VOLUME ONE HUNDRED AND FIFTY SEVEN

INTERNATIONAL REVIEW OF NEUROBIOLOGY

Neurobiology of Addiction and Co-Morbid Disorders

INTERNATIONAL REVIEW OF NEUROBIOLOGY

VOLUME 157

SERIES EDITOR

PATRICIA JANAK
Janak Lab, Department of Neuroscience
Dunning Hall, John Hopkins University
Baltimore, MD, USA

PETER JENNER
Division of Pharmacology and Therapeutics
GKT School of Biomedical Sciences
King's College, London, UK

EDITORIAL BOARD

VOLUME ONE HUNDRED AND FIFTY SEVEN

INTERNATIONAL REVIEW OF NEUROBIOLOGY

Neurobiology of Addiction and Co-Morbid Disorders

Edited by

ERIN S. CALIPARI, Ph.D.
Department of Pharmacology; Department of Molecular Physiology and Biophysics; Department of Psychiatry and Behavioral Sciences; Vanderbilt Center for Addiction Research; Vanderbilt Brain Institute, Vanderbilt University, Nashville, TN, United States

NICHOLAS W. GILPIN, Ph.D.
Department of Physiology, Alcohol & Drug Abuse Center of Excellence, LSUHSC; Southeast Louisiana V.A. Healthcare System (SLVHCS), New Orleans, LA, United States

ELSEVIER

ACADEMIC PRESS
An imprint of Elsevier

Academic Press is an imprint of Elsevier
50 Hampshire Street, 5th Floor, Cambridge, MA 02139, United States
525 B Street, Suite 1650, San Diego, CA 92101, United States
The Boulevard, Langford Lane, Kidlington, Oxford OX5 1GB, United Kingdom
125 London Wall, London, EC2Y 5AS, United Kingdom

First edition 2021

Notices
Knowledge and best practice in this field are constantly changing. As new research and experience broaden our understanding, changes in research methods, professional practices, or medical treatment may become necessary.

Practitioners and researchers must always rely on their own experience and knowledge in evaluating and using any information, methods, compounds, or experiments described herein. In using such information or methods they should be mindful of their own safety and the safety of others, including parties for whom they have a professional responsibility.

To the fullest extent of the law, neither the Publisher nor the authors, contributors, or editors, assume any liability for any injury and/or damage to persons or property as a matter of products liability, negligence or otherwise, or from any use or operation of any methods, products, instructions, or ideas contained in the material herein.

ISBN: 978-0-12-820355-2
ISSN: 0074-7742

For information on all Academic Press publications
visit our website at https://www.elsevier.com/books-and-journals

Publisher: Zoe Kruze
Acquisitions Editor: Sam Mahfoudh
Editorial Project Manager: Shellie Bryant
Production Project Manager: Abdulla Sait
Cover Designer: Alan Studholme

Typeset by SPi Global, India

Contents

Contributors

Ream Al-Hasani
Department of Anesthesiology, Washington University in St. Louis; Washington University in St. Louis, Pain Center; Washington University in St. Louis, School of Medicine; Department of Pharmaceutical and Administrative Sciences, St. Louis College of Pharmacy; Center for Clinical Pharmacology, St. Louis College of Pharmacy and Washington University in St. Louis School of Medicine, St. Louis, MO, United States

Catherine M. Cahill
Department of Psychiatry and Biobehavioural Sciences, University of California; Shirley and Stefan Hatos Center for Neuropharmacology; Jane & Terry Semel Institute for Neuroscience and Human Behavior, University of California Los Angeles, Los Angeles, CA, United States

Cali A. Calarco
Department of Anatomy and Neurobiology, University of Maryland School of Medicine, Baltimore, MD, United States

Erika R. Carlson
Intramural Research Program, National Institute on Drug Abuse, National Institutes of Health, Baltimore, MD, United States

Hannah N. Carlson
Department of Physiology and Pharmacology, Wake Forest School of Medicine, Winston-Salem, NC, United States

Meaghan Creed
Department of Anesthesiology; Department of Neuroscience; Department of Psychiatry; Department of Biomedical Engineering, Washington University in St. Louis; Washington University in St. Louis, Pain Center; Washington University in St. Louis, School of Medicine, St. Louis, MO, United States

Jessica A. Cucinello-Ragland
Department of Physiology and Comprehensive Alcohol-HIV/AIDS Research Center, LSU Health Sciences Center, New Orleans, LA, United States

Scott Edwards
Department of Physiology and Comprehensive Alcohol-HIV/AIDS Research Center, LSU Health Sciences Center, New Orleans, LA, United States

Elizabeth A. Fucich
Department of Physiology; Alcohol and Drug Abuse Center of Excellence, Louisiana State University Health Sciences Center, New Orleans, LA, United States

Alejandra Jacotte-Simancas
Department of Physiology; Alcohol and Drug Abuse Center of Excellence, Louisiana State University Health Sciences Center, New Orleans, LA, United States

Drew D. Kiraly
Nash Family Department of Neuroscience; Department of Psychiatry, Icahn School of Medicine at Mount Sinai, New York, NY, United States

Veronika Kondev
Department of Psychiatry and Behavioral Sciences, Vanderbilt University Medical Center, Nashville, TN, United States

George F. Koob
Intramural Research Program, National Institute on Drug Abuse, National Institutes of Health, Baltimore, MD, United States

Mary Kay Lobo
Department of Anatomy and Neurobiology, University of Maryland School of Medicine, Baltimore, MD, United States

Kelsey E. Lucerne
Nash Family Department of Neuroscience, Icahn School of Medicine at Mount Sinai, New York, NY, United States

Renata C.N. Marchette
Intramural Research Program, National Institute on Drug Abuse, National Institutes of Health, Baltimore, MD, United States

Tamara Markovic
Department of Anesthesiology, Washington University in St. Louis; Washington University in St. Louis, Pain Center; Washington University in St. Louis, School of Medicine, St. Louis, MO, United States

Nicolas Massaly
Department of Anesthesiology, Washington University in St. Louis; Washington University in St. Louis, Pain Center; Washington University in St. Louis, School of Medicine, St. Louis, MO, United States

M. Adrienne McGinn
Intramural Research Program, National Institute on Drug Abuse, National Institutes of Health, Baltimore, MD, United States

Patrick R. Melugin
Department of Pharmacology, Vanderbilt Brain Institute, Vanderbilt Center for Addiction Research, Vanderbilt University, Nashville, TN, United States

Patricia E. Molina
Department of Physiology; Alcohol and Drug Abuse Center of Excellence, Louisiana State University Health Sciences Center, New Orleans, LA, United States

Jose A. Moron
Department of Anesthesiology; Department of Psychiatry; Department of Biomedical Engineering; Washington University in St. Louis, Pain Center; Washington University in St. Louis, School of Medicine, St. Louis, MO, United States

Suzanne O. Nolan
Department of Pharmacology, Vanderbilt Brain Institute, Vanderbilt Center for Addiction Research, Vanderbilt University, Nashville, TN, United States

Caroline B. Pantazis
Intramural Research Program, National Institute on Drug Abuse, National Institutes of Health, Baltimore, MD, United States

Sachin Patel
Department of Psychiatry and Behavioral Sciences, Vanderbilt University Medical Center; Department of Molecular Physiology & Biophysics; Department of Pharmacology; Vanderbilt Center for Addiction Research, Vanderbilt University, Nashville, TN, United States

Nadia Said
Intramural Research Program, National Institute on Drug Abuse, National Institutes of Health, Baltimore, MD, United States

Cody A. Siciliano
Department of Pharmacology, Vanderbilt Brain Institute, Vanderbilt Center for Addiction Research, Vanderbilt University, Nashville, TN, United States

Zachary F. Stielper
Department of Physiology; Alcohol and Drug Abuse Center of Excellence, Louisiana State University Health Sciences Center, New Orleans, LA, United States

Brendan J. Tunstall
Intramural Research Program, National Institute on Drug Abuse, National Institutes of Health, Baltimore, MD, United States

Leandro F. Vendruscolo
Intramural Research Program, National Institute on Drug Abuse, National Institutes of Health, Baltimore, MD, United States

Jeff L. Weiner
Department of Physiology and Pharmacology, Wake Forest School of Medicine, Winston-Salem, NC, United States

Nathan Winters
Department of Psychiatry and Behavioral Sciences, Vanderbilt University Medical Center; Department of Pharmacology, Vanderbilt University, Nashville, TN, United States

The critical importance of understanding comorbidities to effectively treat drug addiction

Nicholas W. Gilpin[a,b,c,d] **and Erin S. Calipari**[e,f,g,h,i]
[a]Physiology, LSUHSC, New Orleans, LA, United States
[b]Neuroscience, LSUHSC, New Orleans, LA, United States
[c]Alcohol & Drug Abuse Center of Excellence, LSUHSC, New Orleans, LA, United States
[d]Southeast Louisiana V.A. Healthcare System (SLVHCS), New Orleans, LA, United States
[e]Department of Pharmacology, Vanderbilt University, Nashville, TN, United States
[f]Department of Molecular Physiology and Biophysics, Vanderbilt University, Nashville, TN, United States
[g]Department of Psychiatry and Behavioral Sciences, Vanderbilt University Medical Center, Nashville, TN, United States
[h]Vanderbilt Center for Addiction Research, Vanderbilt University, Nashville, TN, United States
[i]Vanderbilt Brain Institute, Vanderbilt University, Nashville, TN, United States

Many individuals that live with one mental or neurological disorder are also living with at least one comorbid disorder. This is particularly true in the case of addiction, such that people with an addictive disorder are more likely to be diagnosed with other mental disorders (National Institute of Drug Abuse, n.d.; Petrakis, Gonzalez, Rosenheck, & Krystal, 2002), as well as health conditions not mediated by the central nervous system (American Addiciton Centers, 2020; National Institute of Drug Abuse (NIDA), 2020; National Institute on Alcohol Abuse and Alcoholism (NIAAA), 1991). Given the high rate of comorbidity between addiction and other disorders, effectively treating addiction hinges on our understanding of the complex relationships between addiction and comorbid disorders.

In some cases, misuse of drugs and alcohol contributes to the onset, maintenance, progression, and acceleration of other disorders. In other cases, drug use is driven by physiological and psychotropic drug effects that mitigate the effects or dull the symptoms of other disorders (i.e., self-medication). In many cases, both of these statements may be true. Importantly, the physiological and behavioral effects of drugs vary widely across classes, which is likely why specific drug classes tend to be more highly comorbid with specific disorders, some of which are discussed in this volume. Drug and alcohol use can exacerbate symptoms or lead to the development of other health conditions, and conversely, preexisting conditions can make individuals more likely to develop addiction. Understanding the causes and consequences of these comorbidities is critical because the interaction between two comorbid disorders may alter the neurobiological mediators that should be targeted by treatments.

Even within the realm of addictive disorders, individuals often meet the DSM-V criteria for multiple substance use disorders simultaneously. While it is tempting to classify drug addiction as a singular disorder, drugs act in the brain through different mechanisms of action. Drugs also interact in the body and brain, such that the effects of one drug can be altered by another, even at relatively low doses. Furthermore, the trajectory and expression of a substance use disorder can be altered by use or misuse of a second drug. In some cases, combinations of co-administered drugs can result in the synthesis of a completely different drug in the body with distinct pharmacological actions from the original ingested compounds. Therefore, moving beyond studying individual drugs in isolation will be a critical next step in our understanding of the complex biology and behavior that defines drug addiction in humans.

Despite the fact that addictive disorders are frequently comorbid with other disorders, the overwhelming majority of preclinical research examines drug-taking, drug-seeking, and drug effects on the organism in isolation (i.e., in the absence of a second disorder). Even in studies focused on addiction, the overwhelming majority of preclinical research examines individual drugs in isolation (i.e., in the absence of a second drug). The reasons for this are likely varied, but we speculate it may be explained by some combination of the following factors: (1) these experiments are difficult to design and execute due to a high number of experimental factors and large scope of work, and (2) it may be difficult to determine where to seek funding for such projects that seemingly float between the portfolios of separate funding agencies.

The purpose of this volume is to present the state-of-the-science on *Neurobiology of Addiction and Comorbid Disorders*. This was achieved through the generous and scholarly contributions of international experts in this area. These chapters also discuss challenges, barriers, opportunities, and solutions as they relate to increasing the focus of future research on addiction and comorbid disorders. Specifically, the following chapters focus on the neurobiology underlying interactions between addictive disorders and pain, depression, immune dysfunction, cognitive disorders, and traumatic brain injury and traumatic stress disorders.

At least in part, this volume is a call to scientists to design and perform experiments that represent the highly comorbid nature of addiction and other central and peripheral disorders. It is also a call to funding agencies to prioritize this important, challenging, and understudied area. Our hope is that this volume will shine a light on advances and gaps in basic neuroscience research aimed at investigating a human condition (i.e., addiction) that is often characterized by comorbid disease states.

References

American Addiciton Centers. (2020). *Excessive use of alcohol is frequently associated with comorbid medical issues in the body*. https://www.alcohol.org/comorbid/.

National Institute of Drug Abuse (NIDA). (2020). *Part 2: Co-occurring substance use disorder and physical comorbidities*. https://www.drugabuse.gov/publications/research-reports/common-comorbidities-substance-use-disorders/part-2-co-occurring-substance-use-disorder-physical-comorbidities.

National Institute of Drug Abuse (NIDA). Health consequences of drug misuse mental health effects. https://www.drugabuse.gov/drug-topics/health-consequences-drug-misuse/mental-health-effects.

National Institute on Alcohol Abuse and Alcoholism (NIAAA). (1991). *Alcoholism and co-occurring disorders. No. 14 PH 302*. https://pubs.niaaa.nih.gov/publications/aa14.htm.

Petrakis, I. L., Gonzalez, G., Rosenheck, R., & Krystal, J. H. (2002). *Comorbidity of alcoholism and psychiatric disorders: An overview*. https://pubs.niaaa.nih.gov/publications/arh26-2/81-89.htm.

CHAPTER ONE

Neurobiological aspects of pain in the context of alcohol use disorder

Jessica A. Cucinello-Ragland and Scott Edwards*
Department of Physiology and Comprehensive Alcohol-HIV/AIDS Research Center, LSU Health Sciences Center, New Orleans, LA, United States
*Corresponding author: e-mail address: sedwa5@lsuhsc.edu

Contents

Abstract

Alcohol is an effective and widely utilized analgesic. However, the chronic use of alcohol can actually facilitate nociceptive sensitivity over time, a condition known as hyperalgesia. Excessive and uncontrollable alcohol drinking is also a hallmark feature of alcohol use disorder (AUD). Both AUD and chronic pain are typically accompanied by negative affective states that may underlie reinforcement mechanisms contributing to AUD maintenance or progression. Frequent utilization of alcohol to relieve pain in individuals suffering from AUD or other chronic pain conditions may thus represent a powerful negative reinforcement construct. This chapter will describe ties between

International Review of Neurobiology, Volume 157
ISSN 0074-7742
https://doi.org/10.1016/bs.irn.2020.09.001

alcohol-mediated pain relief and potential exacerbation of AUD. We describe neurobiological systems engaged in alcohol analgesia as well as systems recruited in the development and maintenance of AUD and hyperalgesia. Although few effective therapies exist for either chronic pain or AUD, the common interaction of these conditions will likely lead the way for promising new discoveries of more effective and even simultaneous treatment of AUD and co-morbid hyperalgesia. An abundance of neurobiological findings from multiple laboratories has implicated a potentiation of central amygdala (CeA) signaling in both pain and AUD, and these data also suggest that attenuation of stress-related systems (including corticotropin-releasing factor, vasopressin, and glucocorticoid receptor activity) would be particularly effective and comprehensive therapeutic strategies targeting the critical intersection of somatic and motivational mechanisms driving AUD, including alcohol-induced hyperalgesia.

1. Alcohol use disorder and chronic pain

1.1 Pain as a central symptom of AUD

Alcohol use disorder (AUD, ranging from mild to severe) is characterized by an escalation of alcohol drinking and gradual emergence of negative affective symptoms (e.g., depression, anxiety) over time (Edwards & Koob, 2010). Such negative emotional states may promote escalated drinking as well as relapse during attempted abstinence periods. Consequently, AUD typically represents a chronic psychiatric condition requiring close management over the patient's lifespan (Koob & Volkow, 2010). AUD affects roughly 10–13% of the population in the United States at any given time, and up to one-third of adults in the United States will meet criteria for AUD in their lifetime (Grant et al., 2017). In addition to the negative affective symptoms described above, both clinical and preclinical evidence demonstrates that nociceptive hypersensitivity emerges alongside chronic or excessive alcohol exposure (Dina et al., 2000; Edwards, Guerrero, et al., 2012; Edwards, Vendruscolo, et al., 2012; Edwards, Vendruscolo, Gilpin, Wojnar, & Witkiewitz, 2020; Fu et al., 2015; Gatch & Lal, 1999; Jochum, Boettger, Burkhardt, Juckel, & Bar, 2010). Facilitation of pain by alcohol is evident in that the nocifensive response to painful (noxious) stimuli is enhanced (termed hyperalgesia) and also in that innocuous stimuli (normally non-painful) become noxious (termed allodynia). Consequently, pain represents a negative subjective experience that can have a powerful influence on reward and reinforcement mechanisms, possibly facilitating the transition to substance use disorders in vulnerable individuals (McGinn & Edwards, 2016).

1.2 The negative affective dimension of chronic pain

Chronic pain affects approximately 20% of adults worldwide (Goldberg & McGee, 2011), a number that will likely increase over the next several decades given the aging global population. In the United States alone, over 50 million individuals report chronic pain at a cost of over $600 billion in healthcare expenses and lost productivity (Dahlhamer et al., 2018). While these numbers appear to eclipse the impact of other chronic diseases such as heart disease and cancer, the devastating impact of chronic pain is related more to its effects on morbidity (compared to mortality), especially with regard to the potentiation of negative affect in association with unrelieved pain. Negative affective mood states (anxiety, sadness, anger, and irritability) are considered highly aversive psychodynamic constructs that are distinguishing features of anxiety and depression (Watson & Tellegen, 1985). In general, negative affective disorders represents an umbrella term encompassing a number of mood disorders, including neuroticism, major depressive disorder, general anxiety disorder, and panic disorder.

Chronic pain and negative affective disorders have an alarming rate of comorbidity, likely because of the bidirectional relationship between the two (Elman, Borsook, & Volkow, 2013). Indeed, patients with chronic pain and high negative affect report higher pain severity and pain interference compared to patients with only one of the two disorders (Arnow et al., 2006; Naliboff, Chang, Munakata, & Mayer, 2000). In both children and adults, multiple types of pain disorders are associated with increased likelihood of a positive diagnosis for negative affective disorders, with higher pain ratings elevating the likelihood of affective disorder diagnoses (Campo et al., 2004; Means-Christensen, Roy-Byrne, Sherbourne, Craske, & Stein, 2008). It is important to note that the prevalence of comorbid chronic pain and negative affect varies depending on the type of negative affective disorder. For example, 35–85% of chronic pain patients experience comorbid depression, while 16–60% experience comorbid anxiety disorders (Bair, Robinson, Katon, & Kroenke, 2003; Fishbain, Cutler, Rosomoff, & Rosomoff, 1998; Fishbain, Goldberg, Meagher, Steele, & Rosomoff, 1986; Lee et al., 2018; McWilliams, Cox, & Enns, 2003; Williams et al., 2003).

1.3 Chronic pain and depression

Because chronic pain has a higher incidence of comorbidity with depression than anxiety, and because depression is a major global health burden

(Liu et al., 2019), much of the research into the relationship between chronic pain and negative affect has revolved around depressive disorders. Clinical studies have investigated the development of depression across a number of different chronic pain conditions (Bair et al., 2003), including low back pain (France, Houpt, Skott, Krishnan, & Varia, 1986), orofacial pain (Cox et al., 2016; Feinmann, 1983), and pelvic pain (Magni, Salmi, de Leo, & Ceola, 1984; Siqueira-Campos, Da Luz, de Deus, Martinez, & Conde, 2019). Additionally, both clinical (Aguera-Ortiz, Failde, Mico, Cervilla, & Lopez-Ibor, 2011; Lee et al., 2009; von Knorring, Perris, Eisemann, Eriksson, & Perris, 1983) and preclinical (Goncalves et al., 2008; Kontinen, Kauppila, Paananen, Pertovaara, & Kalso, 1999; Schwartz et al., 2014) studies have found that chronic pain precedes and induces symptoms of depression in a variety of chronic pain conditions and models.

1.4 Chronic pain and anxiety

Although most of the research on chronic pain and negative affect has traditionally revolved around depression, there has been a rapid growth in research regarding comorbid chronic pain and anxiety disorders (Asmundson & Katz, 2009). While clinical reports of comorbid chronic pain and anxiety are highly variable across populations, chronic pain-induced anxiety-like behavior is readily observed in both neuropathic and non-neuropathic animal models of pain (Bahaaddini, Khatamsaz, Esmaeili-Mahani, Abbasnejad, & Raoof, 2016; Dimitrov, Tsuda, Cameron, & Usdin, 2014; Ji, Fu, Ruppert, & Neugebauer, 2007; Matsuzawa-Yanagida et al., 2008; Narita et al., 2006). Additionally, chronic pain can generate pain-related anxiety in humans, in which the patient experiences fear and anxiety in the anticipation of experiencing future pain (McCracken & Dhingra, 2002; McCracken, Zayfert, & Gross, 1992).

1.5 Does hyperalgesia and associated negative affect facilitate AUD severity?

An emerging question is whether and how the negative experience of chronic, unrelieved pain contributes to the maintenance of AUD. Hyperalgesia may drive both continuous alcohol drinking and relapse propensity as a powerful negative reinforcement mechanism based on the strong affective valence associated with persistent and unrelieved nociceptive hypersensitivity (Egli, Koob, & Edwards, 2012; Zale, Maisto, & Ditre, 2015).

Within this context, negative reinforcement would occur whenever the perceived or real effects of alcohol drinking reduce hyperalgesia symptoms. At the basic science level, evidence for this relationship comes from a recent study where hyperalgesia symptoms in alcohol-dependent animals were attenuated by alcohol self-administration (Roltsch Hellard, Impastato, & Gilpin, 2017). At the clinical level, pain prospectively predicts AUD development (McDermott, Joyner, Hakes, Okey, & Cougle, 2018) and relapse propensity (Witkiewitz et al., 2015). A host of neurobiological and psychosocial mechanisms are shared between chronic pain and alcohol reinforcement (Ditre, Zale, & LaRowe, 2019; LeBlanc, McGinn, Itoga, & Edwards, 2015). As described within this chapter, the dysregulation of such systems may become central in the progression of AUD, while targeting these aberrant processes may represent valuable therapeutic strategies for the treatment of pain in the context of AUD.

2. Analgesic actions of alcohol

2.1 History of alcohol as an analgesic

For centuries, people have consumed alcohol for its analgesic properties, dating as far back as ancient Egypt and Greece (Crocq, 2007; McGovern, Mirzoian, & Hall, 2009). The medicinal properties of alcohol were so numerous that 14th and 15th century European physicians referred to alcohol as "aqua vitae," or "water of life," reporting that alcohol "eases the pain in the teeth,…[and] in the breasts when swollen" (Roueche, 1963). Many 19th and 20th century physicians prescribed and administered oral or intravenous alcohol prior to medical procedures due to its analgesic and anesthetic properties (Brown & Cutter, 1977; Chapman & William, 1951; Crocq, 2007; Hanson, 1995). In the 1930s and 1940s, scientists began investigating the analgesic effects of alcohol in the laboratory setting. Indeed, various studies have found that alcohol dampens pain in a variety of pain tests in both humans and animals. Orally consumed alcohol decreases both mechanical and thermal pain sensitivity in both humans and rodents (Gatch, 2009; Gatch & Lal, 1999; Mullin & Luckhardt, 1934; Wolff, Hardy, & Goodell, 1942; Woodrow & Eltherington, 1988). Although there is conflicting evidence regarding the analgesic efficacy of intravenous (i.v.) ethanol in rodents (Bukusoglu, Thalhammer, & Krieger, 1993), in humans i.v. alcohol has been shown to produce analgesia comparable to i.v. morphine (Saddler, James, & Harington, 1985) as well as anti-nociceptive effects against pain due to electrical stimulation (Perrino et al., 2008; Ralevski et al., 2010) and capsaicin-induced hyperalgesia (Arout et al., 2016).

2.2 Alcohol analgesia in the context of AUD

Seventy-nine percent of high-risk drinkers and 61% of individuals suffering from alcohol dependence consume alcohol to treat pain, compared to only 38% of heavy drinkers and 35% of unhealthy drinkers (Alford et al., 2016). There is a strong mutual relationship between pain and alcohol in which alcohol affects nociceptive sensitivity and pain motivates alcohol consumption (Beasley, Macfarlane, & Macfarlane, 2016; Zale et al., 2015). This shared symptomatology can often make co-morbid directionalities difficult to assess (Macfarlane & Beasley, 2015). We will next discuss these interactions, beginning with a discussion of the analgesic effects of alcohol based on drinking history and on chronic pain status.

Early studies investigating the analgesic effects of alcohol in people with alcoholism found that whiskey dose-dependently reduces and abolishes reported pain in alcoholics but has no analgesic effects in non-alcoholics, regardless of dose (Cutter, Maloof, Kurtz, & Jones, 1976). Later studies found that alcohol reduces pain ratings in heavy social drinkers but increases pain ratings in people who regularly consume alcohol in a home setting, suggesting that the analgesic effects of alcohol may vary based on drinking history and pattern of drinking (Brown & Cutter, 1977). From a motivational perspective, alcohol is frequently self-administered for both its stress-reducing and analgesic effects (Egli et al., 2012; Thompson, Oram, Correll, Tsermentseli, & Stubbs, 2017). Self-reports of alcohol use specifically for pain management are common (e.g., Riley & King, 2009). Problem drinkers at risk for AUD not only report more severe pain symptoms compared to non-drinkers, but also report a higher incidence of using alcohol to manage their pain (Brennan, Schutte, & Moos, 2005).

As people with AUD experience a higher incidence of pain (Brennan, Schutte, SooHoo, & Moos, 2011; Zale et al., 2015) and appear to be more sensitive to the analgesic effects of alcohol, it is not surprising that pain severity and experience affects AUD treatment outcomes. Increases in both pain interference and intensity significantly increase the risk for relapse both during and after AUD treatment, and, more specifically, pain significantly predicts heavy drinking lapses both during and after AUD treatment (Witkiewitz et al., 2015). Likewise, decreases in physical pain from the start of AUD treatment are correlated with a decreased likelihood of relapse

(Jakubczyk et al., 2016). Additionally, persistent pain is associated with increased heavy alcohol use 24 months following detoxification (Larson et al., 2007).

Chronic pain also affects the analgesic efficacy of alcohol. Adults with chronic pain, regardless of age, drink alcohol more heavily than the general population (Ditre et al., 2019), and 25% of people with chronic pain report using alcohol to manage their pain symptoms (Riley & King, 2009). Although one study found that a higher percentage of younger participants self-medicated their chronic pain with alcohol compared to older participants (Riley & King, 2009), alcohol consumption is self-reported as one of the most effective pain-relief methods in elderly persons who require assistance with activities of daily living (Jakobsson, Rahm Hallberg, & Westergren, 2004). Both human and animal studies have found that chronic pain is associated with and may drive increased alcohol consumption (Butler et al., 2017; Lawton & Simpson, 2009; Yu, Hwa, Makhijani, Besheer, & Kash, 2019).

Genetic vulnerability for high alcohol drinking may also play a role in the analgesic effects of alcohol. One study found that men with a high familial-genetic risk for alcoholism report higher pain ratings and experience greater reduction of pain in response to alcohol compared to men with no familial-genetic risk for alcoholism (Stewart, Finn, & Pihl, 1995). However, more recent studies have found that the analgesic effects of alcohol do not differ based on family history (Perrino et al., 2008; Ralevski et al., 2010). Rodents selectively bred for alcohol preference display decreased thermal pain thresholds in examinations of either spinal or supraspinal nociceptive behaviors (Kampov-Polevoy et al., 1996; Kimpel, Brown, & Froehlich, 2003). Several genes have been proposed to link alcohol dependence and pain, including the catechol-O-methyl-transferase (COMT) and mu opioid receptor 1 (OPRM1) genes (Egli et al., 2012).

2.3 Neurobiological mechanisms of alcohol analgesia

Despite the surplus of evidence that alcohol produces analgesia, the spinal and supraspinal neurobiological mechanisms behind this phenomenon are drastically understudied (Thompson et al., 2017). Three current hypotheses for this mechanism of action are discussed here, but it is important to note that alcohol likely produces analgesia via interactions between all of the mechanisms described below, as well as through other undiscovered mechanisms.

2.3.1 Opioid receptor systems

As the endogenous opioid neuropeptide system is integral to the production of analgesia, many scientists have examined the hypothesis that alcohol produces its analgesic effects by interacting with opioid receptor pharmacology. Interestingly, mice with high activity (HA) opioid systems display significantly greater ethanol-induced analgesia compared to controls, while their low activity (LA) counterparts display significantly less ethanol-induced analgesia (Mogil et al., 1993). However, other studies have produced mixed results. Depending on the dose of alcohol, pain model, pain test, and experimental species, the non-selective opioid receptor antagonist naloxone has been shown to both prevent and not affect alcohol-induced analgesia (Boada, Feria, & Sanz, 1981; Campbell, Taylor, & Tizabi, 2006, 2007; Cutter & O'Farrell, 1987; Jorgensen & Hole, 1981; Saddler et al., 1985). The specific opioid receptors responsible for the analgesic effects of alcohol are yet to be definitively determined. Genetic and pharmacological manipulation has implicated the delta-opioid receptor in the differences seen between the aforementioned HA and LA mice (Poznanski et al., 2017; Sacharczuk et al., 2014). However, a recent study found that the selective delta-opioid receptor antagonist naltrindole did not affect ethanol-induced anti-nociception (Neddenriep et al., 2019). This same study found that both high doses of naltrexone and the selective kappa-opioid receptor antagonist nor-BNI prevented ethanol-induced anti-nociception in a mouse model of chronic neuropathic pain, suggesting that the analgesic effects of alcohol are also mediated through both mu- and kappa-opioid receptors (Neddenriep et al., 2019). Complementary to their role in analgesia, it is important to note that all three opioid receptor systems represent targets for treating hyperalgesia in the context of substance use disorders (Delery & Edwards, 2020).

2.3.2 GABAergic and glutamatergic signaling

Alcohol is a positive allosteric modulator of the $GABA_A$ receptor, and GABAergic signaling likely plays a major role in the analgesic effects of drinking (Davies, 2003). The benzodiazepine antagonist flumazenil abolishes the anti-nociceptive effects of alcohol in thermal pain tests in rodents (Gatch, 1999), suggesting that GABA receptors are involved in mediating alcohol-induced analgesia. Additional studies are warranted to determine the contributions of regional (both spinal and supraspinal) activation of GABAergic neurons and interneurons in regulating pain sensitivity. Interestingly, infusion of ethanol directly into the medial prefrontal cortex of rats increases mechanical pain sensitivity by activating $GABA_A$ receptors there (Geng et al., 2016). In addition to the $GABA_A$ receptor, alcohol-induced analgesia may also be mediated via interactions with the *N*-methyl-D-aspartate (NMDA) glutamate receptor. The NMDA receptor antagonist MK-801 decreases anti-nociception produced by alcohol, and systemic administration of both MK-801 and naloxone completely abolishes alcohol-induced anti-nociception in mice, suggesting that the analgesic effects of alcohol are mediated by both NMDA and opioid receptors (Mogil et al., 1993).

2.3.3 GIRK2 potassium channels

The G protein-coupled inwardly rectifying potassium 2 (GIRK2) channels play a major role in substance-induced analgesia and have also been implicated in alcohol-induced analgesia (Bodhinathan & Slesinger, 2014; Ikeda et al., 2002; Lewohl et al., 1999). Ethanol directly interacts with and opens GIRK2 channels (Kobayashi et al., 1999). Mice with a mutated GIRK2 channel display reduced ethanol analgesia (Kobayashi et al., 1999), while the analgesic effects of ethanol are fully abolished in GIRK2 knockout mice (Blednov, Stoffel, Alva, & Harris, 2003).

2.3.4 Additional systems and considerations

Additional research is needed to examine the contribution of other neurobiological systems to alcohol analgesia. In addition to opioid signaling, it is likely that another endogenous analgesic system, the endocannabinoid system, may be intimately involved. Endocannabinoids regulate not only pain sensitivity (Woodhams, Chapman, Finn, Hohmann, & Neugebauer, 2017), but also act as a buffer of stress and negative affect (Hillard, 2014; Morena, Patel, Bains, & Hill, 2016; Tasker et al., 2015). Extensive research has also implicated endocannabinoid signaling in alcohol reward and reinforcement (Kirson, Oleata, Parsons, Ciccocioppo, & Roberto, 2018; Serrano et al., 2018; Varodayan et al., 2016).

Another consideration for future research is whether and how tolerance to alcohol analgesia develops over time across a range of disease contexts. Damaj and colleagues (Neddenriep et al., 2019) recently demonstrated that the analgesic efficacy of repeated alcohol diminished over time in an animal model of chronic neuropathic pain. Such findings presage a possible rationale for escalation of drinking in individuals using alcohol to manage pain, possibly also driving AUD. Further research is needed to determine the neuroanatomical and molecular sites of action for alcohol analgesia. It will also be important to understand which analgesic systems or mechanisms become compromised with chronic alcohol use, as supporting these systems may represent a beneficial pharmacological strategy for targeting alcohol-associated hyperalgesia. Such treatments may also support additional strategies described in the next section, which correspond to known neuroadaptations shared between excessive alcohol use and chronic pain conditions that underlie hyperalgesia.

3. Medication strategies for treating pain in the context of AUD

3.1 Shared neuroadaptations between chronic pain, negative affect, and AUD

AUD has been hypothesized to stem from functional alterations in the neurobiological substrates of supraspinal pain processing (Egli et al., 2012), including a key contribution from the nociceptive central amygdala (CeA; Neugebauer, 2015). Importantly, the sensitization of pain-driven negative affect (Ji et al., 2007) is also thought to be mediated by the CeA and to closely interact with similar neuronal mechanisms related to the development of AUD (Edwards et al., 2020). Both alcohol dependence and persistent pain produce multiple neurophysiological adaptations within the CeA (e.g., Carrasquillo & Gereau, 2008; Gilpin, Herman, & Roberto, 2015; Ji & Neugebauer, 2009; Roberto et al., 2010; Zhang et al., 2014). The CeA receives functionally distinct projections from the pontine parabrachial area (PB, nociceptive coding) and basolateral amygdala (BLA, sensory-affective coding) that are sensitized in the context of chronic pain (Ikeda, Takahashi, Inoue, & Kato, 2007; Neugebauer, Li, Bird, Bhave, & Gereau, 2003). Chronic pain-induced activation of the CeA is accompanied by dysregulation of amygdala-driven prefrontal cortex function and resultant cognitive deficits (Ji & Neugebauer, 2011; Ji et al., 2010; Sun & Neugebauer, 2011). Importantly, executive system dysfunction is thought to play a central role in the heavily compromised decision-making that accompanies the transition from recreational drug use to substance use disorder (George & Koob, 2010), which may contribute to alcohol misuse in individuals suffering from chronic pain. Alcohol withdrawal produces hyperalgesia that is mediated by altered activity of CeA projections to the periaqueductal gray (PAG), a region that is critically involved in descending pain modulation (Avegno et al., 2018).

Elucidation of chronic alcohol-induced neuroadaptations within brain stress and nociceptive systems may provide valuable insights into potential mechanisms underlying the transition to AUD in vulnerable individuals. In these regions, elements of brain stress signaling, such as the corticotropin-releasing factor (CRF), vasopressin, and glucocorticoid receptor (GR) systems, are recruited (Edwards, Little, Richardson, & Vendruscolo, 2015; Koob & Le Moal, 2008) during the transition to dependence, as described below.

3.2 Corticotropin-releasing factor receptor systems

As a key mediator of the heightened negative affective response that manifests during alcohol withdrawal, the functional potentiation of CRF signaling is evident in the establishment and progression of AUD-related behaviors (including escalated alcohol drinking) observed in preclinical animal models (Edwards & Koob, 2010; Funk, Zorrilla, Lee, Rice, & Koob, 2007; Roberto et al., 2010; Schreiber & Gilpin, 2018). CRF is a neuropeptide synthesized by the paraventricular nucleus (PVN) of the hypothalamus where it initiates the hypothalamic-pituitary-adrenal (HPA) axis response to stress. CRF stimulates synthesis of adrenocorticotropic hormone (ACTH) in the anterior pituitary, which then facilitates the secretion of glucocorticoids from the adrenal gland to mediate adaptive stress responses and restore homeostasis. High concentrations of CRF are also present in basal forebrain and brainstem areas (Swanson, Sawchenko, Rivier, & Vale, 1983), where the neuropeptide regulates autonomic and behavioral responses to stress. Importantly, extra-hypothalamic CRF1 receptors (CRF1Rs) are widely expressed throughout the brain, and activation of these receptors potentiates central stress responsiveness. However, activation of complementary CRF2 receptors mediates a contrasting (but not necessarily opposing) effect on anxiety-like behavior (Zhao et al., 2007) and CeA plasticity (Varodayan et al., 2017).

Chronic pain produces a number of physiological changes throughout the body, including increases in CRF levels in the CeA (Rouwette et al., 2012). Complementary to this neuroadaptation, the anti-hyperalgesic effects of CRF1R antagonists (administered either systemically or in some cases intra-CeA) have been demonstrated in both mice and rats across a variety of pain models, including neuropathic pain, inflammatory pain, and visceral pain (e.g., Hummel et al., 2010; Ji & Neugebauer, 2007; Nijsen, Ongenae, Meulemans, & Coulie, 2005; Schwetz et al., 2004). CRF would indeed appear to play a key role in processing an array of pain modalities (e.g., mechanical, thermal, visceral), suggesting that its role is not modality-specific but may generalize to all classes of pain. Importantly, CRF1R antagonists do not alter baseline measures of various nociception-related indices (e.g., mechanical or thermal paw withdrawal thresholds, audible or ultrasonic vocalizations) in non-injured animals, indicating the absence of a direct analgesic effect (e.g., Baiamonte et al., 2014; Cohen et al., 2015; Fu & Neugebauer, 2008; Roltsch et al., 2014). In comparison, CRF infusion into the CeA produces hyperalgesia in otherwise naïve animals (Itoga et al., 2016). CRF also directly alters the electrophysiological properties of CeA neurons to drive pain sensitization. For example, rodents in a state of arthritic inflammatory pain (Neugebauer, Han, Adwanikar, Fu, & Ji, 2007) exhibit enhanced CeA excitability that is attenuated by CRF1R antagonism (Ji et al., 2007). The authors also found that the normally inhibitory role of CRF2R signaling in this region was lost in the arthritis model, further demonstrating the complementary roles of CRF1 vs CRF2 receptors in neurophysiology, including pain signaling (Ji & Neugebauer, 2008). Interestingly, arthritis increases anxiety-like behavior and this effect is attenuated by either systemic or intra-CeA CRF1R antagonism (Ji et al., 2007), suggesting that CRF signaling drives the intersection of pain and negative affect within the CeA (Egli et al., 2012; Neugebauer, Li, Bird, & Han, 2004). Stress- and alcohol-related factors other than pain also produce an enhancement of CRF activity in the CeA (Gilpin, 2012; Roberto et al., 2010), and it was also demonstrated that non-pain-related activation of CRF1R signaling in the CeA augments nociceptive sensitivity (Ji, Fu, Adwanikar, & Neugebauer, 2013).

3.3 Vasopressin receptor systems

Several emerging lines of evidence have implicated the neuropeptide vasopressin in the pathophysiology of stress- and emotion-related behaviors, often in a sexually dimorphic fashion (Bisagno & Cadet, 2014). In addition to the well-characterized role of hypothalamic vasopressin acting in conjunction with CRF to stimulate the HPA axis (Lolait, Stewart, Jessop, Young, & O'Carroll, 2007), vasopressin-synthesizing neurons are also localized in the bed nucleus of the stria terminalis (BNST) and medial amygdala (De Vries & Buijs, 1983), project extensively throughout the limbic system (Veinante & Freund-Mercier, 1997), and play a significant role in regulating various complex behaviors, including aggression and social affiliation.

Vasopressin signaling within the brain is thought to underlie various aspects of emotional processing (Caldwell, Lee, Macbeth, & Young, 2008), and early studies established a specific role for central vasopressin in aversive learning and memory mechanisms (Koob & Bloom, 1982). More recently, the generation of small molecule, receptor subtype-selective antagonists has greatly advanced the characterization of the complex behavior regulated by V1b receptors (V1bR), which are thought to play a role in generalized negative affective-like behaviors, as demonstrated by the anxiolytic-like, antidepressant-like, and anti-hyperalgesic-like effects of the small molecule V1bR antagonist SSR149415 (Bradesi, Martinez, Lao, Larsson, & Mayer, 2009; Griebel et al., 2002). Studies utilizing regional microinjections of SSR149415 have established a particular role for negative affective-like V1bR signaling in the amygdala, as SSR149415 reduces both anxiety- and depression-like behavior (Salome, Stemmelin, Cohen, & Griebel, 2006). Vasopressin gene (*Avp*) expression levels are higher in the CeA of male Sardinian alcohol-preferring (sP) relative to alcohol-non-preferring (sNP) rats. Interestingly, the excessive drinking of sP rats reduces elevated *Avp* levels, while systemic SSR149415 administration is effective in reducing drinking in this population (Zhou et al., 2011). Independent lines of preclinical evidence strongly point to a role for vasopressin/V1bR signaling in the transition to alcohol dependence (Edwards, Guerrero, et al., 2012; Zhou & Kreek, 2014).

Given the literature suggesting that V1bR antagonists have substantial efficacy in reducing negative affective-like profiles and the fact that vasopressin and its central receptors (V1bR and V1aR) are highly expressed in extended amygdala and cortical areas (Stemmelin, Lukovic, Salome, & Griebel, 2005; Tribollet, Raufaste, Maffrand, & Serradeil-Le Gal, 1999), vasopressin systems could play a central role in the increased alcohol intake and hyperalgesia associated with dependence. These effects may occur via the ability of vasopressin to activate the HPA axis (Volpi, Rabadan-Diehl, & Aguilera, 2004) and/or via direct central (i.e., extra-hypothalamic) actions. Interestingly, the effects of V1bR blockade on excessive alcohol self-administration (Edwards, Guerrero, et al., 2012) closely resemble inhibition of the CRF1 receptor system on this behavior (Richardson et al., 2008), and CRF1R antagonism has already been demonstrated to alleviate symptoms of mechanical hyperalgesia in alcohol-dependent rats (Edwards, Vendruscolo, et al., 2012). Accordingly, either in addition to or in cooperation with CRF, vasopressin V1bR activity may represent another mechanism whereby negative reinforcement mechanisms regulate alcohol dependence-related symptomatology. Perhaps most importantly, recent evidence has emerged for the safety and clinical efficacy of V1bR antagonists to reduce HPA axis activation and increase abstinence in alcohol-dependent patients (Katz, Liu, Locke, Dutta, & Tracy, 2016; Ryan et al., 2017; *ClinicalTrials.gov Identifier NCT01613014*).

In comparison to V1bRs, V1aRs are expressed at even higher levels in the brain (Tribollet et al., 1999), and are important for mediating social behavior. Systemic and brain administration of the V1aR antagonist SR49059 improves neurological and neurobehavioral outcomes in animal models of traumatic brain injury (TBI; Krieg, Sonanini, Plesnila, & Trabold, 2015; Manaenko et al., 2011; Marmarou et al., 2014). Interestingly, pain-related affective responses are enhanced via vasopressin signaling through V1a receptors in the CeA (Cragg, Ji, & Neugebauer, 2016), an effect that is blocked via intra-CeA administration of SR49059. These findings make V1aR antagonism a promising therapeutic strategy for both TBI and pain (which are themselves highly comorbid), although the contribution of this receptor system to excessive drinking or hyperalgesia in the context of alcohol dependence has remained unexplored.

3.4 Glucocorticoid receptor system

Both alcohol and alcohol withdrawal elevate systemic glucocorticoid levels due to activation of the endocrine HPA axis (Rivier, 2014). In addition to profound dysregulation of HPA axis activity (Richardson, Lee, O'Dell, Koob, & Rivier, 2008), alcohol-dependent animals exhibit a functional increase in central brain GR signaling, a key mediator of stress responsiveness (Edwards et al., 2015). Importantly, GR antagonism via mifepristone reduces escalated drinking in both preclinical animal models and early clinical trials (Vendruscolo et al., 2015; *ClinicalTrials.gov Identifier NCT01548417*). As a potential biochemical correlate of sensitized central GR signaling that may underlie these promising clinical directions, previous work describes increases in GR phosphorylation in the CeA of alcohol-dependent animals (Vendruscolo et al., 2015). At the molecular level, the transcriptional activity of GR is directly controlled via phosphorylation at serine 232 (Adzic et al., 2009), and two key downstream targets of GR transcriptional activity are *Crh* and *Crhr1* (Yao & Denver, 2007). Importantly, systemic administration of either the GR antagonist mifepristone (Dina et al., 2008) or a CRF1R antagonist (Edwards, Vendruscolo, et al., 2012) alleviates hyperalgesia in binge drinking and alcohol-dependent animals (respectively). It is also important to note that in contrast to the classic

GR-mediated negative feedback regulation of CRF at the level of the HPA axis, GR activity actually facilitates CRF gene expression in certain central brain regions (Edwards et al., 2015) including the CeA (Makino, Gold, & Schulkin, 1994). Differential expression and activity of steroid receptor coactivator-1 (SRC-1) isoforms between the PVN and CeA may account for this dichotomy (Lachize et al., 2009). As described above, intra-CeA administration of CRF1R antagonists is effective in alleviating hyperalgesia in both pain and drug dependence models (McGinn & Edwards, 2016). Unfortunately, evidence for the clinical efficacy of CRF1 receptor antagonists to reduce excessive drinking has been slow to develop, with some early failures in clinical trials (e.g., Kwako et al., 2015). However, development of novel antagonists with different brain pharmacokinetics and/or a more precise selection of specific treatment populations (e.g., those suffering from co-morbid hyperalgesia) may unmask the true utility of this medication class (Spierling & Zorrilla, 2017). In addition to the CRF system, alternative stress-regulated targets either independent from or even transcriptionally regulated by GR, as in the case of V1aR (Watters, Swank, Wilkinson, & Dorsa, 1996) and V1bR (Aguilera & Rabadan-Diehl, 2000), would appear to offer a very promising path forward for treating severe AUD and associated pain conditions.

4. Summary

Pain represents a uniquely subjective experience that may exert a powerful influence on reinforcement processes, possibly facilitating the transition to substance use disorders (Miller & Gold, 2007; Shurman, Koob, & Gutstein, 2010), including AUD (Edwards et al., 2020). Recent insights into the mechanisms that mediate the analgesic efficacy and abuse-related properties of alcohol have provided a basis for a conceptualization of the neurobiological intersection of pain and AUD. The analgesic effects of alcohol have been known anecdotally for centuries and have been scientifically investigated for almost 100 years. Drinking history, alcohol dependence, chronic pain status, and genetic risk for AUD all affect the analgesic efficacy of alcohol (Edwards et al., 2020). The mechanisms behind alcohol-induced analgesia are understudied, but the primary mechanisms proposed include the endogenous opioid system, GABA/NMDA signaling, and GIRK2 channel activity. It is likely that alcohol interacts with each of these systems to produce its analgesic effects, while future studies are needed to understand the possible role of other endogenous analgesia systems, such as the endocannabinoid system.

Neurophysiological processes that counter the anxiolytic and analgesic properties of alcohol may drive AUD progression through the generation of hyperalgesia and closely associated negative affective states. In turn, given that chronic pain itself often produces emotional distress, alcohol-related hyperalgesia may be closely associated with the transition to AUD via facilitation of negative reinforcement mechanisms. Strong evidence suggests that the neural substrates associated with addiction may overlap with substrates of emotional aspects of pain processing in brain regions where ascending nociceptive circuitry terminates (Egli et al., 2012). Specifically, the affective component of pain is strongly regulated by the CeA (Neugebauer, 2015). Numerous preclinical models have demonstrated a functional role for the CeA in regulating persistent hyperalgesia associated with a range of neurological and psychiatric conditions ranging from arthritis (Ji et al., 2007) to opioid dependence (McNally & Akil, 2002). This neurocircuitry may promote crosstalk among a variety of psychiatric disorders and account for the high degree of comorbidity between affective disorders, AUD and chronic pain (Edwards et al., 2020). Medications such as CRF1 receptor antagonists, vasopressin receptor antagonists, and GR antagonists that act on these shared systems would likely be effective in treating a range of affective pain-related conditions, including AUD (Fig. 1). Moreover, dependence on other

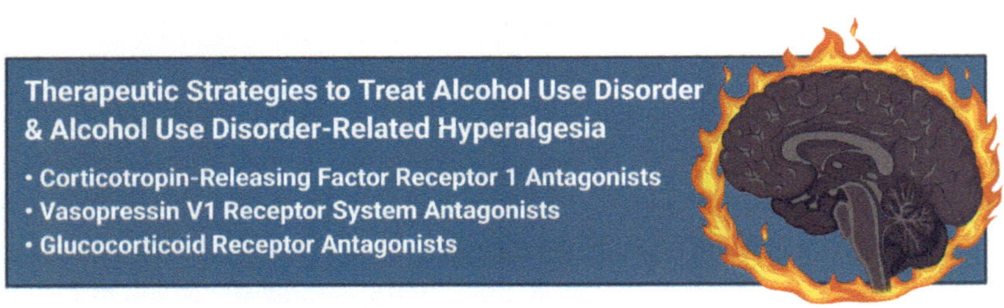

Fig. 1 There is an urgent need to understand the neurobiological mechanisms of pain and AUD, as well as a continuing need for preclinical medication development targeted to novel systems to treat these devastating conditions. As hyperalgesia represents a form of chronic stress and negative affect, we expect that this condition plays a critical role in negative reinforcement mechanisms underlying excessive drinking in the context of AUD. Following from this conceptualization, the interrogation of brain stress systems would likely represent a prolific avenue for medication development. Neurobiological changes across both hypothalamic and extra-hypothalamic stress systems (including changes in CRF, vasopressin, and GR signaling) have been observed in preclinical AUD animal models, with the central amygdala (CeA) playing a critical role in mediating the interaction of pain and negative affective-related behaviors expected to contribute to AUD.

abused drugs commonly taken with alcohol (e.g., nicotine) is associated with a CRF1R-dependent hyperalgesia (Baiamonte et al., 2014; Cohen et al., 2015; Edwards, Vendruscolo, et al., 2012). As a result, patients with a history of alcohol or other substance use disorders may be susceptible to the activation of a common set of nociceptive factors that could drive relapse and abuse of any or all of these substances. Thus, we and others have hypothesized that CeA neuroadaptations facilitate pain-driven motivated behaviors that underlie dependence-induced hyperalgesia and excessive drinking. It is our hope that the functional integration of pain- and addiction-related research objectives across the National Institutes of Health will promote further collaboration among addiction and pain neuroscientists, leading to broader and more integrative therapeutic strategies for these devastating conditions.

Acknowledgments

Preparation of this review was supported by National Institute on Alcohol Abuse and Alcoholism grants AA028445, AA025996, AA007577, and AA009803.

Disclosure

S.E. is a consultant for Avanos Medical, Inc., a private company focused on the development of medical devices for pain management. J.A.C.R. has no conflicts related to this work.

References

Adzic, M., Djordjevic, J., Djordjevic, A., Niciforovic, A., Demonacos, C., Radojcic, M., et al. (2009). Acute or chronic stress induce cell compartment-specific phosphorylation of glucocorticoid receptor and alter its transcriptional activity in Wistar rat brain. *The Journal of Endocrinology*, *202*(1), 87–97. https://doi.org/10.1677/JOE-08-0509.

Aguera-Ortiz, L., Failde, I., Mico, J. A., Cervilla, J., & Lopez-Ibor, J. J. (2011). Pain as a symptom of depression: Prevalence and clinical correlates in patients attending psychiatric clinics. *Journal of Affective Disorders*, *130*(1–2), 106–112. https://doi.org/10.1016/j.jad.2010.10.022.

Aguilera, G., & Rabadan-Diehl, C. (2000). Regulation of vasopressin V1b receptors in the anterior pituitary gland of the rat. *Experimental Physiology*, *85*(s1), 19S–26S. https://doi.org/10.1111/j.1469-445x.2000.tb00004.x.

Alford, D. P., German, J. S., Samet, J. H., Cheng, D. M., Lloyd-Travaglini, C. A., & Saitz, R. (2016). Primary care patients with drug use report chronic pain and self-medicate with alcohol and other drugs. *Journal of General Internal Medicine*, *31*(5), 486–491. https://doi.org/10.1007/s11606-016-3586-5.

Arnow, B. A., Hunkeler, E. M., Blasey, C. M., Lee, J., Constantino, M. J., Fireman, B., et al. (2006). Comorbid depression, chronic pain, and disability in primary care. *Psychosomatic Medicine*, *68*(2), 262–268. https://doi.org/10.1097/01.psy.0000204851.15499.fc.

Arout, C. A., Perrino, A. C., Jr., Ralevski, E., Acampora, G., Koretski, J., Limoncelli, D., et al. (2016). Effect of intravenous ethanol on capsaicin-induced hyperalgesia in human subjects. *Alcoholism, Clinical and Experimental Research*, *40*(7), 1425–1429. https://doi.org/10.1111/acer.13095.

Asmundson, G. J., & Katz, J. (2009). Understanding the co-occurrence of anxiety disorders and chronic pain: State-of-the-art. *Depression and Anxiety, 26*(10), 888–901. https://doi.org/10.1002/da.20600.

Avegno, E. M., Lobell, T. D., Itoga, C. A., Baynes, B. B., Whitaker, A. M., Weera, M. M., et al. (2018). Central amygdala circuits mediate hyperalgesia in alcohol-dependent rats. *The Journal of Neuroscience, 38*(36), 7761–7773. https://doi.org/10.1523/JNEUROSCI.0483-18.2018.

Bahaaddini, M., Khatamsaz, S., Esmaeili-Mahani, S., Abbasnejad, M., & Raoof, M. (2016). The role of trigeminal nucleus caudalis orexin 1 receptor in orofacial pain-induced anxiety in rat. *Neuroreport, 27*(15), 1107–1113. https://doi.org/10.1097/WNR.0000000000000660.

Baiamonte, B. A., Valenza, M., Roltsch, E. A., Whitaker, A. M., Baynes, B. B., Sabino, V., et al. (2014). Nicotine dependence produces hyperalgesia: Role of corticotropin-releasing factor-1 receptors (CRF1Rs) in the central amygdala (CeA). *Neuropharmacology, 77*, 217–223. https://doi.org/10.1016/j.neuropharm.2013.09.025.

Bair, M. J., Robinson, R. L., Katon, W., & Kroenke, K. (2003). Depression and pain comorbidity: A literature review. *Archives of Internal Medicine, 163*(20), 2433–2445. https://doi.org/10.1001/archinte.163.20.2433.

Beasley, M. J., Macfarlane, T. V., & Macfarlane, G. J. (2016). Is alcohol consumption related to likelihood of reporting chronic widespread pain in people with stable consumption? Results from UK biobank. *Pain, 157*(11), 2552–2560. https://doi.org/10.1097/j.pain.0000000000000675.

Bisagno, V., & Cadet, J. L. (2014). Stress, sex, and addiction: Potential roles of corticotropin-releasing factor, oxytocin, and arginine-vasopressin. *Behavioural Pharmacology, 25*(5–6), 445–457. https://doi.org/10.1097/FBP.0000000000000049.

Blednov, Y. A., Stoffel, M., Alva, H., & Harris, R. A. (2003). A pervasive mechanism for analgesia: Activation of GIRK2 channels. *Proceedings of the National Academy of Sciences of the United States of America, 100*(1), 277–282. https://doi.org/10.1073/pnas.012682399.

Boada, J., Feria, M., & Sanz, E. (1981). Inhibitory effect of naloxone on the ethanol-induced antinociception in mice. *Pharmacological Research Communications, 13*(7), 673–678. https://doi.org/10.1016/s0031-6989(81)80055-0.

Bodhinathan, K., & Slesinger, P. A. (2014). Alcohol modulation of G-protein-gated inwardly rectifying potassium channels: From binding to therapeutics. *Frontiers in Physiology, 5*, 76. https://doi.org/10.3389/fphys.2014.00076.

Bradesi, S., Martinez, V., Lao, L., Larsson, H., & Mayer, E. A. (2009). Involvement of vasopressin 3 receptors in chronic psychological stress-induced visceral hyperalgesia in rats. *American Journal of Physiology. Gastrointestinal and Liver Physiology, 296*(2), G302–G309. https://doi.org/10.1152/ajpgi.90557.2008.

Brennan, P. L., Schutte, K. K., & Moos, R. H. (2005). Pain and use of alcohol to manage pain: Prevalence and 3-year outcomes among older problem and non-problem drinkers. *Addiction, 100*(6), 777–786. https://doi.org/10.1111/j.1360-0443.2005.01074.x.

Brennan, P. L., Schutte, K. K., SooHoo, S., & Moos, R. H. (2011). Painful medical conditions and alcohol use: A prospective study among older adults. *Pain Medicine, 12*(7), 1049–1059. https://doi.org/10.1111/j.1526-4637.2011.01156.x.

Brown, R. A., & Cutter, H. S. G. (1977). Alcohol, customary drinking behavior, and pain. *Journal of Abnormal Psychology, 86*(2), 179–188. https://doi.org/10.1037/0021-843X.86.2.179.

Bukusoglu, C., Thalhammer, J. G., & Krieger, N. R. (1993). Analgesia with anesthetic steroids and ethanol. *Analgesia and Anesthesia, 77*(1), 27–31.

Butler, R. K., Knapp, D. J., Ulici, V., Longobardi, L., Loeser, R. F., & Breese, G. R. (2017). A mouse model for chronic pain-induced increase in ethanol consumption. *Pain, 158*(3), 457–462.

Caldwell, H. K., Lee, H. J., Macbeth, A. H., & Young, W. S., 3rd. (2008). Vasopressin: Behavioral roles of an "original" neuropeptide. *Progress in Neurobiology, 84*(1), 1–24. https://doi.org/10.1016/j.pneurobio.2007.10.007.

Campbell, V. C., Taylor, R. E., & Tizabi, Y. (2006). Antinociceptive effects of alcohol and nicotine: Involvement of the opioid system. *Brain Research, 1097*(1), 71–77. https://doi.org/10.1016/j.brainres.2006.04.054.

Campbell, V. C., Taylor, R. E., & Tizabi, Y. (2007). Effects of selective opioid receptor antagonists on alcohol-induced and nicotine-induced antinociception. *Alcoholism, Clinical and Experimental Research, 31*(8), 1435–1440. https://doi.org/10.1111/j.1530-0277.2007.00432.x.

Campo, J. V., Bridge, J., Ehmann, M., Altman, S., Lucas, A., Birmaher, B., et al. (2004). Recurrent abdominal pain, anxiety, and depression in primary care. *Pediatrics, 113*(4), 817–824. https://doi.org/10.1542/peds.113.4.817.

Carrasquillo, Y., & Gereau, R. W. (2008). Hemispheric lateralization of a molecular signal for pain modulation in the amygdala. *Molecular Pain, 4*, 24. https://doi.org/10.1186/1744-8069-4-24.

Chapman, E. R., & William, P. T. (1951). Intravenous alcohol as an obstetrical analgesia. *American Journal of Obstetrics and Gynecology, 61*(3), 676–679.

Cohen, A., Treweek, J., Edwards, S., Leao, R. M., Schulteis, G., Koob, G. F., et al. (2015). Extended access to nicotine leads to a CRF1 receptor dependent increase in anxiety-like behavior and hyperalgesia in rats. *Addiction Biology, 20*(1), 56–68. https://doi.org/10.1111/adb.12077.

Cox, D. R., Ashby, S., DeConde, A. S., Mace, J. C., Orlandi, R. R., Smith, T. L., et al. (2016). Dyad of pain and depression in chronic rhinosinusitis. *International Forum of Allergy & Rhinology, 6*(3), 308–314. https://doi.org/10.1002/alr.21664.

Cragg, B., Ji, G., & Neugebauer, V. (2016). Differential contributions of vasopressin V1A and oxytocin receptors in the amygdala to pain-related behaviors in rats. *Molecular Pain, 12*. https://doi.org/10.1177/1744806916676491.

Crocq, M. A. (2007). Historical and cultural aspects of man's relationship with addictive drugs. *Dialogues in Clinical Neuroscience, 9*(4), 355–361. Retrieved from https://www.ncbi.nlm.nih.gov/pubmed/18286796.

Cutter, H. S., Maloof, B., Kurtz, N. R., & Jones, W. C. (1976). "Feeling no pain" differential responses to pain by alcoholics and nonalcoholics before and after drinking. *Journal of Studies on Alcohol, 37*(3), 273–277. https://doi.org/10.15288/jsa.1976.37.273.

Cutter, H. S., & O'Farrell, T. J. (1987). Experience with alcohol and the endogenous opioid system in ethanol analgesia. *Addictive Behaviors, 12*(4), 331–343. https://doi.org/10.1016/0306-4603(87)90047-5.

Dahlhamer, J., Lucas, J., Zelaya, C., Nahin, R., Mackey, S., DeBar, L., et al. (2018). Prevalence of chronic pain and high-impact chronic pain among adults—United States, 2016. *MMWR: Morbidity and Mortality Weekly Report, 67*(36), 1001–1006. https://doi.org/10.15585/mmwr.mm6736a2.

Davies, M. (2003). The role of GABAA receptors in mediating the effects of alcohol in the central nervous system. *Journal of Psychiatry & Neuroscience, 28*(4), 263–274. Retrieved from https://www.ncbi.nlm.nih.gov/pubmed/12921221.

De Vries, G. J., & Buijs, R. M. (1983). The origin of the vasopressinergic and oxytocinergic innervation of the rat brain with special reference to the lateral septum. *Brain Research, 273*(2), 307–317. https://doi.org/10.1016/0006-8993(83)90855-7.

Delery, E. C., & Edwards, S. (2020). Neuropeptide and cytokine regulation of pain in the context of substance use disorders. *Neuropharmacology, 174*, 108153. https://doi.org/10.1016/j.neuropharm.2020.108153.

Dimitrov, E. L., Tsuda, M. C., Cameron, H. A., & Usdin, T. B. (2014). Anxiety- and depression-like behavior and impaired neurogenesis evoked by peripheral neuropathy persist following resolution of prolonged tactile hypersensitivity. *The Journal of Neuroscience*, *34*(37), 12304–12312. https://doi.org/10.1523/JNEUROSCI.0312-14.2014.

Dina, O. A., Barletta, J., Chen, X., Mutero, A., Martin, A., Messing, R. O., et al. (2000). Key role for the epsilon isoform of protein kinase C in painful alcoholic neuropathy in the rat. *Journal of Neuroscience*, *20*(22), 8614–8619. Retrieved from https://www.ncbi.nlm.nih.gov/pubmed/11069970.

Dina, O. A., Khasar, S. G., Alessandri-Haber, N., Green, P. G., Messing, R. O., & Levine, J. D. (2008). Alcohol-induced stress in painful alcoholic neuropathy. *The European Journal of Neuroscience*, *27*(1), 83–92. https://doi.org/10.1111/j.1460-9568.2007.05987.x.

Ditre, J. W., Zale, E. L., & LaRowe, L. R. (2019). A reciprocal model of pain and substance use: Transdiagnostic considerations, clinical implications, and future directions. *Annual Review of Clinical Psychology*, *15*, 503–528. https://doi.org/10.1146/annurev-clinpsy-050718-095440.

Edwards, S., Guerrero, M., Ghoneim, O. M., Roberts, E., & Koob, G. F. (2012). Evidence that vasopressin V1b receptors mediate the transition to excessive drinking in ethanol-dependent rats. *Addiction Biology*, *17*(1), 76–85. https://doi.org/10.1111/j.1369-1600.2010.00291.x.

Edwards, S., & Koob, G. F. (2010). Neurobiology of dysregulated motivational systems in drug addiction. *Future Neurology*, *5*(3), 393–401. https://doi.org/10.2217/fnl.10.14.

Edwards, S., Little, H. J., Richardson, H. N., & Vendruscolo, L. F. (2015). Divergent regulation of distinct glucocorticoid systems in alcohol dependence. *Alcohol*, *49*(8), 811–816. https://doi.org/10.1016/j.alcohol.2015.04.004.

Edwards, S., Vendruscolo, L. F., Gilpin, N. W., Wojnar, M., & Witkiewitz, K. (2020). Alcohol and pain: A translational review of preclinical and clinical findings to inform future treatment strategies. *Alcoholism, Clinical and Experimental Research*, *44*(2), 368–383. https://doi.org/10.1111/acer.14260.

Edwards, S., Vendruscolo, L. F., Schlosburg, J. E., Misra, K. K., Wee, S., Park, P. E., et al. (2012). Development of mechanical hypersensitivity in rats during heroin and ethanol dependence: Alleviation by CRF(1) receptor antagonism. *Neuropharmacology*, *62*(2), 1142–1151. https://doi.org/10.1016/j.neuropharm.2011.11.006.

Egli, M., Koob, G. F., & Edwards, S. (2012). Alcohol dependence as a chronic pain disorder. *Neuroscience and Biobehavioral Reviews*, *36*(10), 2179–2192. https://doi.org/10.1016/j.neubiorev.2012.07.010.

Elman, I., Borsook, D., & Volkow, N. D. (2013). Pain and suicidality: Insights from reward and addiction neuroscience. *Progress in Neurobiology*, *109*, 1–27. https://doi.org/10.1016/j.pneurobio.2013.06.003.

Feinmann, C. (1983). Psychogenic facial pain: Presentation and treatment. *Journal of Psychosomatic Research*, *27*(5), 403–410. https://doi.org/10.1016/0022-3999(83)90076-4.

Fishbain, D. A., Cutler, R. B., Rosomoff, H. L., & Rosomoff, R. S. (1998). Do antidepressants have an analgesic effect in psychogenic pain and somatoform pain disorder? A meta-analysis. *Psychosomatic Medicine*, *60*(4), 503–509. https://doi.org/10.1097/00006842-199807000-00019.

Fishbain, D. A., Goldberg, M., Meagher, B. R., Steele, R., & Rosomoff, H. (1986). Male and female chronic pain patients categorized by DSM-III psychiatric diagnostic criteria. *Pain*, *26*(2), 181–197. https://doi.org/10.1016/0304-3959(86)90074-6.

France, R. D., Houpt, J. L., Skott, A., Krishnan, K. R., & Varia, I. M. (1986). Depression as a psychopathological disorder in chronic low back pain patients. *Journal of Psychosomatic Research*, *30*(2), 127–133. https://doi.org/10.1016/0022-3999(86)90041-3.

Fu, R., Gregor, D., Peng, Z., Li, J., Bekker, A., & Ye, J. (2015). Chronic intermittent voluntary alcohol drinking induces hyperalgesia in Sprague-Dawley rats. *International Journal of Physiology, Pathophysiology and Pharmacology*, 7(3), 136–144. Retrieved from https://www.ncbi.nlm.nih.gov/pubmed/26823962.

Fu, Y., & Neugebauer, V. (2008). Differential mechanisms of CRF1 and CRF2 receptor functions in the amygdala in pain-related synaptic facilitation and behavior. *The Journal of Neuroscience*, *28*(15), 3861–3876. https://doi.org/10.1523/JNEUROSCI.0227-08.2008.

Funk, C. K., Zorrilla, E. P., Lee, M. J., Rice, K. C., & Koob, G. F. (2007). Corticotropin-releasing factor 1 antagonists selectively reduce ethanol self-administration in ethanol-dependent rats. *Biological Psychiatry*, *61*(1), 78–86. https://doi.org/10.1016/j.biopsych.2006.03.063.

Gatch, M. B. (1999). Effects of benzodiazepines on acute and chronic ethanol-induced nociception in rats. *Alcoholism, Clinical and Experimental Research*, *23*(11), 1736–1743. Retrieved from https://www.ncbi.nlm.nih.gov/pubmed/10591589.

Gatch, M. B. (2009). Ethanol withdrawal and hyperalgesia. *Current Drug Abuse Reviews*, *2*(1), 41–50. https://doi.org/10.2174/1874473710902010041.

Gatch, M. B., & Lal, H. (1999). Effects of ethanol and ethanol withdrawal on nociception in rats. *Alcoholism, Clinical and Experimental Research*, *23*(2), 328–333. Retrieved from https://www.ncbi.nlm.nih.gov/pubmed/10069564.

Geng, K. W., He, T., Wang, R. R., Li, C. L., Luo, W. J., Wu, F. F., et al. (2016). Ethanol increases mechanical pain sensitivity in rats via activation of GABAA receptors in medial prefrontal cortex. *Neuroscience Bulletin*, *32*(5), 433–444. https://doi.org/10.1007/s12264-016-0063-4.

George, O., & Koob, G. F. (2010). Individual differences in prefrontal cortex function and the transition from drug use to drug dependence. *Neuroscience and Biobehavioral Reviews*, *35*(2), 232–247. https://doi.org/10.1016/j.neubiorev.2010.05.002.

Gilpin, N. W. (2012). Corticotropin-releasing factor (CRF) and neuropeptide Y (NPY): Effects on inhibitory transmission in central amygdala, and anxiety- & alcohol-related behaviors. *Alcohol*, *46*(4), 329–337. https://doi.org/10.1016/j.alcohol.2011.11.009.

Gilpin, N. W., Herman, M. A., & Roberto, M. (2015). The central amygdala as an integrative hub for anxiety and alcohol use disorders. *Biological Psychiatry*, 77, 859–869.

Goldberg, D. S., & McGee, S. J. (2011). Pain as a global public health priority. *BMC Public Health*, *11*, 770. https://doi.org/10.1186/1471-2458-11-770.

Goncalves, L., Silva, R., Pinto-Ribeiro, F., Pego, J. M., Bessa, J. M., Pertovaara, A., et al. (2008). Neuropathic pain is associated with depressive behaviour and induces neuroplasticity in the amygdala of the rat. *Experimental Neurology*, *213*(1), 48–56. https://doi.org/10.1016/j.expneurol.2008.04.043.

Grant, B. F., Chou, S. P., Saha, T. D., Pickering, R. P., Kerridge, B. T., Ruan, W. J., et al. (2017). Prevalence of 12-month alcohol use, high-risk drinking, and DSM-IV alcohol use disorder in the United States, 2001-2002 to 2012-2013: Results from the national epidemiologic survey on alcohol and related conditions. *JAMA Psychiatry*, *74*(9), 911–923. https://doi.org/10.1001/jamapsychiatry.2017.2161.

Griebel, G., Simiand, J., Serradeil-Le Gal, C., Wagnon, J., Pascal, M., Scatton, B., et al. (2002). Anxiolytic- and antidepressant-like effects of the non-peptide vasopressin V1b receptor antagonist, SSR149415, suggest an innovative approach for the treatment of stress-related disorders. *Proceedings of the National Academy of Sciences of the United States of America*, *99*(9), 6370–6375. https://doi.org/10.1073/pnas.092012099.

Hanson, D. J. (1995). *Preventing alcohol abuse: Alcohol, culture, and control*. Praeger.

Hillard, C. J. (2014). Stress regulates endocannabinoid-CB1 receptor signaling. *Seminars in Immunology*, *26*(5), 380–388. https://doi.org/10.1016/j.smim.2014.04.001.

Hummel, M., Cummons, T., Lu, P., Mark, L., Harrison, J. E., Kennedy, J. D., et al. (2010). Pain is a salient "stressor" that is mediated by corticotropin-releasing factor-1 receptors. *Neuropharmacology*, *59*(3), 160–166. https://doi.org/10.1016/j.neuropharm.2010.05.001.

Ikeda, K., Kobayashi, T., Kumanishi, T., Yano, R., Sora, I., & Niki, H. (2002). Molecular mechanisms of analgesia induced by opioids and ethanol: Is the GIRK channel one of the keys? *Neuroscience Research*, *44*(2), 121–131. https://doi.org/10.1016/s0168-0102(02)00094-9.

Ikeda, R., Takahashi, Y., Inoue, K., & Kato, F. (2007). NMDA receptor-independent synaptic plasticity in the central amygdala in the rat model of neuropathic pain. *Pain*, *127*(1–2), 161–172. https://doi.org/10.1016/j.pain.2006.09.003.

Itoga, C. A., Roltsch Hellard, E. A., Whitaker, A. M., Lu, Y. L., Schreiber, A. L., Baynes, B. B., et al. (2016). Traumatic stress promotes hyperalgesia via corticotropin-releasing factor-1 receptor (CRFR1) signaling in central amygdala. *Neuropsychopharmacology*, *41*, 2463–2472. https://doi.org/10.1038/npp.2016.44.

Jakobsson, U., Rahm Hallberg, I., & Westergren, A. (2004). Pain management in elderly persons who require assistance with activities of daily living: A comparison of those living at home with those in special accommodations. *European Journal of Pain*, *8*(4), 335–344. https://doi.org/10.1016/j.ejpain.2003.10.007.

Jakubczyk, A., Ilgen, M. A., Kopera, M., Krasowska, A., Klimkiewicz, A., Bohnert, A., et al. (2016). Reductions in physical pain predict lower risk of relapse following alcohol treatment. *Drug and Alcohol Dependence*, *158*, 167–171. https://doi.org/10.1016/j.drugalcdep.2015.11.020.

Ji, G., Fu, Y., Adwanikar, H., & Neugebauer, V. (2013). Non-pain-related CRF1 activation in the amygdala facilitates synaptic transmission and pain responses. *Molecular Pain*, *9*, 2. https://doi.org/10.1186/1744-8069-9-2.

Ji, G., Fu, Y., Ruppert, K. A., & Neugebauer, V. (2007). Pain-related anxiety-like behavior requires CRF1 receptors in the amygdala. *Molecular Pain*, *3*, 13. https://doi.org/10.1186/1744-8069-3-13.

Ji, G., & Neugebauer, V. (2007). Differential effects of CRF1 and CRF2 receptor antagonists on pain-related sensitization of neurons in the central nucleus of the amygdala. *Journal of Neurophysiology*, *97*(6), 3893–3904. https://doi.org/10.1152/jn.00135.2007.

Ji, G., & Neugebauer, V. (2008). Pro- and anti-nociceptive effects of corticotropin-releasing factor (CRF) in central amygdala neurons are mediated through different receptors. *Journal of Neurophysiology*, *99*(3), 1201–1212. https://doi.org/10.1152/jn.01148.2007.

Ji, G., & Neugebauer, V. (2009). Hemispheric lateralization of pain processing by amygdala neurons. *Journal of Neurophysiology*, *102*(4), 2253–2264. https://doi.org/10.1152/jn.00166.2009.

Ji, G., & Neugebauer, V. (2011). Pain-related deactivation of medial prefrontal cortical neurons involves mGluR1 and GABA(A) receptors. *Journal of Neurophysiology*, *106*(5), 2642–2652. https://doi.org/10.1152/jn.00461.2011.

Ji, G., Sun, H., Fu, Y., Li, Z., Pais-Vieira, M., Galhardo, V., et al. (2010). Cognitive impairment in pain through amygdala-driven prefrontal cortical deactivation. *The Journal of Neuroscience*, *30*(15), 5451–5464. https://doi.org/10.1523/JNEUROSCI.0225-10.2010.

Jochum, T., Boettger, M. K., Burkhardt, C., Juckel, G., & Bar, K. J. (2010). Increased pain sensitivity in alcohol withdrawal syndrome. *European Journal of Pain*, *14*(7), 713–718. https://doi.org/10.1016/j.ejpain.2009.11.008.

Jorgensen, H. A., & Hole, K. (1981). Does ethanol stimulate brain opiate receptors? Studies on receptor binding and naloxone inhibition of ethanol-induced effects. *European Journal of Pharmacology*, *75*(4), 223–229. https://doi.org/10.1016/0014-2999(81)90548-3.

Kampov-Polevoy, A. B., Kasheffskaya, O. P., Overstreet, D. H., Rezvani, A. H., Viglinskaya, I. V., Badistov, B. A., et al. (1996). Pain sensitivity and saccharin intake in alcohol-preferring and -nonpreferring rat strains. *Physiology & Behavior*, *59*(4–5), 683–688. https://doi.org/10.1016/0031-9384(95)02110-8.

Katz, D. A., Liu, W., Locke, C., Dutta, S., & Tracy, K. A. (2016). Clinical safety and hypothalamic-pituitary-adrenal axis effects of the arginine vasopressin type 1B receptor antagonist ABT-436. *Psychopharmacology*, *233*(1), 71–81. https://doi.org/10.1007/s00213-015-4089-5.

Kimpel, M. W., Brown, M. M., & Froehlich, J. C. (2003). Pain thresholds in alcohol preferring and non-preferring rats: Diurnal and repeated trial line differences. *Alcoholism, Clinical and Experimental Research*, *27*(12), 1921–1928.

Kirson, D., Oleata, C. S., Parsons, L. H., Ciccocioppo, R., & Roberto, M. (2018). CB1 and ethanol effects on glutamatergic transmission in the central amygdala of male and female msP and Wistar rats. *Addiction Biology*, *23*(2), 676–688. https://doi.org/10.1111/adb.12525.

Kobayashi, T., Ikeda, K., Kojima, H., Niki, H., Yano, R., Yoshioka, T., et al. (1999). Ethanol opens G-protein-activated inwardly rectifying K+ channels. *Nature Neuroscience*, *2*(12), 1091–1097. https://doi.org/10.1038/16019.

Kontinen, V. K., Kauppila, T., Paananen, S., Pertovaara, A., & Kalso, E. (1999). Behavioural measures of depression and anxiety in rats with spinal nerve ligation-induced neuropathy. *Pain*, *80*(1–2), 341–346. https://doi.org/10.1016/s0304-3959(98)00230-9.

Koob, G. F., & Bloom, F. E. (1982). Behavioral effects of neuropeptides: Endorphins and vasopressin. *Annual Review of Physiology*, *44*, 571–582. https://doi.org/10.1146/annurev.ph.44.030182.003035.

Koob, G. F., & Le Moal, M. (2008). Addiction and the brain antireward system. *Annual Review of Psychology*, *59*, 29–53. https://doi.org/10.1146/annurev.psych.59.103006.093548.

Koob, G. F., & Volkow, N. D. (2010). Neurocircuitry of addiction. *Neuropsychopharmacology*, *35*(1), 217–238. https://doi.org/10.1038/npp.2009.110.

Krieg, S. M., Sonanini, S., Plesnila, N., & Trabold, R. (2015). Effect of small molecule vasopressin V1a and V2 receptor antagonists on brain edema formation and secondary brain damage following traumatic brain injury in mice. *Journal of Neurotrauma*, *32*(4), 221–227. https://doi.org/10.1089/neu.2013.3274.

Kwako, L. E., Spagnolo, P. A., Schwandt, M. L., Thorsell, A., George, D. T., Momenan, R., et al. (2015). The corticotropin releasing hormone-1 (CRH1) receptor antagonist pexacerfont in alcohol dependence: A randomized controlled experimental medicine study. *Neuropsychopharmacology*, *40*(5), 1053–1063. https://doi.org/10.1038/npp.2014.306.

Lachize, S., Apostolakis, E. M., van der Laan, S., Tijssen, A. M., Xu, J., de Kloet, E. R., et al. (2009). Steroid receptor coactivator-1 is necessary for regulation of corticotropin-releasing hormone by chronic stress and glucocorticoids. *Proceedings of the National Academy of Sciences of the United States of America*, *106*(19), 8038–8042. https://doi.org/10.1073/pnas.0812062106.

Larson, M. J., Paasche-Orlow, M., Cheng, D. M., Lloyd-Travaglini, C., Saitz, R., & Samet, J. H. (2007). Persistent pain is associated with substance use after detoxification: A prospective cohort analysis. *Addiction*, *102*(5), 752–760. https://doi.org/10.1111/j.1360-0443.2007.01759.x.

Lawton, J., & Simpson, J. (2009). Predictors of alcohol use among people experiencing chronic pain. *Psychology, Health & Medicine*, *14*(4), 487–501. https://doi.org/10.1080/13548500902923177.

LeBlanc, D. M., McGinn, M. A., Itoga, C. A., & Edwards, S. (2015). The affective dimension of pain as a risk factor for drug and alcohol addiction. *Alcohol*, *49*(8), 803–809. https://doi.org/10.1016/j.alcohol.2015.04.005.

Lee, H. S., Lee, S., Park, S., Baek, Y., Youn, J. H., Cho, D. B., et al. (2018). The association between somatic and psychological discomfort and health-related quality of life according to the elderly and non-elderly. *Quality of Life Research*, *27*(3), 673–681. https://doi.org/10.1007/s11136-017-1715-1.

Lee, P., Zhang, M., Hong, J. P., Chua, H. C., Chen, K. P., Tang, S. W., et al. (2009). Frequency of painful physical symptoms with major depressive disorder in Asia: Relationship with disease severity and quality of life. *The Journal of Clinical Psychiatry*, *70*(1), 83–91. https://doi.org/10.4088/jcp.08m04114.

Lewohl, J. M., Wilson, W. R., Mayfield, R. D., Brozowski, S. J., Morrisett, R. A., & Harris, R. A. (1999). G-protein-coupled inwardly rectifying potassium channels are targets of alcohol action. *Nature Neuroscience*, *2*(12), 1084–1090. https://doi.org/10.1038/16012.

Liu, C. H., Zhang, G. Z., Li, B., Li, M., Woelfer, M., Walter, M., et al. (2019). Role of inflammation in depression relapse. *Journal of Neuroinflammation*, *16*, 90. https://doi.org/10.1186/s12974-019-1475-7.

Lolait, S. J., Stewart, L. Q., Jessop, D. S., Young, W. S., 3rd, & O'Carroll, A. M. (2007). The hypothalamic-pituitary-adrenal axis response to stress in mice lacking functional vasopressin V1b receptors. *Endocrinology*, *148*(2), 849–856. https://doi.org/10.1210/en.2006-1309.

Macfarlane, G. J., & Beasley, M. (2015). Alcohol consumption in relation to risk and severity of chronic widespread pain: Results from a UK population-based study. *Arthritis Care & Research*, *67*(9), 1297–1303. https://doi.org/10.1002/acr.22604.

Magni, G., Salmi, A., de Leo, D., & Ceola, A. (1984). Chronic pelvic pain and depression. *Psychopathology*, *17*(3), 132–136. https://doi.org/10.1159/000284030.

Makino, S., Gold, P. W., & Schulkin, J. (1994). Corticosterone effects on corticotropin-releasing hormone mRNA in the central nucleus of the amygdala and the parvocellular region of the paraventricular nucleus of the hypothalamus. *Brain Research*, *640*(1–2), 105–112. https://doi.org/10.1016/0006-8993(94)91862-7.

Manaenko, A., Fathali, N., Khatibi, N. H., Lekic, T., Hasegawa, Y., Martin, R., et al. (2011). Arginine-vasopressin V1a receptor inhibition improves neurologic outcomes following an intracerebral hemorrhagic brain injury. *Neurochemistry International*, *58*(4), 542–548. https://doi.org/10.1016/j.neuint.2011.01.018.

Marmarou, C. R., Liang, X., Abidi, N. H., Parveen, S., Taya, K., Henderson, S. C., et al. (2014). Selective vasopressin-1a receptor antagonist prevents brain edema, reduces astrocytic cell swelling and GFAP, V1aR and AQP4 expression after focal traumatic brain injury. *Brain Research*, *1581*, 89–102. https://doi.org/10.1016/j.brainres.2014.06.005.

Matsuzawa-Yanagida, K., Narita, M., Nakajima, M., Kuzumaki, N., Niikura, K., Nozaki, H., et al. (2008). Usefulness of antidepressants for improving the neuropathic pain-like state and pain-induced anxiety through actions at different brain sites. *Neuropsychopharmacology*, *33*(8), 1952–1965. https://doi.org/10.1038/sj.npp.1301590.

McCracken, L. M., & Dhingra, L. (2002). A short version of the Pain Anxiety Symptoms Scale (PASS-20): Preliminary development and validity. *Pain Research & Management*, 7(1), 45–50. https://doi.org/10.1155/2002/517163.

McCracken, L. M., Zayfert, C., & Gross, R. T. (1992). The Pain Anxiety Symptoms Scale: Development and validation of a scale to measure fear of pain. *Pain*, *50*(1), 67–73. https://doi.org/10.1016/0304-3959(92)90113-p.

McDermott, K. A., Joyner, K. J., Hakes, J. K., Okey, S. A., & Cougle, J. R. (2018). Pain interference and alcohol, nicotine, and cannabis use disorder in a national sample of substance users. *Drug and Alcohol Dependence*, *186*, 53–59. https://doi.org/10.1016/j.drugalcdep.2018.01.011.

McGinn, M. A., & Edwards, S. (2016). CRF1 receptor signaling at the intersection of pain and opioid addiction. In V. Preedy (Ed.), *Vol. 1. The neuropathology of drug addictions and substance misuse* (pp. 891–896). Academic Press.

McGovern, P. E., Mirzoian, A., & Hall, G. R. (2009). Ancient Egyptian herbal wines. *Proceedings. National Academy of Sciences. United States of America*, *106*(18), 7361–7366.

McNally, G. P., & Akil, H. (2002). Role of corticotropin-releasing hormone in the amygdala and bed nucleus of the stria terminalis in the behavioral, pain modulatory, and endocrine consequences of opiate withdrawal. *Neuroscience*, *112*(3), 605–617. https://doi.org/10.1016/s0306-4522(02)00105-7.

McWilliams, L. A., Cox, B. J., & Enns, M. W. (2003). Mood and anxiety disorders associated with chronic pain: An examination in a nationally representative sample. *Pain*, *106*(1–2), 127–133. https://doi.org/10.1016/s0304-3959(03)00301-4.

Means-Christensen, A. J., Roy-Byrne, P. P., Sherbourne, C. D., Craske, M. G., & Stein, M. B. (2008). Relationships among pain, anxiety, and depression in primary care. *Depression and Anxiety*, *25*(7), 593–600. https://doi.org/10.1002/da.20342.

Miller, N. S., & Gold, M. S. (2007). Opiate prescription medication dependence and pain perceptions. *Journal of Addictive Diseases*, *26*(Suppl. 1), 65–71. https://doi.org/10.1300/J069v26S01_07.

Mogil, J. S., Marek, P., Yirmiya, R., Balian, H., Sadowski, B., Taylor, A. N., et al. (1993). Antagonism of the non-opioid component of ethanol-induced analgesia by the NMDA receptor antagonist MK-801. *Brain Research*, *602*(1), 126–130. https://doi.org/10.1016/0006-8993(93)90251-h.

Morena, M., Patel, S., Bains, J. S., & Hill, M. N. (2016). Neurobiological interactions between stress and the endocannabinoid system. *Neuropsychopharmacology*, *41*(1), 80–102. https://doi.org/10.1038/npp.2015.166.

Mullin, F. J., & Luckhardt, A. B. (1934). The effect of alcohol on cutaneous tactile and pain sensitivity. *American Journal of Physiology*, *109*, 77–78.

Naliboff, B. D., Chang, L., Munakata, J., & Mayer, E. A. (2000). Towards an integrative model of irritable bowel syndrome. *Progress in Brain Research*, *122*, 413–423. https://doi.org/10.1016/s0079-6123(08)62154-8.

Narita, M., Kaneko, C., Miyoshi, K., Nagumo, Y., Kuzumaki, N., Nakajima, M., et al. (2006). Chronic pain induces anxiety with concomitant changes in opioidergic function in the amygdala. *Neuropsychopharmacology*, *31*(4), 739–750. https://doi.org/10.1038/sj.npp.1300858.

Neddenriep, B., Bagdas, D., Contreras, K. M., Ditre, J. W., Wolstenholme, J. T., Miles, M. F., et al. (2019). Pharmacological mechanisms of alcohol analgesic-like properties in mouse models of acute and chronic pain. *Neuropharmacology*, *160*, 107793. https://doi.org/10.1016/j.neuropharm.2019.107793.

Neugebauer, V. (2015). Amygdala pain mechanisms. *Handbook of Experimental Pharmacology*, *227*, 261–284. https://doi.org/10.1007/978-3-662-46450-2_13.

Neugebauer, V., Han, J. S., Adwanikar, H., Fu, Y., & Ji, G. (2007). Techniques for assessing knee joint pain in arthritis. *Molecular Pain*, *3*, 8. https://doi.org/10.1186/1744-8069-3-8.

Neugebauer, V., Li, W., Bird, G. C., Bhave, G., & Gereau, R. W. T. (2003). Synaptic plasticity in the amygdala in a model of arthritic pain: Differential roles of metabotropic glutamate receptors 1 and 5. *Journal of Neuroscience*, *23*(1), 52–63. Retrieved from https://www.ncbi.nlm.nih.gov/pubmed/12514201.

Neugebauer, V., Li, W., Bird, G. C., & Han, J. S. (2004). The amygdala and persistent pain. *The Neuroscientist*, *10*(3), 221–234. https://doi.org/10.1177/1073858403261077.

Nijsen, M., Ongenae, N., Meulemans, A., & Coulie, B. (2005). Divergent role for CRF1 and CRF2 receptors in the modulation of visceral pain. *Neurogastroenterology and Motility*, *17*(3), 423–432. https://doi.org/10.1111/j.1365-2982.2005.00644.x.

Perrino, A. C., Jr., Ralevski, E., Acampora, G., Edgecombe, J., Limoncelli, D., & Petrakis, I. L. (2008). Ethanol and pain sensitivity: Effects in healthy subjects using an acute pain paradigm. *Alcoholism, Clinical and Experimental Research*, *32*(6), 952–958. https://doi.org/10.1111/j.1530-0277.2008.00653.x.

Poznanski, P., Lesniak, A., Korostynski, M., Szklarczyk, K., Lazarczyk, M., Religa, P., et al. (2017). Delta-opioid receptor antagonism leads to excessive ethanol consumption in

mice with enhanced activity of the endogenous opioid system. *Neuropharmacology, 118*, 90–101. https://doi.org/10.1016/j.neuropharm.2017.03.016.
Ralevski, E., Perrino, A., Acampora, G., Koretski, J., Limoncelli, D., & Petrakis, I. (2010). Analgesic effects of ethanol are influenced by family history of alcoholism and neuroticism. *Alcoholism, Clinical and Experimental Research, 34*(8), 1433–1441. https://doi.org/10.1111/j.1530-0277.2010.01228.x.
Richardson, H. N., Lee, S. Y., O'Dell, L. E., Koob, G. F., & Rivier, C. L. (2008). Alcohol self-administration acutely stimulates the hypothalamic-pituitary-adrenal axis, but alcohol dependence leads to a dampened neuroendocrine state. *The European Journal of Neuroscience, 28*(8), 1641–1653. https://doi.org/10.1111/j.1460-9568.2008.06455.x.
Richardson, H. N., Zhao, Y., Fekete, E. M., Funk, C. K., Wirsching, P., Janda, K. D., et al. (2008). MPZP: A novel small molecule corticotropin-releasing factor type 1 receptor (CRF1) antagonist. *Pharmacology, Biochemistry, and Behavior, 88*(4), 497–510. https://doi.org/10.1016/j.pbb.2007.10.008.
Riley, J. L., 3rd, & King, C. (2009). Self-report of alcohol use for pain in a multi-ethnic community sample. *The Journal of Pain, 10*(9), 944–952. https://doi.org/10.1016/j.jpain.2009.03.005.
Rivier, C. (2014). Role of hypothalamic corticotropin-releasing factor in mediating alcohol-induced activation of the rat hypothalamic-pituitary-adrenal axis. *Frontiers in Neuroendocrinology, 35*(2), 221–233. https://doi.org/10.1016/j.yfrne.2013.10.005.
Roberto, M., Cruz, M. T., Gilpin, N. W., Sabino, V., Schweitzer, P., Bajo, M., et al. (2010). Corticotropin releasing factor-induced amygdala gamma-aminobutyric acid release plays a key role in alcohol dependence. *Biological Psychiatry, 67*(9), 831–839. https://doi.org/10.1016/j.biopsych.2009.11.007.
Roltsch, E. A., Baynes, B. B., Mayeux, J. P., Whitaker, A. M., Baiamonte, B. A., & Gilpin, N. W. (2014). Predator odor stress alters corticotropin-releasing factor-1 receptor (CRF1R)-dependent behaviors in rats. *Neuropharmacology, 79*, 83–89. https://doi.org/10.1016/j.neuropharm.2013.11.005.
Roltsch Hellard, E. A., Impastato, R. A., & Gilpin, N. W. (2017). Intra-cerebral and intra-nasal melanocortin-4 receptor antagonist blocks withdrawal hyperalgesia in alcohol-dependent rats. *Addiction Biology, 22*(3), 692–701. https://doi.org/10.1111/adb.12360.
Roueche, B. (1963). Alcohol in human culture. In L. P. Salvatore (Ed.), *Alcohol and civilization* (pp. 167–182). New York: McGraw-Hill.
Rouwette, T., Vanelderen, P., de Reus, M., Loohuis, N. O., Giele, J., van Egmond, J., et al. (2012). Experimental neuropathy increases limbic forebrain CRF. *European Journal of Pain, 16*(1), 61–71. https://doi.org/10.1016/j.ejpain.2011.05.016.
Ryan, M. L., Falk, D. E., Fertig, J. B., Rendenbach-Mueller, B., Katz, D. A., Tracy, K. A., et al. (2017). A phase 2, double-blind, placebo-controlled randomized trial assessing the efficacy of ABT-436, a novel V1b receptor antagonist, for alcohol dependence. *Neuropsychopharmacology, 42*(5), 1012–1023. https://doi.org/10.1038/npp.2016.214.
Sacharczuk, M., Lesniak, A., Lipkowski, A. W., Korostynski, M., Przewlocki, R., & Sadowski, B. (2014). Association between the A107V substitution in the delta-opioid receptors and ethanol drinking in mice selected for high and low analgesia. *Addiction Biology, 19*(4), 643–651. https://doi.org/10.1111/adb.12030.
Saddler, J. M., James, M. F., & Harington, A. P. (1985). Naloxone does not reverse ethanol analgesia in man. *Clinical and Experimental Pharmacology & Physiology, 12*(4), 359–364. https://doi.org/10.1111/j.1440-1681.1985.tb00883.x.
Salome, N., Stemmelin, J., Cohen, C., & Griebel, G. (2006). Differential roles of amygdaloid nuclei in the anxiolytic- and antidepressant-like effects of the V1b receptor antagonist, SSR149415, in rats. *Psychopharmacology, 187*(2), 237–244. https://doi.org/10.1007/s00213-006-0424-1.

Schreiber, A. L., & Gilpin, N. W. (2018). Corticotropin-releasing factor (CRF) neurocircuitry and neuropharmacology in alcohol drinking. *Handbook of Experimental Pharmacology*, *248*, 435–471. https://doi.org/10.1007/164_2017_86.

Schwartz, N., Temkin, P., Jurado, S., Lim, B. K., Heifets, B. D., Polepalli, J. S., et al. (2014). Chronic pain. Decreased motivation during chronic pain requires long-term depression in the nucleus accumbens. *Science*, *345*(6196), 535–542. https://doi.org/10.1126/science.1253994.

Schwetz, I., Bradesi, S., McRoberts, J. A., Sablad, M., Miller, J. C., Zhou, H., et al. (2004). Delayed stress-induced colonic hypersensitivity in male Wistar rats: Role of neurokinin-1 and corticotropin-releasing factor-1 receptors. *American Journal of Physiology. Gastrointestinal and Liver Physiology*, *286*(4), G683–G691. https://doi.org/10.1152/ajpgi.00358.2003.

Serrano, A., Pavon, F. J., Buczynski, M. W., Schlosburg, J., Natividad, L. A., Polis, I. Y., et al. (2018). Deficient endocannabinoid signaling in the central amygdala contributes to alcohol dependence-related anxiety-like behavior and excessive alcohol intake. *Neuropsychopharmacology*, *43*(9), 1840–1850. https://doi.org/10.1038/s41386-018-0055-3.

Shurman, J., Koob, G. F., & Gutstein, H. B. (2010). Opioids, pain, the brain, and hyperkatifeia: A framework for the rational use of opioids for pain. *Pain Medicine*, *11*(7), 1092–1098. https://doi.org/10.1111/j.1526-4637.2010.00881.x.

Siqueira-Campos, V. M. E., Da Luz, R. A., de Deus, J. M., Martinez, E. Z., & Conde, D. M. (2019). Anxiety and depression in women with and without chronic pelvic pain: prevalence and associated factors. *Journal of Pain Research*, *12*, 1223–1233. https://doi.org/10.2147/JPR.S195317.

Spierling, S. R., & Zorrilla, E. P. (2017). Don't stress about CRF: Assessing the translational failures of CRF1 antagonists. *Psychopharmacology*, *234*(9–10), 1467–1481. https://doi.org/10.1007/s00213-017-4556-2.

Stemmelin, J., Lukovic, L., Salome, N., & Griebel, G. (2005). Evidence that the lateral septum is involved in the antidepressant-like effects of the vasopressin V1b receptor antagonist, SSR149415. *Neuropsychopharmacology*, *30*(1), 35–42. https://doi.org/10.1038/sj.npp.1300562.

Stewart, S. H., Finn, P. R., & Pihl, R. O. (1995). A dose-response study of the effects of alcohol on the perceptions of pain and discomfort due to electric shock in men at high familial-genetic risk for alcoholism. *Psychopharmacology*, *119*(3), 261–267. https://doi.org/10.1007/BF02246289.

Sun, H., & Neugebauer, V. (2011). mGluR1, but not mGluR5, activates feed-forward inhibition in the medial prefrontal cortex to impair decision making. *Journal of Neurophysiology*, *106*(2), 960–973. https://doi.org/10.1152/jn.00762.2010.

Swanson, L. W., Sawchenko, P. E., Rivier, J., & Vale, W. W. (1983). Organization of ovine corticotropin-releasing factor immunoreactive cells and fibers in the rat brain: An immunohistochemical study. *Neuroendocrinology*, *36*(3), 165–186. https://doi.org/10.1159/000123454.

Tasker, J. G., Chen, C., Fisher, M. O., Fu, X., Rainville, J. R., & Weiss, G. L. (2015). Endocannabinoid regulation of neuroendocrine systems. *International Review of Neurobiology*, *125*, 163–201. https://doi.org/10.1016/bs.irn.2015.09.003.

Thompson, T., Oram, C., Correll, C. U., Tsermentseli, S., & Stubbs, B. (2017). Analgesic effects of alcohol: A systematic review and meta-analysis of controlled experimental studies in healthy participants. *The Journal of Pain*, *18*(5), 499–510. https://doi.org/10.1016/j.jpain.2016.11.009.

Tribollet, E., Raufaste, D., Maffrand, J., & Serradeil-Le Gal, C. (1999). Binding of the non-peptide vasopressin V1a receptor antagonist SR-49059 in the rat brain: An in vitro and in vivo autoradiographic study. *Neuroendocrinology*, *69*(2), 113–120. https://doi.org/10.1159/000054409.

Varodayan, F. P., Correia, D., Kirson, D., Khom, S., Oleata, C. S., Luu, G., et al. (2017). CRF modulates glutamate transmission in the central amygdala of naïve and ethanol-dependent rats. *Neuropharmacology, 125*, 418–428. https://doi.org/10.1016/j.neuropharm.2017.08.009.

Varodayan, F. P., Soni, N., Bajo, M., Luu, G., Madamba, S. G., Schweitzer, P., et al. (2016). Chronic ethanol exposure decreases CB1 receptor function at GABAergic synapses in the rat central amygdala. *Addiction Biology, 21*(4), 788–801. https://doi.org/10.1111/adb.12256.

Veinante, P., & Freund-Mercier, M. J. (1997). Distribution of oxytocin- and vasopressin-binding sites in the rat extended amygdala: A histoautoradiographic study. *The Journal of Comparative Neurology, 383*(3), 305–325. Retrieved from https://www.ncbi.nlm.nih.gov/pubmed/9205043.

Vendruscolo, L. F., Estey, D., Goodell, V., Macshane, L. G., Logrip, M. L., Schlosburg, J. E., et al. (2015). Glucocorticoid receptor antagonism decreases alcohol seeking in alcohol-dependent individuals. *The Journal of Clinical Investigation, 125*(8), 3193–3197. https://doi.org/10.1172/JCI79828.

Volpi, S., Rabadan-Diehl, C., & Aguilera, G. (2004). Vasopressinergic regulation of the hypothalamic pituitary adrenal axis and stress adaptation. *Stress, 7*(2), 75–83. https://doi.org/10.1080/10253890410001733535.

von Knorring, L., Perris, C., Eisemann, M., Eriksson, U., & Perris, H. (1983). Pain as a symptom in depressive disorders. II. Relationship to personality traits as assessed by means of KSP. *Pain, 17*(4), 377–384. https://doi.org/10.1016/0304-3959(83)90169-0.

Watson, D., & Tellegen, A. (1985). Toward a consensual structure of mood. *Psychological Bulletin, 98*(2), 219–235. https://doi.org/10.1037//0033-2909.98.2.219.

Watters, J. J., Swank, M. W., Wilkinson, C. W., & Dorsa, D. M. (1996). Evidence for glucocorticoid regulation of the rat vasopressin V1a receptor gene. *Peptides, 17*(1), 67–73. https://doi.org/10.1016/0196-9781(95)02085-3.

Williams, L. S., Jones, W. J., Shen, J., Robinson, R. L., Weinberger, M., & Kroenke, K. (2003). Prevalence and impact of depression and pain in neurology outpatients. *Journal of Neurology, Neurosurgery, and Psychiatry, 74*(11), 1587–1589. https://doi.org/10.1136/jnnp.74.11.1587.

Witkiewitz, K., Vowles, K. E., McCallion, E., Frohe, T., Kirouac, M., & Maisto, S. A. (2015). Pain as a predictor of heavy drinking and any drinking lapses in the COMBINE study and the UK Alcohol Treatment Trial. *Addiction, 110*(8), 1262–1271. https://doi.org/10.1111/add.12964.

Wolff, H. G., Hardy, J. D., & Goodell, H. (1942). Studies on pain: Measurement of the effect of ethyl alcohol on the pain threshold and on the "alarm" reaction. *Journal de Pharmacologie, 75*, 38–49.

Woodhams, S. G., Chapman, V., Finn, D. P., Hohmann, A. G., & Neugebauer, V. (2017). The cannabinoid system and pain. *Neuropharmacology, 124*, 105–120. https://doi.org/10.1016/j.neuropharm.2017.06.015.

Woodrow, K. M., & Eltherington, L. G. (1988). Feeling no pain: Alcohol as an analgesic. *Pain, 32*(2), 159–163. https://doi.org/10.1016/0304-3959(88)90064-4.

Yao, M., & Denver, R. J. (2007). Regulation of vertebrate corticotropin-releasing factor genes. *General and Comparative Endocrinology, 153*(1–3), 200–216. https://doi.org/10.1016/j.ygcen.2007.01.046.

Yu, W., Hwa, L. S., Makhijani, V. H., Besheer, J., & Kash, T. L. (2019). Chronic inflammatory pain drives alcohol drinking in a sex-dependent manner for C57BL/6J mice. *Alcohol, 77*, 135–145. https://doi.org/10.1016/j.alcohol.2018.10.002.

Zale, E. L., Maisto, S. A., & Ditre, J. W. (2015). Interrelations between pain and alcohol: An integrative review. *Clinical Psychology Review, 37*, 57–71. https://doi.org/10.1016/j.cpr.2015.02.005.

Zhang, Z., Tao, W., Hou, Y. Y., Wang, W., Lu, Y. G., & Pan, Z. Z. (2014). Persistent pain facilitates response to morphine reward by downregulation of central amygdala GABAergic function. *Neuropsychopharmacology*, *39*(9), 2263–2271. https://doi.org/10.1038/npp.2014.77.

Zhao, Y., Valdez, G. R., Fekete, E. M., Rivier, J. E., Vale, W. W., Rice, K. C., et al. (2007). Subtype-selective corticotropin-releasing factor receptor agonists exert contrasting, but not opposite, effects on anxiety-related behavior in rats. *The Journal of Pharmacology and Experimental Therapeutics*, *323*(3), 846–854. https://doi.org/10.1124/jpet.107.123208.

Zhou, Y., Colombo, G., Carai, M. A., Ho, A., Gessa, G. L., & Kreek, M. J. (2011). Involvement of arginine vasopressin and V1b receptor in alcohol drinking in Sardinian alcohol-preferring rats. *Alcoholism, Clinical and Experimental Research*, *35*(10), 1876–1883. https://doi.org/10.1111/j.1530-0277.2011.01532.x.

Zhou, Y., & Kreek, M. J. (2014). Alcohol: A stimulant activating brain stress responsive systems with persistent neuroadaptation. *Neuropharmacology*, *87*, 51–58. https://doi.org/10.1016/j.neuropharm.2014.05.044.

CHAPTER TWO

Pain, negative affective states and opioid-based analgesics: Safer pain therapies to dampen addiction

Nicolas Massaly[a,b,c,*], Tamara Markovic[a,b,c], Meaghan Creed[a,b,c,d,e,f], Ream Al-Hasani[a,b,c,g,h], Catherine M. Cahill[i,j,k], and Jose A. Moron[a,b,c,d,e]

[a]Department of Anesthesiology, Washington University in St. Louis, St. Louis, MO, United States
[b]Washington University in St Louis, Pain Center, St. Louis, MO, United States
[c]Washington University in St Louis, School of Medicine, St. Louis, MO, United States
[d]Department of Neuroscience, Washington University in St. Louis, St. Louis, MO, United States
[e]Department of Psychiatry, Washington University in St. Louis, St. Louis, MO, United States
[f]Department of Biomedical Engineering, Washington University in St. Louis, St. Louis, MO, United States
[g]Department of Pharmaceutical and Administrative Sciences, St. Louis College of Pharmacy, St. Louis, MO, United States
[h]Center for Clinical Pharmacology, St. Louis College of Pharmacy and Washington University in St. Louis School of Medicine, St. Louis, MO, United States
[i]Department of Psychiatry and Biobehavioural Sciences, University of California, Los Angeles, CA, United States
[j]Shirley and Stefan Hatos Center for Neuropharmacology, University of California Los Angeles, Los Angeles, CA, United States
[k]Jane & Terry Semel Institute for Neuroscience and Human Behavior, University of California Los Angeles, Los Angeles, CA, United States
*Corresponding author: e-mail address: nmassaly@wustl.edu

Contents

Abstract

Across centuries and civilizations opioids have been used to relieve pain. In our modern societies, opioid-based analgesics remain one of the most efficient treatments for acute pain. However, the long-term use of opioids can lead to the development of analgesic tolerance, opioid-induced hyperalgesia, opioid use disorders, and overdose, which can ultimately produce respiratory depressant effects with fatal consequences. In addition

International Review of Neurobiology, Volume 157
ISSN 0074-7742
https://doi.org/10.1016/bs.irn.2020.09.002

to the nociceptive sensory component of pain, negative affective states arising from persistent pain represent a risk factor for developing an opioid use disorder. Several studies have indicated that the increase in prescribed opioid analgesics since the 1990s represents the root of our current opioid epidemic. In this review, we will present our current knowledge on the endogenous opioid system within the pain neuroaxis and the plastic changes occurring in this system that may underlie the occurrence of pain-induced negative affect leading to misuse and abuse of opioid medications. Dissecting the allostatic neuronal changes occurring during pain is the most promising avenue to uncover novel targets for the development of safer pain medications. We will discuss this along with current and potential approaches to treat pain-induced negative affective states that lead to drug misuse. Moreover, this chapter will provide a discussion on potential avenues to reduce the abuse potential of new analgesic drugs and highlight a basis for future research and drug development based on recent advances in this field.

1. Introduction: Pain and the opioid epidemic

In recent years, North America has faced an opioid epidemic. Over a number of decades, opioid use disorder (OUD), including misuse and abuse of both opioid analgesics and illegal opioid substances, has reached an unprecedented prevalence. In 2018, in the United States, 128 cases of lethal overdose attributed to an opioid (opioid medication, heroin and illegal synthetic opioids) were reported daily. In addition, emergency departments have been dealing with an insurgence of opioid-induced respiratory depression, as well as other opioid use related illnesses such as infectious diseases (Dowell, Haegerich, & Chou, 2016; Volkow & McLellan, 2016). This alarming growth in opioid use and misuse in the North American population correlates with a massive increase in opioid analgesic prescriptions. Currently, more than 178 million Americans experience pain that persists for weeks to years, where the prevalence of chronic pain is 11.2% of the United States adult population (Nahin, 2015).

Since the 19th century following the isolation of morphine from crude opium, many opioid formulations have been designed to engage our endogenous opioid system and provide efficient pain relief. The endogenous opioid system is critical in the modulation of pain signaling, but despite its analgesic properties its engagement also has deleterious side effects such as constipation, nausea, sedation, the development of tolerance, abuse potential, and the induction of respiratory depression. To this day, opioid medications remain the most efficient pain reliever for acute, inflammatory conditions, nerve-injured induced pain and other persistent conditions (less than 16 weeks in duration from pain onset) (Dowell et al., 2016; Furlan, Chaparro, Irvin, & Mailis-Gagnon, 2011). Recently, and as a result of recommendations described by the Institute of Medicine in 2010 on pain management (Institute of Medicine (US) Committee on Advancing Pain Research, Care, and Education, 2011), pain was introduced as the "fifth vital sign" requiring practitioners to provide an effective and sustained pain treatment (Lanser & Gesell, 2001). To provide efficient acute pain management, in 1995 the U.S. Food and Drug Administration (FDA) approved the use of an extended release formulation of oxycodone, Oxycontin®, which was marketed as an opioid with less addictive profile, and potential than previous opioid-based analgesics. The aggressive promotion of Oxycontin® for the treatment of acute, moderate to severe pain relief together with the false belief that opioid-based pain killers possess low abuse liability when used for treating pain led to a vast increase in prescription use (Cochran et al., 2016; Krashin, Murinova, & Sullivan, 2016). In the span of only 15 years, from 1997 to 2012, opioid-based prescriptions rose from 91 million to 255 million units dispensed in the United States. During this period, the number of reported opioid-related lethal overdoses, mainly from prescription opioids, increased exponentially (Dowell et al., 2016). As a consequence, the medical community and federal governments imposed restrictions in the number of opioid based therapy units dispensed from health care providers (Von Korff, Kolodny, Deyo, & Chou, 2011). Following restrictions on opioid analgesic dispensing, along with the threat of litigation facing physicians, and the decision by some general practitioners to stop prescribing opioids, a rapid increase in heroin consumption and illicit opioids has been observed. From 2010 to 2017, deaths from synthetic opioids such as fentanyl increased nearly 10-fold, from 3007 (14.3% of opioid-related death) to 28,466 (59.8%) (https://www.cdc.gov/nchs/data/databriefs/db329_tables-508.pdf). While it is likely that the development of misuse and abuse behaviors often arises from the diversion and recreational use of those opioids, it is been shown that prescription opioid use triggers the development of OUDs in 21–29% of pain patients (Vowles et al., 2015).

To better understand the potential risk for opioid misuse in pain conditions, all aspects of the pain experience must be considered. Pain is comprised of many components including a sensory component (pain intensity and location), an affective component (emotional reaction to the pain) and cognitive component (evaluation of the pain). The intense emotional suffering often present in chronic pain patients correlates with the development of persistent disabilities in social, professional and self-care activities (Elman, Borsook, & Volkow, 2013), as well as impairments or disturbances in memory, cognition and attention, sleep, and reduced physical functioning (Dahan, van Velzen, & Niesters, 2014). The development of chronic pain drives synaptic adaptations in brain structures where nociceptive and pain-associated affective components are integrated, which contributes to the development of negative affective states (Garland, Froeliger, Zeidan, Partin, & Howard, 2013; Krashin et al., 2016; Massaly, Morón, & Al-Hasani, 2016; Taylor, Becker, Schweinhardt, & Cahill, 2016). Importantly, persistent negative affective states can lead to co-morbid psychiatric disease such as depression and represent a trigger for the development of OUD as well as other drugs of abuse (Hogarth, 2020). Despite the sustained effort of the scientific community to identify novel targets and develop innovative and non-addictive treatments for the relief of chronic pain, this has yet to result in safer analgesics. Current approaches for the development of non-addictive analgesics consist of non-opioid drugs, biased opioid ligands devoid of addictive properties and positive allosteric modulators and are further discussed later in this chapter. Minimizing the occurrence of OUDs driven by prescription opiates for chronic pain and the development of safer pain mediation requires a better understanding of the neurocircuitry of pain and its interaction with opioid-based pain killers. It should be noted that multiple non-pharmacological approaches are currently used to relieve pain and improve quality of life of patients experiencing chronic pain such as cognitive, rehabilitation and physical therapies. Indeed, the integration of several above mentioned approaches has shown high efficiency in chronic pain relief and may help the medical community in decreasing chronic use of opioid analgesics and prevent the development of OUD in pain patients (Koele et al., 2014).

In this review, we will describe our current understanding of the pain neurocircuitry and its interaction with the endogenous opioid system. We will then discuss the impact of chronic and persistent pain on the development of negative affective states, how this may impact opioid misuse and abuse potential, the neurocircuitry mediating these conditions and the role of endogenous opioid systems within them. Lastly, we provide insight on the current and future directions for the development of safer opioid-based analgesics devoid of abuse potential and chronic pain treatment.

2. Pain and the endogenous opioid system

As defined by the International Association for the Study of Pain (IASP) pain is "*an unpleasant sensory and emotional experience associated with actual or potential tissue damage or described in term of such damage.*" Acute pain is necessary and protective. It serves as a warning signal allowing our bodies to heal while avoiding further injury that may jeopardize our well-being and survival. Pain, thus, primarily helps organisms, through the initiation of defensive and avoidance mechanisms. Noxious stimuli and tissue damage can activate a selective network of peripheral nociceptors projecting (further detailed in the following section), via a spinal relay network, to central brain regions such as cortex, amygdala, bilateral insula, thalamus, anterior cingulate and the prefrontal cortex (Coghill, McHaffie, & Yen, 2003). Within these brain structures, the nociceptive experience acquires an emotional valence which allows an organism to learn and avoid future similar situations.

Despite this obvious protective role, the occurrence of long term adaptive cellular changes within the pain neuroaxis can ultimately lead to the development of chronic pain. Chronic pain is defined as pain lasting at least 3 months, whether associated with an identifiable cause or not (for review Dowell et al., 2016). Preclinical studies and brain imaging in patients experiencing chronic pain reveal neuronal allostatic changes (the process of achieving stability, or homeostasis, through physiological or behavioral change) along pain transmission pathways such as nociceptors phenotype alteration, transcriptome and translatome changes, adaptations in limbic circuits, and interactions between the nervous and immune systems which represent key factors in the development of these chronic states (Ji, Chamessian, & Zhang, 2016; Price et al., 2018). Correlated with those allostatic adaptations, brain structure activity and connectivity are impacted by the presence of chronic pain states (Apkarian, Baliki, & Farmer, 2013; Price et al., 2018). Furthermore, chronic pain states and their negative affective consequences are exacerbated by stress exposure, emotional states, and several other psychological factors (Denk, McMahon, & Tracey, 2014; Garland, 2012). The interaction between long term plastic changes and psychiatric outcomes during chronic pain may explain, at least in part, the ineffectiveness of common therapies. Thus, it is important to understand how pain is encoded and what alterations occur in nociceptive and limbic circuits during chronic pain in order to develop novel and safer analgesics.

Nociceptive sensation is first integrated by activation of peripheral nociceptors projecting to several superficial layers of the spinal dorsal horn as well as in the deeper layers associated with non-noxious sensations. There, nociceptors can transmit the "pain signal" onto second-order dorsal horn neurons that project to the thalamus and subsequently to the somatosensory cortex. The networks from nociceptors to supraspinal centers is defined collectively as the ascending pain pathway which serves as a first-order integration of the pain experience, allowing protective behavior and avoidance learning to prevent further damage (for detailed review Basbaum, Bautista, Scherrer, & Julius, 2009). During pain processing, the activity and efficiency of the spinal projection tracts that relay to different brain structures can be modulated by the activity of local spinal circuits and descending circuits comprised in part by the periaqueductal gray (PAG) neurons projecting to the rostral ventral medulla (RVM) and the locus coeruleus (LC). The RVM and LC send axons to the dorsal horn of the spinal cord to modulate transmission between primary nociceptors and second-order ascending neurons (Basbaum et al., 2009; Heinricher, Tavares, Leith, & Lumb, 2009; Julius & Basbaum, 2001). In addition, several other cortical and subcortical structures play a critical role in pain modulation through descending circuits. The prefrontal cortex (PFC), the anterior cingulate cortex (ACC), the hypothalamus, the insula, the mesolimbic pathway and the amygdala are all involved in the integration of the aversive and emotional component of pain (Tracey & Mantyh, 2007).

Components of the endogenous opioid system are highly expressed CNS sites important for pain processing/transmission, and endogenous opioid signaling is crucial in the modulation of pain. Indeed, both endogenous and exogenous opioid agonists possess potent analgesic properties by inhibiting nociceptive transmission (Besson, 1999; Tracey, 2010). The opioid system is composed of four main receptors: the delta-opioid receptor (DOR), the mu-opioid receptor (MOR) and the kappa-opioid receptor (KOR) and the nociceptin (orphanin FQ) receptor (NOR). These receptors are members of the superfamily of transmembrane G-protein coupled receptors (GPCRs). The MOR and DOR can be activated by the endogenous opioid peptides enkephalins and endorphins, while dynorphin peptide is the main endogenous agonist for the KOR. There are almost 30 peptides that bind to opioid receptors in the central nervous system that are derived from four peptide precursors (pre-prodynorphin, pre-proenkephalin, pre-pronociceptin, and prooopiomelanocortin).

The role of opioid systems in modulating nociception has been investigated using preclinical models during, acute, inflammatory and chronic pain. Until very recently, nociception was assessed using reflexive measures of withdrawal latencies to noxious pressure, thermal, cold and touch stimuli. The presence of hyperalgesia, characterized by an increased response to a normally painful stimulus, and allodynia, defined as painful response to a stimulus that is not normally noxious, represent the main measurable outcomes in animal pain models. To date, engagement of MOR signaling using endogenous or synthetic agonists represent the most efficient analgesic in reversing hyperalgesia and allodynia. Along these lines, most of the current opioid analgesics derived from the *papaver somniferum* opium poppy (morphine, codeine), semi-synthetics based on the structure of thebaine from the opium poppy (oxycodone, hydrocodone), and synthetics with no similarity in chemical structure with any of the poppy alkaloids (methadone, fentanyl) target the MOR to alleviate pain. Interestingly, opioid analgesics do not only modulate nociceptive and reflexive pain signals, but they also alleviate affective and emotional dimensions of pain (Garland et al., 2013). Indeed, motivational pain assays such as conditioned place preference (CPP), Real Time Place Paradigm (RTPP), sucrose preference test, progressive ratio schedule of reinforcement in self-administration approach helped uncovering the role of MOR in the emotional component of pain experience in preclinical models. However, activation of opioid receptors in the peripheral and central nervous systems causes numerous undesirable side effects including vomiting, urinary retention, bradycardia, hypotension, sedation, pruritis, and respiratory depression, which restrict their long-term use and contribute to non-compliance (Arvidsson, Dado, et al., 1995; Arvidsson, Riedl, et al., 1995; Basbaum et al., 2009; Cahill et al., 2014; Mansour, Khachaturian, Lewis, Akil, & Watson, 1988; Zhu, Hsu, & Pintar, 1998).

A key goal of the opioid research field is to dissociate the mechanisms underlying opioid analgesia from those responsible for adverse effects produced by MOR activation. At the cellular level, activation of MOR by endogenous or exogenous agonists engages intracellular recruitment of inhibitory guanosine-5′-triphosophate (GTP) binding proteins (Gi/o) and signaling through the beta-arrestin pathway. Upon opioid receptor recruitment, the alpha and beta-gamma subunits of the G-coupled protein can activate intracellular kinases, such as c-Jun and ERK, and inwardly rectifying potassium channels, respectively. Dissociated beta-gamma subunits can bind and inhibit Ca^{2+} channels, decreasing presynaptic calcium currents and presynaptic neurotransmitter release (Rusin, Giovannucci, Stuenkel, & Moises, 1997; Williams, Christie, & Manzoni, 2001). These signaling cascades subsequently lead to neuronal hyperpolarization and produce analgesia. In addition, activation of the beta-arrestin pathway is sufficient to mediate MOR agonist-induced side effects such as tolerance, respiratory depression and constipation (Bohn et al., 1999; Manglik et al., 2016; Schmid et al., 2017). Thus, the development of G-protein signaling biased-agonists for the MOR, engaging the G-protein pathway with little or no beta-arrestin pathway signaling, have been a main objective to develop safer opioid-based pain medications (Grim et al., 2020; Manglik et al., 2016; Raehal, Walker, & Bohn, 2005). However, the lack of translational efficacy and safety in the development of MOR biased-agonists in human studies and recent preclinical evidence have challenged this model and demonstrated that tolerance and respiratory depression are still observed in beta-2 arrestin knockout mice, suggesting that beta-arrestin pathway may not be necessary for MOR agonist-induced side effects (Kliewer et al., 2019, 2020). Future research will be necessary to further dissect the contribution of G-protein and beta-arrestin pathways to undesirable side effects of MOR agonists.

As nociceptors express MOR peripherally and on their axon terminals within the spinal cord, the development of peripherally restricted MOR agonists, sparing centrally mediated side effects, such as abuse potential and respiratory depression, represent another promising avenue for safer analgesics. Several research groups have been working along those lines on the development of such MOR agonists to provide analgesics that are devoid of misuse potential and lethal side effects. Several studies have demonstrated that topical and subcutaneous administration of loperamide, a peripherally restricted MOR agonist, or its analogs is sufficient to reverse hyperalgesia induced by inflammatory pain (DeHaven-Hudkins et al., 2002; DeHaven-Hudkins & Dolle, 2004; Vadivelu, Mitra, & Hines, 2011). Systemic and local administration of peripherally restricted MOR agonists can also dose-dependently attenuate neuropathic pain-induced hyperalgesia and allodynia (Chung et al., 2012; Guan et al., 2008; Obara et al., 2009; Tiwari et al., 2016), and this analgesic effect is reversed when drugs are co-administered with methyl-naltrexone, a peripherally acting MOR antagonist (Guan et al., 2008). Interestingly, deletion of MOR in nociceptors blocks morphine antinociceptive tolerance and morphine-induced hyperalgesia in uninjured and neuropathic pain conditions, without altering acute morphine antinociception (Corder et al., 2017).

Altogether, these studies establish that peripheral MOR engagement is sufficient but not necessary for opioid-induced changes in sensory nociception. However, if peripherally restricted opioid receptor ligands were effective for pain management, loperamide (Imodium®) would be more widely used, yet there is no firm evidence of loperamide use for pain management. With the objective of blunting centrally mediated reinforcing properties of MOR agonists, a novel category of opioid agonists with actions restricted to the inflammatory site has been recently developed (Spahn et al., 2017). After tissue damage or lesion, the recruitment of inflammatory mediators at the injury site creates a localized acidic milieu (Rodriguez-Gaztelumendi, Spahn, Labuz, Machelska, & Stein, 2018), and a low pKa fentanyl derivate was recently developed to selectively target low pH milieu (Spahn et al., 2017). More specifically, the novel opioid agonist, (±)-*N*-(3-fluoro-1-phenethylpiperidine-4-yl)-*N*-phenyl propionamide (NFEPP), relieves hyperalgesia and allodynia induced by inflammatory and neuropathic pain without inducing commonly observed opioid-induced side effects such as constipation, sedation, and abuse liability (Spahn et al., 2017). While further studies are needed to assess the sufficiency of NFEPP for relief of ongoing non-evoked pain and to assess analgesic tolerance during chronic use, targeting opioids selectively in the inflammatory site represents a novel promising strategy for pain treatment.

MORs are widely distributed throughout the brainstem, midbrain, and forebrain structures where endogenous MOR peptides mediate analgesia. Several studies have used locally targeted microinjections of MOR agonists to uncover brain structures involved in analgesia and the negative affective states arising as consequence of persistent pain. The descending pathway, composed of ventrolateral (vl) PAG to RVM projections, represents a main site of action for the MOR system to modulate nociceptive inputs. MORs are widely expressed in the vlPAG and local infusions of MOR agonists into this structure produces analgesia (Bobeck, McNeal, & Morgan, 2009; Buntin-Mushock, Phillip, Moriyama, & Palmer, 2005; Jacquet & Lajtha, 1974). MOR signaling within the vlPAG is necessary for the development of analgesic tolerance to long-term opioid treatment (Lane, Patel, & Morgan, 2005). The RVM possesses different cell types characterized as ON, OFF, or neutral cells. They differ in their analgesic or pro-nociceptive properties and their response to opioid agonists (Basbaum & Fields, 1984; Fang, Haws, Drasner, Williamson, & Fields, 1989). Briefly, MOR agonists inhibit ON cells while they disinhibit OFF cells to produce antinociception (Fang et al., 1989). The selective engagement of MOR in a cell type-specific fashion remains an unrealistic task for human chronic pain therapies, with perhaps the exception of intrathecal (spinal) pumps for cancer pain treatment. Our current therapies mostly rely on chronic systemic (primarily oral) administration of MOR agonists, which can lead to aberrant plastic events and produce opioid-induced-hyperalgesia (OIH); a phenomenon that may or may not be a consequence of opioid-induced physical withdrawal (Hayhurst & Durieux, 2016). This particular issue represents a trigger for the development of misuse behavior and relapse to opioid use to relieve the symptoms of OIH, as well as the negative affective states associated with pain and opioid withdrawal.

Opioids, at least in part, exert reinforcing properties by inhibiting GABA release onto dopamine neurons in the ventral tegmental area (VTA) thus leading to increased release of dopamine in various brain regions (Devine, Leone, Pocock, & Wise, 1993; Giuliano, Robbins, Wille, Bullmore, & Everitt, 2013; Le Merrer, Becker, Befort and Kieffer, 2009; Xiao & Ye, 2008). In the VTA, MORs are expressed on GABAergic interneurons and GABAergic terminals arising from the rostromedial tegmental nucleus, and activation of these MORs leads to inhibition of GABA release, thereby removing their tonic inhibition, and increasing dopaminergic transmission from the VTA to efferent structures, including the nucleus accumbens (NAc) (Johnson & North, 1992; Matsui, Jarvie, Robinson, Hentges, & Williams, 2014; Siuda et al., 2015). Interestingly, dopamine is necessary for opioid reward only when animals are opioid-dependent (Hnasko, Sotak, & Palmiter, 2005; Laviolette, Gallegos, Henriksen, & van der Kooy, 2004). Thus, the initial learning of opioid reward as expressed as a conditioned place preference does not require dopamine. However, dopamine is critical for the reinforcing effects of opioids in operant paradigms and changes in dopaminergic neuronal activity represent a basis for the misuse and abuse potential of MOR agonists (Corre et al., 2018; Fields & Margolis, 2015; Wise, 2006). The engagement of mesolimbic dopamine is required for motivational learning and important in assigning salience and valence to the stimuli. Interestingly, local activation of MOR in the NAc core and shell can also increase dopamine release, suggesting both regulation of dopamine terminals by local striatal circuits in this target structure (Hipólito et al., 2015), and a feedback mechanism for MOR signaling onto D1 receptor-expressing medium spiny neurons in the NAc that project back to the VTA (Cui et al., 2014). Stimulation of MORs in the NAc is associated with reinforcement behaviors in rodents (Castro & Berridge, 2014). More specifically, MOR engagement in NAc shell substructures can drive either rewarding (hot spot) or aversive effects (cold spot) suggesting that there is further complexity in exactly how these receptors can regulate striatal function (Castro & Berridge, 2014). Overall, MOR engagement in the VTA and "hotspot" sub-regions of the NAc shell drive opioid reinforcement.

In contrast to the analgesic and rewarding properties of acute activation of the MOR system (Matsui et al., 2014; Siuda, Carr, Rominger, & Violin, 2017; Wade & Fairbanks, 2014; Wise & Koob, 2014), stimulation of the KOR system leads to dysphoria, anhedonia, and aversive behaviors (Al-Hasani et al., 2015; Bruchas, Land, & Chavkin, 2010; Land et al., 2008, 2009; Leitl, Onvani, et al., 2014; Muschamp & Carlezon, 2013; Shippenberg, Stein, Huber, Millan, & Herz, 1988). Because of these opposing actions, the MOR and the KOR are often referred as opponent systems. However, stimulation of the KOR system by its endogenous opioid ligand dynorphin (Chavkin, James, & Goldstein, 1982) can efficiently relieve itch and injury-induced hyperalgesia (Cowan, Kehner, & Inan, 2015; Ho, Mannes, Dubner, & Caudle, 1997; Kivell & Prisinzano, 2010; Schäfer, Brack, & Stein, 2000; Snyder et al., 2018; Vanderah, 2010). Indeed, KOR agonists applied peripherally reduce nociception in several persistent pain models such as formalin and inflammatory pain models (Antonijevic, Mousa, Schafer, & Stein, 1995; Auh & Ro, 2012; Binder et al., 2001; Cunha et al., 2012; Keïta, Kayser, & Guilbaud, 1995; Obara et al., 2009; Stein, Millan, Shippenberg, Peter, & Herz, 1989). In conditions of chronic pain, a subset of dorsal horn interneurons was shown to increase dynorphin expression (Boyle et al., 2017; Lai, Luo, Chen, & Porreca, 2008; Podvin, Yaksh, & Hook, 2016; Xu et al., 2004), suggesting endogenous adaptations within the KOR system. Peripherally restricted KOR agonists have also demonstrated efficacy in persistent pain and incision models (a more acute pain model) as well as the acetic writhing and itch models (Barber et al., 1994; Binder & Walker, 1998; Caram-Salas et al., 2007; Inan & Cowan, 2004; Snyder et al., 2018; Vanderah et al., 2008, 2004). In addition, recruitment of KORs in several discrete brain structures, such as the dorsal raphe nucleus (DRN) can produce antinociception (Land et al., 2009). While activation of KOR system may potentially represent a promising approach for the treatment of pain, several potential limitations must be considered. Indeed, as mentioned above, enhanced KOR signaling produces aversion, impairs stress coping mechanisms, reduces the reinforcing properties of stimuli and promotes drug reinstatement (Bruchas & Chavkin, 2010; Crowley et al., 2016; Lalanne, Ayranci, Kieffer, & Lutz, 2014; Land et al., 2008; Leitl, Onvani, et al., 2014; Massaly et al., 2019; McCall et al., 2017;

Shippenberg et al., 1988; Shippenberg, Zapata, & Chefer, 2007; Tejeda et al., 2017; Van't Veer & Carlezon, 2013). In addition, KOR signaling is necessary for the development of heroin escalation (Schlosburg et al., 2013) suggesting that engagement of this system may promote misuse and abuse of opioids. Thus, KOR agonists possess limited potential as analgesics due to their negative centrally mediated effects and their potential to increase drug misuse and abuse (Chavkin & Koob, 2016; Schlosburg et al., 2013).

Several recent studies demonstrated that the central KOR system is actively recruited by pain to drive negative affective states and that the use of KOR antagonists may represent a potential avenue for the treatment of pain-induced negative affective states (Liu et al., 2019; Massaly et al., 2019; Narita et al., 2005; Nation et al., 2018; Navratilova et al., 2019; Page et al., 2019; Phelps, Navratilova, Dickenson, Porreca, & Bannister, 2019). Within the NAc, a structure important for the integration of hedonic value of both rewarding and aversive stimuli, the KOR system plays an important role in mediating reward-seeking behaviors (Al-Hasani et al., 2015; Castro & Berridge, 2014; Massaly et al., 2019). In addition, evidence has demonstrated that both inflammatory and chronic pain can recruit KOR signaling in the NAc to drive negative affective states and decrease the reinforcing properties of opioids (Liu et al., 2019; Massaly et al., 2019; Narita et al., 2005). We propose that KOR antagonists may represent a functional substrate to alleviate pain-induced negative states and improve patient well-being. In this way, the development of novel selective, potent and non-addictive KOR antagonists to dampen the negative affect associated with pain could (1) reduce the likelihood of co-morbid psychiatric disorders and (2) avoid the drug misuse potential associated with most opioid analgesics (Hogarth, 2020). Along these lines, buprenorphine is usually administered to treat both pain conditions and OUDs. Buprenorphine is a partial MOR agonist but possesses antagonistic properties at the KOR and DOR. However, despite being considered a safer analgesic than morphine and fentanyl (due to its ability to dampen KOR-induced negative affective states), buprenorphine still retains misuse potential (Jones et al., 2015). Thus, the development of selective KOR antagonists may represent a safer strategy for the development of pain treatments. KOR antagonists such as norBNI and JDTic possess a delayed onset of action that ranges from hours to days, but even minimally effective doses produce antagonist action that can last several weeks. Therefore, the development of KOR antagonists with a shorter "medication-like" duration of action is currently undergoing. While a few short-acting KOR antagonists have been developed

(Guerrero et al., 2019; Page et al., 2019), a thorough characterization of these ligands for pain treatment and OUD, as well as their efficacy in a clinical setting still needs to be determined.

The DOR system represents another potential opioid target for the management of pain states. DORs are highly expressed in forebrain regions (Mansour et al., 1988) and while DOR activation exhibits low analgesic properties in acute pain models, activation of the DOR system provides substantial analgesia in chronic pain conditions (Cahill, Holdridge, & Morinville, 2007; Pradhan, Befort, Nozaki, Gavériaux-Ruff, & Kieffer, 2011). Peripheral application of DOR agonists promotes analgesia in animal models of inflammatory pain (Gendron et al., 2006; Gendron, Mittal, Beaudry, & Walwyn, 2015; Pradhan et al., 2009; Zhang & Bao, 2012) and neuropathic pain (Zhang & Bao, 2012). Unlike opioid-based MOR analgesics, DOR agonists possess fewer side effects (Bodnar, 2020; Gavériaux-Ruff & Kieffer, 2011; Gendron et al., 2015) and most importantly, exhibit no abuse potential (Negus, Gatch, Mello, Zhang, & Rice, 1998). In chronic pain states (i.e., neuropathic pain models or migraine), DOR stimulation demonstrates high analgesic potential (Cahill, Morinville, Hoffert, O'Donnell, & Beaudet, 2003; Fraser, Pradhan, Clarke, & Wahlestedt, 2000; Gavériaux-Ruff, Karchewski, Hever, Matifas, & Kieffer, 2008; Gendron et al., 2006; Hurley & Hammond, 2000; Nadal, Baños, Kieffer, & Maldonado, 2006; Pradhan et al., 2009). Further supporting the role of DORs as analgesics, genetic ablation of these receptors enhances injury-induced hyperalgesia and allodynia (Gavériaux-Ruff et al., 2008; Nadal et al., 2006). Moreover, chronic use of DOR agonists does not induce physical dependence or withdrawal upon treatment cessation (Brandt, Furness, Rice, Fischer, & Negus, 2001). The anxiolytic properties of DOR activation also make it a promising target (similar to KOR antagonism) for improving mood of chronic pain patients. It is well known that negative mood states are not only predictors of opioid misuse, but it also impacts the pain experience, such that co-morbid anxiety or depression exaggerates pain intensity and negatively impacts quality of life (Jamison & Edwards, 2013; Martel, Dolman, Edwards, Jamison, & Wasan, 2014). In addition, DOR agonists have great potential as a therapeutic treatment for chronic migraine (Pradhan, Smith, Zyuzin, & Charles, 2014). However, DOR agonists have been associated with convulsions, which has terminated many clinical development studies (Broom et al., 2002a, 2002b; Chung et al., 2015, Comer et al., 1993; Danielsson et al., 2006; Dykstra, Schoenbaum, Yarbrough, McNutt, & Chang, 1993;

Jutkiewicz, Baladi, Folk, Rice, & Woods, 2006). These negative consequences are an important factor to consider for their use as analgesics in vivo and in clinical settings, although there is evidence that non-internalizing ligands do not share this convulsive side effect (Vicente-Sanchez et al., 2018).

As mentioned above, DOR activation has little or no effect on nociception in pain-naïve and acute phasic pain, as compared to MOR activation (Gallantine & Meert, 2005). Yet, there is still interest in a combination of peripherally restricted agonists for both MOR and DOR as a potential pain relief strategy (Bruce et al., 2019; Günther et al., 2018). In addition to its analgesic properties, DOR signaling alters the development of tolerance to the analgesic effects of MOR agonists. For example, tolerance to the analgesic properties of MOR agonists is abolished in DOR knockout animals or when combined with the use of DOR antagonists (Abdelhamid, Sultana, Portoghese, & Takemori, 1991; Beaudry, Gendron, & Morón, 2015; Váradi et al., 2016; Zhu et al., 1999). Furthermore, DOR antagonists decrease the reinforcing properties of MOR agonist-based analgesics (Abdelhamid et al., 1991; Billa, Xia, & Morón, 2010; Morón et al., 2010; Váradi et al., 2016). Despite this promise, an argument has been made that since DOR is required for hippocampal-dependent learning, ablation or antagonism of DOR could impair associative learning that contributes to the analgesic tolerance of morphine (Cahill & Ong, 2018).

3. Pain-induced alterations in opioid consumption

Pain and pleasure have opposite hedonic value, and thus can both guide our motivational decisions and learning to either avoid or seek emotional experiences. From a behavioral standpoint, reinforcing stimuli such as pleasant odors, music, palatable food and sexual behavior can decrease pain sensitivity and negative affect (Leknes & Tracey, 2008). While pain generates aversive states that disrupt homeostatic balance, pain relief produces negative reinforcement (King et al., 2009; Navratilova & Porreca, 2014). Interestingly, clinical reports suggest that opioids produce significant reinforcement by the relief of pain-induced negative affect; for example, patients with chronic pain have reported reduction in their experience of affective aversive states following morphine use while the nociceptive pain sensation was still present (Price, Von der Gruen, Miller, Rafii, & Price, 1985). This provides further evidence that negative affect is an important part of the pain experience and plays a major role in patients' quality of life.

The persistence of pain-induced negative affective states can trigger allostatic changes in limbic and emotion-related brain structures (Baliki & Apkarian, 2015; Leknes & Tracey, 2008) which over time can lead to the development of anxiety and/or major depression (Elman et al., 2013; Yalcin & Barrot, 2014). Additionally, pain itself can impair reward processing and the patient's ability to enjoy daily activities, characteristic of an anhedonic state (Elman et al., 2013; Leknes & Tracey, 2008).

Preclinical evidence demonstrates that both acute and chronic pain indeed impact reward-induced positive reinforcement (Cahill et al., 2013; Elvemo, Landrø, Borchgrevink, & Håberg, 2015; Hipólito et al., 2015; Leitl, Onvani, et al., 2014; Leknes & Tracey, 2008; Martin & Ewan, 2008; Navratilova & Porreca, 2014; Porreca & Navratilova, 2017; Taylor et al., 2016). Given the fact that opioid-based analgesics possess positive reinforcing properties that can lead to OUD, it is critical to understand the role of pain on the abuse potential of these medications.

Clinical studies have reported that a subset of patients with chronic pain develop misuse and overuse patterns of opioid consumption (Passik, Messina, Golsorkhi, & Xie, 2011). Thus, several preclinical behavioral paradigms have been developed to assess the impact of pain on the reinforcing properties of opioids and the neuronal mechanisms underlying pain-induced negative affect. Many models of pain and voluntary consumption of opioid in animal models have been used to uncover those mechanisms. Unfortunately, the plethora of models led to discrepancy and conflicting results on the consequences of pain on opioid consumption. One of the first reports on the interaction between pain and opioid consumption investigated the reinforcing properties of fentanyl in rats experiencing inflammatory pain (induced by *Mycobacterium butyricum* injection in the tail base) and reported that ongoing inflammatory pain increases voluntary drinking of fentanyl when compared to control animals (Colpaert, Meert, De Witte, & Schmitt, 1982; Colpaert et al., 2001). These results have been replicated in rats by the same group (Colpaert et al., 2001) and others (Kupers & Gybels, 1995). However, using a similar voluntary drinking operant task in, another group reported that inflammatory, chemotherapy-induced and neuropathic pain reduced motivation to seek and consume fentanyl in mice (Wade et al., 2013). It should be noted that in the latter study conducted in mice, the overall locomotor activity of pain animals tended to be lower than controls, which may confound interpretation of the results. To this end, a separate experiment in the same study showed that pain does not impact reinforcing properties of natural rewards as compared to opioids

(Wade et al., 2013). Since then, additional reports have demonstrated that pain can impact the hedonic value of opioids, without changing food seeking and consumption behaviors (Hipólito et al., 2015; Taylor et al., 2015).

To further assess the potential rewarding properties of opioids under pain conditions other approaches such as conditioned place preference (CPP) and intravenous drug self-administration (IVSA) have been used. Conditioned place preference consists of repeated pairing of a specific context with drug, such that the animal associates the interoceptive cues of the drug with the contextual cues. Using this model in adult rats, inflammatory pain has been shown to lower the number of morphine-context conditioning sessions necessary to elicit a preference for the drug context (Zhang et al., 2014), as well as leftward shift in the dose response curve for morphine-induced CPP for male rats in a model of neuropathic pain (Cahill et al., 2013) and inflammatory pain (Sufka, 1994). However, many other reports demonstrate that the rewarding properties of morphine are decreased in both adult rats and mice experiencing inflammatory, neuropathic and, cancer pain (Betourne et al., 2008; Narita et al., 2005; Niikura et al., 2008; Ozaki, Narita, Iino, Miyoshi, & Suzuki, 2003; Ozaki et al., 2002; Taylor et al., 2015). While CPP is a widely used model to assess rewarding properties of stimuli, it has some limitations, especially when studying opioids in pain conditions as central opioid actions in the brain or spinal cord that are important in mediating anti-hyperalgesia and analgesia (i.e., negative reinforcement). As mentioned above, the analgesic properties of opioids may be sufficient to trigger preference for the drug-paired context, thus introducing a confound between the negative reinforcement of pain relief and the rewarding properties of the drug itself. Furthermore, CPP is dependent on non-voluntary drug exposure and thus, despite its translational value, possesses a limited face validity when compared to patients voluntarily consuming opioids to relieve the nociceptive and negative emotional aspects of pain. In this regard, the operant intracranial self-stimulation (ICSS) model can be used to integrate voluntary seeking of reinforcing stimulations during pain conditions (Moerke & Negus, 2019). ICSS is an operant paradigm in which animals press a lever triggering a contingent electrical stimulation. This approach can be used to test abuse liability of drugs where electrical stimuli are delivered in the medial forebrain bundle of the lateral hypothalamus or the VTA, driving excitatory inputs to the mesolimbic dopaminergic neurons in the VTA and thus promoting DA release in nucleus accumbens (Negus & Moerke, 2019).

Acute oxycodone or morphine administration increases the frequency of responding for ICSS, whereas natural or naloxone precipitated opioid withdrawal suppressed ICSS responding (Altarifi & Negus, 2011; Wiebelhaus, Walentiny, & Beardsley, 2016). Acute pain induced by intraperitoneal administration of dilute acid did not alter medial forebrain bundle ICSS frequency (Miller, Altarifi, & Negus, 2015), nor altered medial forebrain bundle ICSS in male or female rats affected by chemotherapy-induced (paclitaxel) neuropathic pain (Legakis & Negus, 2018). However, treatment of the neuropathic pain induced by the chemotherapeutic drug showed that as morphine tolerance developed to the anti-allodynic effects, reward increased as evidenced by changes in ICSS frequency (Legakis & Negus, 2018). Although, morphine and cocaine produced similar effects on ICSS responding in the medial forebrain bundle, but paraventricular nucleus of the hypothalamus ICSS was facilitated to a greater extent by morphine than cocaine, and the effects of either drug were unaltered in neuropathic pain rats induced by spinal nerve ligation (Ewan & Martin, 2012). Interestingly, VTA ICSS was not altered by the presence of neuropathic pain in rats, but oxycodone, methadone, fentanyl and hydromorphine facilitation of VTA ICSS in controls rats was absent in neuropathic pain rats (Ewan & Martin, 2011). Although, post-surgical pain induced by paw incision initially decreased VTA ICSS for the first 2 days after injury (Ewan & Martin, 2014). Taken together, a significant number of studies agree that the occurrence of chronic pain does not alter ICSS stimulation frequency, independent of the brain region of stimulation, however, it does alter the responses to opioids.

Intravenous self-administration represents the gold standard model to assess the reinforcing properties and abuse potential of drugs. This approach allows assessment of drug-seeking and drug intake in an operant paradigm. Self-titer, rate of drug intake and escalation of drug intake can be measured during single sessions or across days and weeks (Tsibulsky & Norman, 2005), allowing studies of opioid analgesic intake during the development and maintenance of pain states. One study demonstrated that morphine consumption is not altered during the first 10 days of IVSA after pain induction ($1\,mg\,kg^{-1}$ per injection). Interestingly, after a period of withdrawal (assessed up to 13 days after termination of morphine access), animals experiencing neuropathic pain demonstrated significantly higher rates of morphine-seeking and morphine consumption than control animals, when the amount of morphine obtained per injection was lowered to $0.2\,mg\,kg^{-1}$ (Hou, Cai, & Pan, 2015). In contrast and using the same IVSA approach, another group demonstrated that morphine self-administration was lower in animals experiencing inflammatory pain (Lyness, Smith, Heavner, Iacono, & Garvin, 1989). While these results again seem contradictory, it is important to note that the in the study by Hou and collaborators, animals were trained to self-administer morphine before the induction of pain, while the study by Lyness and collaborators tested the acquisition of morphine self-administration after pain induction. Importantly, it has been shown that heroin, morphine, methadone, fentanyl, and hydromorphone consumption (using the IVSA procedure) is lower in rats with neuropathic pain induced by sciatic nerve ligation (SNL) compared to sham animals (Martin, Kim, Buechler, Porreca, & Eisenach, 2007). In this study, full pharmacological characterization of the impact of pain on the reinforcing properties of opioids showed that while IVSA of low doses of each opioid (heroin, morphine, methadone, fentanyl, and hydromorphone) compound was lower in pain conditions, the reinforcing properties of higher doses were unaltered—with pain animals consuming a total amount of opioid per self-administration session similar to that achieved by control animals (Martin et al., 2007). This result is similar to another recent report showing that induction of inflammatory pain following initial heroin self-administration, decreased the number of low-dose heroin reinforcers consumed per self-administration session. Interestingly, in this study by Hipólito and collaborators, IVSA of heroin was significantly increased when rats were exposed to very high doses of the opioid (Hipólito et al., 2015). Together, these studies demonstrate that during inflammatory and chronic pain states, the opioid dose-response relationship for its reinforcing properties is shifted to the right (such as animals in pain doubled the intake of a very high dose of heroin, whereas control animals didn't alter the intake of the same dose. Conversely, animals in pain significantly decreased the intake of the lower dose of heroin). While this can lead to lower drug consumption for low-titer opioid analgesics in the short run (acute and short opioid exposure to treat perioperative pain, for example), it highlights an increased risk for the development of OUD in patients that develop tolerance and are exposed to higher concentrations of opioid analgesics.

4. Circuits underlying pain-induced negative affect and OUD

There is an increased risk for the development of OUD in patients with long-term persistent pain and opioid access, which is likely related to plastic changes occurring within the central nervous system that drive the emergence of negative affective states (Garland et al., 2013; Martin & Ewan, 2008; Massaly et al., 2016). Clinical and preclinical studies have pinpointed several brain structures involved in the affective dimension of pain: for example, the ACC is highly involved in processing the affective and emotional components of pain (Becerra, Navratilova, Porreca, & Borsook, 2013; Vogt, 2005). Functional Magnetic Resonance Imaging (fMRI) and Positron Emission Tomography (PET) studies have revealed that the endogenous MOR system is engaged in the ACC during persistent pain and affective pain relief states (Borras et al., 2004; Zubieta et al., 2005). Early studies have uncovered that pain unpleasantness rating in patients is tightly correlated with the activity of the ACC (Rainville, Duncan, Price, Carrier, & Bushnell, 1997). Additionally, preclinical studies confirmed that MOR signaling within the ACC is necessary to alleviate ongoing pain in rodents (Navratilova et al., 2015). In line with these results, microinfusions of MOR agonists locally in the ACC reduce pain-induced negative affect without altering the expression of mechanical allodynia (LaGraize, Borzan, Peng, & Fuchs, 2006). Altogether, these findings point toward a major role for the ACC in the integration and maintenance of negative affective states induced by pain.

The mesolimbic pathway is a key brain circuit involved in the development of opioid addiction and, because of its alterations during chronic and persistent pain states, it represents an ideal candidate for mechanistic studies on pain and opioid misuse co-morbidities (Cui et al., 2014; Fields & Margolis, 2015). The mesolimbic dopamine pathway is composed of dopaminergic projections from the VTA projecting to the striatum, including the NAc (Morales & Margolis, 2017). The dopaminergic projections to the NAc are involved in reward and aversion processing (Becerra & Borsook, 2008) and also mediates pain-induced negative affect. However, it should be noted that dopaminergic neurons co-release other neurotransmitters such as glutamate and GABA, also involved in integration of reward and aversion. In addition, non-dopaminergic neurons originating from the VTA are also involved in these processes (see Morales & Margolis, 2017 for extensive review). It is well established that neuroplastic changes in mesolimbic circuitry, including molecular and cellular changes within the mesolimbic dopamine system, lead to the genesis of negative affect (Koob & Le Moal, 2001; Koob & Volkow, 2010, 2016; Wise & Koob, 2014). Both natural rewards and drugs of abuse, including opioids, activate this pathway and trigger accumbal dopamine release (Devine et al., 1993;

Giuliano et al., 2013; Hipólito et al., 2015; Taylor et al., 2015). Interestingly, those drug-induced dopaminergic responses are blunted in detoxified abusers (Volkow & Morales, 2015; Volkow et al., 1997). Patients experiencing chronic pain show reduced NAc activity and alterations in reward evaluation, decision making, and motivation tasks (Apkarian et al., 2004; Verdejo-García, López-Torrecillas, Calandre, Delgado-Rodríguez, & Bechara, 2009; Walteros et al., 2011). Furthermore, the resolution of pain is negatively associated with NAc activity in chronic pain patients (Baliki, Geha., Fields, & Apkarian, 2010), suggesting a dysregulation of dopaminergic function during persistent pain states. As noted above, positive stimuli and opioids cause release of dopamine in the NAc via inhibition of GABA release onto VTA dopamine neurons. Thus, the observed decrease in MOR function in GABAergic projections onto the VTA during pain impairs opioid-induced disinhibition of dopaminergic neurons, which in turn decreases opioid reinforcing properties (Hipólito et al., 2015; Niikura, Narita, Butelman, Kreek, & Suzuki, 2010; Ozaki et al., 2002). Similarly, a positive correlation has been observed between alterations in opioid reinforcing properties during pain states and dopaminergic transmission within the NAc (Hipólito et al., 2015; Narita et al., 2005; Niikura et al., 2008; Ozaki et al., 2002; Taylor et al., 2015). Furthermore, pain relief produces MOR-dependent dopamine release within the NAc, further corroborating the interactions between opioid systems and dopamine-dependent reinforcement processes (Navratilova et al., 2012). In addition to MOR, KORs can also locally alter dopamine release within the NAc (Muschamp & Carlezon, 2013; Nestler & Carlezon, 2006; Van't Veer & Carlezon, 2013). Recent work demonstrated that neuropathic pain recruits KOR signaling in the NAc to reduce dopaminergic activity and induce negative affective states (Liu et al., 2019). Another laboratory demonstrated that not only KOR recruitment but also dynorphin-containing neurons in the NAc were sufficient and necessary to drive pain-induced depression of motivational states (Massaly et al., 2019). These observations support the notion that dynorphin acting on KORs in the NAc is responsible for blunting morphine-induced dopamine release and for the development of morphine CPP under inflammatory pain conditions (Narita et al., 2005). Altogether, these studies demonstrate that KOR signaling represents a potential target to prevent negative affective states induced by pain which could ultimately reduce or mitigate the emergence of OUDs (Cahill et al., 2014; Chavkin & Koob, 2016; Liu et al., 2019; Massaly et al., 2019, 2016). It must also be noted that despite its high potential to treat these different disorders, former studies have demonstrated that subcutaneous injection of the KOR antagonist, norBNI, does not reverse pain-induced decrease in lever presses during ICSS procedure (Leitl, Potter, et al., 2014; Leitl, Onvani, et al., 2014). This lack of effect may be explained by the simultaneous analgesic and aversive properties of systemically administered KOR agonists that readily cross the blood-brain barrier. Indeed, if KOR antagonists relieve negative affective centrally, their peripheral action may blunt KOR-mediated endogenous analgesia and thus mask negative affect relief, that should be reflected in altered ICSS responding. In addition, KOR stimulation in discrete substructures of the NAc drive opposite hedonic effects (Castro & Berridge, 2014). Overall, we posit that targeted inhibition of KOR signaling in pain states represents a promising strategy to dampen negative affective states and reduce the development of OUD in patients using opioid-based analgesics (Cahill et al., 2014; Chavkin, 2011; Liu et al., 2019; Massaly et al., 2019, 2016).

In addition to the mesolimbic pathway, the amygdala is another brain area that undergoes neuroplastic changes during persistent pain states and in turn drives alterations in reinforcing and rewarding properties of opioids (Carrasquillo & Gereau, 2007; Crock et al., 2012; Janak & Tye, 2015; Narita et al., 2006). Both the central nucleus of the amygdala and the basolateral amygdala (BLA) are key players in the integration of affect, mood, fear disorders as well as reinforcement (Pare & Duvarci, 2012; Veinante, Yalcin, & Barrot, 2013). Recent studies have elegantly uncovered neuronal ensembles in the amygdala that are necessary for the expression of the emotional component of noxious stimuli (Corder et al., 2019). This study holds promising potential for future treatment of negative affective states that accompany ongoing pain and thus decrease both the development of co-morbid psychiatric disorders and opioid drug misuse potential. This is especially interesting as projections from the BLA to the NAc modulate dopamine-mediated neuronal responses to both stress and opioid reward salience (Lintas et al., 2011, 2012; Namburi et al., 2015; Stuber et al., 2011). Blocking MOR in the lateral amygdala using MOR antagonist beta-funaltrexamine reverses pain-induced reductions in heroin intake in rats with neuropathic pain (Martin et al., 2011). Further, inhibition of amygdala GABAergic neurons with MOR agonists reduces foot-shock aversive conditioning reactions and disinhibits the descending brainstem pain modulation pathway (Han, Soleiman, Soden, Zweifel, & Palmiter, 2015; Namburi et al., 2015; Winters et al., 2017). These studies further implicate the amygdala as an important regulation center for pain and negative affect. Other work has shown that inflammatory pain promotes KOR signaling in the amygdala to drive negative affective states (Nation et al., 2018; Navratilova et al., 2019). Hence, manipulating MOR and KOR signaling in the amygdala holds promising therapeutic potential for the treatment of pain-induced negative affect and anhedonia.

5. Development of safer analgesics approaches for chronic pain

The pain research community aims to develop the Holy Grail of analgesics—a drug with significant analgesic properties that is devoid of abuse potential. Current pain treatments with opioid medications are mostly limited by their abuse potential and the respiratory depression events that may arise from misuse or overdose. There are several ongoing drug development projects funded by the NIH Heal Initiative to develop new pharmaceutical compounds devoid of those deleterious side effects. Several research groups have been focusing on the development of MOR and KOR biased agonists. This strategy leverages the fact that signal transduction from opioid GPCRs through either G-protein or beta-arrestin may lead to distinct behavioral outcomes. For example, mice lacking beta-arrestin protein do eexhibit an analgesic response to opioid treatment but not respiratory depression, tolerance and opioid reinforcement (Bohn et al., 1999). This finding prompted the design of MOR agonists biased toward G-protein signaling to produce efficient analgesics devoid of side effects (Gupta et al., 2016; Kingwell, 2015; Manglik et al., 2016; White et al., 2014). While these biased MOR compounds represent a promising approach for the improvement of pain treatment, recent evidence using beta-arrestin knockout mice demonstrate that both morphine and fentanyl still retain respiratory depressant potential (Kliewer et al., 2020). This underscores the necessity for more in-depth examination on the characterization of these biased compounds, their mechanism of action and their full set of effects.

Another exciting strategy for pain treatment points toward the use of opioid receptor heterodimers as drug targets to reach maximal analgesic efficiency. High-throughput screening approaches have identified many compounds that are being tested for their analgesic properties as well as their safety profile, mostly as agonists of the MOR/DOR heterodimers (Bruce et al., 2019; Fujita, Gomes, & Devi, 2014, 2015). The development of selective milieu MOR agonist has also been investigated: for example, as mentioned above, a fentanyl derivate with decreased pKA allows selective activity on MOR in a low pH milieu (Rodriguez-Gaztelumendi et al., 2018; Spahn et al., 2017). The authors took advantage of the acidic nature of the inflammatory milieu and demonstrate that NFEPP has higher efficiency at low pH and selectively reverses hyperalgesia and allodynia without inducing commonly observed opioid side effects (Rosas, Huang, Roth, & Dockendorff, 2019; Spahn et al., 2017). Lastly, the development of KOR antagonists to decrease negative affective states induced by persistent pain represents a potential promising avenue (Cahill et al., 2014; Chavkin, 2011; Liu et al., 2019; Massaly et al., 2019; Navratilova et al., 2019). Improving the treatment of prolonged negative affective states may decrease the incidence or severity of psychiatric co-morbidities associated with pain, and may also reduce opioid abuse potential.

In addition to opioid pharmacology, chronic pain treatment still employs a combination of nonsteroidal anti-inflammatory drugs (NSAIDs), antidepressants, anticonvulsants, local anesthetics, alpha-2 adrenoreceptors agonists, NMDA antagonists and more recently, in certain states, cannabinoids (Derry, Derry, & Moore, 2013; Eilers, Philip, Bickler, McKay, & Schumacher, 2001; Graff-Radford, Shaw, & Naliboff, 2000; Loftus et al., 2010; Lunn, Hughes, & Wiffen, 2014; Moore, Derry, Aldington, Cole, & Wiffen, 2015; Pirbudak, Sevinç, Maralcan, & Kiliç, 2014; Smith, Deshpande, Collins, Katz, & Losina, 2016; Wang et al., 2015; Whiting et al., 2015; Wiffen et al., 2017). All these pharmacological agents are routinely used to try and dampen the nociceptive and emotional component of acute and chronic pain without the use of opioid-based analgesics. Some treatments such as ketamine and cannabinoids have been shown to also decrease opioid intake during perioperative care (Boehnke, Litinas, & Clauw, 2016; Cengiz et al., 2014; Cichewicz, 2004; Zakine et al., 2008). However, despite their "safer" profiles and analgesic properties, cannabinoids and NMDA antagonists (Ketamine) still retain misuse potential and need to be closely monitored by health care providers even if their life-threatening potential remains lower than opioid-based analgesics.

6. Conclusions

The widespread use of opioid-based analgesics has been a major factor in the emergence of the opioid epidemic. While the diversion of these pills for recreational use by pain-free individuals may have contributed in this process, one-fifth of chronic pain patients treated with opioid-based medications develop OUD. Even more than the sensory component of chronic pain conditions, the persistence of negative affective states associated with chronic pain represents an important trigger for abuse potential in patients treated with increasing doses of opioids over the long-term. Thus, the development of non-addictive pharmacological analgesics and non-pharmacological approaches is critical for the treatment of the emotional component of pain. Decreasing the occurrence and persistence of negative affect associated with chronic pain conditions would indeed dampen both the development of co-morbid neuropsychiatric diseases and the abuse potential of opioid-based analgesics. In parallel, the development of pain treatments devoid of abuse potential will be necessary to reduce pain analgesics diversion and reduce the proportion of chronic pain patients with a high risk of abuse that are exposed to opioid medications. Overall, these efforts are necessary for optimizing and developing novel safe and effective strategies for treating chronic pain.

References

Abdelhamid, E. E., Sultana, M., Portoghese, P. S., & Takemori, A. E. (1991). Selective blockage of delta opioid receptor prevents the development of morphine tolerance and dependence in mice. *The Journal of Pharmacology and Experimental Therapeutics*, *258*, 299–303.

Al-Hasani, R., McCall, J. G., Shin, G., Gomez, A. M., Schmitz, G. P., Bernardi, J. M., et al. (2015). Distinct subpopulations of nucleus accumbens dynorphin neurons drive aversion and reward. *Neuron*, *87*, 1063–1077.

Altarifi, A. A., & Negus, S. S. (2011). Some determinants of morphine effects on intracranial self-stimulation in rats: Dose, pretreatment time, repeated treatment, and rate dependence. *Behavioural Pharmacology*, *22*, 663–673.

Antonijevic, I., Mousa, S., Schafer, M., & Stein, C. (1995). Perineurial defect and peripheral opioid analgesia in inflammation. *The Journal of Neuroscience*, *15*, 165–172.

Apkarian, A. V., Baliki, M. N., & Farmer, M. A. (2013). Predicting transition to chronic pain. *Current Opinion in Neurology*, *26*, 360–367.

Apkarian, A. V., Sosa, Y., Krauss, B. R., Thomas, P. S., Fredrickson, B. E., Levy, R. E., et al. (2004). Chronic pain patients are impaired on an emotional decision making task. *Pain*, ***108***, 129–136.

Arvidsson, U., Dado, R., Riedl, M., Lee, J., Law, P., Loh, H., et al. (1995). Delta-opioid receptor immunoreactivity: Distribution in brainstem and spinal cord, and relationship to biogenic amines and enkephalin. *The Journal of Neuroscience*, *15*, 1215–1235.

Arvidsson, U., Riedl, M., Chakrabarti, S., Lee, J., Nakano, A., Dado, R., et al. (1995). Distribution and targeting of a mu-opioid receptor (MOR1) in brain and spinal cord. *The Journal of Neuroscience*, *15*, 3328–3341.

Auh, Q.-S., & Ro, J. Y. (2012). Effects of peripheral κ opioid receptor activation on inflammatory mechanical hyperalgesia in male and female rats. *Neuroscience Letters*, *524*, 111–115.

Baliki, M. N., & Apkarian, A. V. (2015). Nociception, pain, negative moods, and behavior selection. *Neuron*, *87*, 474–491.

Baliki, M. N., Geha, P. Y., Fields, H. L., & Apkarian, A. V. (2010). Predicting value of pain and analgesia: Nucleus accumbens response to noxious stimuli changes in the presence of chronic pain. *Neuron*, ***66***, 149–160.

Barber, A., Bartoszyk, G. D., Bender, H. M., Gottschlich, R., Greiner, H. E., Harting, J., et al. (1994). A pharmacological profile of the novel, peripherally-selective kappa-opioid receptor agonist, EMD 61753. *British Journal of Pharmacology*, *113*, 1317–1327.

Basbaum, A. I., Bautista, D. M., Scherrer, G., & Julius, D. (2009). Cellular and molecular mechanisms of pain. *Cell*, *139*, 267–284.

Basbaum, A. I., & Fields, H. L. (1984). Endogenous pain control systems: Brainstem spinal pathways and endorphin circuitry. *Annual Review of Neuroscience*, 7, 309–338.

Beaudry, H., Gendron, L., & Morón, J. A. (2015). Implication of delta opioid receptor subtype 2 but not delta opioid receptor subtype 1 in the development of morphine analgesic tolerance in a rat model of chronic inflammatory pain. *The European Journal of Neuroscience*, *41*, 901–907.

Becerra, L., & Borsook, D. (2008). Signal valence in the nucleus accumbens to pain onset and offset. *European Journal of Pain*, *12*, 866–869.

Becerra, L., Navratilova, E., Porreca, F., & Borsook, D. (2013). Analogous responses in the nucleus accumbens and cingulate cortex to pain onset (aversion) and offset (relief) in rats and humans. *Journal of Neurophysiology*, *110*, 1221–1226.

Besson, J. M. (1999). The neurobiology of pain. *Lancet*, *353*, 1610–1615.

Betourne, A., Familiades, J., Lacassagne, L., Halley, H., Cazales, M., Ducommun, B., et al. (2008). Decreased motivational properties of morphine in mouse models of cancerous- or inflammatory-chronic pain: Implication of supraspinal neuropeptide FF2 receptors. *Neuroscience*, *157*, 12–21.

Billa, S. K., Xia, Y., & Morón, J. A. (2010). Disruption of morphine conditioned place preference by delta-2-opioid receptor antagonist: Study of mu- and delta-opioid receptor expression at the synapse. *The European Journal of Neuroscience*, *32*, 625–631.

Binder, W., Machelska, H., Mousa, S., Schmitt, T., Rivière, P. J., Junien, J. L., et al. (2001). Analgesic and antiinflammatory effects of two novel kappa-opioid peptides. *Anesthesiology*, *94*, 1034–1044.

Binder, W., & Walker, J. S. (1998). Effect of the peripherally selective κ-opioid agonist, asimadoline, on adjuvant arthritis. *British Journal of Pharmacology*, *124*, 647–654.

Bobeck, E. N., McNeal, A. L., & Morgan, M. M. (2009). Drug dependent sex differences in periaqueducatal gray mediated antinociception in the rat. *Pain*, *147*, 210.

Bodnar, R. J. (2020). Endogenous opiates and behavior: 2017. *Peptides*, *124*, 170223.

Boehnke, K. F., Litinas, E., & Clauw, D. J. (2016). Medical cannabis use is associated with decreased opiate medication use in a retrospective cross-sectional survey of patients with chronic pain. *The Journal of Pain*, *17*, 739–744.

Bohn, L. M., Lefkowitz, R. J., Gainetdinov, R. R., Peppel, K., Caron, M. G., & Lin, F. T. (1999). Enhanced morphine analgesia in mice lacking beta-arrestin 2. *Science*, *286*, 2495–2498.

Borras, M. C., Becerra, L., Ploghaus, A., Gostic, J. M., DaSilva, A., Gonzalez, R. G., et al. (2004). fMRI measurement of CNS responses to naloxone infusion and subsequent mild noxious thermal stimuli in healthy volunteers. *Journal of Neurophysiology*, *91*, 2723–2733.

Boyle, K. A., Gutierrez-Mecinas, M., Polgár, E., Mooney, N., O'Connor, E., Furuta, T., et al. (2017). A quantitative study of neurochemically defined populations of inhibitory interneurons in the superficial dorsal horn of the mouse spinal cord. *Neuroscience*, *363*, 120–133.

Brandt, M. R., Furness, M. S., Rice, K. C., Fischer, B. D., & Negus, S. S. (2001). Studies of tolerance and dependence with the delta-opioid agonist SNC80 in rhesus monkeys responding under a schedule of food presentation. *Journal of Pharmacology and Experimental Therapeutics*, ***299***, 629–637.

Broom, D. C., Jutkiewicz, E. M., Folk, J. E., Traynor, J. R., Rice, K. C., & Woods, J. H. (2002a). Nonpeptidic delta-opioid receptor agonists reduce immobility in the forced swim assay in rats. *Neuropsychopharmacology*, ***26***, 744–755.

Broom, D. C., Nitsche, J. F., Pintar, J. E., Rice, K. C., Woods, J. H., & Traynor, J. R. (2002b). Comparison of receptor mechanisms and efficacy requirements for delta-agonist-induced convulsive activity and antinociception in mice. *Journal of Pharmacology and Experimental Therapeutics*, ***303***, 723–729.

Bruce, D. J., Peterson, C. D., Kitto, K. F., Akgün, E., Lazzaroni, S., Portoghese, P. S., et al. (2019). Combination of a δ-opioid receptor agonist and loperamide produces peripherally-mediated analgesic synergy in mice. *Anesthesiology*, *131*, 649–663.

Bruchas, M. R., & Chavkin, C. (2010). Kinase cascades and ligand-directed signaling at the kappa opioid receptor. *Psychopharmacology*, *210*, 137–147.

Bruchas, M. R., Land, B. B., & Chavkin, C. (2010). The dynorphin/kappa opioid system as a modulator of stress-induced and pro-addictive behaviors. *Brain Research*, *1314*, 44–55.

Buntin-Mushock, C., Phillip, L., Moriyama, K., & Palmer, P. P. (2005). Age-dependent opioid escalation in chronic pain patients. *Anesthesia and Analgesia*, *100*, 1740–1745.

Cahill, C. M., Holdridge, S. V., & Morinville, A. (2007). Trafficking of delta-opioid receptors and other G-protein-coupled receptors: Implications for pain and analgesia. *Trends in Pharmacological Sciences*, *28*, 23–31.

Cahill, C. M., Morinville, A., Hoffert, C., O'Donnell, D., & Beaudet, A. (2003). Up-regulation and trafficking of delta opioid receptor in a model of chronic inflammation: Implications for pain control. *Pain*, *101*, 199–208.

Cahill, C. M., & Ong, E. (2018). Evidence and function relevance of native DOR-MOR heteromers. *Handbook of Experimental Pharmacology*, *247*, 115–127.

Cahill, C. M., Taylor, A. M. W., Cook, C., Ong, E., Morón, J. A., & Evans, C. J. (2014). Does the kappa opioid receptor system contribute to pain aversion? *Frontiers in Pharmacology*, *5*, 253.

Cahill, C. M., Xue, L., Grenier, P., Magnussen, C., Lecour, S., & Olmstead, M. C. (2013). Changes in morphine reward in a model of neuropathic pain. *Behavioural Pharmacology*, *24*, 207–213.

Caram-Salas, N. L., Reyes-García, G., Bartoszyk, G. D., Araiza-Saldaña, C. I., Ambriz-Tututi, M., Rocha-González, H. I., et al. (2007). Subcutaneous, intrathecal and periaqueductal grey administration of asimadoline and ICI-204448 reduces tactile allodynia in the rat. *European Journal of Pharmacology*, *573*, 75–83.

Carrasquillo, Y., & Gereau, R. W. (2007). Activation of the extracellular signal-regulated kinase in the amygdala modulates pain perception. *The Journal of Neuroscience*, *27*, 1543–1551.

Castro, D. C., & Berridge, K. C. (2014). Opioid hedonic hotspot in nucleus accumbens shell: Mu, delta, and kappa maps for enhancement of sweetness "liking" and "wanting". *The Journal of Neuroscience*, *34*, 4239–4250.

Cengiz, P., Gokcinar, D., Karabeyoglu, I., Topcu, H., Cicek, G. S., & Gogus, N. (2014). Intraoperative low-dose ketamine infusion reduces acute postoperative pain following total knee replacement surgery: A prospective, randomized double-blind placebo-controlled trial. *Journal of the College of Physicians and Surgeons–Pakistan*, *24*, 299–303.

Chavkin, C. (2011). The therapeutic potential of κ-opioids for treatment of pain and addiction. *Neuropsychopharmacology*, *36*, 369–370.

Chavkin, C., James, I. F., & Goldstein, A. (1982). Dynorphin is a specific endogenous ligand of the kappa opioid receptor. *Science*, *215*, 413–415.

Chavkin, C., & Koob, G. F. (2016). Dynorphin, dysphoria, and dependence: The stress of addiction. *Neuropsychopharmacology*, *41*, 373–374.

Chung, C., Carteret, A. F., McKelvy, A. D., Ringkamp, M., Yang, F., Hartke, T. V., et al. (2012). Analgesic properties of loperamide differ following systemic and local administration to rats after spinal nerve injury. *European Journal of Pain*, *16*, 1021–1032.

Chung, P. C. S., Keyworth, H. L., Martin-Garcia, E., Charbogne, P., Darcq, E., Bailey, A., et al. (2015). A novel anxiogenic role for the delta opioid receptor expressed in GABAergic forebrain neurons. *Biological Psychiatry*, ***77***, 404–415.

Cichewicz, D. L. (2004). Synergistic interactions between cannabinoid and opioid analgesics. *Life Sciences*, *74*, 1317–1324.

Cochran, G., Bacci, J. L., Ylioja, T., Hruschak, V., Miller, S., Seybert, A. L., et al. (2016). Prescription opioid use: Patient characteristics and misuse in community pharmacy. *Journal of the American Pharmaceutical Association*, *56*, 248–256.e6.

Coghill, R. C., McHaffie, J. G., & Yen, Y.-F. (2003). Neural correlates of interindividual differences in the subjective experience of pain. *Proceedings of the National Academy of Sciences of the United States of America*, *100*, 8538–8542.

Colpaert, F. C., Meert, T., De Witte, P., & Schmitt, P. (1982). Further evidence validating adjuvant arthritis as an experimental model of chronic pain in the rat. *Life Sciences*, *31*, 67–75.

Colpaert, F. C., Tarayre, J. P., Alliaga, M., Bruins Slot, L. A., Attal, N., & Koek, W. (2001). Opiate self-administration as a measure of chronic nociceptive pain in arthritic rats. *Pain*, *91*, 33–45.

Comer, S. D., Hoenicke, E. M., Sable, A. I., McNutt, R. W., Chang, K. J., De Costa, B. R., et al. (1993). Convulsive effects of systemic administration of the delta opioid agonist BW373U86 in mice. *Journal of Pharmacology and Experimental Therapeutics*, ***267***, 888–895.

Corder, G., Ahanonu, B., Grewe, B. F., Wang, D., Schnitzer, M. J., & Scherrer, G. (2019). An amygdalar neural ensemble that encodes the unpleasantness of pain. *Science*, *363*, 276–281.

Corder, G., Tawfik, V. L., Wang, D., Sypek, E. I., Low, S. A., Dickinson, J. R., et al. (2017). Loss of μ opioid receptor signaling in nociceptors, but not microglia, abrogates morphine tolerance without disrupting analgesia. *Nature Medicine*, *23*, 164–173.

Corre, J., van Zessen, R., Loureiro, M., Patriarchi, T., Tian, L., Pascoli, V., et al. (2018). Dopamine neurons projecting to medial shell of the nucleus accumbens drive heroin reinforcement. *eLife*, 7, e39945.

Cowan, A., Kehner, G. B., & Inan, S. (2015). Targeting itch with ligands selective for κ opioid receptors. *Handbook of Experimental Pharmacology*, *226*, 291–314.

Crock, L. W., Kolber, B. J., Morgan, C. D., Sadler, K. E., Vogt, S. K., Bruchas, M. R., et al. (2012). Central amygdala metabotropic glutamate receptor 5 in the modulation of visceral pain. *The Journal of Neuroscience*, *32*, 14217–14226.

Crowley, N. A., Bloodgood, D. W., Hardaway, J. A., Kendra, A. M., McCall, J. G., Al-Hasani, R., et al. (2016). Dynorphin controls the gain of an amygdalar anxiety circuit. *Cell Reports*, *14*, 2774–2783.

Cui, Y., Ostlund, S. B., James, A. S., Park, C. S., Ge, W., Roberts, K. W., et al. (2014). Targeted expression of μ-opioid receptors in a subset of striatal direct-pathway neurons restores opiate reward. *Nature Neuroscience*, *17*, 254–261.

Cunha, T. M., Souza, G. R., Domingues, A. C., Carreira, E. U., Lotufo, C. M., Funez, M. I., et al. (2012). Stimulation of peripheral kappa opioid receptors inhibits inflammatory hyperalgesia via activation of the PI3Kγ/AKT/nNOS/NO signaling pathway. *Molecular Pain*, *8*, 10.

Dahan, A., van Velzen, M., & Niesters, M. (2014). Comorbidities and the complexities of chronic pain. *Anesthesiology*, *121*, 675–677.

Danielsson, I., Gasior, M., Stevenson, G. W., Folk, J. E., Rice, K. C., & Negus, S. S. (2006). Electroencephalographic and convulsant effects of the delta opioid agonist SNC80 in rhesus monkeys. *Pharmacology Biochemistry and Behavior*, ***85***, 428–434.

DeHaven-Hudkins, D. L., Cowan, A., Cortes Burgos, L., Daubert, J. D., Cassel, J. A., DeHaven, R. N., et al. (2002). Antipruritic and antihyperalgesic actions of loperamide and analogs. *Life Sciences*, *71*, 2787–2796.

DeHaven-Hudkins, D. L., & Dolle, R. E. (2004). Peripherally restricted opioid agonists as novel analgesic agents. *Current Pharmaceutical Design, 10*, 743–757.

Denk, F., McMahon, S. B., & Tracey, I. (2014). Pain vulnerability: A neurobiological perspective. *Nature Neuroscience, 17*, 192–200.

Derry, C. J., Derry, S., & Moore, R. A. (2013). Single dose oral ibuprofen plus paracetamol (acetaminophen) for acute postoperative pain. *The Cochrane Database of Systematic Reviews, 2013*, CD010210.

Devine, D. P., Leone, P., Pocock, D., & Wise, R. A. (1993). Differential involvement of ventral tegmental mu, delta and kappa opioid receptors in modulation of basal mesolimbic dopamine release: In vivo microdialysis studies. *The Journal of Pharmacology and Experimental Therapeutics, 266*, 1236–1246.

Dowell, D., Haegerich, T. M., & Chou, R. (2016). CDC guideline for prescribing opioids for chronic pain—United States, 2016. *JAMA, 315*, 1624–1645.

Dykstra, L. A., Schoenbaum, G. M., Yarbrough, J., McNutt, R., & Chang, K. J. (1993). A novel delta opioid agonist, BW373U86, in squirrel monkeys responding under a schedule of shock titration. *Journal of Pharmacology and Experimental Therapeutics,* ***267***, 875–882.

Eilers, H., Philip, L. A., Bickler, P. E., McKay, W. R., & Schumacher, M. A. (2001). The reversal of fentanyl-induced tolerance by administration of "small-dose" ketamine. *Anesthesia and Analgesia, 93*, 213–214.

Elman, I., Borsook, D., & Volkow, N. D. (2013). Pain and suicidality: Insights from reward and addiction neuroscience. *Progress in Neurobiology, 109*, 1–27.

Elvemo, N. A., Landrø, N. I., Borchgrevink, P. C., & Håberg, A. K. (2015). Reward responsiveness in patients with chronic pain. *European Journal of Pain, 19*, 1537–1543.

Ewan, E. E., & Martin, T. J. (2011). Rewarding electrical brain stimulation in rats after peripheral nerve injury: Decreased facilitation by commonly abused prescription opioids. *Anesthesiology, 115*, 1271–1280.

Ewan, E. E., & Martin, T. J. (2012). Intracranial self-stimulation of the paraventricular nucleus of the hypothalamus: Increased faciliation by morphine compared to cocaine. *Anesthesiology, 116*, 1116–1123.

Ewan, E. E., & Martin, T. J. (2014). Differential suppression of intracranial self-stimulation, food-maintained operant responding, and open field activity by paw incision and spinal nerve ligation in rats. *Anesthesia and Analgesia, 118*, 854–862.

Fang, F. G., Haws, C. M., Drasner, K., Williamson, A., & Fields, H. L. (1989). Opioid peptides (DAGO-enkephalin, dynorphin A(1-13), BAM 22P) microinjected into the rat brainstem: Comparison of their antinociceptive effect and their effect on neuronal firing in the rostral ventromedial medulla. *Brain Research, 501*, 116–128.

Fields, H. L., & Margolis, E. B. (2015). Understanding opioid reward. *Trends in Neurosciences, 38*, 217–225.

Fraser, G. L., Pradhan, A. A., Clarke, P. B., & Wahlestedt, C. (2000). Supraspinal antinociceptive response to [D-Pen(2,5)]-enkephalin (DPDPE) is pharmacologically distinct from that to other delta-agonists in the rat. *The Journal of Pharmacology and Experimental Therapeutics, 295*, 1135–1141.

Fujita, W., Gomes, I., & Devi, L. A. (2014). Revolution in GPCR signalling: Opioid receptor heteromers as novel therapeutic targets: IUPHAR review 10. *British Journal of Pharmacology, 171*, 4155–4176.

Fujita, W., Gomes, I., & Devi, L. A. (2015). Heteromers of μ-δ opioid receptors: New pharmacology and novel therapeutic possibilities. *British Journal of Pharmacology, 172*, 375–387.

Furlan, A., Chaparro, L. E., Irvin, E., & Mailis-Gagnon, A. (2011). A comparison between enriched and nonenriched enrollment randomized withdrawal trials of opioids for chronic noncancer pain. *Pain Research & Management, 16*, 337–351.

Gallantine, E. L., & Meert, T. F. (2005). A comparison of the antinociceptive and adverse effects of the mu-opioid agonist morphine and the delta-opioid agonist SNC80. *Basic & Clinical Pharmacology & Toxicology*, ***97***, 39–51.

Garland, E. L. (2012). Pain processing in the human nervous system: A selective review of nociceptive and biobehavioral pathways. *Primary Care*, *39*, 561–571.

Garland, E. L., Froeliger, B., Zeidan, F., Partin, K., & Howard, M. O. (2013). The downward spiral of chronic pain, prescription opioid misuse, and addiction: Cognitive, affective, and neuropsychopharmacologic pathways. *Neuroscience and Biobehavioral Reviews*, *37*, 2597–2607.

Gavériaux-Ruff, C., Karchewski, L. A., Hever, X., Matifas, A., & Kieffer, B. L. (2008). Inflammatory pain is enhanced in delta opioid receptor-knockout mice. *The European Journal of Neuroscience*, *27*, 2558–2567.

Gavériaux-Ruff, C., & Kieffer, B. L. (2011). Delta opioid receptor analgesia: Recent contributions from pharmacology and molecular approaches. *Behavioural Pharmacology*, *22*, 405–414.

Gendron, L., Lucido, A. L., Mennicken, F., O'Donnell, D., Vincent, J.-P., Stroh, T., et al. (2006). Morphine and pain-related stimuli enhance cell surface availability of somatic δ-opioid receptors in rat dorsal root ganglia. *The Journal of Neuroscience*, *26*, 953–962.

Gendron, L., Mittal, N., Beaudry, H., & Walwyn, W. (2015). Recent advances on the δ opioid receptor: From trafficking to function. *British Journal of Pharmacology*, *172*, 403–419.

Giuliano, C., Robbins, T. W., Wille, D. R., Bullmore, E. T., & Everitt, B. J. (2013). Attenuation of cocaine and heroin seeking by μ-opioid receptor antagonism. *Psychopharmacology*, *227*, 137–147.

Graff-Radford, S. B., Shaw, L. R., & Naliboff, B. N. (2000). Amitriptyline and fluphenazine in the treatment of postherpetic neuralgia. *The Clinical Journal of Pain*, *16*, 188–192.

Grim, T. W., Schmid, C. L., Stahl, E. L., Pantouli, F., Ho, J.-H., Acevedo-Canabal, A., et al. (2020). A G protein signaling-biased agonist at the μ-opioid receptor reverses morphine tolerance while preventing morphine withdrawal. *Neuropsychopharmacology*, *45*, 416–425.

Guan, Y., Johanek, L. M., Hartke, T. V., Shim, B., Tao, Y.-X., Ringkamp, M., et al. (2008). Peripherally acting mu-opioid receptor agonist attenuates neuropathic pain in rats after L5 spinal nerve injury. *Pain*, *138*, 318–329.

Guerrero, M., Urbano, M., Kim, E.-K., Gamo, A. M., Riley, S., Abgaryan, L., et al. (2019). Design and synthesis of a novel and selective kappa opioid receptor (KOR) antagonist (BTRX-335140). *Journal of Medicinal Chemistry*, *62*, 1761–1780.

Günther, T., Dasgupta, P., Mann, A., Miess, E., Kliewer, A., Fritzwanker, S., et al. (2018). Targeting multiple opioid receptors—Improved analgesics with reduced side effects? *British Journal of Pharmacology*, *175*, 2857–2868.

Gupta, A., Gomes, I., Bobeck, E. N., Fakira, A. K., Massaro, N. P., Sharma, I., et al. (2016). Collybolide is a novel biased agonist of κ-opioid receptors with potent antipruritic activity. *Proceedings of the National Academy of Sciences of the United States of America*, *113*, 6041–6046.

Han, S., Soleiman, M. T., Soden, M. E., Zweifel, L. S., & Palmiter, R. D. (2015). Elucidating an affective pain circuit that creates a threat memory. *Cell*, *162*, 363–374.

Hayhurst, C. J., & Durieux, M. E. (2016). Differential opioid tolerance and opioid-induced hyperalgesia: A clinical reality. *Anesthesiology*, *124*, 483–488.

Heinricher, M. M., Tavares, I., Leith, J. L., & Lumb, B. M. (2009). Descending control of nociception: Specificity, recruitment and plasticity. *Brain Research Reviews*, *60*, 214–225.

Hipólito, L., Wilson-Poe, A., Campos-Jurado, Y., Zhong, E., Gonzalez-Romero, J., Virag, L., et al. (2015). Inflammatory pain promotes increased opioid self-administration: Role of dysregulated ventral tegmental area μ opioid receptors. *The Journal of Neuroscience*, *35*, 12217–12231.

Hnasko, T. S., Sotak, B. N., & Palmiter, R. D. (2005). Morphine reward in dopamine-deficient mice. *Nature*, *438*, 854–857.

Ho, J., Mannes, A. J., Dubner, R., & Caudle, R. M. (1997). Putative kappa-2 opioid agonists are antihyperalgesic in a rat model of inflammation. *The Journal of Pharmacology and Experimental Therapeutics*, *281*, 136–141.

Hogarth, L. (2020). Addiction is driven by excessive goal-directed drug choice under negative affect: Translational critique of habit and compulsion theory. *Neuropsychopharmacology*, *45*, 720–735.

Hou, Y.-Y., Cai, Y.-Q., & Pan, Z. Z. (2015). Persistent pain maintains morphine-seeking behavior after morphine withdrawal through reduced MeCP2 repression of Glua1 in rat central amygdala. *The Journal of Neuroscience*, *35*, 3689–3700.

Hurley, R. W., & Hammond, D. L. (2000). The analgesic effects of supraspinal μ and δ opioid receptor agonists are potentiated during persistent inflammation. *The Journal of Neuroscience*, *20*, 1249–1259.

Inan, S., & Cowan, A. (2004). Kappa opioid agonists suppress chloroquine-induced scratching in mice. *European Journal of Pharmacology*, *502*, 233–237.

Institute of Medicine (US) Committee on Advancing Pain Research, Care, and Education. (2011). *Relieving pain in America: A blueprint for transforming prevention, care, education, and research*. Washington (DC), USA: National Academies Press.

Jacquet, Y. F., & Lajtha, A. (1974). Paradoxical effects after microinjection of morphine in the periaqueductal gray matter in the rat. *Science*, *185*, 1055–1057.

Jamison, R. N., & Edwards, R. R. (2013). Risk factor assessment for problematic use of opioids for chronic pain. *The Clinical Neuropsychologist*, *27*, 60–80.

Janak, P. H., & Tye, K. M. (2015). From circuits to behaviour in the amygdala. *Nature*, *517*, 284–292.

Ji, R.-R., Chamessian, A., & Zhang, Y.-Q. (2016). Pain regulation by non-neuronal cells and inflammation. *Science*, *354*, 572–577.

Johnson, S. W., & North, R. A. (1992). Opioids excite dopamine neurons by hyperpolarization of local interneurons. *The Journal of Neuroscience*, ***12***, 483–488.

Jones, J. D., Sullivan, M. A., Vosburg, S. K., Manubay, J. M., Mogali, S., Metz, V., et al. (2015). Abuse potential of intranasal buprenorphine versus buprenorphine/naloxone in buprenorphine-maintained heroin users. *Addiction Biology*, *20*, 784–798.

Julius, D., & Basbaum, A. I. (2001). Molecular mechanisms of nociception. *Nature*, *413*, 203–210.

Jutkiewicz, E. M., Baladi, M. G., Folk, J. E., Rice, K. C., & Woods, J. H. (2006). The convulsive and electroencephalographic changes produced by nonpeptidic delta-opioid agonists in rats: Comparison with pentylenetetrazol. *Journal of Pharmacology and Experimental Therapeutics*, ***317***, 1337–1348.

Keïta, H., Kayser, V., & Guilbaud, G. (1995). Antinociceptive effect of a kappa-opioid receptor agonist that minimally crosses the blood-brain barrier (ICI 204448) in a rat model of mononeuropathy. *European Journal of Pharmacology*, *277*, 275–280.

King, T., Vera-Portocarrero, L., Gutierrez, T., Vanderah, T. W., Dussor, G., Lai, J., et al. (2009). Unmasking the tonic-aversive state in neuropathic pain. *Nature Neuroscience*, *12*, 1364–1366.

Kingwell, K. (2015). Pioneering biased ligand offers efficacy with reduced on-target toxicity. *Nature Reviews Drug Discovery*, *14*, 809–810.

Kivell, B., & Prisinzano, T. E. (2010). Kappa opioids and the modulation of pain. *Psychopharmacology*, *210*, 109–119.

Kliewer, A., Schmiedel, F., Sianati, S., Bailey, A., Bateman, J. T., Levitt, E. S., et al. (2019). Phosphorylation-deficient G-protein-biased μ-opioid receptors improve analgesia and diminish tolerance but worsen opioid side effects. *Nature Communications*, ***10***, 367.

Kliewer, A., Gillis, A., Hill, R., Schmidel, F., Bailey, C., Kelly, E., et al. (2020). Morphine-induced respiratory depression is independent of β-arrestin2 signalling. *British Journal of Pharmacology*, *177*, 2923–2931.

Koele, R., Volker, G., van Vree, F., van Gestel, M., Köke, A., & Vliet Vlieland, T. (2014). Multidisciplinary rehabilitation for chronic widespread musculoskeletal pain: Results from daily practice. *Musculoskeletal Care*, *12*, 210–220.

Koob, G. F., & Le Moal, M. (2001). Drug addiction, dysregulation of reward, and allostasis. *Neuropsychopharmacology*, *24*, 97–129.

Koob, G. F., & Volkow, N. D. (2010). Neurocircuitry of addiction. *Neuropsychopharmacology*, *35*, 217–238.

Koob, G. F., & Volkow, N. D. (2016). Neurobiology of addiction: A neurocircuitry analysis. *Lancet Psychiatry*, *3*, 760–773.

Krashin, D., Murinova, N., & Sullivan, M. (2016). Challenges to treatment of chronic pain and addiction during the "opioid crisis". *Current Pain and Headache Reports*, *20*, 65.

Kupers, R., & Gybels, J. (1995). The consumption of fentanyl is increased in rats with nociceptive but not with neuropathic pain. *Pain*, *60*, 137–141.

LaGraize, S. C., Borzan, J., Peng, Y. B., & Fuchs, P. N. (2006). Selective regulation of pain affect following activation of the opioid anterior cingulate cortex system. *Experimental Neurology*, *197*, 22–30.

Lai, J., Luo, M., Chen, Q., & Porreca, F. (2008). Pronociceptive actions of dynorphin via bradykinin receptors. *Neuroscience Letters*, *437*, 175–179.

Lalanne, L., Ayranci, G., Kieffer, B. L., & Lutz, P.-E. (2014). The kappa opioid receptor: From addiction to depression, and back. *Frontiers in Psychiatry*, *5*, 170.

Land, B. B., Bruchas, M. R., Lemos, J. C., Xu, M., Melief, E. J., & Chavkin, C. (2008). The dysphoric component of stress is encoded by activation of the dynorphin kappa-opioid system. *The Journal of Neuroscience*, *28*, 407–414.

Land, B. B., Bruchas, M. R., Schattauer, S., Giardino, W. J., Aita, M., Messinger, D., et al. (2009). Activation of the kappa opioid receptor in the dorsal raphe nucleus mediates the aversive effects of stress and reinstates drug seeking. *Proceedings of the National Academy of Sciences of the United States of America*, *106*, 19168–19173.

Lane, D. A., Patel, P. A., & Morgan, M. M. (2005). Evidence for an intrinsic mechanism of antinociceptive tolerance within the ventrolateral periaqueductal gray of rats. *Neuroscience*, *135*, 227–234.

Lanser, P., & Gesell, S. (2001). Pain management: The fifth vital sign. *Healthcare Benchmarks*, *8*(68–70), 62.

Laviolette, S. R., Gallegos, R. A., Henriksen, S. J., & van der Kooy, D. (2004). Opiate state controls bi-directional reward signaling via GABAA receptors in the ventral tegmental area. *Nature Neuroscience*, 7, 160–169.

Le Merrer, J., Becker, J. A. J., Befort, K., & Kieffer, B. L. (2009). Reward processing by the opioid system in the brain. *Physiological Reviews*, ***89***, 1379–1412.

Legakis, L. P., & Negus, S. S. (2018). Repeated morphine produces sensitization to reward and tolerance to anti-allodynia in male and female rats with chemotherapy-induced neuropathy. *The Journal of Pharmacology and Experimental Therapeutics*, *365*, 9–19.

Leitl, M. D., Onvani, S., Bowers, M. S., Cheng, K., Rice, K. C., Carlezon, W. A., et al. (2014). Pain-related depression of the mesolimbic dopamine system in rats: Expression, blockade by analgesics, and role of endogenous κ-opioids. *Neuropsychopharmacology*, *39*, 614–624.

Leitl, M. D., Potter, D. N., Cheng, K., Rice, K. C., Carlezon, W. A., & Negus, S. S. (2014). Sustained pain-related depression of behavior: Effects of intraplantar formalin and complete freund's adjuvant on intracranial self-stimulation (ICSS) and endogenous kappa opioid biomarkers in rats. *Molecular Pain*, *10*, 62.

Leknes, S., & Tracey, I. (2008). A common neurobiology for pain and pleasure. *Nature Reviews. Neuroscience*, *9*, 314–320.
Lintas, A., Chi, N., Lauzon, N. M., Bishop, S. F., Gholizadeh, S., Sun, N., et al. (2011). Identification of a dopamine receptor-mediated opiate reward memory switch in the basolateral amygdala-nucleus accumbens circuit. *The Journal of Neuroscience*, ***31***, 11172–11183.
Lintas, A., Chi, N., Lauzon, N. M., Bishop, S. F., Sun, N., Tan, H., et al. (2012). Inputs from the basolateral amygdala to the nucleus accumbens shell control opiate reward magnitude via differential dopamine D1 or D2 receptor transmission. *European Journal of Neuroscience*, ***35***, 279–290.
Liu, S. S., Pickens, S., Burma, N. E., Ibarra-Lecue, I., Yang, H., Xue, L., et al. (2019). Kappa opioid receptors drive a tonic aversive component of chronic pain. *The Journal of Neuroscience*, *39*, 4162–4178.
Loftus, R. W., Yeager, M. P., Clark, J. A., Brown, J. R., Abdu, W. A., Sengupta, D. K., et al. (2010). Intraoperative ketamine reduces perioperative opiate consumption in opiate-dependent patients with chronic back pain undergoing back surgery. *Anesthesiology*, *113*, 639–646.
Lunn, M. P. T., Hughes, R. A. C., & Wiffen, P. J. (2014). Duloxetine for treating painful neuropathy, chronic pain or fibromyalgia. *Cochrane Database of Systematic Reviews*, *4*, CD007115.
Lyness, W. H., Smith, F. L., Heavner, J. E., Iacono, C. U., & Garvin, R. D. (1989). Morphine self-administration in the rat during adjuvant-induced arthritis. *Life Sciences*, *45*, 2217–2224.
Manglik, A., Lin, H., Aryal, D. K., McCorvy, J. D., Dengler, D., Corder, G., et al. (2016). Structure-based discovery of opioid analgesics with reduced side effects. *Nature*, *537*, 185–190.
Mansour, A., Khachaturian, H., Lewis, M. E., Akil, H., & Watson, S. J. (1988). Anatomy of CNS opioid receptors. *Trends in Neurosciences*, *11*, 308–314.
Martel, M. O., Dolman, A. J., Edwards, R. R., Jamison, R. N., & Wasan, A. D. (2014). The association between negative affect and prescription opioid misuse in patients with chronic pain: The mediating role of opioid craving. *The Journal of Pain*, *15*, 90–100.
Martin, T. J., Buechler, N. L., Kim, S. A., Ewan, E. E., Xiao, R., & Childers, S. R. (2011). Involvement of the lateral amygdala in the antiallodynic and reinforcing effects of heroin in rats after peripheral nerve injury. *Anesthesiology*, *114*, 633–642.
Martin, T. J., & Ewan, E. (2008). Chronic pain alters drug self-administration: Implications for addiction and pain mechanisms. *Experimental and Clinical Psychopharmacology*, *16*, 357–366.
Martin, T. J., Kim, S. A., Buechler, N. L., Porreca, F., & Eisenach, J. C. (2007). Opioid self-administration in the nerve-injured rat relevance of antiallodynic effects to drug consumption and effects of intrathecal analgesics. *Anesthesiology*, *106*, 312–322.
Massaly, N., Copits, B. A., Wilson-Poe, A. R., Hipólito, L., Markovic, T., Yoon, H. J., et al. (2019). Pain-induced negative affect is mediated via recruitment of the nucleus accumbens kappa opioid system. *Neuron*, *102*, 564–573.e6.
Massaly, N., Morón, J. A., & Al-Hasani, R. (2016). A trigger for opioid misuse: Chronic pain and stress dysregulate the mesolimbic pathway and kappa opioid system. *Frontiers in Neuroscience*, *10*, 480.
Matsui, A., Jarvie, B. C., Robinson, B. G., Hentges, S. T., & Williams, J. T. (2014). Separate GABA afferents to dopamine neurons mediate acute action of opioids, development of tolerance, and expression of withdrawal. *Neuron*, *82*, 1346–1356.
McCall, J. G., Siuda, E. R., Bhatti, D. L., Lawson, L. A., McElligott, Z. A., Stuber, G. D., et al. (2017). Locus coeruleus to basolateral amygdala noradrenergic projections promote anxiety-like behavior. *eLife*, *6*, e18247.

Miller, L. L., Altarifi, A. A., & Negus, S. S. (2015). Effects of repeated morphine on intracranial self-stimulation in male rats in the absence or presence of a noxious pain stimulus. *Experimental and Clinical Psychopharmacology*, *23*, 405–414.

Moerke, M. J., & Negus, S. S. (2019). Interactions between pain states and opioid reward assessed with intracranial self-stimulation in rats. *Neuropharmacology*, *160*, 107689.

Moore, R. A., Derry, S., Aldington, D., Cole, P., & Wiffen, P. J. (2015). Amitriptyline for neuropathic pain in adults. *Cochrane Database of Systematic Reviews*, 7, CD008242.

Morales, M., & Margolis, E. B. (2017). Ventral tegmental area: Cellular heterogeneity, connectivity and behaviour. *Nature Reviews Neuroscience*, *18*, 73–85.

Morón, J. A., Gullapalli, S., Taylor, C., Gupta, A., Gomes, I., & Devi, L. A. (2010). Modulation of opiate-related signaling molecules in morphine-dependent conditioned behavior: Conditioned place preference to morphine induces CREB phosphorylation. *Neuropsychopharmacology*, *35*, 955–966.

Muschamp, J. W., & Carlezon, W. A. (2013). Roles of nucleus accumbens CREB and dynorphin in dysregulation of motivation. *Cold Spring Harbor Perspectives in Medicine*, *3*, a012005.

Nadal, X., Baños, J.-E., Kieffer, B. L., & Maldonado, R. (2006). Neuropathic pain is enhanced in delta-opioid receptor knockout mice. *The European Journal of Neuroscience*, *23*, 830–834.

Nahin, R. L. (2015). Estimates of pain prevalence and severity in adults: United States, 2012. *The Journal of Pain*, *16*, 769–780.

Namburi, P., Beyeler, A., Yorozu, S., Calhoon, G. G., Halbert, S. A., Wichmann, R., et al. (2015). A circuit mechanism for differentiating positive and negative associations. *Nature*, *520*, 675–678.

Narita, M., Kaneko, C., Miyoshi, K., Nagumo, Y., Kuzumaki, N., Nakajima, M., et al. (2006). Chronic pain induces anxiety with concomitant changes in opioidergic function in the amygdala. *Neuropsychopharmacology*, *31*, 739–750.

Narita, M., Kishimoto, Y., Ise, Y., Yajima, Y., Misawa, K., & Suzuki, T. (2005). Direct evidence for the involvement of the mesolimbic kappa-opioid system in the morphine-induced rewarding effect under an inflammatory pain-like state. *Neuropsychopharmacology*, *30*, 111–118.

Nation, K. M., De Felice, M., Hernandez, P. I., Dodick, D. W., Neugebauer, V., Navratilova, E., et al. (2018). Lateralized kappa opioid receptor signaling from the amygdala central nucleus promotes stress-induced functional pain. *Pain*, *159*, 919–928.

Navratilova, E., Ji, G., Phelps, C., Qu, C., Hein, M., Yakhnitsa, V., et al. (2019). Kappa opioid signaling in the central nucleus of the amygdala promotes disinhibition and aversiveness of chronic neuropathic pain. *Pain*, *160*, 824–832.

Navratilova, E., & Porreca, F. (2014). Reward and motivation in pain and pain relief. *Nature Neuroscience*, *17*, 1304–1312.

Navratilova, E., Xie, J. Y., Meske, D., Qu, C., Morimura, K., Okun, A., et al. (2015). Endogenous opioid activity in the anterior cingulate cortex is required for relief of pain. *The Journal of Neuroscience*, *35*, 7264–7271.

Navratilova, E., Xie, J. Y., Okun, A., Qu, C., Eyde, N., Ci, S., et al. (2012). Pain relief produces negative reinforcement through activation of mesolimbic reward-valuation circuitry. *Proceedings of the National Academy of Sciences of the United States of America*, *109*, 20709–20713.

Negus, S. S., Gatch, M. B., Mello, N. K., Zhang, X., & Rice, K. (1998). Behavioral effects of the delta-selective opioid agonist SNC80 and related compounds in rhesus monkeys. *The Journal of Pharmacology and Experimental Therapeutics*, *286*, 362–375.

Negus, S. S., & Moerke, M. J. (2019). Determinants of opioid abuse potential: Insights using intracranial self-stimulation. *Peptides*, *112*, 23–31.

Nestler, E. J., & Carlezon, W. A. (2006). The mesolimbic dopamine reward circuit in depression. *Biological Psychiatry*, *59*, 1151–1159.

Niikura, K., Narita, M., Butelman, E. R., Kreek, M. J., & Suzuki, T. (2010). Neuropathic and chronic pain stimuli downregulate central mu-opioid and dopaminergic transmission. *Trends in Pharmacological Sciences*, *31*, 299–305.
Niikura, K., Narita, M., Narita, M., Nakamura, A., Okutsu, D., Ozeki, A., et al. (2008). Direct evidence for the involvement of endogenous β-endorphin in the suppression of the morphine-induced rewarding effect under a neuropathic pain-like state. *Neuroscience Letters*, *435*, 257–262.
Obara, I., Parkitna, J. R., Korostynski, M., Makuch, W., Kaminska, D., Przewlocka, B., et al. (2009). Local peripheral opioid effects and expression of opioid genes in the spinal cord and dorsal root ganglia in neuropathic and inflammatory pain. *Pain*, *141*, 283–291.
Ozaki, S., Narita, M., Iino, M., Miyoshi, K., & Suzuki, T. (2003). Suppression of the morphine-induced rewarding effect and G-protein activation in the lower midbrain following nerve injury in the mouse: Involvement of G-protein-coupled receptor kinase 2. *Neuroscience*, *116*, 89–97.
Ozaki, S., Narita, M., Narita, M., Iino, M., Sugita, J., Matsumura, Y., et al. (2002). Suppression of the morphine-induced rewarding effect in the rat with neuropathic pain: Implication of the reduction in μ-opioid receptor functions in the ventral tegmental area. *Journal of Neurochemistry*, *82*, 1192–1198.
Page, S., Mavrikaki, M. M., Lintz, T., Puttick, D., Roberts, E., Rosen, H., et al. (2019). Behavioral pharmacology of novel kappa opioid receptor antagonists in rats. *The International Journal of Neuropsychopharmacology*, *22*, 735–745.
Pare, D., & Duvarci, S. (2012). Amygdala microcircuits mediating fear expression and extinction. *Current Opinion in Neurobiology*, *22*, 717–723.
Passik, S. D., Messina, J., Golsorkhi, A., & Xie, F. (2011). Aberrant drug-related behavior observed during clinical studies involving patients taking chronic opioid therapy for persistent pain and fentanyl buccal tablet for breakthrough pain. *Journal of Pain and Symptom Management*, *41*, 116–125.
Phelps, C. E., Navratilova, E., Dickenson, A. H., Porreca, F., & Bannister, K. (2019). Kappa opioid signaling in the right central amygdala causes hind paw specific loss of diffuse noxious inhibitory controls in experimental neuropathic pain. *Pain*, *160*, 1614–1621.
Pirbudak, L., Sevinç, A., Maralcan, G., & Kiliç, E. (2014). Pain management with intrathecal clonidine in a colon cancer patient with opioid hyperalgesia: Case presentation. *Ağrı*, *26*, 93–96.
Podvin, S., Yaksh, T., & Hook, V. (2016). Emerging role of spinal dynorphin in chronic pain, a therapeutic perspective. *Annual Review of Pharmacology and Toxicology*, *56*, 511–533.
Porreca, F., & Navratilova, E. (2017). Reward, motivation and emotion of pain and its relief. *Pain*, *158*, S43–S49.
Pradhan, A. A. A., Becker, J. A. J., Scherrer, G., Tryoen-Toth, P., Filliol, D., Matifas, A., et al. (2009). In vivo delta opioid receptor internalization controls behavioral effects of agonists. *PLoS One*, *4,5*. https://doi.org/10.1371/journal.pone.0005425. e5425.
Pradhan, A. A., Befort, K., Nozaki, C., Gavériaux-Ruff, C., & Kieffer, B. L. (2011). The delta opioid receptor: An evolving target for the treatment of brain disorders. *Trends in Pharmacological Sciences*, *32*, 581–590.
Pradhan, A. A., Smith, M. L., Zyuzin, J., & Charles, A. (2014). δ-Opioid receptor agonists inhibit migraine-related hyperalgesia, aversive state and cortical spreading depression in mice. *British Journal of Pharmacology*, *171*, 2375–2384.
Price, T. J., Basbaum, A. I., Bresnahan, J., Chambers, J. F., De Koninck, Y., Edwards, R. R., et al. (2018). Transition to chronic pain: Opportunities for novel therapeutics. *Nature Reviews. Neuroscience*, *19*, 383–384.
Price, D. D., Von der Gruen, A., Miller, J., Rafii, A., & Price, C. (1985). A psychophysical analysis of morphine analgesia. *Pain*, *22*, 261–269.

Raehal, K. M., Walker, J. K. L., & Bohn, L. M. (2005). Morphine side effects in beta-arrestin 2 knockout mice. *The Journal of Pharmacology and Experimental Therapeutics, 314*, 1195–1201.

Rainville, P., Duncan, G. H., Price, D. D., Carrier, B., & Bushnell, M. C. (1997). Pain affect encoded in human anterior cingulate but not somatosensory cortex. *Science, 277*, 968–971.

Rodriguez-Gaztelumendi, A., Spahn, V., Labuz, D., Machelska, H., & Stein, C. (2018). Analgesic effects of a novel pH-dependent μ-opioid receptor agonist in models of neuropathic and abdominal pain. *Pain, 159*, 2277–2284.

Rosas, R., Huang, X.-P., Roth, B. L., & Dockendorff, C. (2019). β-Fluorofentanyls are pH-sensitive Mu opioid receptor agonists. *ACS Medicinal Chemistry Letters, 10*, 1353–1356.

Rusin, K. I., Giovannucci, D. R., Stuenkel, E. L., & Moises, H. C. (1997). Kappa-opioid receptor activation modulates Ca2+ currents and secretion in isolated neuroendocrine nerve terminals. *The Journal of Neuroscience, 17*, 6565–6574.

Schäfer, M., Brack, A., & Stein, C. (2000). Dynorphin a peptides. *Molecular Diagnosis & Therapy, 13*, 161–166.

Schlosburg, J. E., Whitfield, T. W., Park, P. E., Crawford, E. F., George, O., Vendruscolo, L. F., et al. (2013). Long-term antagonism of κ opioid receptors prevents escalation of and increased motivation for heroin intake. *The Journal of Neuroscience, 33*, 19384–19392.

Schmid, C. L., Kennedy, N. M., Ross, N. C., Lovell, K. M., Yue, Z., Morgenweck, J., et al. (2017). Bias factor and therapeutic window correlate to predict safer opioid analgesics. *Cell, 171*, 1165–1175.e13.

Shippenberg, T. S., Stein, C., Huber, A., Millan, M. J., & Herz, A. (1988). Motivational effects of opioids in an animal model of prolonged inflammatory pain: Alteration in the effects of kappa- but not of mu-receptor agonists. *Pain, 35*, 179–186.

Shippenberg, T. S., Zapata, A., & Chefer, V. I. (2007). Dynorphin and the pathophysiology of drug addiction. *Pharmacology & Therapeutics, 116*, 306–321.

Siuda, E. R., Copits, B. A., Schmidt, M. J., Baird, M. A., Al-Hasani, R., Planer, W. J., et al. (2015). Spatiotemporal control of opioid signaling and behavior. *Neuron, **86***, 923–935.

Siuda, E. R., Carr, R., Rominger, D. H., & Violin, J. D. (2017). Biased mu-opioid receptor ligands: A promising new generation of pain therapeutics. *Current Opinion in Pharmacology, 32*, 77–84.

Smith, S. R., Deshpande, B. R., Collins, J. E., Katz, J. N., & Losina, E. (2016). Comparative pain reduction of oral non-steroidal anti-inflammatory drugs and opioids for knee osteoarthritis: Systematic analytic review. *Osteoarthritis and Cartilage, 24*, 962–972.

Snyder, L. M., Chiang, M. C., Loeza-Alcocer, E., Omori, Y., Hachisuka, J., Sheahan, T. D., et al. (2018). Kappa opioid receptor distribution and function in primary afferents. *Neuron, 99*, 1274–1288.e6.

Spahn, V., Del Vecchio, G., Labuz, D., Rodriguez-Gaztelumendi, A., Massaly, N., Temp, J., et al. (2017). A nontoxic pain killer designed by modeling of pathological receptor conformations. *Science, 355*, 966–969.

Stein, C., Millan, M. J., Shippenberg, T. S., Peter, K., & Herz, A. (1989). Peripheral opioid receptors mediating antinociception in inflammation. Evidence for involvement of mu, delta and kappa receptors. *The Journal of Pharmacology and Experimental Therapeutics, 248*, 1269–1275.

Stuber, G. D., Sparta, D. R., Stamatakis, A. M., van Leeuwen, W. A., Hardjoprajitno, J. E., Cho, S., et al. (2011). Excitatory transmission from the amygdala to nucleus accumbens facilitates reward seeking. *Nature, **475***, 377–380.

Sufka, K. J. (1994). Conditioned place preference paradigm: A novel approach for analgesic drug assessment against chronic pain. *Pain, 58*, 355–366.

Taylor, A. M. W., Becker, S., Schweinhardt, P., & Cahill, C. (2016). Mesolimbic dopamine signaling in acute and chronic pain: Implications for motivation, analgesia, and addiction. *Pain*, *157*, 1194–1198.

Taylor, A. M. W., Castonguay, A., Taylor, A. J., Murphy, N. P., Ghogha, A., Cook, C., et al. (2015). Microglia disrupt mesolimbic reward circuitry in chronic pain. *The Journal of Neuroscience*, *35*, 8442–8450.

Tejeda, H. A., Wu, J., Kornspun, A. R., Pignatelli, M., Kashtelyan, V., Krashes, M. J., et al. (2017). Pathway- and cell-specific kappa-opioid receptor modulation of excitation-inhibition balance differentially gates D1 and D2 accumbens neuron activity. *Neuron*, *93*, 147–163.

Tiwari, V., Yang, F., He, S.-Q., Shechter, R., Zhang, C., Shu, B., et al. (2016). Activation of peripheral μ-opioid receptors by dermorphin [D-Arg2, Lys4] (1-4) amide leads to modality-preferred inhibition of neuropathic pain. *Anesthesiology*, *124*, 706–720.

Tracey, I. (2010). Getting the pain you expect: Mechanisms of placebo, nocebo and reappraisal effects in humans. *Nature Medicine*, *16*, 1277–1283.

Tracey, I., & Mantyh, P. W. (2007). The cerebral signature for pain perception and its zmodulation. *Neuron*, *55*, 377–391.

Tsibulsky, V. L., & Norman, A. B. (2005). Real time computation of in vivo drug levels during drug self-administration experiments. *Brain Research. Brain Research Protocols*, *15*, 38–45.

Vadivelu, N., Mitra, S., & Hines, R. L. (2011). Peripheral opioid receptor agonists for analgesia: A comprehensive review. *Journal of Opioid Management*, 7, 55–68.

Vanderah, T. W. (2010). Delta and kappa opioid receptors as suitable drug targets for pain. *The Clinical Journal of Pain*, *26*(Suppl. 10), S10–S15.

Vanderah, T. W., Largent-Milnes, T., Lai, J., Porreca, F., Houghten, R. A., Menzaghi, F., et al. (2008). Novel D-amino acid tetrapeptides produce potent antinociception by selectively acting at peripheral kappa-opioid receptors. *European Journal of Pharmacology*, *583*, 62–72.

Vanderah, T. W., Schteingart, C. D., Trojnar, J., Junien, J.-L., Lai, J., & Riviere, P. J.-M. (2004). FE200041 (D-Phe-D-Phe-D-Nle-D-Arg-NH2): A peripheral efficacious kappa opioid agonist with unprecedented selectivity. *The Journal of Pharmacology and Experimental Therapeutics*, *310*, 326–333.

Van't Veer, A., & Carlezon, W. A. (2013). Role of kappa-opioid receptors in stress and anxiety-related behavior. *Psychopharmacology*, *229*, 435–452.

Váradi, A., Marrone, G. F., Palmer, T. C., Narayan, A., Szabó, M. R., Le Rouzic, V., et al. (2016). Mitragynine/corynantheidine pseudoindoxyls as opioid analgesics with Mu agonism and delta antagonism, which do not recruit β-arrestin-2. *Journal of Medicinal Chemistry*, *59*, 8381–8397.

Veinante, P., Yalcin, I., & Barrot, M. (2013). The amygdala between sensation and affect: A role in pain. *Journal of Molecular Psychiatry*, *1*, 9. https://doi.org/10.1186/2049-9256-1-9.

Verdejo-García, A., López-Torrecillas, F., Calandre, E. P., Delgado-Rodríguez, A., & Bechara, A. (2009). Executive function and decision-making in women with fibromyalgia. *Archives of Clinical Neuropsychology*, ***24***, 113–122.

Vicente-Sanchez, A., Dripps, I. J., Tipton, A. F., Akbari, H., Akbari, A., Jutkiewicz, E. M., et al. (2018). Tolerance to high-internalizing δ opioid receptor agonist is critically mediated by arrestin 2. *British Journal of Pharmacology*, *175*, 3050–3059.

Vogt, B. A. (2005). Pain and emotion interactions in subregions of the cingulate gyrus. *Nature Reviews. Neuroscience*, *6*, 533–544.

Volkow, N. D., & McLellan, A. T. (2016). Opioid abuse in chronic pain—Misconceptions and mitigation strategies. *The New England Journal of Medicine*, *374*, 1253–1263.

Volkow, N. D., & Morales, M. (2015). The brain on drugs: From reward to addiction. *Cell, 162*, 712–725.

Volkow, N. D., Wang, G.-J., Fowler, J. S., Logan, J., Gatley, S. J., Hitzemann, R., et al. (1997). Decreased striatal dopaminergic responsiveness in detoxified cocaine-dependent subjects. *Nature, 386*, 830–833.

Von Korff, M., Kolodny, A., Deyo, R. A., & Chou, R. (2011). Long-term opioid therapy reconsidered. *Annals of Internal Medicine, 155*, 325–328.

Vowles, K. E., McEntee, M. L., Julnes, P. S., Frohe, T., Ney, J. P., & van der Goes, D. N. (2015). Rates of opioid misuse, abuse, and addiction in chronic pain: A systematic review and data synthesis. *Pain, 156*, 569–576.

Wade, C. L., & Fairbanks, C. A. (2014). The self-administration of analgesic drugs in experimentally induced chronic pain. *Current Topics in Behavioral Neurosciences, 20*, 217–232.

Wade, C. L., Krumenacher, P., Kitto, K. F., Peterson, C. D., Wilcox, G. L., & Fairbanks, C. A. (2013). Effect of chronic pain on fentanyl self-administration in mice. *PLoS One, 8*, e79239.

Walteros, C., Sánchez-Navarro, J. P., Muñoz, M. A., Martínez-Selva, J. M., Chialvo, D., & Montoya, P. (2011). Altered associative learning and emotional decision making in fibromyalgia. *The Journal of Psychosomatic Research*, ***70***, 294–301.

Wang, Z. Y., Shi, S. Y., Li, S. J., Chen, F., Chen, H., Lin, H. Z., et al. (2015). Efficacy and safety of duloxetine on osteoarthritis knee pain: A meta-analysis of randomized controlled trials. *Pain Medicine, 16*, 1373–1385.

White, K. L., Scopton, A. P., Rives, M.-L., Bikbulatov, R. V., Polepally, P. R., Brown, P. J., et al. (2014). Identification of novel functionally selective κ-opioid receptor scaffolds. *Molecular Pharmacology, 85*, 83–90.

Whiting, P. F., Wolff, R. F., Deshpande, S., Di Nisio, M., Duffy, S., Hernandez, A. V., et al. (2015). Cannabinoids for medical use: A systematic review and meta-analysis. *JAMA, 313*, 2456–2473.

Wiebelhaus, J. M., Walentiny, D. M., & Beardsley, P. M. (2016). Effects of acute and repeated administration of oxycodone and naloxone-precipitated withdrawal on intracranial self-stimulation in rats. *The Journal of Pharmacology and Experimental Therapeutics, 356*, 43–52.

Wiffen, P. J., Derry, S., Bell, R. F., Rice, A. S., Tölle, T. R., Phillips, T., et al. (2017). Gabapentin for chronic neuropathic pain in adults. *Cochrane Database of Systematic Reviews, 6*, CD007938.

Williams, J. T., Christie, M. J., & Manzoni, O. (2001). Cellular and synaptic adaptations mediating opioid dependence. *Physiological Reviews, 81*, 299–343.

Winters, B. L., Gregoriou, G. C., Kissiwaa, S. A., Wells, O. A., Medagoda, D. I., Hermes, S. M., et al. (2017). Endogenous opioids regulate moment-to-moment neuronal communication and excitability. *Nature Communications, 8*, 14611.

Wise, R. A. (2006). Role of brain dopamine in food reward and reinforcement. *Philosophical Transactions of the Royal Society of London. Series B, Biological Sciences, 361*, 1149–1158.

Wise, R. A., & Koob, G. F. (2014). The development and maintenance of drug addiction. *Neuropsychopharmacology, 39*, 254–262.

Xiao, C., & Ye, J.-H. (2008). Ethanol dually modulates GABAergic synaptic transmission onto dopaminergic neurons in ventral tegmental area: Role of mu-opioid receptors. *Neuroscience*, ***153***, 240–248.

Xu, M., Petraschka, M., McLaughlin, J. P., Westenbroek, R. E., Caron, M. G., Lefkowitz, R. J., et al. (2004). Neuropathic pain activates the endogenous κ opioid system in mouse spinal cord and induces opioid receptor tolerance. *The Journal of Neuroscience, 24*, 4576–4584.

Yalcin, I., & Barrot, M. (2014). The anxiodepressive comorbidity in chronic pain. *Current Opinion in Anaesthesiology, 27*, 520–527.

Zakine, J., Samarcq, D., Lorne, E., Moubarak, M., Montravers, P., Beloucif, S., et al. (2008). Postoperative ketamine administration decreases morphine consumption in major abdominal surgery: A prospective, randomized, double-blind, controlled study. *Anesthesia and Analgesia*, *106*, 1856–1861.

Zhang, X., & Bao, L. (2012). Interaction and regulatory functions of μ- and δ-opioid receptors in nociceptive afferent neurons. *Neuroscience Bulletin*, *28*, 121–130.

Zhang, Z., Tao, W., Hou, Y.-Y., Wang, W., Lu, Y.-G., & Pan, Z. Z. (2014). Persistent pain facilitates response to morphine reward by downregulation of central amygdala GABAergic function. *Neuropsychopharmacology*, *39*, 2263–2271.

Zhu, Y., Hsu, M.-S., & Pintar, J. E. (1998). Developmental expression of the μ, κ, and δ opioid receptor mRNAs in mouse. *The Journal of Neuroscience*, *18*, 2538–2549.

Zhu, Y., King, M. A., Schuller, A. G. P., Nitsche, J. F., Reidl, M., Elde, R. P., et al. (1999). Retention of supraspinal delta-like analgesia and loss of morphine tolerance in δ opioid receptor knockout mice. *Neuron*, *24*, 243–252.

Zubieta, J.-K., Bueller, J. A., Jackson, L. R., Scott, D. J., Xu, Y., Koeppe, R. A., et al. (2005). Placebo effects mediated by endogenous opioid activity on μ-opioid receptors. *The Journal of Neuroscience*, *25*, 7754–7762.

CHAPTER THREE

The neural, behavioral, and epidemiological underpinnings of comorbid alcohol use disorder and post-traumatic stress disorder

Hannah N. Carlson and Jeff L. Weiner*
Department of Physiology and Pharmacology, Wake Forest School of Medicine, Winston-Salem, NC, United States
*Corresponding author: e-mail address: jweiner@wakehealth.edu

Contents

Abstract

Alcohol use disorder (AUD) and (PTSD) frequently co-occur and individuals suffering from this dual diagnosis often exhibit increased symptom severity and poorer treatment outcomes than those with only one of these diseases. Although there have been

International Review of Neurobiology, Volume 157
ISSN 0074-7742
https://doi.org/10.1016/bs.irn.2020.09.006

significant advances in our understanding of the neurobiological mechanisms underlying each of these disorders, the neural underpinnings of the comorbid condition remain poorly understood. This chapter summarizes recent epidemiological findings on comorbid AUD and PTSD, with a focus on vulnerable populations, the temporal relationship between these disorders, and the clinical consequences associated with the dual diagnosis. We then review animal models of the comorbid condition and emerging human and non-human animal research that is beginning to identify maladaptive neural changes common to both disorders, primarily involving functional changes in brain reward and stress networks. We end by proposing a neural framework, based on the emerging field of affective valence encoding, that may better explain the epidemiological and neural findings on AUD and PTSD.

Abbreviations

AMY	amygdala
ACC	anterior cingulate cortex
aHC	anterior hippocampus
AUD	alcohol use disorder
BLA	basolateral amygdala
BOLD	blood-oxygen-dependent-level
CEA	central amygdala
Cort	corticosterone
DMN	default mode network
DSM	diagnostic and statistical manual of mental disorders
DA	dopamine
dACC	dorsal anterior cingulate cortex
dlPFC	dorsal lateral prefrontal cortex
dmPFC	dorsal medial prefrontal cortex
ELS	early life stress
EEG	electroencephalography
ECA	epidemiological catchment area
FHN	family history negative
FHP	family history positive
fMRI	functional magnetic resonance imaging
GAD	generalized anxiety disorder
HC	hippocampus
ICD	international classification of disease
KOR	kappa opioid receptor
LTP	long term potentiation
MDD	major depressive disorder
mPFC	medial prefrontal cortex
mCC	middle cingulate cortex
NESARC	national epidemiological survey on alcohol and related conditions
NMDA	*n*-methyl-D-aspartate
NE	norepinephrine
NAc	nucleus accumbens
OCD	obsessive compulsive disorder
OFC	orbital frontal cortex

PET	positron emission tomography
pACC	posterior anterior cingulate cortex
PFC	prefrontal cortex
pgACC	pregenual anterior cingulate cortex
PTSD	post-traumatic stress disorder
PKA	protein kinase A
Rt-fMRI NF	real-time fMRI neurofeedback
RSFC	resting state functional connectivity
RIP	resident intruder procedures
SEFL	Stress enhanced fear learning
SN	salience network
SPECT	single-photon emission computed tomography
SPS	single prolonged stress
TMS	transcranial magnetic stimulation
VN	valence network
vHC	ventral hippocampus
vmPFC	ventral medial prefrontal cortex
VS	ventral striatum
VTA	ventral tegmental area

1. Introduction

The co-occurrence of alcohol use disorder (AUD) and post-traumatic stress disorder (PTSD) represents one of the most common neuropsychiatric dual diagnoses (Smith & Cottler, 2018). Individuals afflicted with both disorders frequently experience increased symptom severity and poorer treatment outcomes than those diagnosed with only one of these illnesses. Despite the major global socio-economic and public health impact of comorbid AUD and PTSD, our understanding of the neural substrates responsible for the frequent co-occurrence of these disorders is relatively limited and effective treatments are lacking. The goal of this article is to review the epidemiology and major risk factors associated with comorbid AUD and PTSD. We then discuss animal models of these disorders and outline the maladaptive changes in key neural circuits and networks that may contribute to this dual diagnosis, centered primarily on data from human functional imaging of brain reward and stress networks. Finally, we outline a more holistic neural framework for the comorbid condition based on emerging findings from both human imaging work and non-human animal circuit-mapping research.

2. The diagnosis of AUD and PTSD

AUD and PTSD are clinically defined by the Diagnostic and Statistical Manual of Mental Disorders (DSM) and the International Classification of Disease (ICD) (American Psychiatric Association, 2013; World Health Organization, 2018). In both classifications, AUD is defined by the presence of a series of diagnostic criteria or guidelines focused primarily on harmful drinking patterns, loss of control over drinking behavior, and the emergence of a negative affective state during withdrawal, with symptoms persisting for at least 12 months. The DSM-V describes AUD as a single condition that varies in severity based on the number of criteria exhibited (mild, moderate, severe). Similarly, the ICD-11 differentiates alcohol dependence (three or more symptom clusters) from harmful patterns of alcohol use (less than three symptom clusters).

With regard to PTSD, both classification systems require acute and/or chronic trauma exposure and also employ a core symptom cluster-based diagnostic approach. These latest iterations of both diagnostic systems have taken steps to distinguish PTSD from other affective disorders. In fact, the DSM-V no longer classifies PTSD as an anxiety disorder but rather incorporates it into a new class of trauma and stressor-related disorders that also includes Adjustment Disorders. There are now 20 PTSD symptoms contained within four core categories: (1) intrusion or re-experiencing, (2) avoidant symptoms, (3) negative alterations in mood or cognition, and (4) alterations in arousal and reactivity. Similarly, the ICD11 includes PTSD in a category for "Disorders specifically associated with stress." In this system, PTSD is diagnosed by three symptom clusters including (1) re-experiencing of the trauma, (2) avoidance of trauma reminders, and (3) a persistent sense of current threat that manifests as hypervigilance or enhanced startle reactivity. The ICD11 also includes a complex PTSD diagnosis with three additional symptom clusters that reflect "disturbances in self-organization; (1) affect dysregulation, (2) negative self-concept and (3) disturbances in relationships."

Although there are important differences in the symptom clusters used to diagnose AUD and PTSD, it is hard to overlook important similarities in the behavioral phenotypes associated with both disorders. For example, many of the symptoms used to diagnose AUD are centered around alcohol itself (e.g., loss of control over alcohol intake, persistent and unsuccessful attempts at reducing alcohol consumption, tolerance to behavioral and cognitive effects of alcohol) whereas with PTSD, exposure to a traumatic or stressful event (e.g., death, threatened death, actual/threatened serious injury of sexual violence) is a requisite criterion. However, both disorders are strongly associated with the emergence of negative affective mood and behaviors, a profound decrease in natural interest in social, occupational, and/or recreational activities, as well as major alterations in normal sleep patterns.

AUD is often considered a chronic, relapsing disorder characterized by significant fluctuations in symptom severity over the course of the illness (NIAAA, 2020). While PTSD is generally not viewed in a similar manner, it is worth noting that individuals with this disorder often experience a similar variation in symptoms over the course of their recovery. For example, a 5-year longitudinal study of PTSD in primary care patients found that the likelihood of PTSD recovery was 38% but that symptoms reoccurred in almost 30% of these individuals (Perez Benitez et al., 2012). There may also be coordinated fluctuations in the symptoms of both disorders. One study found that, in a cohort of individuals with the comorbid condition receiving outpatient treatment with sertraline, AUD symptoms improved before or simultaneously with PTSD symptoms (Back, Brady, Sonne, & Verduin, 2006). Similarly, an inpatient study with Vietnam veterans being treated for PTSD found that the onset of alcohol use coincided with the development of PTSD symptoms and the alcohol consumption levels increased as did PTSD symptoms (Bremner, Southwick, Darnell, & Charney, 1996). If one begins to consider the severity of the stress experienced by many individuals living with AUD, the overlap in both the symptoms of these disorders, and their frequent recurrence over the trajectory of these illnesses, becomes easier to understand.

3. Epidemiology of comorbid AUD and PTSD in the general population

Despite the fact that the diagnostic criteria for AUD and PTSD have evolved considerably, and that there are significant criterion differences between the DSM and ICD, epidemiological evidence for the frequent comorbidity between these two disorders has remained robust. Although clinicians have anecdotally recognized the frequent co-occurrence of AUD and PTSD for many years, it was only after the Vietnam War and the recognition of PTSD as a unique affective disorder, that the frequency of the comorbid condition began to be formally documented. In the United States, three major surveys have tracked data on neuropsychiatric illnesses in the general population, the Epidemiological Catchment Area (ECA) program, the National Comorbidity Survey (NCS), and the National Epidemiological Survey on Alcohol and Related Conditions (NESARC) and all three have provided data highlighting the frequent co-occurrence of AUD and PTSD. For example, some of the earliest evidence linking AUD and PTSD came from an analysis of ECA data, finding that a DSM-III diagnosis of PTSD was associated with a near twofold increase in risk of developing alcoholism in men and an almost threefold increase in women (Helzer, Robins, & McEvoy, 1987). Similarly, a large review of NCS data that also used DSM-III diagnostic criteria found that a diagnosis of PTSD was associated with a two- to threefold increase in risk of developing DSM-III alcohol abuse or dependence (Kessler, Sonnega, Bromet, Hughes, & Nelson, 1995). More recent studies, using data from the NESARC also found significant, albeit more modest, associations between lifetime PTSD and DSM-IV alcohol abuse/alcohol dependence (1.7 odds ratio) (Pietrzak, Goldstein, Southwick, & Grant, 2011) and DSM-V AUD (1.2 odds ratio) (Goldstein et al., 2016).

Despite major differences in how AUD and PTSD are diagnosed around the world, the striking frequency of this dual diagnosis has been reported in numerous countries. For instance, in Europe, where the prevalence of PTSD in the general population is in the range of 0.6–7% (Burri & Maercker, 2014), these numbers jump to 15–25% in treatment-seeking alcohol dependent patients (Preuss et al., 2018). Similarly, in Australia, where lifetime prevalence of PTSD is around 4–5%, one study reported that 34% of individuals with PTSD were also diagnosed with a substance use disorder, most commonly AUD (24%) (Mills, Teesson, Ross, & Peters, 2006). Even in Brazil, where the overall PTSD rate is already relatively high (8–9%) and likely underestimated (da Silva et al., 2019), the prevalence of PTSD is in excess of 30% in those with an AUD diagnosis (Castillo-Carniglia, Keyes, Hasin, & Cerda, 2019). Collectively, these epidemiological studies provide compelling evidence for the frequent comorbidity between AUD and PTSD in the United States and many other countries.

4. Epidemiology of AUD and PTSD in vulnerable populations

4.1 Military and veterans

Although the incidence of AUD and PTSD in the general population is high, there are certain populations that are particularly vulnerable to develop both disorders. One such group are members of the military and veterans.

Indeed, as already noted, clinical recognition of PTSD and the comorbid condition emerged from epidemiological data collected on veterans of the Vietnam War. It is now well established that the prevalence of comorbid AUD and PTSD is significantly higher among members of the military, relative to civilians, and therefore it is not surprising that the comorbid condition is also more prevalent in military personnel (Dworkin, Bergman, Walton, Walker, & Kaysen, 2018; Richardson et al., 2017; Stein et al., 2017). For example, a recent study on mental disorders in a representative sample of National Guard members found that AUD was the most common disorder in this cohort, with a lifetime prevalence of 44% (Fink et al., 2016) and another found almost the same incidence in US veterans (Fuehrlein et al., 2016). With regard to PTSD, one study that examined the 30-day prevalence of DSM-IV mental disorders among non-deployed US soldiers found that 15% met criteria for an affective disorder, with PTSD being the most common (Kessler et al., 2014). Another study on past year prevalence of mental disorders in members of the Canadian Armed Forces reported a similar 16.5% had at least one of six disorders examined, with major depressive disorder and PTSD being most common. This study also reported a significant increase in the incidence of PTSD and other anxiety disorders between 2002 and 2013 in this population (Zamorski et al., 2016). When looking at the comorbid condition among members of the military and veterans, the data are particularly striking. In a study on veterans of the Iraq and Afghanistan wars, 63% of individuals with AUD also met criteria for PTSD (Seal et al., 2011). More recent data from the National Health and Resilience in Veterans Study reported that approximately 15% of this cohort had probable current AUD or PTSD. Of those with AUD, 20% met criteria for PTSD and for those with PTSD, approximately 17% also met criteria for AUD (Norman, Haller, Hamblen, Southwick, & Pietrzak, 2018). Another recent study on new U.S. recruits found that binge drinking, heavy drinking, and AUD increased the odds of meeting criteria for all mental disorders, with lifetime prevalence of co-occurring PTSD being one of the highest (23%) (Stein et al., 2017).

4.2 Early-life trauma and stress

Another demographic group that is highly vulnerable to AUD, PTSD, and their co-occurrence, are individuals exposed to early life stressors (ELS) such as sexual, physical, and emotional abuse or neglect (Nemeroff, 2016). Over the past three decades, a vast body of epidemiological data has shown an almost dose-response like relationship between the number of childhood/adolescent adverse life events and the likelihood of developing AUD and/or other affective disorders, including PTSD (Dube, Anda, Felitti, Edwards, & Croft, 2002; Felitti et al., 1998; Kaysen, Rosen, Bowman, & Resick, 2010; Khoury, Tang, Bradley, Cubells, & Ressler, 2010; Weber et al., 2008). This relationship is particularly disconcerting given the very high prevalence of ELS. For example, a large epidemiological study found that almost 65% of people in the United States experience at least one, and upward of 12% experience as many as four ELS events (Middlebrooks & Audage, 2008). The vulnerability for developing these disorders conferred by ELS likely stems from the fact that the nervous system undergoes profound developmental changes not only during childhood but throughout adolescence and well into early adulthood. Exposure to trauma and stress during these critical periods can alter normal neuronal maturation in key stress and reward circuits (Arain et al., 2013; Gogtay et al., 2004), leading to a marked increase in risk of developing AUD and a wide range of neuropsychiatric disorders, including PTSD (Nemeroff, 2016).

As just a few striking examples of the strong association between ELS and comorbid AUD/PTSD, one study on children removed from their parents' care due to allegations of abuse or neglect (average age: 12.5 years) reported that PTSD was the most common neuropsychiatric diagnosis at baseline, with 50% meeting criteria for this disorder. Moreover, early onset alcohol use is a major predictor of AUD and almost 30% of maltreated children reported using alcohol, a rate sevenfold higher than demographically matched controls (Kaufman et al., 2007). Indeed, another study of 587 patients recruited from an urban primary care population found a very high rate of alcohol dependence in this cohort (39%) and notably, the number of childhood traumas experienced was significantly correlated with alcohol use and the severity of PTSD symptoms (Khoury et al., 2010).

Although prospective studies documenting the impact of ELS on AUD and PTSD are less common, data from the 10-year longitudinal Youth Emotion Project revealed that the severity of adolescent (but not childhood) ELS exposure predicted poorer health outcomes and onset of substance use disorders (including alcohol) in late adolescence/early adulthood (Wolitzky-Taylor et al., 2017). Interestingly, the relationship between adolescent ELS and substance use disorder in this study was uniquely driven by emotional abuse symptoms and not those associated with sexual or physical abuse or loss/separation of caregiver. Finally, data from a large population-based birth cohort study in Australia found that approximately 10% of participants had a history of substantiated ELS and that both internalizing and externalizing disorders were strongly associated with ELS in adolescence (Mills et al., 2013) and early adulthood (Kisely et al., 2018), with PTSD being the most strongly associated affective disorder.

4.3 Impact of gender

There is an extensive and growing body of literature suggesting that gender is an important moderating factor in the comorbid condition. In the general U.S. population, being female is associated with a twofold increase in the risk of developing PTSD (Goldstein et al., 2016; Kilpatrick et al., 2013; Pietrzak et al., 2011) whereas being male increases the likelihood of an AUD diagnosis by a similar margin (Grant et al., 2015). Reasons for these prevalence differences are complex and likely involve both biological and sociological factors. For example, the impact of gender on the prevalence of AUD is likely influenced by the fact that men typically consume more alcohol than women (Wilsnack, Wilsnack, Kristjanson, Vogeltanz-Holm, & Gmel, 2009). Interestingly, although it is often suggested that the higher prevalence of PTSD in women can be attributed to gender differences in the likelihood of being exposed to traumatic events, one recent study that sought to control for this variable actually concluded that women may also be more vulnerable to the neurological consequences of traumatic events (Blanco et al., 2018). Other data suggest that women may also be more sensitive to the deleterious effects of alcohol on the brain (although evidence is mixed) (Verplaetse, Cosgrove, Tanabe, & McKee, 2020), a finding that is particularly disconcerting given that the longstanding gender gap in alcohol consumption (Grant et al., 2017; Keyes, Grant, & Hasin, 2008; McHugh, Votaw, Sugarman, & Greenfield, 2018) and AUD (Grant et al., 2017) appears to be shrinking. Together, these studies reveal important gender differences in comorbid PTSD and AUD. The complexity of these associations highlights the need for further studies as a deeper understanding of the neurobiology underlying these differences will likely have significant implications with regard to the development of effective treatments for these disorders in women and men.

5. Which comes first?

It is important to understand the temporal relationship between PTSD and AUD as such knowledge may provide mechanistic insight into the underlying causes of the comorbid condition. Four models have been proposed to explain the functional associations between these disorders (Straus, Haller, Lyons, & Norman, 2018). The Self-Medication model posits that PTSD precedes AUD and that pathological alcohol use develops in an attempt to mitigate the negative affective symptoms of PTSD. The High Risk and Susceptibility models propose that excessive alcohol use can increase the likelihood of experiencing traumatic events or exacerbate emotional and/or neurobiological consequences associated with experiencing a traumatic event. Finally, the Shared Vulnerability model argues that both disorders arise as a result of common biological and/or environmental risk factors. In general, the majority of epidemiological studies suggest that PTSD most often develops before AUD, and, not surprisingly, there is a large body of evidence in support of the Self-Medication model (see Gilpin & Weiner, 2017; Straus et al., 2018). For example, a comprehensive review of large cross-sectional, prospective, and survival analysis datasets found consistent evidence that pre-existing PTSD increased the risk of drug abuse/dependence with little evidence in support of the "High Risk" hypothesis (Chilcoat & Breslau, 1998). A more recent study that sought to evaluate all four hypotheses using longitudinal data from a community sample also found the strongest support for the Self-Medication hypothesis, with PTSD symptoms being the strongest predictor of higher levels of alcohol and drug problems, when compared with pre-trauma family risk factors, pre-trauma substance use problems, trauma exposure, and demographic variables (Haller & Chassin, 2014). This study also found no support for the Susceptibility or Shared Vulnerability models but did show that adolescent substance use problems had a modest but significant effect on risk of assaultive violence exposure, but not overall risk for trauma, partially supporting the High Risk hypothesis. Finally, a study using data from the NESARC found that PTSD may be the anxiety disorder with the highest rates of self-medication with either alcohol or other drugs and that men were twice as likely to seek out alcohol or drugs to alleviate their PTSD symptoms (Leeies, Pagura, Sareen, & Bolton, 2010).

One final comment on the temporal patterns associated with this dual diagnosis is that far fewer studies have actually sought out evidence for the high risk or susceptibility models in which AUD precedes PTSD. One recent study that also used NESARC data and directly addressed the order of onset between these disorders actually found that the pathway from PTSD to AUD was significantly stronger than the reverse, but only in women (Berenz et al., 2017). Moreover, another NESARC study that used wave I and II data to conduct what may be the first longitudinal investigation for the direct effects of past-year AUD on the development of PTSD, found that prior alcohol dependence significantly increased risk of trauma exposure and the development of PTSD, relative to matched controls (Balachandran, Cohen, Le Foll, Rehm, & Hassan, 2020).

6. Clinical consequences of comorbid AUD and PTSD

Although many factors influence the temporal relationship between AUD and PTSD, the physical and psychological burden associated with this dual diagnosis is significantly greater than that of either disorder alone. To that end, a large number of studies that have compared individuals with either AUD or PTSD alone with subjects diagnosed with both disorders have found that the latter group often experiences greater symptom severity of both disorders (Dworkin, Wanklyn, Stasiewicz, & Coffey, 2018; Longo et al., 2020; Somohano, Rehder, Dingle, Shank, & Bowen, 2019), exhibits higher rates of other psychiatric comorbidities (e.g., bipolar disorder, depression, anxiety, schizophrenia) (Bowe & Rosenheck, 2015; Norman et al., 2018; Norman, Tate, Anderson, & Brown, 2007; Read, Brown, & Kahler, 2004; Terhaag et al., 2019), suffers greater physical health and psychosocial problems (Bowe & Rosenheck, 2015; Hoge, Terhakopian, Castro, Messer, & Engel, 2007; Ouimette, Goodwin, & Brown, 2006; Tate, Norman, McQuaid, & Brown, 2007), as well as higher rates of homelessness and poor life conditions (Bowe & Rosenheck, 2015; Najavits, Weiss, & Shaw, 1999). For example, a study of a nationally representative sample of U.S. veterans with AUD, PTSD, or the comorbid condition found that, relative to those with AUD alone, the dual diagnosis cohort scored significantly lower on measures of cognitive, mental and physical functioning as well as quality of life (Norman et al., 2018). The comorbid group also had a 5–10-fold increased incidence of major depression, generalized anxiety disorder, suicidal ideation and suicide attempts. Indeed, many other studies have reported very high levels of suicidal ideation, suicide attempts and increased mortality among the comorbid population (Afzali, Sunderland, Batterham, Carragher, & Slade, 2017; Kachadourian, Gandelman, Ralevski, & Petrakis, 2018; Shorter, Hsieh, & Kosten, 2015; Stefanovics & Rosenheck, 2019) and, although literature on gender differences is sparse, the dual diagnosis may represent an even greater risk factor for suicide attempts in women (Gradus et al., 2017). Not surprisingly, there is also extensive evidence that the dual diagnosis has a profoundly negative impact on treatment and recovery (Flanagan, Jones, Jarnecke, & Back, 2018; Kingston, Marel, & Mills, 2017; Verplaetse, McKee, & Petrakis, 2018).

7. Animal models of comorbid AUD and PTSD

The preceding epidemiological review clearly indicates that comorbid AUD and PTSD represents a significant global public health concern. The co-occurrence of these disorders is high in many countries and there is compelling evidence that this dual diagnosis represents a major treatment challenge. Despite overwhelming data highlighting the clinical significance of this dual diagnosis, there has been surprisingly little human research directed at the neurobiological mechanisms responsible for the comorbid condition. As reviewed in subsequent sections of this chapter, there is a rich literature on the neural mechanisms that are thought to contribute to each disorder alone. However, studies on the neurobiology underlying the comorbid condition in human subjects is sparse. To address this gap in our knowledge, a number of preclinical animal models have been developed that engender symptoms of both disorders. To provide a background for a discussion of some of the neurobiological findings obtained using these paradigms, in this section we provide a brief description of the more commonly used models. Although numerous animal regimens have been developed to model each disorder (for recent reviews see: Deslauriers, Toth, Der-Avakian, & Risbrough, 2018; Flandreau & Toth, 2018; Goltseker, Hopf, & Barak, 2019; Kenny, Hoyer, & Koob, 2018; Pinna, 2019; Richter-Levin, Stork, & Schmidt, 2019; Torok, Sipos, Pivac, & Zelena, 2019), here we focus exclusively on studies that have explicitly sought to model the comorbid condition. All of these models involve some element of stress-induced escalation of alcohol consumption. Briefly, subjects are exposed to some form of stress, such as predator odor (Edwards et al., 2013) or adolescent social isolation (Locci & Pinna, 2019), among others. Then, alcohol intake in stress-exposed animals can be compared to control subjects. Alternatively, the animals may be trained to drink prior to stressor presentation, to monitor for individual changes.

The efficacy of an animal model is based on its adherence to certain metrics of validity (Willner, 1986). Predictive validity refers to the propensity of the dependent variable of the model to predict translational outcomes; in other words, using animal performance to predict human outcomes. These predicable outcomes may be treatments or behavioral or physiological markers of the disorder. Face validity refers to the extent to which the model *appears* to measure what it purports to. For example, a model of AUD would achieve face validity if it were capable of producing behavioral signs of dependence, like tolerance, withdrawal, craving, or binging.

Finally, construct validity involves the relationship between the model and the theory underlying its conception, namely the extent to which a given model represents the construct it is intended to. This representation should be both empirical and theoretical. For example, if traumatic experience increases alcohol use in the human population, a translational model with construct validity would have to demonstrate a similar pattern in animals. Furthermore, the underlying cause of the change in alcohol drinking in this example should be homologous. In other words, if the traumatic condition in the animal model were prolonged water restriction, a subsequent increase in drinking cannot be attributed without doubt to trauma, because thirst is also likely a cause. Opinions differ as to which of these criteria are primary, or even necessary, for a viable animal model (Geyer & Markou, 1995; Koob, Heinrichs, & Britton, 1998; Robbins, 1998; Sarter & Bruno, 2002).

In terms of creating and using valid animal models of PTSD, it is crucial to consider the differences between modeling stress and modeling trauma. Though Generalized Anxiety Disorder (GAD) and PTSD do frequently co-occur in humans and are both highly comorbid with AUD, they are distinct diagnoses. The aforementioned stress-induced drinking models involve acute, low level stressors and may produce a phenotype more consistent with GAD than PTSD. Indeed, GAD may be a risk factor for the development of PTSD following exposure to a traumatic event. However, symptoms that differentiate PTSD involve re-experiencing of the specific trauma, which includes flashbacks, avoidance of related stimuli, and hyper rousal. Models of PTSD are characterized by high intensity stressors and often aim to mirror the conditions in which traumatic events occur in the human population.

Every animal model that aims to represent the comorbid condition of AUD and PTSD necessitates application of a stressor and access to alcohol. A multitude of stressors have been utilized, simulating the variety of events in the human population which may lead to a PTSD diagnosis. In animal models, the stressors may be physical or non-physical. A physical stressor entails physiological threat like shock, restraint, or forced submersion in water. Non-physical stressors includes both social/psychological and pharmacological, where psychological involves manipulation of an external factor, like social isolation or vicarious trauma, and pharmacological involves the induction of autonomic stress reactivity with a drug compound, like yohimbine (Lamb, 1979; Verbitsky, Dopfel, & Zhang, 2020). The vast breadth of these models highlights an especially important consideration: it may not be feasible to use a single model to reliably reproduce a human condition that can be induced by qualitatively distinct stressors. Importantly, though no existing model captures the full symptomatology of both disorders, those reviewed here are capable of reliably producing escalation in alcohol intake as well as at least two core PTSD symptoms.

One early translational model of PTSD is the single prolonged stress (SPS) procedure, which has been performed in many iterations and involves a series of different assortments and schedules of stressor application in a single day (Dal-Zotto, Marti, & Armario, 2003; Lisieski, Eagle, Conti, Liberzon, & Perrine, 2018; Richter-Levin, 1998). Briefly, SPS involves a single period of intense stress exposure, often in the form of multiple physical stressors like restraint, forced swim, and foot shock, exposure to ether until loss of consciousness. The amalgamation of multiple, varied stressors occurring over a relatively short (less than a day) period of time is intended to simulate a translationally relevant traumatic event. Such models do reliably produce behavioral symptoms associated with PTSD, including reduced fear extinction, enhanced startle response, hyperarousal, and heightened anxiety-like behavior (mood alteration), as well as an escalation of ethanol consumption indicative of AUD (Knox et al., 2012; Lisieski et al., 2018; Souza, Noble, & McIntyre, 2017).

Similar models, involving varied schedules of repeated stress have also been performed: these include chronic variable stress (Lopez, Doremus-Fitzwater, & Becker, 2011; McGuire, Herman, Horn, Sallee, & Sah, 2010; Rompala, Simons, Kihle, & Homanics, 2018) and intermittent unpredictable stress (Peay et al., 2020; Sequeira-Cordero, Salas-Bastos, Fornaguera, & Brenes, 2019). These models capitalize on mercurial stressor application, in order to simulate a context in which the stressor may appear at any time. If the stressor is unpredictable, it can neither be avoided nor adapted to, which leaves the animal in a state of hypervigilance or prolonged anxiety, which may disrupt extinction of conditioned fear. Another related model that reliably produces alterations in fear learning is aptly referred to as Stress-Induced Enhancement of Fear Learning (SEFL) (Rajbhandari, Gonzalez, & Fanselow, 2018). In this procedure, prior exposure to repeated footshock is sufficient to produce fear generalization. In other words, animals that have been subjected to electric footshock in the past are more sensitive to similar, but less intense stressors at a later time point. Translationally, this phenotype tracks with resurgence of symptoms in response to related, but nonspecific stressors as well as enhanced acquisition of novel fears in patients with PTSD. Interestingly, this model increases voluntary alcohol drinking in alcohol-naïve rats but not in those with a prior history of alcohol self-administration (Meyer, Long, Fanselow, & Spigelman, 2013).

Given the prevalence of PTSD in combat-exposed populations, some models have sought to more directly mimic the conditions of war. One recent model employed a combination of physical and psychological stressors, including repeated footshock, distressing auditory tones, sleep deprivation and forced exercise to represent the onslaught of stressors experienced by soldiers (Kim, Kim, Koo, Heo, & Cheon, 2017). This model did increase anxiety-like behavior, as well as producing meaningful alterations in cognitive and immune functions. However, because exposed animals displayed no alteration of hypothalamic-pituitary-adrenal (HPA) function, the authors reasoned that the model may not fully represent PTSD. Yet, further validation of this model may prove fruitful, given its robust behavioral effects, despite the lack of histopathological changes. If future studies were to demonstrate either a subsequent increase in voluntary alcohol intake or alterations in reward/stress circuitry, the combat-exposure animal model may have significant construct validity for the comorbid condition, given that HPA-specific changes in PTSD are often inconclusive (de Kloet et al., 2006).

Another popular model involves exposing animal subjects to the odor of a natural predator in order to simulate a life threatening situation, without causing actual harm to the animal (Edwards et al., 2013; Manjoch et al., 2016). In such models, adult rodents will be trained to self-administer alcohol prior to exposure to predator odor (commonly bobcat, feline, or fox urine). Subjects are then categorized by their immediate response to the given stressor and once again allowed to consume alcohol. Rodents that displayed high reactivity to the predator odor display behavioral phenotypes associated with alcohol-dependence, including escalated drinking post-stress, compulsive-like responding, and aversion resistant drinking. This model is unique in its replication of the avoidance symptom of PTSD.

As alluded to previously, some animal models of the comorbid condition have focused more so on socially-derived stressors that may act as a correlate for early life adversity, parental neglect, and/or psychosocial trauma. One such model is adolescent social isolation, during which rodents are separated early after weaning (typically PD 28) into two conditions: socially isolated animals housed singly in standard cages and group housed animals, housed 3–4 in larger, guinea pig cages for 3–6 weeks. Following this housing manipulation, socially isolated rats display enduring increases in alcohol intake and preference, increased anxiety-like behavior, and notably, impaired fear extinction (Skelly, Chappell, Carter, & Weiner, 2015) compared to their group housed counterparts. Importantly, this manipulation only produces the aforementioned phenotypes when performed during adolescence and in males; animals that reach adulthood before being separated do not develop AUD/PTSD-like symptoms, nor do female rodents (Butler, Carter, & Weiner, 2014). Because this housing manipulation leads to the development of both AUD- and PTSD-like symptoms, it has been suggested as a viable model of the comorbid condition (Butler, Karkhanis, Jones, & Weiner, 2016; Locci & Pinna, 2019; Pinna, 2019). This procedure may model early life adversity or neglect, which have been explored as a potential cause or correlate of PTSD and AUD in human populations.

Other social stressor models rely not on the anxiogenic effect of isolation, but on the propensity of negative social interaction to produce trauma. Indeed, psychosocial stress in various forms has been linked to PTSD. Social defeat paradigms are commonly used as models of depressive-like behavior. However, they may also produce behavioral phenotypes that encapsulate the depression, anxiety, and social avoidance symptomology of PTSD and have gained popularity in this field (Ishikawa, Uchida, Kitaoka, Furuyashiki, & Kida, 2019; Sial, Warren, Alcantara, Parise, & Bolanos-Guzman, 2016; Verbitsky et al., 2020). In resident-intruder (RIP) procedures, a male "intruder" rodent is introduced to the homecage of an older, dominant "resident" rodent. The resident acts aggressively, in defense of the homecage, by attacking the intruder. In RIP models, the intruder animal is assessed for symptoms of trauma exposure. Some variations have been assessed in recent years. In a study of adolescent animals, Long-Evans residents were socially defeated by Sprague Dawley intruders (Manz, Levine, Seckler, Iskander, & Reich, 2018). This finding, a reversal of the classic outcome in which the resident is socially dominant, prompted the use of reverse RIP (rRIP) as a model of adolescent stress exposure or bullying. Another variation involves assessment not of the intruder animal, but of a spectator animal (Sial et al., 2016). This model, termed "vicarious" social defeat, seeks to characterize the impact of witnessing trauma. A major limitation for social defeat models is sex specificity; resident-intruder pairings typically involve animals of the same sex. Aggression is differentially expressed in male and female rodents, females tending to attack less frequently than males. Furthermore, males tend not to attack female intruders as readily as other males, which limits the available data on the impact of social defeat on behavior in females (Verbitsky et al., 2020). However, one model had adolescent rodents as the spectator of social defeat of their mother (Liu, Patki, Salvi, Kelly, & Salim, 2018) and did see that both male and female pups developed depressive-like behaviors after exposure to vicarious maternal trauma. All of the social defeat procedures are capable of producing PTSD-like symptoms in the target rodent, including heightened anxiety-like behavior, social avoidance, and alterations of stress-related biomarkers including corticosterone (CORT) as well as increased in alcohol self-administration (Deal, Konstantopoulos, Weiner, & Budygin, 2018; Macedo et al., 2018; Newman, Albrechet-Souza, et al., 2018).

Finally, a necessary component of any model for *comorbid* PTSD/AUD is an escalation of alcohol drinking-related behaviors. Several of the models discussed above have produced considerable effects on varied aspects of alcohol-directed behavior. The adolescent social isolation model produces a robust drinking phenotype in males (Butler et al., 2016; McCool & Chappell, 2009). SI animals consume more alcohol during intermittent two bottle choice procedures, where bottles containing 10% ethanol or water are offered side by side in the home cage for 12 h periods three times per week. The aSI manipulation also produces behavioral changes in operant alcohol self administration. Importantly, aSI animals display a higher level of both appetitive (seeking) and consummatory (drinking) behavior compared to GH rats. Importantly, this effect is alcohol-specific; responding is not similarly altered toward sucrose. Predator odor also engenders vulnerability to drinking in a subset of animals, which is representative of the clinical condition in that *some* but not *all* individuals will be affected. Specifically, animals that avoided the odor paired chamber (Avoiders) exhibited prolonged increases in operant responding and compulsive-like responding for alcohol compared to non-Avoiders (Edwards et al., 2013).

A critical consideration in the stress-induction models discussed above is whether they actually succeed in *traumatizing* the animal. As discussed previously, a diagnostic criterion for PTSD is direct or indirect exposure to a traumatic event leading to symptoms that persist for at least 1-month. Thus, enduring changes in core PTSD-like phenotypes may be the most viable indicator that animals have experienced actual trauma. Ethologically relevant stressors, like predator odor, are perhaps the most appropriate given these criteria. Social defeat paradigms have successfully produced behavioral alteration up to a month following exposure. While some animal models of the comorbid condition, like adolescent social isolation, engender escalations in alcohol drinking and anxiety-like behaviors lasting up to 8 weeks (Skelly et al., 2015; Yorgason, Espana, Konstantopoulos, Weiner, & Jones, 2013), it is not clear if all of the symptoms that manifest early after the housing manipulation last as long. Additionally, paradigms like adolescent social isolation, that focus on stress exposure during specific developmental periods, may not model PTSD that results from trauma exposure during adulthood. An ideal model would involve exposure to a stressor that is capable of producing AUD- and PTSD-like symptoms when applied at any time during the life course. Finally, in an ideal model, only a subset of the animals exposed to the traumatic stimulus or event should manifest symptoms associated with PTSD. This criterion is met in models like the predator odor procedure in which only animals that avoid odor-paired stimuli go on to develop AUD/PTSD-like phenotypes (Edwards et al., 2013).

8. Neurobiological substrates of comorbid AUD and PTSD

As reviewed above, there is extensive epidemiological evidence documenting both the frequent co-occurrence of AUD and PTSD as well as the many negative clinical consequences associated with this dual diagnosis. This evidence has led to the hypothesis that these disorders may involve maladaptive changes in common neural substrates. Surprisingly, relatively few studies have directly examined the neurobiology of the comorbid condition. However, there is a rich literature on the biological basis of each disorder (e.g., Fenster, Lebois, Ressler, & Suh, 2018; Fritz, Klawonn, & Zahr, 2019; Koob, 2014; Koob & Volkow, 2016; Malikowska-Racia & Salat, 2019; Yehuda et al., 2015). Indeed, the available evidence strongly supports the notion that many of the behavioral phenotypes associated with both disorders manifest as a result of a profound dysregulation of brain networks that process reward and stress. In this section, we review recent findings from human studies that highlight the maladaptive effects of AUD and PTSD on these canonical brain reward and stress circuits, with an emphasis on the impact of these illnesses on measures of functional connectivity. We also review findings from preclinical studies using animal models of the comorbid condition that provide convergent evidence that alterations in reward and stress pathways may play a causal role in the AUD/PTSD-like phenotypes promoted by these models. We then advance a novel conceptual framework for the comorbid condition that borrows concepts from the emerging field of affective valence processing. This framework proposes that reward and stress circuits may be viewed as a functionally integrated valence network that assigns some degree of positivity or negativity to all sensory and emotional stimuli, encodes the relative salience or value of this information, and then orchestrates appropriate behavioral responding to these stimuli. Using this model, we propose that AUD and PTSD be viewed as examples of valence network imbalance disorders and outline how this heuristic may provide a better framework for interpreting the epidemiological and neurobiological findings on the comorbid condition. Readers interested in reviews focused on more traditional anatomical, structural, and functional adaptations and treatments for the comorbid condition are encouraged to read these other recent reviews (Blaine & Sinha, 2017; Dunlop & Wong, 2019; Fritz et al., 2019; Gilpin & Weiner, 2017; Goode & Maren, 2019; Suh & Ressler, 2018).

9. Reward circuitry

Many studies over the years have identified neural systems that comprise a "brain reward" circuitry, with the dopaminergic (DAergic) projection from the ventral tegmental area (VTA) to the nucleus accumbens (NAc) receiving the most attention (Fig. 1). Indeed, it is now well established that acute exposure to all drugs of abuse, including alcohol, engages the VTA-NAc pathway to increase striatal DA levels and that activation of this pathway reinforces drug/alcohol taking behaviors (Cooper, Robison, & Mazei-Robison, 2017; Kim, Ham, Hong, Moon, & Im, 2016; Volkow & Morales, 2015). Moreover, as reviewed below, dysregulation of this pathway, and many other elements of the reward network, play an integral role in the development and progression of AUD and PTSD.

9.1 Reward circuit and AUD—Human studies

Numerous imaging studies have demonstrated that AUD is associated with profound alterations in VTA-NAc DA signaling. Early PET studies from Volkow and colleagues reported that individuals diagnosed with AUD had lower D2 receptor availability in the striatum, relative to healthy controls (Volkow et al., 1996), a finding also observed in people with SUDs (Volkow et al., 1990; Wang et al., 1997). These findings, coupled with evidence of decreased methamphetamine-stimulated striatal DA release in AUD (Volkow et al., 2007), led to the hypothesis that AUD was associated with a hypoDAergic state and decreased function of the brain reward circuitry (although see Hirth et al., 2016 for a contrarian view).

While it remains well-established that AUD dampens mesolimbic DA signaling, and that this maladaptive change contributes to the negative affective symptoms of this disorder, more recent work using functional imaging techniques, like fMRI, suggests that AUD has more complex effects on the canonical VTA-NAc DA pathway and many other elements of the reward

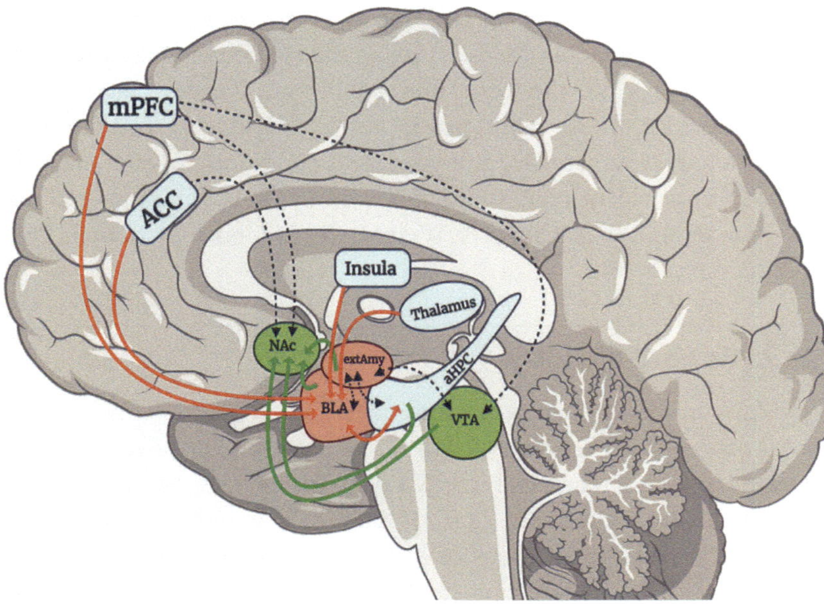

Fig. 1 Brain reward and stress circuitry in the human brain. Diagrammatic illustration of the predominant brain regions and circuits generally though to mediate the processing of reward (green) and stress/fear (red) in the human brain. The nucleus accumbens (NAc) is usually featured as the central node of reward circuit diagrams, which also typically include the DAergic input from the VTA and excitatory input from other limbic structures, like the anterior hippocampus (aHPC). The stress circuit is generally centered around the basolateral AMY (BLA) which has dense, and often reciprocal connectivity with numerous cortical structures including the medial prefrontal cortex, anterior cingulate cortex, insula, the thalamus and aHC as well as nodes of the extended amygdala (CEA, BNST). Illustrated by dashed lines are some of the many additional connections that link elements of both the reward and stress networks that are often omitted from these circuit diagrams.

circuitry. For example, several studies have shown that, relative to healthy controls, individuals with AUD respond more to alcohol-associated cues in the ventral striatum and other reward-related areas like the amygdala (AMY), anterior cingulate cortex (ACC), and medial prefrontal cortex (mPFC) (Huang, Mohan, De Ridder, Sunaert, & Vanneste, 2018; Schacht, Anton, & Myrick, 2013; Vollstadt-Klein et al., 2012). One recent study found that binge drinkers with a heightened risk of developing AUD display heightened NAc BOLD responses to monetary rewards (Crane et al., 2017). These findings are consistent with the incentive salience theory of addiction which suggests that repeated drug exposure actually sensitizes the brain reward system to drug-associated cues, leading to increased craving and relapse (Robinson & Berridge, 1993). However, other fMRI studies have shown that positive cue-evoked BOLD responses in the ventral striatum are lower in individuals with AUD, particularly for non-alcohol-related cues (Fritz et al., 2019). For example, one study found that, while visual alcohol-associated cue reactivity in the NAc was elevated in AUD subjects, relative to controls, the AUD cohort actually had lower NAc BOLD responses to non-alcohol-related monetary cues (Wrase et al., 2007). Notably, both of these changes were correlated with alcohol craving in the AUD group but not the controls.

Over the past decade or so, many groups have begun to use functional imaging to go beyond measurements of task-relate activation of discrete brain regions to examine alterations in functional connectivity across complex brain networks. Most early work focused on task-free assessments of resting state functional connectivity and these studies identified a distributed default mode network (DMN) that was preferentially activated in the resting state. Although there is no universally accepted map of the DMN, numerous regions are consistently identified and can be separated into three interconnected subnetworks: a midline core DMN, a medial temporal DMN, and a dorsal medial prefrontal DMN (Zhang & Volkow, 2019). It is now appreciated that, while these DMN networks are active at rest, they can also be engaged by a variety of tasks, included some that are related to emotional processing. Importantly, the DMN is integrally associated with brain reward circuitry and many studies have now shown that key elements of the DMN are disrupted in AUD. Although a detailed review of this literature is beyond the scope of this chapter (please see Zhang & Volkow, 2019 for a comprehensive recent review), AUD has been shown to be associated with a shift in DMN connectivity from higher order structures (e.g., posterior cingulate cortex to middle cingulate cortex, (MCC)) to more cue-reactive circuits (e.g., increased connectivity between midbrain and MCC). Such changes could contribute to alterations in reward processing for alcohol and non-alcohol-related cues.

Interestingly a few recent studies have explicitly examined NAc connectivity in "at risk" populations and AUD subjects. Given that early onset drinking is a risk factor for AUD, one study explored the relationship between NAc resting state functional connectivity (RSFC) and lifetime drinking behavior in 199 healthy non-alcohol dependent male adolescents and asked whether this measure could predict alcohol consumption during a 1-year follow-up (Veer et al., 2019). The authors reported that greater past alcohol consumption, particularly during the most recent year, was associated with a modest but significant weakening of functional connectivity between the left NAc and dlPFC and bilateral IFF, extending into the left dorsal anterior insula. Most notably, lower connectivity between the NAc and these cortical regions predicted alcohol drinking behavior during the 1-year follow-up. Another study examined NAc RSFC in cohorts of adolescents with a family history of AUD (FHP), another well-established risk factor for this disorder (Dawson, Harford, & Grant, 1992; Sher, 1991), with a control group of family history negative individuals (FHN). Numerous group differences in resting state synchrony were observed, including decreased segregation between the NAc and fronto-parietal networks as well as less integration of the right NAc and left OFC, both changes that may be related to impaired reward processing (Cservenka, Casimo, Fair, & Nagel, 2014). Finally, another group combined resting state EEG and task-based fMRI to examine neural correlates of craving in the NAc and other elements of the reward circuit in a cohort of 11 AUD patients 24 h into abstinence (Huang et al., 2018). This study found convergent evidence, using both imaging techniques, of a "craving network" characterized by increased BOLD activation in response to visual alcohol cues in the NAc, AMY, dACC, pgACC, PCC and parahippocampus as well as increased beta-band activity in the dACC and pgACC in resting-state EEG. These changes were accompanied by hyperconnectivity in resting-state EEG functional connectivity and dysconnectivity in cue-based fMRI functional connectivity across this densely connected network of alcohol-related brain areas. It will be important, in future studies, to conduct similar assessments in healthy controls and AUD patients in long-term abstinence to gain a better understanding of the specific role of these measures of network connectivity in the etiology of the disorder.

9.2 Reward circuit and PTSD—Human studies

As with AUD, negative alterations in mood and anhedonia are core features of PTSD (Eskelund, Karstoft, & Andersen, 2018; Fenster et al., 2018; Vujanovic, Wardle, Smith, & Berenz, 2017). For example, symptoms of anhedonia are observed in approximately two-thirds of PTSD patients, independent of comorbid major depressive disorder (Carmassi et al., 2014; Franklin & Zimmerman, 2001). Moreover, a recent systematic review of 29 studies comparing measures of reward functioning in healthy controls and individuals with PTSD found that most reported decreases in reward anticipation, motivation to obtain rewards, and hedonic responding to reward in PTSD patients (Nawijn et al., 2015). Despite these findings, PET imaging studies of the DAergic system in PTSD are surprisingly scarce. A meta-analysis of PET/SPECT data from individuals with OCD, GAD, phobia, and PTSD found a significant decrease in D2 receptor binding in the ventral striatum, similar to that observed in individuals with AUD (Nikolaus, Antke, Beu, & Muller, 2010). One other study used SPECT to assess striatal DAT levels in asymptomatic trauma-exposed subjects and PTSD patients who had been exposed to urban violence. It reported significantly greater DAT binding density in both the left and right striatum in the PTSD cohort (Hoexter et al., 2012). The authors suggested that this change may reflect a compensation for increased dopamine (DA) signaling in fear circuits. However, an alternative interpretation is that higher striatal DAT may lead to lower DA levels which could contribute to the anhedonia and negative mood associated with this disorder.

Although direct evidence of DA receptor changes in PTSD is limited, numerous functional imaging studies have demonstrated that dysregulation of brain reward circuitry is a central feature of this disease. One early fMRI study examined decision-making in trauma-exposed women with PTSD and healthy controls using a reward processing task in which subjects could maximize their gains by learning an advantageous response pattern (Sailer et al., 2008). While both groups increased their performance and reduced their reaction time across this task, controls improved faster and performed better in the late phase, relative to PTSD patients. These behavioral differences were associated with a significant decrease in activation of the NAc and PFC during the late phase of the task in the PTSD cohort. Another study examined BOLD activation of reward circuit responsivity using a Wheel-of-Fortune-like monetary reward task in a cohort of 28 civilians diagnosed with PTSD and mentally healthy controls (Elman et al., 2009). Using both an unbiased voxel-wise analysis and region-of-interest approach, PTSD subjects exhibited smaller bilateral striatal activation during the passive experiencing of monetary gains vs. losses. Moreover, in the PTSD group, striatal activation was negatively correlated with self-reported motivational and social deficits. Interestingly, other reward-related regions, like the AMY, pallidum, insula, cingulate, orbitofrontal, and frontal cortices, were all activated during the task in control subjects but not those with PTSD. These findings suggest not only a blunting but a narrowing of the elements of the reward circuitry in individuals with PTSD. It should be noted that most, if not all, of these brain regions are also involved in the processing of stressful and aversive stimuli. To that end, one recent study sought to examine the intriguing possibility that PTSD-associated deficits in reward processing may involve over-recruitment of reward-related brain regions by aversive stimuli, making them less available to respond to stimuli with positive valence (Elman et al., 2018). In one experiment, PTSD patients and healthy controls were exposed to pleasant, aversive, or neutral visual stimuli. Individuals with PTSD exhibited decreased subjective positive ratings of pleasant images and decreased activation of the striatum and several cortical regions. Interestingly, although no group differences in subjective ratings of unpleasant visual stimuli were observed, PTSD subjects showed decreased activation in the AMY and thalamus. In a second experiment, the same subjects were exposed to a pain-inducing thermal stimulus. Although PTSD patients and controls reported similar subjective pain ratings in response to this aversive sensory stimulus, this stimulus was associated with increased activation of the striatum, AMY, HC, and mPFC in PTSD patients. These findings reveal that PTSD may be associated with decreased activation of the striatum and other areas that process rewarding (and aversive) psychosocial stimuli but that these same brain regions may be hyper-responsive to aversive sensory stimuli.

As observed in patients with AUD, there is considerable evidence that PTSD has profound effects on brain-wide network connectivity. While most of these studies have focused on network connectivity and fear/stress-related symptoms of PTSD, some alterations in functional connectivity likely impact reward processing and may contribute to the anhedonic symptoms of PTSD. There is a large circuitry called the salience network (SN) which is comprised of numerous cortical and subcortical structures that serve to integrate emotional and sensory input, particularly in the context of threatening stimuli (Peters, Dunlop, & Downar, 2016; Szeszko & Yehuda, 2019). However, many of the brain regions that comprise the SN are also known to play an integral role in reward processing, such as the insula, the AMY, the dorsal ACC, and the OFC (Koch et al., 2016; Kunimatsu, Yasaka, Akai, Kunimatsu, & Abe, 2020) as well as the striatum and brainstem DAergic nuclei (Peters et al., 2016). Although the SN has received relatively little attention in the AUD field (Peters et al., 2016; Zhang & Volkow, 2019), numerous studies have examined intrinsic and extrinsic changes between DMN and SN connectivity in PTSD patients (for reviews see: Koch et al., 2016; Kunimatsu et al., 2020; Peters et al., 2016; Szeszko & Yehuda, 2019). One of the most consistent findings has been a decrease in intrinsic connectivity within nodes of the DMN in PTSD (Akiki et al., 2018; Bluhm et al., 2009; Chen & Etkin, 2013; DiGangi et al., 2016;

Sripada, King, Garfinkel, et al., 2012). In contrast, most studies have observed increased intrinsic SN connectivity in PTSD patients between the insula and both the AMY and HC (Chen & Etkin, 2013; Nicholson et al., 2015; Rabinak et al., 2011; Sripada, King, Welsh, et al., 2012). Although, both increases (Brown et al., 2014) and decreases (Sripada, King, Garfinkel, et al., 2012) in connectivity have been reported between the AMY and dACC. Interestingly, increased cross-network connectivity has been seen in combat veterans with PTSD (Sripada, King, Welsh, et al., 2012) but decreased coupling was observed in PTSD patients with early-life trauma (Bluhm et al., 2009), indicating that the timing and nature of the traumatic event(s) that trigger PTSD may have a profound impact on how this disorder disrupts brain-wide functional connectivity. Nevertheless, these findings seem to suggest that PTSD may lead to a relative dominance of the SN over the DMN in the resting state. While this maladaptive change likely plays an integral role in the hypervigilance and hyperarousal symptoms of PTSD, it may also contribute to the anhedonia and dysregulation of reward processing that is so prevalent in this disorder.

9.3 Reward circuits—Preclinical animal studies

Although human data on the neurobiology of comorbid AUD and PTSD are sparse, findings in animal models that generate symptoms associated with the comorbid condition lend support to the notion that both disorders result in profound dysregulation of key nodes of the reward circuitry. As observed in human studies that have characterized each disorder alone, there is convergent evidence that animal models that promote symptoms of both AUD and PTSD can lead to a sensitization of phasic mesolimbic DA release, particularly in response to alcohol and other drugs of abuse. These adaptations would be expected to enhance the reinforcing properties of these drugs, particularly during the early stages of self-administration. For example, male Sprague Dawley rats exposed to predator odor for a single 15 min session that go on to develop PTSD-like behaviors (e.g., increased anxiety measures on the elevated plus-maze, odor context avoidance) exhibit greater cocaine-stimulated NAc DA release than control subjects as well as odor-exposed rats that did not develop PTSD-like phenotypes (Brodnik et al., 2017). Moreover, susceptible odor-exposed rats also showed greater motoric responses to cocaine and heightened motivation to self-administer this drug, relative to the resilient and control groups. Similarly, another study in male Sprague Dawley rats found that repeated social defeat stress led to an enduring NMDA receptor-dependent LTP of glutamatergic synaptic transmission onto VTA DA cells (Stelly, Pomrenze, Cook, & Morikawa, 2016). They went on to show that this stress-induced plasticity was mediated by a PKA-dependent enhancement in the potency of inositol 1,4,5-triphosphate-induced Ca^{2+} signaling in VTA DA cells. Moreover, blockade of glucocorticoids with mifepristone mitigated these neurobiological adaptations as well as a sensitization of cocaine conditioned place preference that developed after the social defeat paradigm. Another study from our group used ex vivo, fast-scan cyclic voltammetry to evaluate DA release dynamics in DA terminal fields in the NAc of Long-Evans rats (Deal et al., 2018). We found that a relatively mild (3-day) social defeat procedure led to a significant increase in the magnitude of electrically evoked DA transients as well as an increase in the rate of DA uptake. Interestingly, both of these effects were occluded in a separate cohort of rats with a history of intermittent ethanol self-administration, possibly because the drinking procedure may have already engaged the stress systems underlying these changes in DA release dynamics. Finally, a fascinating study revealed that VTA DA neurons in mice susceptible to social defeat stress show increased intrinsic excitability relative to controls, due to an increase in the hyperpolarization-activated cation channel (I_h) conductance (Friedman et al., 2014). Surprisingly, when stress-resilient mice were examined, these subjects had an even strong I_h current. Through an elegant series of experiments, the authors discovered that, in resilient mice, the elevated levels of I_h drove a compensatory increase in an inhibitory potassium channel conductance that normalized VTA DA cell firing. Remarkably, pharmacologically enhancing I_h or optogenetically stimulating DA cells in susceptible mice reversed the maladaptive behavioral phenotypes promoted by this model.

We have also conducted a series of studies characterizing adaptations in the mesolimbic DA system using an adolescent social isolation model of the comorbid condition. As observed following social defeat, relative to rats that were group housed during adolescence, we have consistently observed increases in measures of DA release and uptake in socially isolated rats in adulthood. These changes have been observed in the NAc core, NAc shell, and dorsal striatum and are accompanied by an increase in the expression of the DAT (Karkhanis et al., 2019; Yorgason et al., 2016). Using in vivo microdialysis, we have also reported that adolescent social isolation increases ethanol-stimulated DA release in the NAc (Karkhanis, Locke, McCool, Weiner, & Jones, 2014) and BLA (Karkhanis, Alexander, McCool, Weiner, & Jones, 2015). Notably, we also examined NE release in response to acute ethanol in these studies. Surprisingly, acute ethanol administration had no effect on NE release in either the NAc or BLA but in socially isolated rats, this same treatment resulted in a robust increase in NE release in both brain regions (Karkhanis et al., 2014, 2015).

In more recent work, we have sought to explore possible mechanisms underlying these dramatic effects of adolescent social isolation on the mesolimbic DA system. In the NAc, kappa opioid receptors (KORs) are expressed on presynaptic terminals of DAergic synapses and activation of these receptors inhibits DA release (Svingos, Chavkin, Colago, & Pickel, 2001; Werling, Frattali, Portoghese, Takemori, & Cox, 1988). Since a variety of stressors increase KOR expression in the NAc and KOR system dysregulation has been implicated in the etiology of AUD and PTSD (Karkhanis & Al-Hasani, 2020; Van't Veer & Carlezon Jr, 2013), we examined the effect of adolescent social isolation on KOR signaling in this brain region (Karkhanis, Rose, Weiner, & Jones, 2016). We confirmed that adolescent social isolation led to an increase in both electrically-evoked and ethanol-stimulated DA release and also discovered that, despite this sensitization of DA release dynamics, baseline levels of DA were lower in these animals. This decrease in ambient DA is similar to the hypoDAergic state observed in individuals with AUD and likely contributes to the negative affective state promoted by this model of the comorbid condition (Butler et al., 2016). These changes were accompanied by a significant increase in KOR-mediated inhibition of DA release and pretreatment with a KOR antagonist, normalized baseline DA levels, blocked the increase in ethanol-evoked DA release, and reduced the escalation in alcohol intake observed in socially isolated animals. Collectively, these findings suggest that increased mesolimbic KOR signaling may contribute to the DAergic dysregulation and maladaptive behaviors promoted by this model of the comorbid condition.

10. Stress circuitry

The other major brain system that underlies many of the symptoms of AUD and PTSD is a complex network of cortical and subcortical structures that translates stress-associated stimuli into appropriate behavioral and emotional responses to stressors. Key nodes of this stress circuitry include the hypothalamus as well as the PFC, the basolateral AMY, the ventral hippocampus, the locus coeruleus, and the extended AMY (e.g., central AMY, BNST, nucleus accumbens; Flook et al., 2020; Koob, 2008;

Koob & Volkow, 2010) (Fig. 1). An astute reader will immediately appreciate that there is considerable overlap between the core regions of the stress and reward circuits, a critical issue that will be addressed in the final section of this chapter. While we have already discussed how AUD and PTSD are associated with reward-related dysregulation of some of these brain regions (e.g., NAc, AMY, PFC, ACC), in this section, we focus on functional human imaging studies that have examined these, and other key nodes of this circuitry in the context of stress/fear-related emotional processing. A growing literature has revealed that many brain regions associated with aberrant reward processing in AUD and PTSD are also dysregulated during the processing of negative/aversive stimuli and data from animal models of the comorbid condition suggests that these pathways may play a causal role in many of the behavioral phenotypes associated with both disorders (Gilpin & Weiner, 2017; Suh & Ressler, 2018; Whitaker, Gilpin, & Edwards, 2014).

The lateral and basolateral nuclei of the AMY (BLA) are arguably the central hub of the stress network. The BLA is largely comprised of excitatory pyramidal neurons which are the primary target of all sensory input into the amygdala complex, which mostly originates from the thalamus and cortex (Davis & Whelan, 2001; Paxinos, 2003). These cells in turn project back to many cortical regions as well as numerous elements of the extended AMY and reciprocal connections between these brain regions are thought to encode our behavioral and emotional responses to sensory and emotional stimuli (Janak & Tye, 2015; Sharp, 2017; Zhang et al., 2018). The BLA also receives strong glucocorticoid input and serves as a key interface between the hypothalamic-pituitary-adrenal axis and the stress network (Duvarci & Pare, 2007; Nathan, Griffith, McReynolds, Hahn, & Roozendaal, 2004; Roozendaal, Griffith, Buranday, De Quervain, & McGaugh, 2003; Segev & Akirav, 2016; Stringfield, Higginbotham, & Fuchs, 2016; Taslimi, Sarihi, & Haghparast, 2018). It should also be noted that the afferent and efferent circuitry of the BLA are highly conserved across species (Lanuza, Belekhova, Martinez-Marcos, Font, & Martinez-Garcia, 1998; LeDoux, 2012; Moreno & Gonzalez, 2007; Pabba, 2013). Moreover, the BLA is the largest subregion of the AMY complex, and its relative contribution to the overall size of this structure, both in terms of volume and number of neurons, has increased more than other major AMY nuclei across evolution (Avino et al., 2018; Barger et al., 2012; Chareyron, Banta Lavenex, Amaral, & Lavenex, 2011). For example, the volume of the BLA is 30–40 times larger in monkeys than rats whereas the central and medial nuclei are only 4–8 times larger in monkeys (Chareyron et al., 2011). The fact that the BLA is the primary site of sensory input into the AMY complex and that it represents that largest AMY subregion are important to keep in mind while reviewing human AMY functional imaging studies. While most of these studies do not distinguish between AMY nuclei, given its relatively large size and that it receives the majority of afferent input into the AMY complex, it seems reasonable to assume that the BLA is the predominant contributor to AMY BOLD signals.

10.1 Stress circuitry and AUD—Human studies

Although the stress network plays an integral role in many behavioral and emotional processes, it is most frequently studied in the context of fear and anxiety. For the purposes of this review, fear is defined as an adaptive emotional response to a real or perceived threat that typically dissipates once the threat is no longer present whereas anxiety is a more temporally diffuse, negative emotional state, often evoking arousal and vigilance, that manifests in anticipation of these stressors (Davis, Walker, Miles, & Grillon, 2010; Dias, Banerjee, Goodman, & Ressler, 2013). One of the most influential conceptual frameworks for the neurobiology of AUD posits that, while activation of the canonical reward circuitry likely drives the early stages of this disorder, the stress network plays an increasingly important role in its latter stages (Koob, 2013a, 2013b; Koob et al., 2014). The idea is that, during the initial stages of AUD, alcohol is consumed primarily for its rewarding or pleasurable effects and that abstinence periods are generally associated with a neutral affective state. However, after repeated periods of binge intoxication and withdrawal, a negative affective state develops during abstinence and it is relief from this negative affective state that becomes the primary driver of the maladaptive drinking behaviors associated with AUD. While anhedonia is a key element of this aversive state, excessive anxiety is also an integral component. This anxiety is often linked with craving and together these affective states represent a major trigger for relapse (Huang et al., 2018; Karch et al., 2015; Seo et al., 2013; Sinha, 2012; Tanabe, Regner, Sakai, Martinez, & Gowin, 2019). Not surprisingly, numerous human imaging studies have demonstrated that AUD leads to maladaptive alterations in many elements of the stress circuitry. Moreover, more recent human and animal research has revealed that strikingly similar maladaptive changes in the stress network are associated with relapse in AUD and deficits in fear extinction, a hallmark phenotype of PTSD (Goode & Maren, 2019; Peters, Kalivas, & Quirk, 2009).

Over the past decade, high resolution functional imaging studies have demonstrated that AMY hyperactivity during the processing of anxiogenic or fearful stimuli is a highly reproducible core feature of anxiety and stressor-related disorders (Etkin, Prater, Hoeft, Menon, & Schatzberg, 2010; Etkin & Wager, 2007; Nitschke et al., 2009). While data on AUD and AMY activity has been more sparse, there is now a growing body of evidence revealing that similar alterations may be linked with AUD. One of the first of these studies used fMRI to compare AMY reactivity in individuals with AUD and healthy controls in response to positively and negatively arousing images and found that the AUD cohort showed greater overall AMY activation to negative vs. positive stimuli. While AMY reactivity was greater in AUD subjects, this was the case for both positive and negative images (Gilman & Hommer, 2008). Other early functional imaging studies found mixed evidence of a positive link between AMY hyperactivity and AUD. For example, one study concluded that low AMY responsiveness may be a risk factor for AUD, as they reported decreased threat-related AMY reactivity in individuals at high familial risk for AUD (Glahn, Lovallo, & Fox, 2007) whereas another study found that AMY hyperactivity may be protective against stress-related problem drinking in at least some university students (Nikolova & Hariri, 2012). Despite these seemingly disparate findings, recent evidence suggests that both hypo- and hyperreactivity of the AMY may be linked with AUD, and that these contrasting findings may be explained by examining the balance between overlapping elements of the reward and stress networks.

For example, one large study of 759 undergraduate students used fMRI to image individual differences in reward and threat processing and examined whether these measures predicted stress-related problem drinking (Nikolova, Knodt, Radtke, & Hariri, 2016). Surprisingly, they found that stress-related problem drinking was associated with one of two neural phenotypes: relatively high reward-related ventral striatum (VS) activation and relatively low threat-related AMY activation *or* relatively low reward-related VS activation and high threat-associated AMY activation. Importantly, they also showed that the relationship between stress and problem drinking was mediated by impulsivity (steeper delayed discounting curve) for individuals with high VS-low AMY activation and by anxious/depressive symptoms for those with low VS-high AMY activation. Across both neural phenotypes, they discovered that, regardless of the direction, the greater the VS-AMY imbalance, the higher the risk for problem drinking. Moreover, for the low VS-high AMY risk phenotype, stress not only predicted the presence of AUD at the time the study was conducted but also self-reported problem drinking at a 3-month follow-up.

Although the AMY is a central element of the stress network, its activity is strongly regulated by numerous cortical structures, particularly the insula, PFC, ACC and OFC. There are numerous reciprocal connections between the AMY and many of these cortical and subcortical (e.g., straitum, hippocampus, extended AMY) regions. Emerging network connectivity studies are beginning to shed light on how dysregulation of this cortico-AMY-subcortical stress circuitry contributes to the etiology of AUD. Notably, a series of seminal studies by Rajita Sinha and colleagues have used individualized guided imagery scripts to study stress-related engagement of the stress network in subjects with AUD. Their initial studies, conducted in healthy controls, used fMRI to measure differences in brain-wide activation in response to stressful vs. neutral imagery scripts and found increased stress-related activation in the mPFC and vACC as well as numerous mesolimbic structures (striatum, thalamus, hippocampus), all regions intimately connected with the AMY (Sinha, Lacadie, Skudlarski, & Wexler, 2004). Further analyses were conducted to identify brain regions whose activity correlated with subjective stress ratings. A positive correlation between self-reported distress ratings and caudate and thalamic regions was observed and a negative relationship emerged between distress ratings and the ACC and insula. Together, these data provided early empirical evidence that many elements of the stress network are indeed engaged by stressful stimuli and provided critical insight into how these circuits could modulate negative emotional processing. As just a couple of examples, activation of areas like the ACC may serve to reduce emotional distress whereas thalamic and striatal regions may actually be engaged during the subjective experience of stress.

Guided by these and other findings, the Sinha group, and others, have conducted numerous additional studies in subjects with AUD (for reviews see: Blaine, Milivojevic, Fox, & Sinha, 2016; Sinha, 2012; Sinha & Li, 2007; Tanabe et al., 2019). One important finding that has emerged from these studies is that many nodes of the stress circuitry (e.g., mPFC, OFC, insula) are not only engaged by stressful stimuli but also by alcohol-associated visual cues. These findings suggest that, although alcohol-associated cues clearly engage elements of the reward circuitry, they also activate nodes that are more often considered part of the stress circuitry, likely due to stress-evoked craving. These findings also serve to reinforce the inarguable fact that the reward and stress networks are highly inter-related and largely comprised of overlapping brain regions.

One particularly notable study by the Sinha group used the customized script-driven imagery to examine how engagement of stress and reward circuits by stress and alcohol cues influenced craving and relapse in 45 treatment engaged individuals at 4–8 weeks of abstinence (Seo et al., 2013). Relative to neutral scripts, both the stress and alcohol cues significantly increased self-reported anxiety and craving ratings in the AUD group. Similarly, whole-brain analyses revealed that AUD subjects showed increased activation in corticolimbic and striatal regions during stress and alcohol cues relative to neutral stimuli, including the mPFC, ACC, AMY, and HC. Importantly both stress and alcohol cue-induced craving scores were positively correlated with hyperactivity in the vmPFC and ACC during neutral-relaxing trials with stress-induced craving also being correlated with hyperactivity of the VS and precuneus. Furthermore, using a covariate analysis to exclude behavioral factors that predict relapse (e.g., years of alcohol use, higher stress-induced alcohol craving), they found that vmPFC/ACC, VS, and precuneus hyperactivation during neutral trials was actually predictive of a shorter time to relapse. Notably, a secondary analysis revealed that vmPFC/ACC hyperactivity on neutral trials was the most accurate classifier of relapsers vs. non-relapsers. These findings reveal that key elements of the stress network are profoundly disrupted during early abstinence and suggest that these alterations may contribute to relapse. The mPFC and ACC provide powerful top-down control of many limbic elements of the stress network, including the AMY, and play a key role in emotional processing. The observed blunting of mPFC/ACC engagement in response to stress and alcohol cues in the AUD cohort is consistent with affect dysregulation associated with these brain areas. Moreover, these measures may serve as potential biomarkers for relapse and may represent viable targets for novel treatments, like TMS, that could potentially restore normal neural activity in these circuits.

A number of imaging studies have examined functional connectivity among elements of the stress network in individuals with AUD. One common finding has been that connectivity between the AMY and prefrontal regions is weakened in individuals with AUD, both in the resting state and during the processing of aversive stimuli (Muller-Oehring, Jung, Pfefferbaum, Sullivan, & Schulte, 2015; O'Daly et al., 2012; Wade et al., 2017). One recent study conducted a whole-brain analysis, using the AMY as a seed, to examine the relationship between resting state AMY connectivity and problem drinking in 83 non-dependent alcohol drinkers (Hu et al., 2018). This brain-wide analysis found that, in the resting state, the AMY is positively connected to many cortical and subcortical structures including the HC, insula, and vmPFC. Notably, they reported only one significant relationship between problem drinking and resting state AMY connectivity to any of these nodes, a negative correlation between problem drinking scores and AMY-dACC connectivity and this correlation remained significant even when controlling for depression and other substance use. In addition, they examined this relationship for the four major AMY subregions, including the BLA and CeA. Only weakened BLA-dACC connectivity was significantly correlated with problem drinking scores.

Another study examined task-related functional connectivity using a voxel-wise method, called the intrinsic connectivity distribution, which permits an unbiased analysis of brain-wide changes in connectivity without the need for *a priori* seeds or regions of interest. The first part of this study examined 45 recovering AUD patients and examined whole-brain connectivity during presentation of scripted stress and alcohol imagery cues (Zakiniaeiz, Scheinost, Seo, Sinha, & Constable, 2017). Interestingly, this unbiased approach also identified cingulate cortex as a major locus of connectivity changes. They found reduced voxel-wise connectivity to stress vs. neutral cues in the ACC, MCC, and PCC and reduced PCC connectivity to alcohol cues. They also showed that weaker PCC response to alcohol vs. neutral cues was associated with a longer time to relapse and greater connectivity of the PCC during alcohol- vs. stress-cues was also associated with better recovery outcomes. In the second phase of the study, a subset of 30 AUD patients were demographically matched with 30 healthy controls to permit group comparisons. Using the same unbiased analytical approach, they found reduced MCC and PCC connectivity in the AUD cohort during stress and alcohol cues compared to the healthy controls. These findings extend those from Hu et al., revealing that excessive alcohol drinking is associated with decreased cingulate cortex connectivity in the resting state and during presentation of aversive stimuli known to elicit craving and anxiety. The finding that cingulate cortex connectivity changes correlated with recovery outcomes suggests that this alteration may represent yet another potential biomarker for relapse.

Finally, a related paper by the Sinha group used a seed-based approach to examine neural correlates and connectivity patterns associated with stress-related impulse control problems in 37 recovering AUD patients and 37 demographically matched healthy controls (Seo, Lacadie, & Sinha, 2016). This study employed their alcohol and stress cue imagery scripts and also assessed impulse control difficulties using IMPULSE, a scale previously shown by these investigators to be predictive of addictive disorders and relapse vulnerability (Fox, Axelrod, Paliwal, Sleeper, & Sinha, 2007; Fox, Hong, & Sinha, 2008). As in their prior studies, AUD patients reported higher anxiety ratings during alcohol cues and elevated craving during both stress and alcohol conditions. Interestingly, impulsivity measures did not correlate with anxiety or craving ratings in either the AUD or control groups. A whole-brain analysis of IMPULSE scores during stress vs. neutral and alcohol cue vs. neutral conditions revealed a number of significant correlations in the AUD cohort, but only for the stress vs. neutral comparison. For that condition, activity in the vmPFC, caudate, and dlPFC was negatively correlated with impulsivity ratings. In the healthy control group, no significant correlations emerged between impulsivity and neural activity in any brain region. However, there were also no significant differences in any regional correlational coefficients between the AUD and control groups, suggesting that the neural network encoding impulsivity may be similar in both cohorts but that elements of this network may be strengthened in individuals with AUD. In a final analysis, the authors used the three brain regions identified above as being involved in impulse control difficulties as seeds to compare functional connectivity patterns in AUD and control subjects. A number of significant group differences in connectivity were observed for each seed (both increases and decreases). Perhaps most notably, with the vmPFC as a seed, AUD patients showed significantly reduced connectivity with the rACC compared with the control group. When an analysis of the relationships behavioral ratings of anxiety and craving and all of these functional connectivity patterns was conducted, the only significant finding was a negative correlation between stress-induced anxiety and functional connectivity between the dlPFC and dmPFC in AUD subjects. No correlations were detected between behavioral ratings and functional connectivity patterns in controls. Together, these findings reveal that AUD is associated with reduced stress reactivity of numerous nodes of the stress network and that altered connectivity between several cortical nodes of this network may contribute to greater stress-related impulse control problems in these individuals.

10.2 Stress circuit and PTSD—Human studies

As discussed earlier, hypervigilance, hyperarousal, and deficits in fear learning represent core symptoms of PTSD and it is now widely appreciated that dysregulation of the brain stress network contributes to these maladaptive phenotypes (Akiki, Averill, & Abdallah, 2017; Dossi, Delvecchio, Prunas, Soares, & Brambilla, 2020; Kunimatsu et al., 2020; Prasad, Chaichi, Kelley, Francis, & Gartia, 2019; Shalev, Liberzon, & Marmar, 2017). Indeed, it is perhaps a review of the imaging literature on PTSD that provides the greatest empirical support for considering the AMY (and arguably the BLA) as a core node of the stress circuitry. Decades of animal research has revealed that fear learning is mediated by a strengthening of synaptic activity in the BLA and that fear memories can reside as engrams in this brain region (Goode & Maren, 2019; Ressler & Maren, 2019; Sun, Gooch, & Sah, 2020). Moreover, as reviewed below, AMY hyperexcitability is, perhaps, one of the most replicable neural phenotypes associated with PTSD.

As with most nascent fields, initial functional imaging studies examining AMY responsivity in PTSD were somewhat mixed, with some reporting increases (Rauch et al., 2000; Shin et al., 2005; Williams et al., 2006), decreases (Britton, Phan, Taylor, Fig, & Liberzon, 2005; Phan, Britton, Taylor, Fig, & Liberzon, 2006), or no change in activity (Bremner et al., 1999; Lanius et al., 2001) relative to controls. However, one of the first large meta-analyses of this literature provided compelling evidence that increased AMY reactivity to aversive stimuli represents one of the most robust and reproducible neural phenotypes of PTSD. This review employed an elegant voxel-wise approach to examine fMRI and PET studies of PTSD, social anxiety, and specific phobia (Etkin & Wager, 2007). Their analysis included 40 studies that contrasted a negative emotional condition with a neutral or positive emotional condition or a resting baseline. Importantly, they also included data on fear conditioning in healthy subjects. The AMY was one of only two brain regions, along with the insula, that showed hyperactivity across all three disorders and these two areas were also the only structures that showed increased activity during fear conditioning in controls. This study also identified numerous areas of hypoactivation within the stress network, but only in PTSD subjects, including the mPFC, ACC, MCC, OFC, and anterior hippocampus (aHC). Of note, lower mPFC activity was associated with greater PTSD symptom severity. This was also one of the first studies to examine co-activation across brain regions of interest. They reported consistent co-hypoactivation among cortical areas and a significant negative co-activation between the ACC and the AMY, drawing attention to the now well-established link between hypoactivity of frontal regions and AMY hyperactivity. The authors also speculated that AMY hyperactivity was like driven by the BLA, as it receives the majority of sensory cortical input and they actually observed a PTSD-associated decrease in dorsal AMY activity, which comprises the central AMY.

In the past decade, a number of additional meta-analyses have confirmed and extended these findings, showing that AMY hyperactivity and dysregulation of many nodes of the stress network at rest, and during the processing of fear/stress-related stimuli, are not only a common feature of PTSD but can also be associated with vulnerability to the disorder, symptom severity and treatment outcome (Akiki et al., 2017; Hayes, Hayes, & Mikedis, 2012; Koch et al., 2016; Patel, Spreng, Shin, & Girard, 2012; Prasad et al., 2019; Wang et al., 2016).

One major advance in functional imaging over the past decade has been that improvements in fMRI resolution have made it possible to parse subregions of heterogeneous brain regions like the AMY. In fact, a number of recent studies have taken advantage of these imaging advances to analyze reactivity of the BLA, CEA, and other large AMY subregions in PTSD patients and controls. For example, one study examined AMY subregion activation in response to brief presentations of consciously or unconsciously viewed fearful faces, using a backward masking procedure (Neumeister et al., 2018). This study included three cohorts of patients diagnosed with PTSD, Generalized Anxiety Disorder (GAD) or Panic Disorder (PD) as well as a large sample of healthy controls. When viewing conscious stimuli, all participants (including healthy controls), showed increased activity in several AMY nuclei, including the BLA. However, hyper-responsivity to these stimuli was not significantly greater in any of the patient cohorts and healthy controls. In marked contrast, a similar analysis during the unconscious condition revealed a selective increase in BLA activity in the PTSD patients relative to the PD and GAD groups as well as the healthy controls. These striking findings highlight the central role of BLA dysfunction in the etiology of PTSD and provide evidence of differential neural processing of fear-related stimuli in individuals with stressor and anxiety disorders.

Many recent resting state fMRI studies of PTSD have also focused on differentiating changes in functional connectivity between distinct AMY subregions and other nodes of the stress circuitry. These studies have all shown that the BLA and CEA, the two largest AMY nuclei, have distinct patterns of functional connectivity within the stress network (Aghajani et al., 2016; Brown et al., 2014; Nicholson et al., 2015) and, although not all data concur (Brown et al., 2014), most support the idea, first proposed by Etkin and Wager, that the frequently reported hypoconnectivity between the AMY and cortical areas like the PFC and ACC, likely reflects a reduction in normal cortical regulation of BLA activity. For example, a study of adolescent sexually abused PTSD patients and health controls used intrinsic functional connectivity analysis to examine the BLA and CEA (Aghajani et al., 2016). They found that adolescent PTSD patients showed decreased connectivity between the BLA and a cluster of cortical areas that included the ACC and mPFC. In contrast, they reported increased CEA connectivity with another cortical cluster comprised of the OFC and subcallosal cortices, with the latter change being correlated to increased stress and anxiety symptoms in PTSD patients. Another interesting study examined patients with PTSD alone, PTSD and comorbid MDD, and healthy controls that were all recruited from a cohort who survived the same large earthquake (Yuan et al., 2019). A seed-based approach was employed to compare BLA and CEA functional connectivity across groups. Major findings from this study were that PTSD + MDD patients had lower BLA connectivity with the ACC, SMA and bilateral putamen/pallidum, with the latter negatively correlating with depressive symptoms. Again, functional CEA connectivity, which was predominantly to the ACC/SMA in this study, was actually increased in the PTSD + MDD group. Notably, an analysis of functional connectivity that merged both AMY subregions found no group differences. Collectively, these findings are consistent with the notion that PTSD is associated with a loss of top-down cortical control of BLA activity and that, perhaps, cortical connectivity with the CEA may be increased in this disorder. These findings underscore the distinct connectivity patterns of key AMY subregions and the need to consider, where possible, the possibility that PTSD may have distinct effects on activity patterns between these subregions and other nodes of the stress circuitry.

Finally, several exciting advances suggest that our growing understanding of stress network dysregulation in PTSD may be starting to inform on novel treatments for this disorder. For example, several studies have explored the utility of using real-time fMRI neurofeedback (rt-fMRI NF) to normalize PTSD-associated perturbations in stress network functional connectivity. Rt-fMRI NF is a relatively new technique in which participants receive contingent visual or auditory feedback about ongoing activity in a specific brain region. Remarkably, numerous studies have shown that this procedure allows individuals to exert volitional control over neurophysiological activity within discrete brain regions and networks and that this neurofeedback can be harnessed to alleviate behavioral and neurophysiological symptoms of a number of psychiatric and neurological conditions (Linhartova et al., 2019; Watanabe, Sasaki, Shibata, & Kawato, 2017; Young et al., 2018).

To date, three studies have been published using rt-fMRI NF to target the AMY in PTSD patients. The first conducted an unblinded pilot feasibility study on three veterans with chronic PTSD (Gerin et al., 2016). After a baseline fMRI scan, participants were trained over three sessions to suppress AMY activity after exposure to personalized trauma scripts. All three of the subjects showed an improvement of PTSD symptoms and a normalization of changes in AMY resting-state functional connectivity, with a strengthening of AMY-ACC and AMY-OFC connectivity and a dampening of AMY connectivity to elements of the salience network, including the insula. A slightly larger study on 10 PTSD patients used a similar approach to train subjects to suppress AMY activity across three training sessions (Nicholson et al., 2017). A final scan was also conducted without rt-fMRI NF to determine if the training had any lasting effects. Notably, subjects were able to significantly reduce AMY activation, even during the final scan without neurofeedback. In addition, activation in the PFC, ACC, and insula were negatively correlated with dissociative PTSD symptoms during the final scan. Task-based functional connectivity was also increased between the AMY and dmPFC and ACC during the training sessions.

The largest rt-fMRI NF PTSD study to date was conducted on 40 male veterans with PTSD and 22 veteran controls (Misaki et al., 2018). This study actually sought to increase AMY responsivity in the participants. While this may seem paradoxical, the authors noted that the AMY is activated by both positive and negative emotional stimuli and that training the AMY to increase its activity in response to positive stimuli can actual decrease responsivity to negative cues. Thus, they used rt-fMRI NF to train 25 of the PTSD patients and all of the control subjects to increase AMY responsivity during recall of positive autobiographical memories. The other 15 PTSD patients were also asked to increase an fMRI signal but one derived from activity within the intraparietal sulcus, a region not thought to be involved in emotion regulation. Subjects received three training sessions followed by a final session without feedback approximately 1-week later. PTSD symptoms were significantly improved in the AMY rt-fMRI cohort but not the group receiving feedback from the intraparietal sulcus. Importantly, PTSD subjects had significantly lower AMY-PFC connectivity than controls at baseline and the PTSD group receiving AMY rt-fMRI NF showed an increase in this connectivity over training sessions. Interestingly, this change did not correlate with PTSD symptom improvement. Rather, a connectome-wide analysis revealed that neurofeeback directed at the

AMY normalized other PTSD-associated alterations in functional connectivity. For example, baseline connectivity between the ACC and supplementary motor area (SMA) was lower in PTSD patients vs. controls and AMY rt-fMRI NF increased ACC-SMA connectivity in PTSD patients. Moreover, a positive correlation was noted between the strengthening of this connectivity and an improvement of depressive symptoms in the PTSD group. These findings demonstrate that PTSD patients can be trained to modulate activity in key brain areas impacted by PTSD. In addition, these data reveal that rt-fMRI NF can have widespread effects on network connectivity beyond the targeted region. Though fairly preliminary, the therapeutic implications of these findings are bolstered by other evidence demonstrating that some the beneficial effects of rt-fMRI NF, like the strengthening of AMY-PFC connectivity, are also associated with resilience to PTSD (Bolsinger, Seifritz, Kleim, & Manoliu, 2018) and positive outcomes with other neuromodulatory treatments like transcranial magnetic stimulation (Philip et al., 2018).

10.3 Stress circuitry—Preclinical animal studies

As noted in our review of the reward circuitry in AUD and PTSD, few human studies have directly addressed stress network dysregulation associated with this dual diagnosis. However, animal studies have provided compelling evidence that key nodes of this circuitry likely mediate the stress-related behavioral phenotypes associated with these disorders. Moreover, studies using animal models of the comorbid condition have identified several stress circuit adaptations that are similar to those observed in human subjects with each disorder.

For example, optogenetic circuit manipulations have been elegantly employed to demonstrate that central nodes of the stress network, like the BLA and elements of the extended AMY, can bidirectionally modulate anxiety-like behaviors in mice and rats (Daviu, Bruchas, Moghaddam, Sandi, & Beyeler, 2019; Janak & Tye, 2015; Lalumiere, 2014). In addition, deficits in associative fear learning, particularly extinction of fear-related memories, are a hallmark feature of PTSD (Norrholm & Jovanovic, 2018). While there is only limited evidence linking AUD and alterations in fear learning (Muench et al., 2019; Stephens et al., 2005), extensive studies in animals have shown that stress network regions like the BLA, CeA, BNST, and HC are critical for the acquisition, consolidation, expression and extinction of fear memories (for reviews see: Fenster et al., 2018;

Ressler & Maren, 2019). Early rodent work revealed that the acquisition of fear memories was associated with a strengthening of glutamatergic synapses on BLA projection neurons (Maren & Fanselow, 1996). More recent work, integrating elegant molecular and optogenetic approaches, has demonstrated that there are sparse clusters of cells in the BLA and other stress network nodes like the HC, termed engrams, that are activated during the formation and extinction of fear memories (Davis & Reijmers, 2018; Josselyn, Kohler, & Frankland, 2015; Zhang, Kim, & Tonegawa, 2020). Optogenetic activation of these engrams can elicit freezing behavior or induce extinction in a context-specific fashion. It is also worth noting that animal models of AUD and vulnerability to AUD and PTSD both engender deficits in extinction of fear memories, therefore these cellular populations could play an important role in the disorders (Holmes et al., 2012; Skelly et al., 2015).

Studies in our lab have shown that adolescent social isolation leads to an increase in the intrinsic excitability of glutamatergic projection neurons in the BLA (Rau, Chappell, Butler, Ariwodola, & Weiner, 2015). Similar increases in the excitability of these cells are seen following chronic stress exposure in preweaning (Guadagno, Wong, & Walker, 2018) and adult (Rosenkranz, Venheim, & Padival, 2010) rats. This enhanced BLA excitability is mediated, at least in part, by an increase in the expression and function of calcium-activated small conductance (SK) potassium channels (Atchley, Hankosky, Gasparotto, & Rosenkranz, 2012; Rau et al., 2015) and inhibition of BLA SK channels can significantly attenuates anxiogenic phenotypes promoted by this model (Atchley et al., 2012; Rau et al., 2015). More recently, we have shown that adolescent social isolation is also associated with an increase in synaptic excitation in the ventral hippocampus (analogous to the human aHC), a brain region that receives strong innervation from the BLA (Almonte et al., 2017). In addition, we have confirmed, using chemogenetic approaches, that silencing BLA-vHC synapses reduces anxiety-like behaviors and have also discovered that inhibiting this input reduces alcohol seeking-related behaviors (Ewin et al., 2019).

The Gilpin laboratory has established a predator odor model of comorbid AUD/PTSD in which a subset of rats exposed to bobcat urine exhibit enduring avoidance of odor-paired stimuli (Albrechet-Souza & Gilpin, 2019). This model is particularly powerful as it mirrors findings in humans that only a subset of trauma-exposed subjects go on to develop PTSD. Relative to non-avoiders and control subjects not exposed to the predator odor, avoiders show increased operant alcohol self-administration and aversion resistant drinking. Neurobiological studies using this model have shown that avoiders exhibit greater increases in c-Fos expression in the CeA after odor exposure than non-avoiders and, notably, higher avoidance behavior correlates with increased c-Fos and CRF immunoreactivity in the CeA. In addition, infusion of a CRF1 antagonist into the CeA reduced alcohol self-administration and avoidance behavior in avoiders, while having no effect on these measures in non-avoiders (Weera, Schreiber, Avegno, & Gilpin, 2020).

Social defeat stress, although often used as a model of depression (Knowland & Lim, 2018), also leads to a number of behavioral phenotypes associated with AUD and PTSD, including increases in anxiety-like behaviors, deficits in fear extinction, and escalations in alcohol intake (Lisboa et al., 2018; Narayanan et al., 2011; Newman, Leonard, Arena, de Almeida, & Miczek, 2018). Many studies have shown that this model strongly activates the BLA and dysregulates numerous nodes of the stress network, including the PFC and extended AMY (Diaz & Lin, 2020; Patel, Kas, Chattarji, & Buwalda, 2019). Finally single prolonged stress, in which subjects are exposed to a series of stressors across a single day, followed by a week-long incubation period, also engenders many key phenotypes of AUD and PTSD (Lisieski et al., 2018; Souza et al., 2017). One recent study employed longitudinal magnetic imaging approaches to scan subjects before and after this procedure (Piggott et al., 2019). Their findings revealed a significant decreased in PFC glutamate levels as well as decreased neural activity in the infralimbic PFC region, along with increased activity in the BLA following sPS with no changes in the control group. These findings are remarkably similar to the commonly reported decreases in PFC-AMY functional connectivity reported in both AUD and PTSD.

11. A novel neural framework for comorbid AUD and PTSD

The preceding section sought to provide a comprehensive review of reward and fear circuit dysregulation in AUD and PTSD. Collectively, this literature reveals striking similarities in many of the functional connections within these networks that are altered in both conditions. Focusing first on the reward circuitry, both disorders are associated with a profound dysregulation of ventral striatal activity and connectivity, at rest and during the processing of positive and negative emotional imagery. However, the direction of these alterations is complex and likely dependent on the nature of the stimuli used to experimentally engage this network. For example, relative to healthy controls, individuals with AUD exhibit stronger behavioral reactivity to alcohol-related cues (Huang et al., 2018; Schacht et al., 2013; Vollstadt-Klein et al., 2012) whereas PTSD patients respond more to trauma-related stimuli (Vujanovic et al., 2017) and both conditions are often associated with decreased reactivity to non-alcohol-related positive stimuli (Fritz et al., 2019; Nawijn et al., 2015; Wrase et al., 2007). These behavioral findings are generally in line with functional imaging data showing that AUD subjects exhibit higher VS activity in response to alcohol cues but often lower VS responsivity to the presentation of non-alcohol-related positive stimuli (Schacht et al., 2013; Wrase et al., 2007). Similarly, PTSD patients exhibit lower VS activation during the presentation of positive cues but display heightened VS activation in response to aversive cues (Elman et al., 2009, 2018).

There are also several convergent adaptations within the stress circuitry in both AUD and PTSD. Notably, extensive fMRI findings in individuals with PTSD, along with more recent evidence from functional imaging studies in AUD cohorts, suggest that both disorders involve a loss of top-down cortical control of AMY activity, mainly due to a weakening of functional connectivity between the mPFC-AMY and ACC-AMY. In PTSD patients, these maladaptive changes are generally associated with AMY hyperactivity at rest and during emotional tasks (Etkin & Wager, 2007). Interestingly, as reviewed earlier, in AUD subjects, increased AMY reactivity to stressful stimuli was only seen in individuals with low VS reactivity to rewarding cues and this relationship was mediated by anxious/depressive symptoms, which are also core symptoms of PTSD (Nikolova et al., 2016). Collectively, these findings are consistent with the notion that individuals who exhibit behavioral phenotypes common to AUD and PTSD share a loss of cortical regulation of AMY reactivity that may contribute to these maladaptive behaviors.

Neurobiological studies in animal models of the comorbid condition are generally supportive of the human literature. For example, adolescent social isolation, which promotes symptoms of AUD and PTSD, results in an enduring dysregulation of the mesolimbic DA system similar to that observed in patients with these disorders (Butler, Ariwodola, & Weiner, 2014). In addition, many animal models of AUD and PTSD engender increases in BLA excitability that are comparable to the AMY hyperexcitability that is frequently observed in individuals afflicted with these disorders (Gilpin & Weiner, 2017). Moreover, emerging preclinical data suggest that normalizing DA signaling or AMY activity may reverse some AUD/PTSD-like symptoms (Karkhanis et al., 2016; Rau et al., 2015).

Together, these findings provide compelling evidence that the development and progression of AUD and PTSD both involve common maladaptive changes in brain reward and stress networks. These shared neural alterations provide a simple, biological explanation for the overlap in many of the core symptoms of these disorders as well as their frequent co-occurrence. Unfortunately, despite these impressive advances in our understanding of the neural substrates that may contribute to the comorbid condition, this knowledge has not yet led to the development of novel treatments for the dual diagnosis (or either disorder alone) or a significant improvement in patient outcomes. Thus, a fundamental and, as yet, unresolved question is: what is holding us back? Notwithstanding the large volume of data showing that AUD and PTSD are associated with a profound dysregulation of similar nodes of the reward and stress circuits, we suggest that viewing these circuits as discrete networks that independently process rewarding and stressful stimuli does not actually reflect how these networks function. We end this review with a brief proposal that an over-reliance on this dichotomous framework to interpret emerging findings on the neural substrates of AUD and PTSD may be hampering efforts to translate these advances into better treatments and patient outcomes and offer an alternative, based on sensory/affective valence encoding.

Perhaps the most important limitation with the current conceptual framework of reward and stress networks is that they do not depict the extensive structural and functional connectivity between many of the brain regions within these circuits. Canonical diagrams of reward and stress circuitry also mask the fact that many, if not all, brain regions within these circuits are comprised of heterogeneous cell populations that play an integral role in the processing of both reward and stress-related stimuli. For example, as illustrated in Fig. 1, while the reward circuit is usually centered around the VTA-NAc DA projection, the VTA contains a diverse array of non-DA cells (Morales & Margolis, 2017) and the NAc also receives dense excitatory input from many other regions, like the PFC and aHC, and all of these pathways are involved in the processing of both reward- and stress-related information (Britt et al., 2012; Daviu et al., 2019; Francis & Lobo, 2017; MacAskill, Little, Cassel, & Carter, 2012). Moreover, DA can be released by both aversive and rewarding stimuli (for reviews see: Morales & Margolis, 2017; Thibeault, Kutlu, Sanders, & Calipari, 2019). Similarly, the BLA, which is commonly illustrated as the core node of stress/fear circuits, is inextricably interconnected with many elements of the reward circuitry, including the NAc, and there is now compelling opto- and chemo-genetic evidence that many of these pathways process reward- as well as stress-related stimuli (Daviu et al., 2019; Janak & Tye, 2015; Sharp, 2017). While this overlap between the brain reward and stress systems is widely appreciated in the literature (Centanni, Bedse, Patel, & Winder, 2019; Koob & Schulkin, 2019; Koob & Volkow, 2016), these heuristics are nevertheless still commonly used to interpret functional connectivity data in the AUD and PTSD fields and their continued use may be limiting our ability to truly appreciate how dysregulation of these two networks, which are actually highly functionally connected, contributes to the etiology of these disorders.

How then might we provide a conceptual framework that more accurately illustrates the integration of canonical reward and stress circuits and, perhaps, better explains how maladaptive changes in these functionally inter-related networks leads to the frequent co-occurrence of AUD and PTSD? The answer may lie in a rapidly growing field of neuroscience that is seeking to understand the critical brain circuitry that allows all organisms to perform the essential function of responding to motivationally salient internal and external stimuli in a manner that is congruent with this input. In this field, the term valence is used to define an internal state that is generated by integrating all relevant incoming sensory and/or emotional stimuli. Internal states with positive valence are generally perceived as pleasurable and usually promote behaviors that seek to maintain these conditions while those with negative valence are typically discerned as aversive and tend to encourage behaviors that relieve these unpleasant states. Thus, fundamentally, this circuitry serves the essential function of reinforcing behaviors that are life-sustaining while discouraging behaviors that may be hazardous to one's health and well-being. It is important to note that, in humans, valence can be measured in the absence of overt behavior, by asking a subject to rate the relative pleasantness or aversiveness of a stimulus. In contrast, valence can only be inferred from behavior in non-human animals.

As just one example of valence circuits, a recent study identified excitatory projections from regions of gustatory cortex that encode sweet and bitter taste which project onto distinct clusters of cells in the AMY (Wang et al., 2018). Using elegant optogenetic approaches, this study demonstrated that these cortico-AMY circuits do not encode sweet or bitter taste but rather the hedonic valence of these tastants. In other words, photoactivation of the sweet pathway could promote appetitive behavior directed at a neutral stimulus or override avoidance directed at a bitter tastant and optical stimulation of the bitter circuit could do the opposite. In the last decade, similar approaches have been used to identify neural circuits that encode the valence of other sensory stimuli as well as many emotion-like states, like anhedonia (Hoflich, Michenthaler, Kasper, & Lanzenberger, 2019) and fear (Tovote, Fadok, & Luthi, 2015).

Although this is a nascent field of neuroscience, the brain regions that have already been shown to contribute to the encoding of hedonic valence include all the major nodes that comprise traditional reward and stress circuits that are known to be dysregulated in AUD and PTSD (Fig. 2A). However, a major advantage of this conceptual framework is that it does not assign discrete brain regions for the processing of positive and negative stimuli. Rather, it envisions this functionally cohesive network working in a highly integrated fashion to process the valence of all sensory and emotional information, thus allowing the common maladaptive alterations in functional connectivity associated with AUD and PTSD to be visualized within a single cohesive circuitry (Fig. 2B). Moreover, as alluded to above, this heuristic is supported by a large and growing volume of preclinical circuit-mapping data demonstrating that there are numerous monosynaptic projections that connect many cortical and subcortical structures within the reward and stress circuits (Beyeler et al., 2018; Namburi, Al-Hasani, Calhoon, Bruchas, & Tye, 2016; Tye, 2018). Together, these circuits form a valence network (VN) that we operationally define as a highly distributed circuitry tasked with assigning some degree of positivity or negativity to all sensory and/or affective input, encoding the relative salience or importance of this input (see below), in order to orchestrate behavioral and/or cognitive changes that are congruent with the incoming stimuli.

Importantly, these preclinical animal findings are actually supported by human functional imaging data. A recent meta-analysis of almost 20 years of human affective valence neuroimaging studies sought to test three competing hypotheses about the brain basis of hedonic valence encoding (Lindquist, Satpute, Wager, Weber, & Barrett, 2016). This study found little evidence for a bipolarity hypothesis, in which a single brain network encodes valence by monotonically increasing or decreasing along the spectrum of positive to negative affect. It also did not support a bivalent hypothesis, analogous with brain reward and stress circuits, in which independent brain networks encode positive and negative affect. Instead, it found evidence for what the authors termed an "affective workspace" hypothesis, in which both positive and negative affect are encoded by a single, flexible

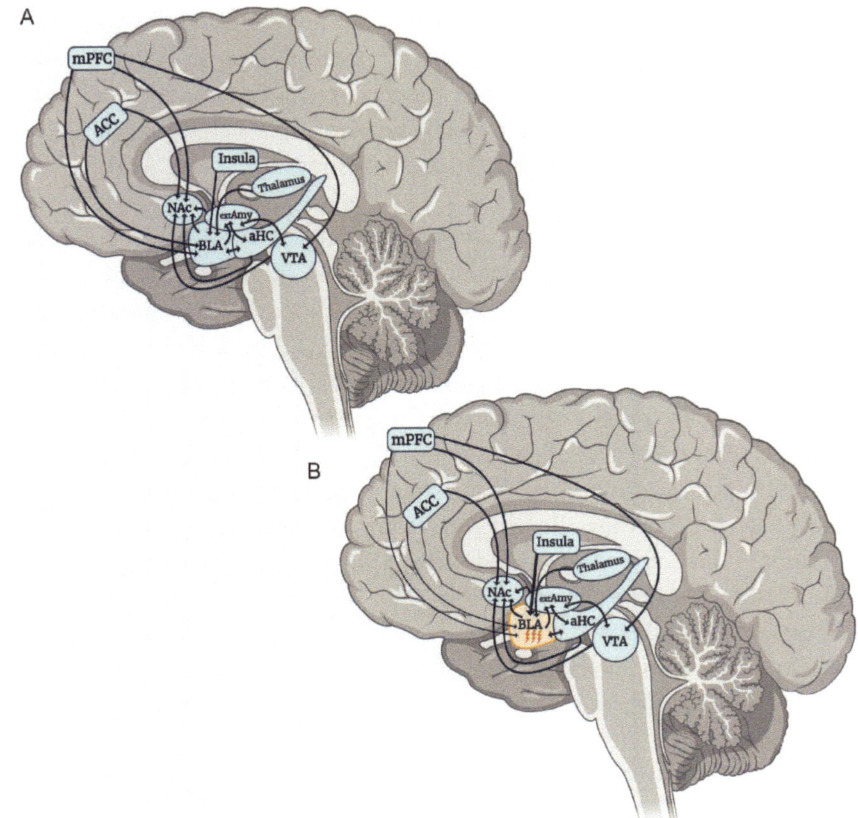

Fig. 2 Using a human valence network model to visualize the primary alterations in functional connectivity that are common to AUD and PTSD. (A) The major cortical and subcortical brain regions and functional connections involved in assigning a degree of positivity or negativity to all sensory and emotional stimuli, encoding the salience of this input, and orchestrating appropriate behavioral and cognitive responses to this information. (B) Illustration of major maladaptive alterations in valence network connectivity common to both AUD and PTSD, based primarily on data from human functional neuroimaging studies. These changes include an increase in BLA excitability driven, in part, by decreased mPFC and ACC connectivity (thinner line), and increased insula connectivity to this brain region (thicker line).

network. In other words, at the level of brain regional activity, there does not appear to be a single brain region or even voxel that uniquely represents positive or negative affect. Rather, there appears to be a valence-general network that engages in the processing of all affective stimuli, regardless of their hedonic valence.

At first, these findings may seem at odds with the preclinical data demonstrating that there do appear to be dedicated monosynaptic circuits within the valence-general network that do indeed encode either positive or negative valence. However, as noted above, many of these studies have also shown that there are direct, non-overlapping connections between elements of the VN that assign both positive and negative valence to incoming stimuli. This can be illustrated by examining the role of the BLA in valence processing. This brain region receives dense input from many sensory and affective brain regions. These inputs synapse primarily onto excitatory BLA pyramidal neurons which encode the positive or negative valence of these stimuli and generate valence-appropriate behavioral and cognitive responses via monosynaptic projections to downstream structures like the CeA, NAc and aHC (Janak & Tye, 2015). Importantly, many studies have demonstrated that discrete projections from the BLA to each of these regions can elicit both positive and negative valence-related behaviors (Kim, Pignatelli, Xu, Itohara, & Tonegawa, 2016; Pi et al., 2020; Zhang et al., 2020). As just one recent example, CCK+ and CCK− BLA pyramidal cells are monosynaptically connected to distinct populations of NAc cells and photoactivation of these circuits can elicit positive or negative affective states (Shen et al., 2019). The reason that dedicated positive and negative valence circuits were not revealed in the human valence processing meta-analysis likely reflects the fact that these neuroimaging studies capture functional, and not structural, connectivity on a time-scale that is orders of magnitude slower than the monosynaptic communication identified in preclinical circuit-mapping studies. Despite these methodological issues, it is noteworthy that the brain regions that comprise the human VN are essentially identical to those identified in animal valence circuit-mapping studies and, perhaps not surprisingly, include many of the structures and functional networks that are altered in AUD and PTSD, including cortical areas like the PFC, ACC, and insula as well as subcortical regions like the AMY and VS (Fig. 2B).

It also warrants discussion that many of the brain regions and functional connections that we refer to as a VN also comprise what is commonly referred to as the salience network (SN) in the human functional imaging literature (Seeley, 2019). The primary function of the SN is to assess the importance (salience) of sensory and/or emotional stimuli, independent of valence, to inform on ongoing behavioral or cognitive processes, often related to threat detection (Kolesar, Bilevicius, Wilson, & Kornelsen, 2019; Szeszko & Yehuda, 2019). However, the authors of the valence functional connectivity meta-analysis correctly note that all of our conscious experience is simultaneously assigned with valence as well as the relative importance of this information. Thus, we suggest that there are two core processing functions of this network: (1) to assign a hedonic valence to all sensory and emotional information and (2) to encode the salience of this input. Since the primary goal of this circuitry is to decipher internal states in manner that promotes the viability of an organism, the term VN may better reflect this role. So, while we refer to this circuitry as the VN, we operationally define it as a distributed brain network that encodes both the valence of all sensory/emotional input as well as the relative salience or importance of this information in order to invigorate behavioral responding for rewarding stimuli and dampen responding for aversive stimuli.

How can this VN framework be used to inform our understanding of the neural network dysregulation in AUD and PTSD? Most importantly, as illustrated in Fig. 2B, the VN provides a way to view the pathophysiology of both disorders within a single integrated circuitry that more accurately depicts the functional role of the brain regions that comprise it. We suggest that, rather than thinking of AUD and PTSD as disorders of reward and/or stress, that they both be considered disorders of hedonic valence processing or VN imbalance disorders. We suggest that, in healthy individuals, the VN is able to reliably encode both the valence of incoming sensory and emotional information as well as the relative salience of that information to appropriately guide behavior and cognition. However, if elements within the VN circuitry are dysregulated, perhaps by exposure to a traumatic event and/or excessive alcohol consumption, incongruous valence encoding may occur which could then lead to the behavioral and cognitive symptoms of disorders like AUD and PTSD.

Indeed, all of the maladaptive behaviors that are central to the diagnosis of AUD and PTSD can be mediated, at least in part, by alterations in VN encoding. For example, many of the symptoms of AUD are associated with excessive alcohol consumption, loss of control over this behavior, and craving. These behaviors could be driven by an increase in the salience of the VN circuitry that encodes the rewarding properties of this drug and/or an amplification of the salience of the negative affect engendered by withdrawal. Similarly, core symptoms of PTSD, like hypervigilance, may arise due to heightened responsivity of VN circuitry to anxiety-provoking stimuli whereas anhedonia and negative mood may reflect hypoactivity of VN nodes that normally encode the positive value of naturally rewarding input.

This VN framework is particularly helpful when trying to interpret the epidemiological and neurobiological data on comorbid AUD and PTSD. First, as already noted, virtually all of the major alterations in functional connectivity that are common to both disorders involve elements of the VN, thus providing a biological basis for the overlap in the many symptoms shared by these diseases. This model may also explain why most epidemiological studies suggest that PTSD more frequently precedes AUD and generally support the self-medication hypothesis for the comorbid condition (Straus et al., 2018). Acute alcohol likely normalizes activity in the valence circuits that are dysregulated and drive the maladaptive symptoms of PTSD. However, repeated cycles of excessive alcohol consumption and withdrawal may actually exacerbate the imbalance in these valence circuits, thus contributing to a worsening of PTSD symptoms and an increased likelihood of developing AUD. For example, animal studies have shown that alcohol effectively decreases BLA excitability and reduces anxiety-like behaviors (McCool, Christian, Diaz, & Lack, 2010; Weiner & Valenzuela, 2006), which may initially quell the negative affective symptoms associated with PTSD. However, repeated cycles of intermittent alcohol exposure result in robust increases in BLA activity that play a causal role in the anxiogenic behaviors that develops during withdrawal (McGinnis, Parrish, Chappell, Alexander, & McCool, 2020; McGinnis, Parrish, & McCool, 2020). These maladaptive changes in the circuits that encode this negative valence information would then promote an even stronger negative affective state during periods of withdrawal, thus exacerbating PTSD symptoms and strengthening motivation to drink to relieve this aversive state.

The VN model also provides a neural framework that may explain the profound heterogeneity in symptom profiles of people with AUD and PTSD as well as individual differences in the severity of these disorders. Given that the VN encodes the valence and salience of all sensory and affective stimuli, there are likely distinct circuits within this network that are responsible for specific symptom clusters of AUD and PTSD, such as anhedonia and anxiety-like behaviors. Indeed, preclinical animal studies have actually identified circuits that encode these behaviors (Hoflich et al., 2019; Janak & Tye, 2015). The clinical presentation of these disorders likely depends on the specific connections within the VN that are dysregulated and may vary depending on the life experience and genetic makeup of each individual. This model can also provide a neural basis for differences in vulnerability and resilience to these disorders. Environmental influences, like early life stress, as well as genetic factors, may compromise or strengthen the resilience of the VN, making it more or less susceptible to the deleterious neural adaptations associated with subsequent exposure to trauma and/or excessive alcohol drinking.

The VN can also reconcile functional neuroimaging datasets that are difficult to interpret using traditional reward and stress circuit constructs. For example, as noted earlier, individuals with AUD often exhibit stronger VS activation in response to alcohol-related cues but show blunted VS reactivity to non-alcohol-related positive stimuli (Schacht et al., 2013; Wrase et al., 2007). Similarly, PTSD patients also commonly show dampened VS responses to cues with positive valence but in some cases, these subjects actually show heightened VS reactivity to aversive stimuli (Elman et al., 2009, 2018). The VN provides a framework to outline simple and testable hypotheses to account for these seemingly discordant findings. While some VN circuits seem to broadly encode the valence of sensory or affective stimuli, others may process a narrower range of this information. It seems reasonable to hypothesize that distinct positive valence circuits may encode the positivity of alcohol-related cues vs. other positive cues that are unrelated to alcohol. If so, AUD may increase salience encoding within the circuitry that processes the positive valence of alcohol-related stimuli while simultaneously decreasing salience encoding in pathways that process non-alcohol-related positively valenced stimuli. Similarly, since it is well-established that many nodes of the VN receive convergent, non-overlapping positive and negative valence information (Tye, 2018), it seems likely that PTSD could diminish the salience of a reward-related signal, while simultaneously increasing the salience of a pathway that encodes aversion. Indeed, recent data suggest that VN neurons that process positive and negative valence are functionally connected by GABAergic interneurons, and thus mutually antagonistic (Kim, Pignatelli, et al., 2016; Zhang et al., 2020).

Finally, we suggest that the VN framework may explain some of the difficulties associated with the development of better treatments for AUD and PTSD and offer new insight into the reasons why novel therapeutic approaches, such as neuromodulation, may prove to be particularly effective in treating these disorders. First, given the complexity of the VN and the broad range of sensory and affective VN circuit imbalances that can potentially give rise to these disorders, it seems highly unlikely that any single treatment could prove effective in all cases. Rather, by gaining a better understanding of the VN elements responsible for specific AUD/PTSD symptoms, we may be able to better tailor treatments to mitigate these distinct neural alterations. More importantly, functional imaging studies have clearly established that the VN processes valence in a highly distributed manner (Lindquist et al., 2016). Although preclinical studies focus primarily on monosynaptic valence circuits, it is clear that these are simply components of a much broader, highly integrated network. Given that AUD and PTSD are associated with widespread alterations in functional connectivity throughout the VN, treatments that could have a distributed effect on VN circuit function may prove to be particularly effective in treating these disorders. As noted earlier, several studies have already demonstrated that real-time fMRI neurofeedback can be used to modulate critical VN nodes, like the AMY, to relieve PTSD symptoms. Indeed, the VN heuristic can readily account for the findings that both decreasing AMY reactivity to aversive stimuli and increasing AMY activity in response to positively valenced cues led to an improvement in PTSD symptoms (Gerin et al., 2016; Misaki et al., 2018; Nicholson et al., 2017). Moreover, the observations that targeting one node of the VN could have distributed effects on VN functional connectivity independent of that node, and that some of these changes actually correlated with improvement of depressive and hyperarousal PTSD symptoms (Misaki et al., 2018), provides encouraging support for the notion that rt-fMRI neurofeedback, and other non-invasive neuromodulation techniques, like exposure therapy and transcranial magnetic stimulation (TMS), may prove particularly effective in treating VN circuit dysregulation in AUD and PTSD. Indeed, TMS is already FDA-approved to treat other affective disorders, like depression and OCD, and there is growing evidence that it may also be an effective treatment for AUD (Diana et al., 2017; Philip, Sorensen, McCalley, & Hanlon, 2020) and PTSD (Kozel, 2018; Trevizol et al., 2016). Future studies are certainly warranted to explore the efficacy of these neuromodulation therapies in the treatment of the comorbid condition.

Despite the advantages that the VN framework offers, there are several major challenges that must be addressed before we are likely to see recent advances in our understanding of the neurobiology of AUD and PTSD translated into better treatments for these disorders. First, there are many discordant findings in the preclinical literature regarding the anatomical segregation of the cells and circuits that encode the valence and salience of sensory and emotional stimuli. While there is a general consensus about the specific brain regions responsible for these functions (Beyeler et al., 2018; Namburi et al., 2016; Peters et al., 2016; Thibeault et al., 2019; Tye, 2018; Zhang & Volkow, 2019), there is considerable disagreement as to the anatomical localization of positive and negative valence encoding cells within these structures. For example, several BLA valence circuit-mapping studies have shown that there are distinct populations of cells that encode positive and negative valence that are largely segregated along the anterior/posterior axis of this brain region (Kim, Pignatelli, et al., 2016; Pi et al., 2020; Wang et al., 2018; Zhang et al., 2020). However, other studies, using similar sensory or affective stimuli, suggest that positive and negative valence cells are intermixed throughout the BLA (Beyeler et al., 2018; Shen et al., 2019). Even more baffling, studies that have reported anatomical segregation of BLA valence encoding cells do not always agree on which subregion encodes positive vs. negative valence (Kim, Pignatelli, et al., 2016; Wang et al., 2018). While there are likely many reasons underlying these disparate findings, such as methodological differences and the inherent difficulty associated with the study of emotion-like behaviors in animals, it will be important to resolve the actual anatomical location of the neurons and circuits responsible for processing sensory and affective valence.

A second challenge is that, although significant progress has been made in identifying changes in VN functional connectivity that contribute to the etiology of AUD and PTSD, our understanding of the neurobiological mechanisms responsible for these alterations is far from complete. While we have advanced the idea that this circuitry be viewed as a VN, we have emphasized that this network not only encodes the valence of sensory and emotional stimuli but also determines the relative salience or importance of this input, likely via modulation of neural activity within these circuits. We have also noted that incongruent encoding of the salience of valence signals likely plays an integral role in the pathophysiology of both disorders. While animal research has identified numerous neurotransmitters and neuropeptides that can modulate the signaling strength of many of the circuits within the VN (Thibeault et al., 2019; Tye, 2018), how such changes actually contribute to alterations in VN functional connectivity in AUD and PTSD is largely unknown. These neuromodulatory systems that govern the salience of VN signals may well represent some of the most promising targets for the development of novel treatments for these disorders.

Finally, for these challenges to be overcome, it is imperative that we improve scientific interaction and communication between researchers that employ human and non-human subjects to elucidate the specific changes in VN functional connectivity that give rise to AUD and PTSD. As alluded to above, animal researchers typically focus on monosynaptic projections that function in the millisecond to second range whereas human neuroimaging researchers study a much slower timescale and characterize networks that are functionally, but not necessarily structurally, connected. While the alterations in monosynaptic circuits being examined in animal models of AUD and PTSD likely give rise to the changes in functional connectivity observed in human studies of individuals afflicted with these conditions, better techniques and methods will be needed to bridge these fields.

12. Conclusions

In summary, this chapter has provided a review of recent literature on the epidemiology and neurobiology of comorbid AUD and PTSD. Collectively, the available data continue to show that there is considerable overlap in the core symptoms associated with these disorders, that these diseases frequently co-occur, and that this dual diagnosis is associated with increased symptom severity and poor treatment outcomes. Although the comorbid condition remains a major global public health concern, groups such as members of the military, war veterans, and individuals exposed to early life stress are particularly vulnerable to these disorders. Focusing primarily on evidence from human functional imaging studies, we have highlighted numerous maladaptive changes in brain reward and stress networks that are common to both disorders. We note, however, that these recent advances have not led to better treatments for the comorbid condition. We suggest that canonical brain reward and stress circuits do not accurately depict the strong, functional integration between these networks and that the continued use these frameworks to interpret clinical and preclinical findings on AUD and PTSD may be limiting our ability to translate recent discoveries into more effective therapies. Instead, we suggest that reward and stress circuits be viewed as an integrated valence network that functions in a cohesive manner to assign a degree of positivity or negativity to all sensory and emotional stimuli. This network also encodes the relative salience or level of importance of these hedonic signals in order to guide behavioral responding that is congruent with this input. Using this heuristic, we suggest that AUD and PTSD may be viewed as valence network imbalance disorders. We highlight a large volume of data from both human functional imaging and animal circuit-mapping studies that provide empirical support for this framework and discuss why this model provides a more parsimonious way of interpreting much of the recent epidemiological and neurobiological data on the comorbid condition. Finally, we propose that emerging non-invasive neuromodulatory therapies, like rt-fMRI NF and TMS, may be particularly effective at restoring the functional perturbations in valence network connectivity that give rise to these disorders.

References

Afzali, M. H., Sunderland, M., Batterham, P. J., Carragher, N., & Slade, T. (2017). Trauma characteristics, post-traumatic symptoms, psychiatric disorders and suicidal behaviours: Results from the 2007 Australian National Survey of Mental Health and Wellbeing. *The Australian and New Zealand Journal of Psychiatry*, *51*(11), 1142–1151. https://doi.org/10.1177/0004867416683815.

Aghajani, M., Veer, I. M., van Hoof, M. J., Rombouts, S. A., van der Wee, N. J., & Vermeiren, R. R. (2016). Abnormal functional architecture of amygdala-centered networks in adolescent posttraumatic stress disorder. *Human Brain Mapping*, *37*(3), 1120–1135. https://doi.org/10.1002/hbm.23093.

Akiki, T. J., Averill, C. L., & Abdallah, C. G. (2017). A network-based neurobiological model of PTSD: Evidence from structural and functional neuroimaging studies. *Current Psychiatry Reports*, *19*(11), 81. https://doi.org/10.1007/s11920-017-0840-4.

Akiki, T. J., Averill, C. L., Wrocklage, K. M., Scott, J. C., Averill, L. A., Schweinsburg, B., et al. (2018). Default mode network abnormalities in posttraumatic stress disorder: A novel network-restricted topology approach. *NeuroImage*, *176*, 489–498. https://doi.org/10.1016/j.neuroimage.2018.05.005.

Albrechet-Souza, L., & Gilpin, N. W. (2019). The predator odor avoidance model of post-traumatic stress disorder in rats. *Behavioural Pharmacology*, *30*(2 and 3-Spec Issue), 105–114. https://doi.org/10.1097/FBP.0000000000000460.

Almonte, A. G., Ewin, S. E., Mauterer, M. I., Morgan, J. W., Carter, E. S., & Weiner, J. L. (2017). Enhanced ventral hippocampal synaptic transmission and impaired synaptic plasticity in a rodent model of alcohol addiction vulnerability. *Scientific Reports*, 7(1). https://doi.org/10.1038/s41598-017-12531-z.

American Psychiatric Association. (2013). *Diagnostic and statistical manual of mental disorders* (5th ed.). Washington, D.C.: American Psychiatric Association.

Arain, M., Haque, M., Johal, L., Mathur, P., Nel, W., Rais, A., et al. (2013). Maturation of the adolescent brain. *Neuropsychiatric Disease and Treatment*, *9*, 449–461. https://doi.org/10.2147/NDT.S39776.

Atchley, D., Hankosky, E. R., Gasparotto, K., & Rosenkranz, J. A. (2012). Pharmacological enhancement of calcium-activated potassium channel function reduces the effects of repeated stress on fear memory. *Behavioural Brain Research*, *232*(1), 37–43. https://doi.org/10.1016/j.bbr.2012.03.037.

Avino, T. A., Barger, N., Vargas, M. V., Carlson, E. L., Amaral, D. G., Bauman, M. D., et al. (2018). Neuron numbers increase in the human amygdala from birth to adulthood, but not in autism. *Proceedings of the National Academy of Sciences of the United States of America*, *115*(14), 3710–3715. https://doi.org/10.1073/pnas.1801912115.

Back, S. E., Brady, K. T., Sonne, S. C., & Verduin, M. L. (2006). Symptom improvement in co-occurring PTSD and alcohol dependence. *The Journal of Nervous and Mental Disease*, *194*(9), 690–696. https://doi.org/10.1097/01.nmd.0000235794.12794.8a.

Balachandran, T., Cohen, G., Le Foll, B., Rehm, J., & Hassan, A. N. (2020). The effect of pre-existing alcohol use disorder on the risk of developing posttraumatic stress disorder: Results from a longitudinal national representative sample. *The American Journal of Drug and Alcohol Abuse*, *46*, 232–240. https://doi.org/10.1080/00952990.2019.1690495.

Barger, N., Stefanacci, L., Schumann, C. M., Sherwood, C. C., Annese, J., Allman, J. M., et al. (2012). Neuronal populations in the basolateral nuclei of the amygdala are differentially increased in humans compared with apes: A stereological study. *The Journal of Comparative Neurology*, *520*(13), 3035–3054. https://doi.org/10.1002/cne.23118.

Berenz, E. C., Roberson-Nay, R., Latendresse, S. J., Mezuk, B., Gardner, C. O., Amstadter, A. B., et al. (2017). Posttraumatic stress disorder and alcohol dependence: Epidemiology and order of onset. *Psychological Trauma*, *9*(4), 485–492. https://doi.org/10.1037/tra0000185.

Beyeler, A., Chang, C. J., Silvestre, M., Leveque, C., Namburi, P., Wildes, C. P., et al. (2018). Organization of valence-encoding and projection-defined neurons in the basolateral amygdala. *Cell Reports*, *22*(4), 905–918. https://doi.org/10.1016/j.celrep.2017.12.097.

Blaine, S. K., Milivojevic, V., Fox, H., & Sinha, R. (2016). Alcohol effects on stress pathways: Impact on craving and relapse risk. *Canadian Journal of Psychiatry-Revue Canadienne De Psychiatrie*, *61*(3), 145–153. https://doi.org/10.1177/0706743716632512.

Blaine, S. K., & Sinha, R. (2017). Alcohol, stress, and glucocorticoids: From risk to dependence and relapse in alcohol use disorders. *Neuropharmacology*, *122*, 136–147. https://doi.org/10.1016/j.neuropharm.2017.01.037.

Blanco, C., Hoertel, N., Wall, M. M., Franco, S., Peyre, H., Neria, Y., et al. (2018). Toward understanding sex differences in the prevalence of posttraumatic stress disorder: Results from the national epidemiologic survey on alcohol and related conditions. *The Journal of Clinical Psychiatry*, *79*(2), 16m11364. https://doi.org/10.4088/JCP.16m11364.

Bluhm, R. L., Williamson, P. C., Osuch, E. A., Frewen, P. A., Stevens, T. K., Boksman, K., et al. (2009). Alterations in default network connectivity in posttraumatic stress disorder related to early-life trauma. *Journal of Psychiatry & Neuroscience*, *34*(3), 187–194. Retrieved from https://www.ncbi.nlm.nih.gov/pubmed/19448848.

Bolsinger, J., Seifritz, E., Kleim, B., & Manoliu, A. (2018). Neuroimaging correlates of resilience to traumatic events—A comprehensive review. *Frontiers in Psychiatry*, *9*, 693. https://doi.org/10.3389/fpsyt.2018.00693.

Bowe, A., & Rosenheck, R. (2015). PTSD and substance use disorder among veterans: Characteristics, service utilization and pharmacotherapy. *Journal of Dual Diagnosis*, *11*(1), 22–32. https://doi.org/10.1080/15504263.2014.989653.

Bremner, J. D., Southwick, S. M., Darnell, A., & Charney, D. S. (1996). Chronic PTSD in Vietnam combat veterans: Course of illness and substance abuse. *The American Journal of Psychiatry*, *153*(3), 369–375. https://doi.org/10.1176/ajp.153.3.369.

Bremner, J. D., Staib, L. H., Kaloupek, D., Southwick, S. M., Soufer, R., & Charney, D. S. (1999). Neural correlates of exposure to traumatic pictures and sound in Vietnam combat veterans with and without posttraumatic stress disorder: A positron emission tomography study. *Biological Psychiatry*, *45*(7), 806–816. https://doi.org/10.1016/S0006-3223(98)00297-2.

Britt, J. P., Benaliouad, F., McDevitt, R. A., Stuber, G. D., Wise, R. A., & Bonci, A. (2012). Synaptic and behavioral profile of multiple glutamatergic inputs to the nucleus accumbens. *Neuron*, *76*(4), 790–803. https://doi.org/10.1016/j.neuron.2012.09.040.

Britton, J. C., Phan, K. L., Taylor, S. F., Fig, L. M., & Liberzon, I. (2005). Corticolimbic blood flow in posttraumatic stress disorder during script-driven imagery. *Biological Psychiatry*, *57*(8), 832–840. https://doi.org/10.1016/j.biopsych.2004.12.025.

Brodnik, Z. D., Black, E. M., Clark, M. J., Kornsey, K. N., Snyder, N. W., & Espana, R. A. (2017). Susceptibility to traumatic stress sensitizes the dopaminergic response to cocaine and increases motivation for cocaine. *Neuropharmacology*, *125*, 295–307. https://doi.org/10.1016/j.neuropharm.2017.07.032.

Brown, V. M., LaBar, K. S., Haswell, C. C., Gold, A. L., Mid-Atlantic, M. W., McCarthy, G., et al. (2014). Altered resting-state functional connectivity of basolateral and centromedial amygdala complexes in posttraumatic stress disorder. *Neuropsychopharmacology*, *39*(2), 351–359. https://doi.org/10.1038/npp.2013.197.

Burri, A., & Maercker, A. (2014). Differences in prevalence rates of PTSD in various European countries explained by war exposure, other trauma and cultural value orientation. *BMC Research Notes*, 7, 407. https://doi.org/10.1186/1756-0500-7-407.

Butler, T. R., Ariwodola, O. J., & Weiner, J. L. (2014). The impact of social isolation on HPA axis function, anxiety-like behaviors, and ethanol drinking. *Frontiers in Integrative Neuroscience*, 7, 102. https://doi.org/10.3389/fnint.2013.00102.

Butler, T. R., Carter, E., & Weiner, J. L. (2014). Adolescent social isolation does not lead to persistent increases in anxiety-like behavior or ethanol intake in female long-evans rats. *Alcoholism, Clinical and Experimental Research*, *38*(8), 2199–2207. https://doi.org/10.1111/acer.12476.

Butler, T. R., Karkhanis, A. N., Jones, S. R., & Weiner, J. L. (2016). Adolescent social isolation as a model of heightened vulnerability to comorbid alcoholism and anxiety disorders. *Alcoholism, Clinical and Experimental Research*, *40*(6), 1202–1214. https://doi.org/10.1111/acer.13075.

Carmassi, C., Akiskal, H. S., Bessonov, D., Massimetti, G., Calderani, E., Stratta, P., et al. (2014). Gender differences in DSM-5 versus DSM-IV-TR PTSD prevalence and criteria comparison among 512 survivors to the L'Aquila earthquake. *Journal of Affective Disorders*, *160*, 55–61. https://doi.org/10.1016/j.jad.2014.02.028.

Castillo-Carniglia, A., Keyes, K. M., Hasin, D. S., & Cerda, M. (2019). Psychiatric comorbidities in alcohol use disorder. *Lancet Psychiatry*, *6*(12), 1068–1080. https://doi.org/10.1016/S2215-0366(19)30222-6.

Centanni, S. W., Bedse, G., Patel, S., & Winder, D. G. (2019). Driving the downward spiral: Alcohol-induced dysregulation of extended amygdala circuits and negative affect. *Alcoholism, Clinical and Experimental Research*, *43*(10), 2000–2013. https://doi.org/10.1111/acer.14178.

Chareyron, L. J., Banta Lavenex, P., Amaral, D. G., & Lavenex, P. (2011). Stereological analysis of the rat and monkey amygdala. *The Journal of Comparative Neurology*, *519*(16), 3218–3239. https://doi.org/10.1002/cne.22677.

Chen, A. C., & Etkin, A. (2013). Hippocampal network connectivity and activation differentiates post-traumatic stress disorder from generalized anxiety disorder. *Neuropsychopharmacology*, *38*(10), 1889–1898. https://doi.org/10.1038/npp.2013.122.

Chilcoat, H. D., & Breslau, N. (1998). Investigations of causal pathways between PTSD and drug use disorders. *Addictive Behaviors*, *23*(6), 827–840. Retrieved from http://www.ncbi.nlm.nih.gov/pubmed/9801719.

Cooper, S., Robison, A. J., & Mazei-Robison, M. S. (2017). Reward circuitry in addiction. *Neurotherapeutics*, *14*(3), 687–697. https://doi.org/10.1007/s13311-017-0525-z.

Crane, N. A., Gorka, S. M., Weafer, J., Langenecker, S. A., de Wit, H., & Phan, K. L. (2017). Preliminary evidence for disrupted nucleus accumbens reactivity and connectivity to reward in binge drinkers. *Alcohol and Alcoholism*, *52*(6), 647–654. https://doi.org/10.1093/alcalc/agx062.

Cservenka, A., Casimo, K., Fair, D. A., & Nagel, B. J. (2014). Resting state functional connectivity of the nucleus accumbens in youth with a family history of alcoholism. *Psychiatry Research*, *221*(3), 210–219. https://doi.org/10.1016/j.pscychresns.2013.12.004.

da Silva, H. C., Furtado da Rosa, M. M., Berger, W., Luz, M. P., Mendlowicz, M., Coutinho, E. S. F., et al. (2019). PTSD in mental health outpatient settings: Highly prevalent and under-recognized. *Brazilian Journal of Psychiatry*, *41*(3), 213–217. https://doi.org/10.1590/1516-4446-2017-0025.

Dal-Zotto, S., Marti, O., & Armario, A. (2003). Glucocorticoids are involved in the long-term effects of a single immobilization stress on the hypothalamic-pituitary-adrenal axis. *Psychoneuroendocrinology*, *28*(8), 992–1009. https://doi.org/10.1016/s0306-4530(02)00120-8.

Davis, P., & Reijmers, L. G. (2018). The dynamic nature of fear engrams in the basolateral amygdala. *Brain Research Bulletin*, *141*, 44–49. https://doi.org/10.1016/j.brainresbull.2017.12.004.

Davis, M., Walker, D. L., Miles, L., & Grillon, C. (2010). Phasic vs sustained fear in rats and humans: Role of the extended amygdala in fear vs anxiety. *Neuropsychopharmacology*, *35*(1), 105–135. https://doi.org/10.1038/npp.2009.109.

Davis, M., & Whelan, P. J. (2001). The amygdala: Vigilance and emotion. *Molecular Psychiatry*, *6*, 13–34. https://doi.org/10.1038/sj.mp.4000812.

Daviu, N., Bruchas, M. R., Moghaddam, B., Sandi, C., & Beyeler, A. (2019). Neurobiological links between stress and anxiety. *Neurobiology of Stress*, *11*. https://doi.org/10.1016/j.ynstr.2019.100191.

Dawson, D. A., Harford, T. C., & Grant, B. F. (1992). Family history as a predictor of alcohol dependence. *Alcoholism, Clinical and Experimental Research*, *16*(3), 572–575. https://doi.org/10.1111/j.1530-0277.1992.tb01419.x.

de Kloet, C. S., Vermetten, E., Geuze, E., Kavelaars, A., Heijnen, C. J., & Westenberg, H. G. (2006). Assessment of HPA-axis function in posttraumatic stress disorder: Pharmacological and non-pharmacological challenge tests, a review. *Journal of Psychiatric Research*, *40*(6), 550–567. https://doi.org/10.1016/j.jpsychires.2005.08.002.

Deal, A. L., Konstantopoulos, J. K., Weiner, J. L., & Budygin, E. A. (2018). Exploring the consequences of social defeat stress and intermittent ethanol drinking on dopamine dynamics in the rat nucleus accumbens. *Scientific Reports*, *8*(1), 332. https://doi.org/10.1038/s41598-017-18706-y.

Deslauriers, J., Toth, M., Der-Avakian, A., & Risbrough, V. B. (2018). Current status of animal models of posttraumatic stress disorder: Behavioral and biological phenotypes, and future challenges in improving translation. *Biological Psychiatry*, *83*(10), 895–907. https://doi.org/10.1016/j.biopsych.2017.11.019.

Diana, M., Raij, T., Melis, M., Nummenmaa, A., Leggio, L., & Bonci, A. (2017). Rehabilitating the addicted brain with transcranial magnetic stimulation. *Nature Reviews. Neuroscience*, *18*(11), 685–693. https://doi.org/10.1038/nrn.2017.113.

Dias, B. G., Banerjee, S. B., Goodman, J. V., & Ressler, K. J. (2013). Towards new approaches to disorders of fear and anxiety. *Current Opinion in Neurobiology*, *23*(3), 346–352. https://doi.org/10.1016/j.conb.2013.01.013.

Diaz, V., & Lin, D. (2020). Neural circuits for coping with social defeat. *Current Opinion in Neurobiology*, *60*, 99–107. https://doi.org/10.1016/j.conb.2019.11.016.

DiGangi, J. A., Tadayyon, A., Fitzgerald, D. A., Rabinak, C. A., Kennedy, A., Klumpp, H., et al. (2016). Reduced default mode network connectivity following combat trauma. *Neuroscience Letters*, *615*, 37–43. https://doi.org/10.1016/j.neulet.2016.01.010.

Dossi, G., Delvecchio, G., Prunas, C., Soares, J. C., & Brambilla, P. (2020). Neural bases of cognitive impairments in post-traumatic stress disorders: A mini-review of functional magnetic resonance imaging findings. *Frontiers in Psychiatry*, *11*, 176. https://doi.org/10.3389/fpsyt.2020.00176.

Dube, S. R., Anda, R. F., Felitti, V. J., Edwards, V. J., & Croft, J. B. (2002). Adverse childhood experiences and personal alcohol abuse as an adult. *Addictive Behaviors*, *27*(5), 713–725. Retrieved from http://www.ncbi.nlm.nih.gov/pubmed/12201379.

Dunlop, B. W., & Wong, A. (2019). The hypothalamic-pituitary-adrenal axis in PTSD: Pathophysiology and treatment interventions. *Progress in Neuro-Psychopharmacology & Biological Psychiatry*, *89*, 361–379. https://doi.org/10.1016/j.pnpbp.2018.10.010.

Duvarci, S., & Pare, D. (2007). Glucocorticoids enhance the excitability of principal basolateral amygdala neurons. *The Journal of Neuroscience*, *27*(16), 4482–4491. https://doi.org/10.1523/JNEUROSCI.0680-07.2007.

Dworkin, E. R., Bergman, H. E., Walton, T. O., Walker, D. D., & Kaysen, D. L. (2018). Co-occurring post-traumatic stress disorder and alcohol use disorder in U.S. military and veteran populations. *Alcohol Research: Current Reviews*, *39*(2), 161–169. Retrieved from https://www.ncbi.nlm.nih.gov/pubmed/31198655.

Dworkin, E. R., Wanklyn, S., Stasiewicz, P. R., & Coffey, S. F. (2018). PTSD symptom presentation among people with alcohol and drug use disorders: Comparisons by substance of abuse. *Addictive Behaviors*, *76*, 188–194. https://doi.org/10.1016/j.addbeh.2017.08.019.

Edwards, S., Baynes, B. B., Carmichael, C. Y., Zamora-Martinez, E. R., Barrus, M., Koob, G. F., et al. (2013). Traumatic stress reactivity promotes excessive alcohol drinking and alters the balance of prefrontal cortex-amygdala activity. *Translational Psychiatry*, *3*, e296. https://doi.org/10.1038/tp.2013.70.

Elman, I., Lowen, S., Frederick, B. B., Chi, W., Becerra, L., & Pitman, R. K. (2009). Functional neuroimaging of reward circuitry responsivity to monetary gains and losses in posttraumatic stress disorder. *Biological Psychiatry*, *66*(12), 1083–1090. https://doi.org/10.1016/j.biopsych.2009.06.006.

Elman, I., Upadhyay, J., Langleben, D. D., Albanese, M., Becerra, L., & Borsook, D. (2018). Reward and aversion processing in patients with post-traumatic stress disorder: Functional neuroimaging with visual and thermal stimuli. *Translational Psychiatry*, *8*(1), 240. https://doi.org/10.1038/s41398-018-0292-6.

Eskelund, K., Karstoft, K. I., & Andersen, S. B. (2018). Anhedonia and emotional numbing in treatment-seeking veterans: Behavioural and electrophysiological responses to reward. *European Journal of Psychotraumatology*, *9*(1). https://doi.org/10.1080/20008198.2018.1446616.

Etkin, A., Prater, K. E., Hoeft, F., Menon, V., & Schatzberg, A. F. (2010). Failure of anterior cingulate activation and connectivity with the amygdala during implicit regulation of emotional processing in generalized anxiety disorder. *The American Journal of Psychiatry*, *167*(5), 545–554. https://doi.org/10.1176/appi.ajp.2009.09070931.

Etkin, A., & Wager, T. D. (2007). Functional neuroimaging of anxiety: A meta-analysis of emotional processing in PTSD, social anxiety disorder, and specific phobia. *The American Journal of Psychiatry*, *164*(10), 1476–1488. https://doi.org/10.1176/appi.ajp.2007.07030504.

Ewin, S. E., Almonte, A. G., Heaney, C. F., Chappell, A. M., Raab-Graham, K. F., & Weiner, J. L. (2019). Chemogenetic inhibition of a monosynaptic projection from the basolateral amygdala to the ventral hippocampus reduces appetitive and consummatory alcohol drinking behaviors. *bioRxiv*.

Felitti, V. J., Anda, R. F., Nordenberg, D., Williamson, D. F., Spitz, A. M., Edwards, V., et al. (1998). Relationship of childhood abuse and household dysfunction to many of the leading causes of death in adults. The Adverse Childhood Experiences (ACE) Study. *American Journal of Preventive Medicine*, *14*(4), 245–258. https://doi.org/10.1016/s0749-3797(98)00017-8.

Fenster, R. J., Lebois, L. A. M., Ressler, K. J., & Suh, J. (2018). Brain circuit dysfunction in post-traumatic stress disorder: From mouse to man. *Nature Reviews. Neuroscience*, *19*(9), 535–551. https://doi.org/10.1038/s41583-018-0039-7.

Fink, D. S., Calabrese, J. R., Liberzon, I., Tamburrino, M. B., Chan, P., Cohen, G. H., et al. (2016). Retrospective age-of-onset and projected lifetime prevalence of psychiatric disorders among U.S. Army National Guard soldiers. *Journal of Affective Disorders*, *202*, 171–177. https://doi.org/10.1016/j.jad.2016.05.025.

Flanagan, J. C., Jones, J. L., Jarnecke, A. M., & Back, S. E. (2018). Behavioral treatments for alcohol use disorder and post-traumatic stress disorder. *Alcohol Research: Current Reviews*, *39*(2), 181–192. Retrieved from https://www.ncbi.nlm.nih.gov/pubmed/31198657.

Flandreau, E. I., & Toth, M. (2018). Animal models of PTSD: A critical review. *Current Topics in Behavioral Neurosciences*, *38*, 47–68. https://doi.org/10.1007/7854_2016_65.

Flook, E. A., Luchsinger, J. R., Silveri, M. M., Winder, D. G., Benningfield, M. M., & Blackford, J. U. (2020). Anxiety during abstinence from alcohol: A systematic review of rodent and human evidence for the anterior insula's role in the abstinence network. *Addiction Biology*, e12861. https://doi.org/10.1111/adb.12861.

Fox, H. C., Axelrod, S. R., Paliwal, P., Sleeper, J., & Sinha, R. (2007). Difficulties in emotion regulation and impulse control during cocaine abstinence. *Drug and Alcohol Dependence*, *89*(2–3), 298–301. https://doi.org/10.1016/j.drugalcdep.2006.12.026.

Fox, H. C., Hong, K. A., & Sinha, R. (2008). Difficulties in emotion regulation and impulse control in recently abstinent alcoholics compared with social drinkers. *Addictive Behaviors, 33*(2), 388–394. https://doi.org/10.1016/j.addbeh.2007.10.002.

Francis, T. C., & Lobo, M. K. (2017). Emerging role for nucleus accumbens medium spiny neuron subtypes in depression. *Biological Psychiatry, 81*(8), 645–653. https://doi.org/10.1016/j.biopsych.2016.09.007.

Franklin, C. L., & Zimmerman, M. (2001). Posttraumatic stress disorder and major depressive disorder: Investigating the role of overlapping symptoms in diagnostic comorbidity. *The Journal of Nervous and Mental Disease, 189*(8), 548–551. Retrieved from http://www.ncbi.nlm.nih.gov/pubmed/11531207.

Friedman, A. K., Walsh, J. J., Juarez, B., Ku, S. M., Chaudhury, D., Wang, J., et al. (2014). Enhancing depression mechanisms in midbrain dopamine neurons achieves homeostatic resilience. *Science, 344*(6181), 313–319. https://doi.org/10.1126/science.1249240.

Fritz, M., Klawonn, A. M., & Zahr, N. M. (2019). Neuroimaging in alcohol use disorder: From mouse to man. *Journal of Neuroscience Research*. https://doi.org/10.1002/jnr.24423.

Fuehrlein, B. S., Mota, N., Arias, A. J., Trevisan, L. A., Kachadourian, L. K., Krystal, J. H., et al. (2016). The burden of alcohol use disorders in US military veterans: Results from the National Health and Resilience in Veterans Study. *Addiction, 111*(10), 1786–1794. https://doi.org/10.1111/add.13423.

Gerin, M. I., Fichtenholtz, H., Roy, A., Walsh, C. J., Krystal, J. H., Southwick, S., et al. (2016). Real-time fMRI neurofeedback with war veterans with chronic PTSD: A feasibility study. *Frontiers in Psychiatry, 7*, 111. https://doi.org/10.3389/fpsyt.2016.00111.

Geyer, M. A., & Markou, A. (1995). Animal models of psychiatric disorders. In F. E. Bloom, & D. J. Kupfer (Eds.), *Psychopharmacology: The fourth generation of progress* (pp. 787–798). New York, NY: Rave Press.

Gilman, J. M., & Hommer, D. W. (2008). Modulation of brain response to emotional images by alcohol cues in alcohol-dependent patients. *Addiction Biology, 13*(3–4), 423–434. https://doi.org/10.1111/j.1369-1600.2008.00111.x.

Gilpin, N. W., & Weiner, J. L. (2017). Neurobiology of comorbid post-traumatic stress disorder and alcohol-use disorder. *Genes, Brain, and Behavior, 16*(1), 15–43. https://doi.org/10.1111/gbb.12349.

Glahn, D. C., Lovallo, W. R., & Fox, P. T. (2007). Reduced amygdala activation in young adults at high risk of alcoholism: Studies from the Oklahoma family health patterns project. *Biological Psychiatry, 61*(11), 1306–1309. https://doi.org/10.1016/j.biopsych.2006.09.041.

Gogtay, N., Giedd, J. N., Lusk, L., Hayashi, K. M., Greenstein, D., Vaituzis, A. C., et al. (2004). Dynamic mapping of human cortical development during childhood through early adulthood. *Proceedings of the National Academy of Sciences of the United States of America, 101*(21), 8174–8179. https://doi.org/10.1073/pnas.0402680101.

Goldstein, R. B., Smith, S. M., Chou, S. P., Saha, T. D., Jung, J., Zhang, H., et al. (2016). The epidemiology of DSM-5 posttraumatic stress disorder in the United States: Results from the National Epidemiologic Survey on Alcohol and Related Conditions-III. *Social Psychiatry and Psychiatric Epidemiology, 51*(8), 1137–1148. https://doi.org/10.1007/s00127-016-1208-5.

Goltseker, K., Hopf, F. W., & Barak, S. (2019). Advances in behavioral animal models of alcohol use disorder. *Alcohol, 74*, 73–82. https://doi.org/10.1016/j.alcohol.2018.05.014.

Goode, T. D., & Maren, S. (2019). Common neurocircuitry mediating drug and fear relapse in preclinical models. *Psychopharmacology, 236*(1), 415–437. https://doi.org/10.1007/s00213-018-5024-3.

Gradus, J. L., Leatherman, S., Curreri, A., Myers, L. G., Ferguson, R., & Miller, M. (2017). Gender differences in substance abuse, PTSD and intentional self-harm among veterans health administration patients. *Drug and Alcohol Dependence, 171*, 66–69. https://doi.org/10.1016/j.drugalcdep.2016.11.012.

Grant, B. F., Chou, S. P., Saha, T. D., Pickering, R. P., Kerridge, B. T., Ruan, W. J., et al. (2017). Prevalence of 12-month alcohol use, high-risk drinking, and DSM-IV alcohol use disorder in the United States, 2001–2002 to 2012–2013: Results from the National Epidemiologic Survey on Alcohol and Related Conditions. *JAMA Psychiatry*, *74*(9), 911–923. https://doi.org/10.1001/jamapsychiatry.2017.2161.

Grant, B. F., Goldstein, R. B., Saha, T. D., Chou, S. P., Jung, J., Zhang, H., et al. (2015). Epidemiology of DSM-5 alcohol use disorder: Results from the National Epidemiologic Survey on Alcohol and Related Conditions III. *JAMA Psychiatry*, *72*(8), 757–766. https://doi.org/10.1001/jamapsychiatry.2015.0584.

Guadagno, A., Wong, T. P., & Walker, C. D. (2018). Morphological and functional changes in the preweaning basolateral amygdala induced by early chronic stress associate with anxiety and fear behavior in adult male, but not female rats. *Progress in Neuro-Psychopharmacology & Biological Psychiatry*, *81*, 25–37. https://doi.org/10.1016/j.pnpbp.2017.09.025.

Haller, M., & Chassin, L. (2014). Risk pathways among traumatic stress, posttraumatic stress disorder symptoms, and alcohol and drug problems: A test of four hypotheses. *Psychology of Addictive Behaviors*, *28*(3), 841–851. https://doi.org/10.1037/a0035878.

Hayes, J. P., Hayes, S. M., & Mikedis, A. M. (2012). Quantitative meta-analysis of neural activity in posttraumatic stress disorder. *Biology of Mood & Anxiety Disorders*, *2*, 9. https://doi.org/10.1186/2045-5380-2-9.

Helzer, J. E., Robins, L. N., & McEvoy, L. (1987). Post-traumatic stress disorder in the general population. Findings of the epidemiologic catchment area survey. *The New England Journal of Medicine*, *317*(26), 1630–1634. https://doi.org/10.1056/NEJM198712243172604.

Hirth, N., Meinhardt, M. W., Noori, H. R., Salgado, H., Torres-Ramirez, O., Uhrig, S., et al. (2016). Convergent evidence from alcohol-dependent humans and rats for a hyperdopaminergic state in protracted abstinence. *Proceedings of the National Academy of Sciences of the United States of America*, *113*(11), 3024–3029. https://doi.org/10.1073/pnas.1506012113.

Hoexter, M. Q., Fadel, G., Felicio, A. C., Calzavara, M. B., Batista, I. R., Reis, M. A., et al. (2012). Higher striatal dopamine transporter density in PTSD: An in vivo SPECT study with [(99m)Tc]TRODAT-1. *Psychopharmacology (Berl)*, *224*(2), 337–345. https://doi.org/10.1007/s00213-012-2755-4.

Hoflich, A., Michenthaler, P., Kasper, S., & Lanzenberger, R. (2019). Circuit mechanisms of reward, anhedonia, and depression. *The International Journal of Neuropsychopharmacology*, *22*(2), 105–118. https://doi.org/10.1093/ijnp/pyy081.

Hoge, C. W., Terhakopian, A., Castro, C. A., Messer, S. C., & Engel, C. C. (2007). Association of posttraumatic stress disorder with somatic symptoms, health care visits, and absenteeism among Iraq war veterans. *The American Journal of Psychiatry*, *164*(1), 150–153. https://doi.org/10.1176/ajp.2007.164.1.150.

Holmes, A., Fitzgerald, P. J., MacPherson, K. P., DeBrouse, L., Colacicco, G., Flynn, S. M., et al. (2012). Chronic alcohol remodels prefrontal neurons and disrupts NMDAR-mediated fear extinction encoding. *Nature Neuroscience*, *15*(10), 1359–1361. https://doi.org/10.1038/nn.3204.

Hu, S., Ide, J. S., Chao, H. H., Zhornitsky, S., Fischer, K. A., Wang, W., et al. (2018). Resting state functional connectivity of the amygdala and problem drinking in non-dependent alcohol drinkers. *Drug and Alcohol Dependence*, *185*, 173–180. https://doi.org/10.1016/j.drugalcdep.2017.11.026.

Huang, Y., Mohan, A., De Ridder, D., Sunaert, S., & Vanneste, S. (2018). The neural correlates of the unified percept of alcohol-related craving: A fMRI and EEG study. *Scientific Reports*, *8*(1), 923. https://doi.org/10.1038/s41598-017-18471-y.

Ishikawa, R., Uchida, C., Kitaoka, S., Furuyashiki, T., & Kida, S. (2019). Improvement of PTSD-like behavior by the forgetting effect of hippocampal neurogenesis enhancer memantine in a social defeat stress paradigm. *Molecular Brain*, *12*(1), 68. https://doi.org/10.1186/s13041-019-0488-6.

Janak, P. H., & Tye, K. M. (2015). From circuits to behaviour in the amygdala. *Nature*, *517*(7534), 284–292. https://doi.org/10.1038/nature14188.

Josselyn, S. A., Kohler, S., & Frankland, P. W. (2015). Finding the engram. *Nature Reviews. Neuroscience*, *16*(9), 521–534. https://doi.org/10.1038/nrn4000.

Kachadourian, L. K., Gandelman, E., Ralevski, E., & Petrakis, I. L. (2018). Suicidal ideation in military veterans with alcohol dependence and PTSD: The role of hostility. *The American Journal on Addictions*, *27*(2), 124–130. https://doi.org/10.1111/ajad.12688.

Karch, S., Keeser, D., Hummer, S., Paolini, M., Kirsch, V., Karali, T., et al. (2015). Modulation of craving related brain responses using real-time fMRI in patients with alcohol use disorder. *PLoS One*, *10*(7). https://doi.org/10.1371/journal.pone.0133034.

Karkhanis, A. N., Alexander, N. J., McCool, B. A., Weiner, J. L., & Jones, S. R. (2015). Chronic social isolation during adolescence augments catecholamine response to acute ethanol in the basolateral amygdala. *Synapse*, *69*(8), 385–395. https://doi.org/10.1002/syn.21826.

Karkhanis, A. N., & Al-Hasani, R. (2020). Dynorphin and its role in alcohol use disorder. *Brain Research*, *1735*. https://doi.org/10.1016/j.brainres.2020.146742.

Karkhanis, A. N., Leach, A. C., Yorgason, J. T., Uneri, A., Barth, S., Niere, F., et al. (2019). Chronic social isolation stress during peri-adolescence alters presynaptic dopamine terminal dynamics via augmentation in accumbal dopamine availability. *ACS Chemical Neuroscience*, *10*(4), 2033–2044. https://doi.org/10.1021/acschemneuro.8b00360.

Karkhanis, A. N., Locke, J. L., McCool, B. A., Weiner, J. L., & Jones, S. R. (2014). Social isolation rearing increases nucleus accumbens dopamine and norepinephrine responses to acute ethanol in adulthood. *Alcoholism, Clinical and Experimental Research*, *38*(11), 2770–2779. https://doi.org/10.1111/acer.12555.

Karkhanis, A. N., Rose, J. H., Weiner, J. L., & Jones, S. R. (2016). Early-life social isolation stress increases kappa opioid receptor responsiveness and downregulates the dopamine system. *Neuropsychopharmacology*, *41*(9), 2263–2274. https://doi.org/10.1038/npp.2016.21.

Kaufman, J., Yang, B. Z., Douglas-Palumberi, H., Crouse-Artus, M., Lipschitz, D., Krystal, J. H., et al. (2007). Genetic and environmental predictors of early alcohol use. *Biological Psychiatry*, *61*(11), 1228–1234. https://doi.org/10.1016/j.biopsych.2006.06.039.

Kaysen, D., Rosen, G., Bowman, M., & Resick, P. A. (2010). Duration of exposure and the dose-response model of PTSD. *Journal of Interpersonal Violence*, *25*(1), 63–74. https://doi.org/10.1177/0886260508329131.

Kenny, P. J., Hoyer, D., & Koob, G. F. (2018). Animal models of addiction and neuropsychiatric disorders and their role in drug discovery: Honoring the legacy of athina markou. *Biological Psychiatry*, *83*(11), 940–946. https://doi.org/10.1016/j.biopsych.2018.02.009.

Kessler, R. C., Heeringa, S. G., Stein, M. B., Colpe, L. J., Fullerton, C. S., Hwang, I., et al. (2014). Thirty-day prevalence of DSM-IV mental disorders among nondeployed soldiers in the US army: Results from the army study to assess risk and resilience in servicemembers (Army STARRS). *JAMA Psychiatry*, *71*(5), 504–513. https://doi.org/10.1001/jamapsychiatry.2014.28.

Kessler, R. C., Sonnega, A., Bromet, E., Hughes, M., & Nelson, C. B. (1995). Posttraumatic stress disorder in the National Comorbidity Survey. *Archives of General Psychiatry*, *52*(12), 1048–1060. Retrieved from http://www.ncbi.nlm.nih.gov/pubmed/7492257.

Keyes, K. M., Grant, B. F., & Hasin, D. S. (2008). Evidence for a closing gender gap in alcohol use, abuse, and dependence in the United States population. *Drug and Alcohol Dependence*, *93*(1–2), 21–29. https://doi.org/10.1016/j.drugalcdep.2007.08.017.

Khoury, L., Tang, Y. L., Bradley, B., Cubells, J. F., & Ressler, K. J. (2010). Substance use, childhood traumatic experience, and posttraumatic stress disorder in an urban civilian population. *Depression and Anxiety*, *27*(12), 1077–1086. https://doi.org/10.1002/da.20751.

Kilpatrick, D. G., Resnick, H. S., Milanak, M. E., Miller, M. W., Keyes, K. M., & Friedman, M. J. (2013). National estimates of exposure to traumatic events and PTSD prevalence using DSM-IV and DSM-5 criteria. *Journal of Traumatic Stress*, *26*(5), 537–547. https://doi.org/10.1002/jts.21848.

Kim, J., Ham, S., Hong, H., Moon, C., & Im, H. I. (2016). Brain reward circuits in morphine addiction. *Molecules and Cells*, *39*(9), 645–653. https://doi.org/10.14348/molcells.2016.0137.

Kim, D., Kim, C. Y., Koo, H., Heo, Y., & Cheon, K. (2017). A novel animal model simulating the beginning of combat exposure. *Neuroimmunomodulation*, *24*(4–5), 211–219. https://doi.org/10.1159/000481914.

Kim, J., Pignatelli, M., Xu, S., Itohara, S., & Tonegawa, S. (2016). Antagonistic negative and positive neurons of the basolateral amygdala. *Nature Neuroscience*, *19*, 1636–1646. https://doi.org/10.1038/nn.4414.

Kingston, R. E. F., Marel, C., & Mills, K. L. (2017). A systematic review of the prevalence of comorbid mental health disorders in people presenting for substance use treatment in Australia. *Drug and Alcohol Review*, *36*(4), 527–539. https://doi.org/10.1111/dar.12448.

Kisely, S., Abajobir, A. A., Mills, R., Strathearn, L., Clavarino, A., & Najman, J. M. (2018). Child maltreatment and mental health problems in adulthood: Birth cohort study. *The British Journal of Psychiatry*, *213*(6), 698–703. https://doi.org/10.1192/bjp.2018.207.

Knowland, D., & Lim, B. K. (2018). Circuit-based frameworks of depressive behaviors: The role of reward circuitry and beyond. *Pharmacology, Biochemistry, and Behavior*, *174*, 42–52. https://doi.org/10.1016/j.pbb.2017.12.010.

Knox, D., George, S. A., Fitzpatrick, C. J., Rabinak, C. A., Maren, S., & Liberzon, I. (2012). Single prolonged stress disrupts retention of extinguished fear in rats. *Learning & Memory*, *19*(2), 43–49. https://doi.org/10.1101/lm.024356.111.

Koch, S. B., van Zuiden, M., Nawijn, L., Frijling, J. L., Veltman, D. J., & Olff, M. (2016). Aberrant resting-state brain activity in posttraumatic stress disorder: A meta-analysis and systematic review. *Depression and Anxiety*, *33*(7), 592–605. https://doi.org/10.1002/da.22478.

Kolesar, T. A., Bilevicius, E., Wilson, A. D., & Kornelsen, J. (2019). Systematic review and meta-analyses of neural structural and functional differences in generalized anxiety disorder and healthy controls using magnetic resonance imaging. *NeuroImage: Clinical*, *24*. https://doi.org/10.1016/j.nicl.2019.102016.

Koob, G. F. (2008). A role for brain stress systems in addiction. *Neuron*, *59*(1), 11–34. https://doi.org/10.1016/j.neuron.2008.06.012.

Koob, G. F. (2013a). Negative reinforcement in drug addiction: The darkness within. *Current Opinion in Neurobiology*, *23*(4), 559–563. https://doi.org/10.1016/j.conb.2013.03.011.

Koob, G. F. (2013b). Theoretical frameworks and mechanistic aspects of alcohol addiction: Alcohol addiction as a reward deficit disorder. *Current Topics in Behavioral Neurosciences*, *13*, 3–30. https://doi.org/10.1007/7854_2011_129.

Koob, G. F. (2014). Neurocircuitry of alcohol addiction: Synthesis from animal models. *Handbook of Clinical Neurology*, *125*, 33–54. https://doi.org/10.1016/B978-0-444-62619-6.00003-3.

Koob, G. F., Buck, C. L., Cohen, A., Edwards, S., Park, P. E., Schlosburg, J. E., et al. (2014). Addiction as a stress surfeit disorder. *Neuropharmacology*, *76*(Pt. B), 370–382. https://doi.org/10.1016/j.neuropharm.2013.05.024.

Koob, G., Heinrichs, S. C., & Britton, K. (1998). Animal models of anxiety disorders. In A. F. Schatzberg, & C. B. Nemeroff (Eds.), *The American psychiatric press textbook of psychopharmacology* (2nd ed., pp. 133–144). Wasington, DC-London: American Psychiatric Press.

Koob, G. F., & Schulkin, J. (2019). Addiction and stress: An allostatic view. *Neuroscience and Biobehavioral Reviews, 106*, 245–262. https://doi.org/10.1016/j.neubiorev.2018.09.008.

Koob, G. F., & Volkow, N. D. (2010). Neurocircuitry of addiction. *Neuropsychopharmacology, 35*(1), 217–238. https://doi.org/10.1038/npp.2009.110.

Koob, G. F., & Volkow, N. D. (2016). Neurobiology of addiction: A neurocircuitry analysis. *Lancet Psychiatry, 3*(8), 760–773. https://doi.org/10.1016/S2215-0366(16)00104-8.

Kozel, F. A. (2018). Clinical repetitive transcranial magnetic stimulation for posttraumatic stress disorder, generalized anxiety disorder, and bipolar disorder. *The Psychiatric Clinics of North America, 41*(3), 433–446. https://doi.org/10.1016/j.psc.2018.04.007.

Kunimatsu, A., Yasaka, K., Akai, H., Kunimatsu, N., & Abe, O. (2020). MRI findings in posttraumatic stress disorder. *Journal of Magnetic Resonance Imaging, 52*, 380–396. https://doi.org/10.1002/jmri.26929.

Lalumiere, R. T. (2014). Optogenetic dissection of amygdala functioning. *Frontiers in Behavioral Neuroscience, 8*, 107. https://doi.org/10.3389/fnbeh.2014.00107.

Lamb, D. H. (1979). On the distinction between physical and psychological stressors. *Motivation and Emotion, 3*, 51–61.

Lanius, R. A., Williamson, P. C., Densmore, M., Boksman, K., Gupta, M. A., Neufeld, R. W., et al. (2001). Neural correlates of traumatic memories in posttraumatic stress disorder: A functional MRI investigation. *American Journal of Psychiatry, 158*(11), 1920–1922. https://doi.org/10.1176/appi.ajp.158.11.1920.

Lanuza, E., Belekhova, M., Martinez-Marcos, A., Font, C., & Martinez-Garcia, F. (1998). Identification of the reptilian basolateral amygdala: An anatomical investigation of the afferents to the posterior dorsal ventricular ridge of the lizard Podarcis hispanica. *The European Journal of Neuroscience, 10*(11), 3517–3534. https://doi.org/10.1046/j.1460-9568.1998.00363.x.

LeDoux, J. E. (2012). Evolution of human emotion: A view through fear. *Progress in Brain Research, 195*, 431–442. https://doi.org/10.1016/B978-0-444-53860-4.00021-0.

Leeies, M., Pagura, J., Sareen, J., & Bolton, J. M. (2010). The use of alcohol and drugs to self-medicate symptoms of posttraumatic stress disorder. *Depression and Anxiety, 27*(8), 731–736. https://doi.org/10.1002/da.20677.

Lindquist, K. A., Satpute, A. B., Wager, T. D., Weber, J., & Barrett, L. F. (2016). The brain basis of positive and negative affect: Evidence from a meta-analysis of the human neuroimaging literature. *Cerebral Cortex, 26*(5), 1910–1922. https://doi.org/10.1093/cercor/bhv001.

Linhartova, P., Latalova, A., Kosa, B., Kasparek, T., Schmahl, C., & Paret, C. (2019). fMRI neurofeedback in emotion regulation: A literature review. *NeuroImage, 193*, 75–92. https://doi.org/10.1016/j.neuroimage.2019.03.011.

Lisboa, S. F., Niraula, A., Resstel, L. B., Guimaraes, F. S., Godbout, J. P., & Sheridan, J. F. (2018). Repeated social defeat-induced neuroinflammation, anxiety-like behavior and resistance to fear extinction were attenuated by the cannabinoid receptor agonist WIN55,212-2. *Neuropsychopharmacology, 43*(9), 1924–1933. https://doi.org/10.1038/s41386-018-0064-2.

Lisieski, M. J., Eagle, A. L., Conti, A. C., Liberzon, I., & Perrine, S. A. (2018). Single-prolonged stress: A review of two decades of progress in a rodent model of post-traumatic stress disorder. *Frontiers in Psychiatry, 9*, 196. https://doi.org/10.3389/fpsyt.2018.00196.

Liu, H., Patki, G., Salvi, A., Kelly, M., & Salim, S. (2018). Behavioral effects of early life maternal trauma witness in rats. *Progress in Neuro-Psychopharmacology & Biological Psychiatry, 81*, 80–87. https://doi.org/10.1016/j.pnpbp.2017.10.013.

Locci, A., & Pinna, G. (2019). Social isolation as a promising animal model of PTSD comorbid suicide: Neurosteroids and cannabinoids as possible treatment options. *Progress in Neuro-Psychopharmacology & Biological Psychiatry*, *92*, 243–259. https://doi.org/10.1016/j.pnpbp.2018.12.014.
Longo, M. S. C., Vilete, L. M. P., Figueira, I., Quintana, M. I., Mello, M. F., Bressan, R. A., et al. (2020). Comorbidity in post-traumatic stress disorder: A population-based study from the two largest cities in Brazil. *Journal of Affective Disorders*, *263*, 715–721. https://doi.org/10.1016/j.jad.2019.11.051.
Lopez, M. F., Doremus-Fitzwater, T. L., & Becker, H. C. (2011). Chronic social isolation and chronic variable stress during early development induce later elevated ethanol intake in adult C57BL/6J mice. *Alcohol*, *45*(4), 355–364. https://doi.org/10.1016/j.alcohol.2010.08.017.
MacAskill, A. F., Little, J. P., Cassel, J. M., & Carter, A. G. (2012). Subcellular connectivity underlies pathway-specific signaling in the nucleus accumbens. *Nature Neuroscience*, *15*(12), 1624–1626. https://doi.org/10.1038/nn.3254.
Macedo, G. C., Morita, G. M., Domingues, L. P., Favoretto, C. A., Suchecki, D., & Quadros, I. M. H. (2018). Consequences of continuous social defeat stress on anxiety- and depressive-like behaviors and ethanol reward in mice. *Hormones and Behavior*, *97*, 154–161. https://doi.org/10.1016/j.yhbeh.2017.10.007.
Malikowska-Racia, N., & Salat, K. (2019). Recent advances in the neurobiology of posttraumatic stress disorder: A review of possible mechanisms underlying an effective pharmacotherapy. *Pharmacological Research*, *142*, 30–49. https://doi.org/10.1016/j.phrs.2019.02.001.
Manjoch, H., Vainer, E., Matar, M., Ifergane, G., Zohar, J., Kaplan, Z., et al. (2016). Predator-scent stress, ethanol consumption and the opioid system in an animal model of PTSD. *Behavioural Brain Research*, *306*, 91–105. https://doi.org/10.1016/j.bbr.2016.03.009.
Manz, K. M., Levine, W. A., Seckler, J. C., Iskander, A. N., & Reich, C. G. (2018). A novel adolescent chronic social defeat model: Reverse-Resident-Intruder Paradigm (rRIP) in male rats. *Stress*, *21*(2), 169–178. https://doi.org/10.1080/10253890.2017.1423285.
Maren, S., & Fanselow, M. S. (1996). The amygdala and fear conditioning: Has the nut been cracked? *Neuron*, *16*(2), 237–240. https://doi.org/10.1016/s0896-6273(00)80041-0.
McCool, B. A., Christian, D. T., Diaz, M. R., & Lack, A. K. (2010). Glutamate plasticity in the drunken amygdala: The making of an anxious synapse. *International Review of Neurobiology*, *91*, 205–233. https://doi.org/10.1016/S0074-7742(10)91007-6.
McCool, B. A., & Chappell, A. M. (2009). Early social isolation in male Long-Evans rats alters both appetitive and consummatory behaviors expressed during operant ethanol self-administration. *Alcoholism: Clinical and Experimental Research*, *33*, 273–282. https://doi.org/10.1111/j.1530-0277.2008.00830.x.
McGinnis, M. M., Parrish, B. C., Chappell, A. M., Alexander, N. J., & McCool, B. A. (2020). Chronic ethanol differentially modulates glutamate release from dorsal and ventral prefrontal cortical inputs onto rat basolateral amygdala principal neurons. *eNeuro*, *7*(2), 1–17. https://doi.org/10.1523/ENEURO.0132-19.2019.
McGinnis, M. M., Parrish, B. C., & McCool, B. A. (2020). Withdrawal from chronic ethanol exposure increases postsynaptic glutamate function of insular cortex projections to the rat basolateral amygdala. *Neuropharmacology*, *172*. https://doi.org/10.1016/j.neuropharm.2020.108129.
McGuire, J., Herman, J. P., Horn, P. S., Sallee, F. R., & Sah, R. (2010). Enhanced fear recall and emotional arousal in rats recovering from chronic variable stress. *Physiology & Behavior*, *101*(4), 474–482. https://doi.org/10.1016/j.physbeh.2010.07.013.

McHugh, R. K., Votaw, V. R., Sugarman, D. E., & Greenfield, S. F. (2018). Sex and gender differences in substance use disorders. *Clinical Psychology Review*, *66*, 12–23. https://doi.org/10.1016/j.cpr.2017.10.012.

Meyer, E. M., Long, V., Fanselow, M. S., & Spigelman, I. (2013). Stress increases voluntary alcohol intake, but does not alter established drinking habits in a rat model of post-traumatic stress disorder. *Alcoholism, Clinical and Experimental Research*, *37*(4), 566–574. https://doi.org/10.1111/acer.12012.

Middlebrooks, J. S., & Audage, N. C. (2008). The effects of childhood stress on health across lifespan. In *National center for injury prevention and control of the centers for disease control and prevention*. Atlanta, GA: National Center for Injury Prevention and Control (U.S.). Centers for Disease Control and Prevention.

Mills, R., Scott, J., Alati, R., O'Callaghan, M., Najman, J. M., & Strathearn, L. (2013). Child maltreatment and adolescent mental health problems in a large birth cohort. *Child Abuse & Neglect*, *37*(5), 292–302. https://doi.org/10.1016/j.chiabu.2012.11.008.

Mills, K. L., Teesson, M., Ross, J., & Peters, L. (2006). Trauma, PTSD, and substance use disorders: Findings from the Australian National Survey of Mental Health and Well-Being. *The American Journal of Psychiatry*, *163*(4), 652–658. https://doi.org/10.1176/appi.ajp.163.4.652.

Misaki, M., Phillips, R., Zotev, V., Wong, C. K., Wurfel, B. E., Krueger, F., et al. (2018). Real-time fMRI amygdala neurofeedback positive emotional training normalized resting-state functional connectivity in combat veterans with and without PTSD: A connectome-wide investigation. *NeuroImage: Clinical*, *20*, 543–555. https://doi.org/10.1016/j.nicl.2018.08.025.

Morales, M., & Margolis, E. B. (2017). Ventral tegmental area: Cellular heterogeneity, connectivity and behaviour. *Nature Reviews. Neuroscience*, *18*(2), 73–85. https://doi.org/10.1038/nrn.2016.165.

Moreno, N., & Gonzalez, A. (2007). Evolution of the amygdaloid complex in vertebrates, with special reference to the anamnio-amniotic transition. *Journal of Anatomy*, *211*(2), 151–163. https://doi.org/10.1111/j.1469-7580.2007.00780.x.

Muench, C., Charlet, K., Balderston, N. L., Grillon, C., Heilig, M., Cortes, C. R., et al. (2019). Fear conditioning and extinction in alcohol dependence: Evidence for abnormal amygdala reactivity. *Addiction Biology*, e12835. https://doi.org/10.1111/adb.12835.

Muller-Oehring, E. M., Jung, Y. C., Pfefferbaum, A., Sullivan, E. V., & Schulte, T. (2015). The resting brain of alcoholics. *Cerebral Cortex*, *25*(11), 4155–4168. https://doi.org/10.1093/cercor/bhu134.

Najavits, L. M., Weiss, R. D., & Shaw, S. R. (1999). A clinical profile of women with post-traumatic stress disorder and substance dependence. *Psychology of Addictive Behaviors*, *13*, 98–104.

Namburi, P., Al-Hasani, R., Calhoon, G. G., Bruchas, M. R., & Tye, K. M. (2016). Architectural representation of valence in the limbic system. *Neuropsychopharmacology*, *41*(7), 1697–1715. https://doi.org/10.1038/npp.2015.358.

Narayanan, V., Heiming, R. S., Jansen, F., Lesting, J., Sachser, N., Pape, H. C., et al. (2011). Social defeat: Impact on fear extinction and amygdala-prefrontal cortical theta synchrony in 5-HTT deficient mice. *PLoS One*, *6*(7), e22600. https://doi.org/10.1371/journal.pone.0022600.

Nathan, S. V., Griffith, Q. K., McReynolds, J. R., Hahn, E. L., & Roozendaal, B. (2004). Basolateral amygdala interacts with other brain regions in regulating glucocorticoid effects on different memory functions. *Annals of the New York Academy of Sciences*, *1032*, 179–182. https://doi.org/10.1196/annals.1314.015.

Nawijn, L., van Zuiden, M., Frijling, J. L., Koch, S. B., Veltman, D. J., & Olff, M. (2015). Reward functioning in PTSD: A systematic review exploring the mechanisms underlying anhedonia. *Neuroscience and Biobehavioral Reviews*, *51*, 189–204. https://doi.org/10.1016/j.neubiorev.2015.01.019.

Nemeroff, C. B. (2016). Paradise lost: The neurobiological and clinical consequences of child abuse and neglect. *Neuron, 89*(5), 892–909. https://doi.org/10.1016/j.neuron.2016.01.019.

Neumeister, P., Feldker, K., Heitmann, C. Y., Buff, C., Brinkmann, L., Bruchmann, M., et al. (2018). Specific amygdala response to masked fearful faces in post-traumatic stress relative to other anxiety disorders. *Psychological Medicine, 48*(7), 1209–1217. https://doi.org/10.1017/S0033291717002513.

Newman, E. L., Albrechet-Souza, L., Andrew, P. M., Auld, J. G., Burk, K. C., Hwa, L. S., et al. (2018). Persistent escalation of alcohol consumption by mice exposed to brief episodes of social defeat stress: Suppression by CRF-R1 antagonism. *Psychopharmacology, 235*(6), 1807–1820. https://doi.org/10.1007/s00213-018-4905-9.

Newman, E. L., Leonard, M. Z., Arena, D. T., de Almeida, R. M. M., & Miczek, K. A. (2018). Social defeat stress and escalation of cocaine and alcohol consumption: Focus on CRF. *Neurobiology of Stress, 9*, 151–165. https://doi.org/10.1016/j.ynstr.2018.09.007.

NIAAA. (2020). Retrieved from https://www.niaaa.nih.gov/alcohol-health/overview-alcohol-consumption/alcohol-use-disorders.

Nicholson, A. A., Densmore, M., Frewen, P. A., Theberge, J., Neufeld, R. W., McKinnon, M. C., et al. (2015). The dissociative subtype of posttraumatic stress disorder: Unique resting-state functional connectivity of basolateral and centromedial amygdala complexes. *Neuropsychopharmacology, 40*(10), 2317–2326. https://doi.org/10.1038/npp.2015.79.

Nicholson, A. A., Rabellino, D., Densmore, M., Frewen, P. A., Paret, C., Kluetsch, R., et al. (2017). The neurobiology of emotion regulation in posttraumatic stress disorder: Amygdala downregulation via real-time fMRI neurofeedback. *Human Brain Mapping, 38*(1), 541–560. https://doi.org/10.1002/hbm.23402.

Nikolaus, S., Antke, C., Beu, M., & Muller, H. W. (2010). Cortical GABA, striatal dopamine and midbrain serotonin as the key players in compulsive and anxiety disorders—Results from in vivo imaging studies. *Reviews in the Neurosciences, 21*(2), 119–139. https://doi.org/10.1515/revneuro.2010.21.2.119.

Nikolova, Y. S., & Hariri, A. R. (2012). Neural responses to threat and reward interact to predict stress-related problem drinking: A novel protective role of the amygdala. *Biology of Mood & Anxiety Disorders, 2*, 19. https://doi.org/10.1186/2045-5380-2-19.

Nikolova, Y. S., Knodt, A. R., Radtke, S. R., & Hariri, A. R. (2016). Divergent responses of the amygdala and ventral striatum predict stress-related problem drinking in young adults: Possible differential markers of affective and impulsive pathways of risk for alcohol use disorder. *Molecular Psychiatry, 21*(3), 348–356. https://doi.org/10.1038/mp.2015.85.

Nitschke, J. B., Sarinopoulos, I., Oathes, D. J., Johnstone, T., Whalen, P. J., Davidson, R. J., et al. (2009). Anticipatory activation in the amygdala and anterior cingulate in generalized anxiety disorder and prediction of treatment response. *The American Journal of Psychiatry, 166*(3), 302–310. https://doi.org/10.1176/appi.ajp.2008.07101682.

Norman, S. B., Haller, M., Hamblen, J. L., Southwick, S. M., & Pietrzak, R. H. (2018). The burden of co-occurring alcohol use disorder and PTSD in U.S. Military veterans: Comorbidities, functioning, and suicidality. *Psychology of Addictive Behaviors, 32*(2), 224–229. https://doi.org/10.1037/adb0000348.

Norman, S. B., Tate, S. R., Anderson, K. G., & Brown, S. A. (2007). Do trauma history and PTSD symptoms influence addiction relapse context? *Drug and Alcohol Dependence, 90*(1), 89–96. https://doi.org/10.1016/j.drugalcdep.2007.03.002.

Norrholm, S. D., & Jovanovic, T. (2018). Fear processing, psychophysiology, and PTSD. *Harvard Review of Psychiatry, 26*(3), 129–141. https://doi.org/10.1097/HRP.0000000000000189.

O'Daly, O. G., Trick, L., Scaife, J., Marshall, J., Ball, D., Phillips, M. L., et al. (2012). Withdrawal-associated increases and decreases in functional neural connectivity associated with altered emotional regulation in alcoholism. *Neuropsychopharmacology*, *37*(10), 2267–2276. https://doi.org/10.1038/npp.2012.77.

Ouimette, P., Goodwin, E., & Brown, P. J. (2006). Health and well being of substance use disorder patients with and without posttraumatic stress disorder. *Addictive Behaviors*, *31*(8), 1415–1423. https://doi.org/10.1016/j.addbeh.2005.11.010.

Pabba, M. (2013). Evolutionary development of the amygdaloid complex. *Frontiers in Neuroanatomy*, 7, 27. https://doi.org/10.3389/fnana.2013.00027.

Patel, D., Kas, M. J., Chattarji, S., & Buwalda, B. (2019). Rodent models of social stress and neuronal plasticity: Relevance to depressive-like disorders. *Behavioural Brain Research*, *369*. https://doi.org/10.1016/j.bbr.2019.111900.

Patel, R., Spreng, R. N., Shin, L. M., & Girard, T. A. (2012). Neurocircuitry models of posttraumatic stress disorder and beyond: A meta-analysis of functional neuroimaging studies. *Neuroscience and Biobehavioral Reviews*, *36*(9), 2130–2142. https://doi.org/10.1016/j.neubiorev.2012.06.003.

Paxinos, G. (2003). *Human nervous system*. San Diego: Academic Press.

Peay, D. N., Saribekyan, H. M., Parada, P. A., Hanson, E. M., Badaruddin, B. S., Judd, J. M., et al. (2020). Chronic unpredictable intermittent restraint stress disrupts spatial memory in male, but not female rats. *Behavioural Brain Research*, *383*. https://doi.org/10.1016/j.bbr.2020.112519.

Perez Benitez, C. I., Zlotnick, C., Stout, R. I., Lou, F., Dyck, I., Weisberg, R., et al. (2012). A 5-year longitudinal study of posttraumatic stress disorder in primary care patients. *Psychopathology*, *45*(5), 286–293. https://doi.org/10.1159/000331595.

Peters, S. K., Dunlop, K., & Downar, J. (2016). Cortico-striatal-thalamic loop circuits of the salience network: A central pathway in psychiatric disease and treatment. *Frontiers in Systems Neuroscience*, *10*, 104. https://doi.org/10.3389/fnsys.2016.00104.

Peters, J., Kalivas, P. W., & Quirk, G. J. (2009). Extinction circuits for fear and addiction overlap in prefrontal cortex. *Learning & Memory*, *16*(5), 279–288. https://doi.org/10.1101/lm.1041309.

Phan, K. L., Britton, J. C., Taylor, S. F., Fig, L. M., & Liberzon, I. (2006). Corticolimbic blood flow during nontraumatic emotional processing in posttraumatic stress disorder. *Archives of General Psychiatry*, *63*(2), 184–192. https://doi.org/10.1001/archpsyc.63.2.184.

Philip, N. S., Barredo, J., van 't Wout-Frank, M., Tyrka, A. R., Price, L. H., & Carpenter, L. L. (2018). Network mechanisms of clinical response to transcranial magnetic stimulation in posttraumatic stress disorder and major depressive disorder. *Biological Psychiatry*, *83*(3), 263–272. https://doi.org/10.1016/j.biopsych.2017.07.021.

Philip, N. S., Sorensen, D. O., McCalley, D. M., & Hanlon, C. A. (2020). Non-invasive brain stimulation for alcohol use disorders: State of the art and future directions. *Neurotherapeutics*, *17*(1), 116–126. https://doi.org/10.1007/s13311-019-00780-x.

Pi, G., Gao, D., Wu, D., Wang, Y., Lei, H., Zeng, W., et al. (2020). Posterior basolateral amygdala to ventral hippocampal CA1 drives approach behaviour to exert an anxiolytic effect. *Nature Communications*, *11*(1), 183. https://doi.org/10.1038/s41467-019-13919-3.

Pietrzak, R. H., Goldstein, R. B., Southwick, S. M., & Grant, B. F. (2011). Prevalence and axis I comorbidity of full and partial posttraumatic stress disorder in the United States: Results from Wave 2 of the National Epidemiologic Survey on Alcohol and Related Conditions. *Journal of Anxiety Disorders*, *25*(3), 456–465. https://doi.org/10.1016/j.janxdis.2010.11.010.

Piggott, V. M., Bosse, K. E., Lisieski, M. J., Strader, J. A., Stanley, J. A., Conti, A. C., et al. (2019). Single-prolonged stress impairs prefrontal cortex control of amygdala and striatum in rats. *Frontiers in Behavioral Neuroscience*, *13*, 18. https://doi.org/10.3389/fnbeh.2019.00018.

Pinna, G. (2019). Animal models of PTSD: The socially isolated mouse and the biomarker role of allopregnanolone. *Frontiers in Behavioral Neuroscience, 13*, 114. https://doi.org/10.3389/fnbeh.2019.00114.

Prasad, A., Chaichi, A., Kelley, D. P., Francis, J., & Gartia, M. R. (2019). Current and future functional imagining techniques for post-traumatic stress disorder. *Royal Society of Chemistry Advances, 9*, 24568.

Preuss, U. W., Gouzoulis-Mayfrank, E., Havemann-Reinecke, U., Schafer, I., Beutel, M., Hoch, E., et al. (2018). Psychiatric comorbidity in alcohol use disorders: Results from the German S3 guidelines. *European Archives of Psychiatry and Clinical Neuroscience, 268*(3), 219–229. https://doi.org/10.1007/s00406-017-0801-2.

Rabinak, C. A., Angstadt, M., Welsh, R. C., Kenndy, A. E., Lyubkin, M., Martis, B., et al. (2011). Altered amygdala resting-state functional connectivity in post-traumatic stress disorder. *Frontiers in Psychiatry, 2*, 62. https://doi.org/10.3389/fpsyt.2011.00062.

Rajbhandari, A. K., Gonzalez, S. T., & Fanselow, M. S. (2018). Stress-enhanced fear learning, a robust rodent model of post-traumatic stress disorder. *Journal of Visualized Experiments, 140*, 58306. https://doi.org/10.3791/58306.

Rau, A. R., Chappell, A. M., Butler, T. R., Ariwodola, O. J., & Weiner, J. L. (2015). Increased basolateral amygdala pyramidal cell excitability may contribute to the anxiogenic phenotype induced by chronic early-life stress. *The Journal of Neuroscience, 35*(26), 9730–9740. https://doi.org/10.1523/JNEUROSCI.0384-15.2015.

Rauch, S. L., Whalen, P. J., Shin, L. M., McInerney, S. C., Macklin, M. L., Lasko, N. B., et al. (2000). Exaggerated amygdala response to masked facial stimuli in posttraumatic stress disorder: A functional MRI study. *Biological Psychiatry, 47*(9), 769–776. https://doi.org/10.1016/s0006-3223(00)00828-3.

Read, J. P., Brown, P. J., & Kahler, C. W. (2004). Substance use and posttraumatic stress disorders: Symptom interplay and effects on outcome. *Addictive Behaviors, 29*(8), 1665–1672. https://doi.org/10.1016/j.addbeh.2004.02.061.

Ressler, R. L., & Maren, S. (2019). Synaptic encoding of fear memories in the amygdala. *Current Opinion in Neurobiology, 54*, 54–59. https://doi.org/10.1016/j.conb.2018.08.012.

Richardson, J. D., Ketcheson, F., King, L., Shnaider, P., Marlborough, M., Thompson, A., et al. (2017). Psychiatric comorbidity pattern in treatment-seeking veterans. *Psychiatry Research, 258*, 488–493. https://doi.org/10.1016/j.psychres.2017.08.091.

Richter-Levin, G. (1998). Acute and long-term behavioral correlates of underwater trauma—Potential relevance to stress and post-stress syndromes. *Psychiatry Research, 79*(1), 73–83. https://doi.org/10.1016/s0165-1781(98)00030-4.

Richter-Levin, G., Stork, O., & Schmidt, M. V. (2019). Animal models of PTSD: A challenge to be met. *Molecular Psychiatry, 24*(8), 1135–1156. https://doi.org/10.1038/s41380-018-0272-5.

Robbins, T. W. (1998). Homology in behavioural pharmacology: An approach to animal models of human cognition. *Behavioural Pharmacology, 9*(7), 509–519. https://doi.org/10.1097/00008877-199811000-00005.

Robinson, T. E., & Berridge, K. C. (1993). The neural basis of drug craving: An incentive-sensitization theory of addiction. *Brain Research. Brain Research Reviews, 18*(3), 247–291. https://doi.org/10.1016/0165-0173(93)90013-p.

Rompala, G. R., Simons, A., Kihle, B., & Homanics, G. E. (2018). Paternal preconception chronic variable stress confers attenuated ethanol drinking behavior selectively to male offspring in a pre-stress environment dependent manner. *Frontiers in Behavioral Neuroscience, 12*, 257. https://doi.org/10.3389/fnbeh.2018.00257.

Roozendaal, B., Griffith, Q. K., Buranday, J., De Quervain, D. J., & McGaugh, J. L. (2003). The hippocampus mediates glucocorticoid-induced impairment of spatial memory retrieval: Dependence on the basolateral amygdala. *Proceedings of the National Academy of Sciences of the United States of America, 100*(3), 1328–1333. https://doi.org/10.1073/pnas.0337480100.

Rosenkranz, J. A., Venheim, E. R., & Padival, M. (2010). Chronic stress causes amygdala hyperexcitability in rodents. *Biological Psychiatry*, *67*(12), 1128–1136. https://doi.org/10.1016/j.biopsych.2010.02.008.

Sailer, U., Robinson, S., Fischmeister, F. P., Konig, D., Oppenauer, C., Lueger-Schuster, B., et al. (2008). Altered reward processing in the nucleus accumbens and mesial prefrontal cortex of patients with posttraumatic stress disorder. *Neuropsychologia*, *46*(11), 2836–2844. https://doi.org/10.1016/j.neuropsychologia.2008.05.022.

Sarter, M., & Bruno, J. P. (2002). Animal models in biological psychiatry. In H. A. H. D'Haenen, J. A. den Boer, & P. Winer (Eds.), *Vol. 1. Biological psychiatry* (pp. 37–44). Chicester: Wiley.

Schacht, J. P., Anton, R. F., & Myrick, H. (2013). Functional neuroimaging studies of alcohol cue reactivity: A quantitative meta-analysis and systematic review. *Addiction Biology*, *18*(1), 121–133. https://doi.org/10.1111/j.1369-1600.2012.00464.x.

Seal, K. H., Cohen, G., Waldrop, A., Cohen, B. E., Maguen, S., & Ren, L. (2011). Substance use disorders in Iraq and Afghanistan veterans in VA healthcare, 2001–2010: Implications for screening, diagnosis and treatment. *Drug and Alcohol Dependence*, *116*(1–3), 93–101. https://doi.org/10.1016/j.drugalcdep.2010.11.027.

Seeley, W. W. (2019). The salience network: A neural system for perceiving and responding to homeostatic demands. *The Journal of Neuroscience*, *39*(50), 9878–9882. https://doi.org/10.1523/JNEUROSCI.1138-17.2019.

Segev, A., & Akirav, I. (2016). Cannabinoids and glucocorticoids in the basolateral amygdala modulate hippocampal-accumbens plasticity after stress. *Neuropsychopharmacology*, *41*(4), 1066–1079. https://doi.org/10.1038/npp.2015.238.

Seo, D., Lacadie, C. M., & Sinha, R. (2016). Neural correlates and connectivity underlying stress-related impulse control difficulties in alcoholism. *Alcoholism, Clinical and Experimental Research*, *40*(9), 1884–1894. https://doi.org/10.1111/acer.13166.

Seo, D., Lacadie, C. M., Tuit, K., Hong, K. I., Constable, R. T., & Sinha, R. (2013). Disrupted ventromedial prefrontal function, alcohol craving, and subsequent relapse risk. *JAMA Psychiatry*, *70*(7), 727–739. https://doi.org/10.1001/jamapsychiatry.2013.762.

Sequeira-Cordero, A., Salas-Bastos, A., Fornaguera, J., & Brenes, J. C. (2019). Behavioural characterisation of chronic unpredictable stress based on ethologically relevant paradigms in rats. *Scientific Reports*, *9*(1). https://doi.org/10.1038/s41598-019-53624-1.

Shalev, A., Liberzon, I., & Marmar, C. (2017). Post-traumatic stress disorder. *New England Journal of Medicine*, *376*(25), 2459–2469. https://doi.org/10.1056/NEJMra1612499.

Sharp, B. M. (2017). Basolateral amygdala and stress-induced hyperexcitability affect motivated behaviors and addiction. *Translational Psychiatry*, 7(8), e1194. https://doi.org/10.1038/tp.2017.161.

Shen, C. J., Zheng, D., Li, K. X., Yang, J. M., Pan, H. Q., Yu, X. D., et al. (2019). Cannabinoid CB1 receptors in the amygdalar cholecystokinin glutamatergic afferents to nucleus accumbens modulate depressive-like behavior. *Nature Medicine*, *25*(2), 337–349. https://doi.org/10.1038/s41591-018-0299-9.

Sher, K. J. (1991). Psychological characteristics of children of alcoholics. Overview of research methods and findings. *Recent Developments in Alcoholism*, *9*, 301–326. Retrieved from https://www.ncbi.nlm.nih.gov/pubmed/1758989.

Shin, L. M., Wright, C. I., Cannistraro, P. A., Wedig, M. M., McMullin, K., Martis, B., et al. (2005). A functional magnetic resonance imaging study of amygdala and medial prefrontal cortex responses to overtly presented fearful faces in posttraumatic stress disorder. *Archives of General Psychiatry*, *62*(3), 273–281. https://doi.org/10.1001/archpsyc.62.3.273.

Shorter, D., Hsieh, J., & Kosten, T. R. (2015). Pharmacologic management of comorbid post-traumatic stress disorder and addictions. *The American Journal on Addictions*, *24*(8), 705–712. https://doi.org/10.1111/ajad.12306.

Sial, O. K., Warren, B. L., Alcantara, L. F., Parise, E. M., & Bolanos-Guzman, C. A. (2016). Vicarious social defeat stress: Bridging the gap between physical and emotional stress. *Journal of Neuroscience Methods*, *258*, 94–103. https://doi.org/10.1016/j.jneumeth.2015.10.012.

Sinha, R. (2012). How does stress lead to risk of alcohol relapse? *Alcohol Research: Current Reviews*, *34*(4), 432–440. Retrieved from https://www.ncbi.nlm.nih.gov/pubmed/23584109.

Sinha, R., Lacadie, C., Skudlarski, P., & Wexler, B. E. (2004). Neural circuits underlying emotional distress in humans. *Annals of the New York Academy of Sciences*, *1032*, 254–257. https://doi.org/10.1196/annals.1314.032.

Sinha, R., & Li, C. S. (2007). Imaging stress- and cue-induced drug and alcohol craving: Association with relapse and clinical implications. *Drug and Alcohol Review*, *26*(1), 25–31. https://doi.org/10.1080/09595230601036960.

Skelly, M. J., Chappell, A. E., Carter, E., & Weiner, J. L. (2015). Adolescent social isolation increases anxiety-like behavior and ethanol intake and impairs fear extinction in adulthood: Possible role of disrupted noradrenergic signaling. *Neuropharmacology*, *97*, 149–159. https://doi.org/10.1016/j.neuropharm.2015.05.025.

Smith, N. D. L., & Cottler, L. B. (2018). The epidemiology of post-traumatic stress disorder and alcohol use disorder. *Alcohol Research: Current Reviews*, *39*(2), 113–120. Retrieved from https://www.ncbi.nlm.nih.gov/pubmed/31198651.

Somohano, V. C., Rehder, K. L., Dingle, T., Shank, T., & Bowen, S. (2019). PTSD symptom clusters and craving differs by primary drug of choice. *Journal of Dual Diagnosis*, *15*(4), 233–242. https://doi.org/10.1080/15504263.2019.1637039.

Souza, R. R., Noble, L. J., & McIntyre, C. K. (2017). Using the single prolonged stress model to examine the pathophysiology of PTSD. *Frontiers in Pharmacology*, *8*, 615. https://doi.org/10.3389/fphar.2017.00615.

Sripada, R. K., King, A. P., Garfinkel, S. N., Wang, X., Sripada, C. S., Welsh, R. C., et al. (2012). Altered resting-state amygdala functional connectivity in men with posttraumatic stress disorder. *Journal of Psychiatry & Neuroscience*, *37*(4), 241–249. https://doi.org/10.1503/jpn.110069.

Sripada, R. K., King, A. P., Welsh, R. C., Garfinkel, S. N., Wang, X., Sripada, C. S., et al. (2012). Neural dysregulation in posttraumatic stress disorder: Evidence for disrupted equilibrium between salience and default mode brain networks. *Psychosomatic Medicine*, 74(9), 904–911. https://doi.org/10.1097/PSY.0b013e318273bf33.

Stefanovics, E. A., & Rosenheck, R. A. (2019). Predictors of post-discharge suicide attempt among veterans receiving specialized intensive treatment for posttraumatic stress disorder. *The Journal of Clinical Psychiatry*, *80*(5), 19m12745. https://doi.org/10.4088/JCP.19m12745.

Stein, M. B., Campbell-Sills, L., Gelernter, J., He, F., Heeringa, S. G., Nock, M. K., et al. (2017). Alcohol misuse and co-occurring mental disorders among new soldiers in the U.S. army. *Alcoholism, Clinical and Experimental Research*, *41*(1), 139–148. https://doi.org/10.1111/acer.13269.

Stelly, C. E., Pomrenze, M. B., Cook, J. B., & Morikawa, H. (2016). Repeated social defeat stress enhances glutamatergic synaptic plasticity in the VTA and cocaine place conditioning. *eLife*, *5*, e15448. https://doi.org/10.7554/eLife.15448.

Stephens, D. N., Ripley, T. L., Borlikova, G., Schubert, M., Albrecht, D., Hogarth, L., et al. (2005). Repeated ethanol exposure and withdrawal impairs human fear conditioning and depresses long-term potentiation in rat amygdala and hippocampus. *Biological Psychiatry*, *58*(5), 392–400. https://doi.org/10.1016/j.biopsych.2005.04.025.

Straus, E., Haller, M., Lyons, R. C., & Norman, S. B. (2018). Functional and psychiatric correlates of comorbid post-traumatic stress disorder and alcohol use disorder. *Alcohol Research: Current Reviews*, *39*(2), 121–129. Retrieved from https://www.ncbi.nlm.nih.gov/pubmed/31198652.

Stringfield, S. J., Higginbotham, J. A., & Fuchs, R. A. (2016). Requisite role of basolateral amygdala glucocorticoid receptor stimulation in drug context-induced cocaine-seeking behavior. *The International Journal of Neuropsychopharmacology, 19*(12), pyw073. https://doi.org/10.1093/ijnp/pyw073.

Suh, J., & Ressler, K. J. (2018). Common biological mechanisms of alcohol use disorder and post-traumatic stress disorder. *Alcohol Research: Current Reviews, 39*(2), 131–145. Retrieved from https://www.ncbi.nlm.nih.gov/pubmed/31198653.

Sun, Y., Gooch, H., & Sah, P. (2020). Fear conditioning and the basolateral amygdala. *F1000Res, 9*, 53. https://doi.org/10.12688/f1000research.21201.1.

Svingos, A. L., Chavkin, C., Colago, E. E., & Pickel, V. M. (2001). Major coexpression of kappa-opioid receptors and the dopamine transporter in nucleus accumbens axonal profiles. *Synapse, 42*(3), 185–192. https://doi.org/10.1002/syn.10005.

Szeszko, P. R., & Yehuda, R. (2019). Magnetic resonance imaging predictors of psychotherapy treatment response in post-traumatic stress disorder: A role for the salience network. *Psychiatry Research, 277*, 52–57. https://doi.org/10.1016/j.psychres.2019.02.005.

Tanabe, J., Regner, M., Sakai, J., Martinez, D., & Gowin, J. (2019). Neuroimaging reward, craving, learning, and cognitive control in substance use disorders: Review and implications for treatment. *The British Journal of Radiology, 92*(1101). https://doi.org/10.1259/bjr.20180942.

Taslimi, Z., Sarihi, A., & Haghparast, A. (2018). Glucocorticoid receptors in the basolateral amygdala mediated the restraint stress-induced reinstatement of methamphetamine-seeking behaviors in rats. *Behavioural Brain Research, 348*, 150–159. https://doi.org/10.1016/j.bbr.2018.04.022.

Tate, S. R., Norman, S. B., McQuaid, J. R., & Brown, S. A. (2007). Health problems of substance-dependent veterans with and those without trauma history. *Journal of Substance Abuse Treatment, 33*(1), 25–32. https://doi.org/10.1016/j.jsat.2006.11.006.

Terhaag, S., Cowlishaw, S., Steel, Z., Brewer, D., Howard, A., Armstrong, R., et al. (2019). Psychiatric comorbidity for veterans with posttraumatic stress disorder (PTSD): A latent profile analysis and implications for treatment. *Psychological Trauma*. https://doi.org/10.1037/tra0000520.

Thibeault, K. C., Kutlu, M. G., Sanders, C., & Calipari, E. S. (2019). Cell-type and projection-specific dopaminergic encoding of aversive stimuli in addiction. *Brain Research, 1713*, 1–15. https://doi.org/10.1016/j.brainres.2018.12.024.

Torok, B., Sipos, E., Pivac, N., & Zelena, D. (2019). Modelling posttraumatic stress disorders in animals. *Progress in Neuro-Psychopharmacology & Biological Psychiatry, 90*, 117–133. https://doi.org/10.1016/j.pnpbp.2018.11.013.

Tovote, P., Fadok, J. P., & Luthi, A. (2015). Neuronal circuits for fear and anxiety. *Nature Reviews. Neuroscience, 16*(6), 317–331. https://doi.org/10.1038/nrn3945.

Trevizol, A. P., Barros, M. D., Silva, P. O., Osuch, E., Cordeiro, Q., & Shiozawa, P. (2016). Transcranial magnetic stimulation for posttraumatic stress disorder: An updated systematic review and meta-analysis. *Trends in Psychiatry and Psychotherapy, 38*(1), 50–55. https://doi.org/10.1590/2237-6089-2015-0072.

Tye, K. M. (2018). Neural circuit motifs in valence processing. *Neuron, 100*(2), 436–452. https://doi.org/10.1016/j.neuron.2018.10.001.

Van't Veer, A., & Carlezon, W. A., Jr. (2013). Role of kappa-opioid receptors in stress and anxiety-related behavior. *Psychopharmacology, 229*(3), 435–452. https://doi.org/10.1007/s00213-013-3195-5.

Veer, I. M., Jetzschmann, P., Garbusow, M., Nebe, S., Frank, R., Kuitunen-Paul, S., et al. (2019). Nucleus accumbens connectivity at rest is associated with alcohol consumption in young male adults. *European Neuropsychopharmacology, 29*(12), 1476–1485. https://doi.org/10.1016/j.euroneuro.2019.10.008.

Verbitsky, A., Dopfel, D., & Zhang, N. (2020). Rodent models of post-traumatic stress disorder: Behavioral assessment. *Translational Psychiatry, 10*(1), 132. https://doi.org/10.1038/s41398-020-0806-x.

Verplaetse, T. L., Cosgrove, K. P., Tanabe, J., & McKee, S. A. (2020). Sex/gender differences in brain function and structure in alcohol use: A narrative review of neuroimaging findings over the last 10 years. *Journal of Neuroscience Research*, 1–15. https://doi.org/10.1002/jnr.24625.

Verplaetse, T. L., McKee, S. A., & Petrakis, I. L. (2018). Pharmacotherapy for co-occurring alcohol use disorder and post-traumatic stress disorder: Targeting the opioidergic, noradrenergic, serotonergic, and GABAergic/glutamatergic systems. *Alcohol Research: Current Reviews, 39*(2), 193–205. Retrieved from https://www.ncbi.nlm.nih.gov/pubmed/31198658.

Volkow, N. D., Fowler, J. S., Wolf, A. P., Schlyer, D., Shiue, C. Y., Alpert, R., et al. (1990). Effects of chronic cocaine abuse on postsynaptic dopamine receptors. *The American Journal of Psychiatry, 147*(6), 719–724. https://doi.org/10.1176/ajp.147.6.719.

Volkow, N. D., & Morales, M. (2015). The brain on drugs: From reward to addiction. *Cell, 162*(4), 712–725. https://doi.org/10.1016/j.cell.2015.07.046.

Volkow, N. D., Wang, G. J., Fowler, J. S., Logan, J., Hitzemann, R., Ding, Y. S., et al. (1996). Decreases in dopamine receptors but not in dopamine transporters in alcoholics. *Alcoholism, Clinical and Experimental Research, 20*(9), 1594–1598. Retrieved from http://www.ncbi.nlm.nih.gov/pubmed/8986209.

Volkow, N. D., Wang, G. J., Telang, F., Fowler, J. S., Logan, J., Jayne, M., et al. (2007). Profound decreases in dopamine release in striatum in detoxified alcoholics: Possible orbitofrontal involvement. *The Journal of Neuroscience, 27*(46), 12700–12706. https://doi.org/10.1523/JNEUROSCI.3371-07.2007.

Vollstadt-Klein, S., Loeber, S., Richter, A., Kirsch, M., Bach, P., von der Goltz, C., et al. (2012). Validating incentive salience with functional magnetic resonance imaging: Association between mesolimbic cue reactivity and attentional bias in alcohol-dependent patients. *Addiction Biology, 17*(4), 807–816. https://doi.org/10.1111/j.1369-1600.2011.00352.x.

Vujanovic, A. A., Wardle, M. C., Smith, L. J., & Berenz, E. C. (2017). Reward functioning in posttraumatic stress and substance use disorders. *Current Opinion in Psychology, 14*, 49–55. https://doi.org/10.1016/j.copsyc.2016.11.004.

Wade, N. E., Padula, C. B., Anthenelli, R. M., Nelson, E., Eliassen, J., & Lisdahl, K. M. (2017). Blunted amygdala functional connectivity during a stress task in alcohol dependent individuals: A pilot study. *Neurobiology of Stress*, 7, 74–79. https://doi.org/10.1016/j.ynstr.2017.05.004.

Wang, L., Gillis-Smith, S., Peng, Y., Zhang, J., Chen, X., Salzman, C. D., et al. (2018). The coding of valence and identity in the mammalian taste system. *Nature, 558*(7708), 127–131. https://doi.org/10.1038/s41586-018-0165-4.

Wang, T., Liu, J., Zhang, J., Zhan, W., Li, L., Wu, M., et al. (2016). Altered resting-state functional activity in posttraumatic stress disorder: A quantitative meta-analysis. *Scientific Reports, 6*. https://doi.org/10.1038/srep27131.

Wang, G. J., Volkow, N. D., Fowler, J. S., Logan, J., Abumrad, N. N., Hitzemann, R. J., et al. (1997). Dopamine D2 receptor availability in opiate-dependent subjects before and after naloxone-precipitated withdrawal. *Neuropsychopharmacology, 16*(2), 174–182. https://doi.org/10.1016/S0893-133X(96)00184-4.

Watanabe, T., Sasaki, Y., Shibata, K., & Kawato, M. (2017). Advances in fMRI real-time neurofeedback. *Trends in Cognitive Sciences, 21*(12), 997–1010. https://doi.org/10.1016/j.tics.2017.09.010.

Weber, K., Rockstroh, B., Borgelt, J., Awiszus, B., Popov, T., Hoffmann, K., et al. (2008). Stress load during childhood affects psychopathology in psychiatric patients. *BMC Psychiatry, 8*, 63. https://doi.org/10.1186/1471-244X-8-63.

Weera, M. M., Schreiber, A. L., Avegno, E. M., & Gilpin, N. W. (2020). The role of central amygdala corticotropin-releasing factor in predator odor stress-induced avoidance behavior and escalated alcohol drinking in rats. *Neuropharmacology*, *166*. https://doi.org/10.1016/j.neuropharm.2020.107979.

Weiner, J. L., & Valenzuela, C. F. (2006). Ethanol modulation of GABAergic transmission: The view from the slice. *Pharmacology & Therapeutics*, *111*(3), 533–554. https://doi.org/10.1016/j.pharmthera.2005.11.002.

Werling, L. L., Frattali, A., Portoghese, P. S., Takemori, A. E., & Cox, B. M. (1988). Kappa receptor regulation of dopamine release from striatum and cortex of rats and guinea pigs. *The Journal of Pharmacology and Experimental Therapeutics*, *246*(1), 282–286. Retrieved from https://www.ncbi.nlm.nih.gov/pubmed/2839666.

Whitaker, A. M., Gilpin, N. W., & Edwards, S. (2014). Animal models of post-traumatic stress disorder and recent neurobiological insights. *Behavioural Pharmacology*, *25*(5–6), 398–409. https://doi.org/10.1097/FBP.0000000000000069.

Williams, L. M., Kemp, A. H., Felmingham, K., Barton, M., Olivieri, G., Peduto, A., et al. (2006). Trauma modulates amygdala and medial prefrontal responses to consciously attended fear. *NeuroImage*, *29*(2), 347–357. https://doi.org/10.1016/j.neuroimage.2005.03.047.

Willner, P. (1986). Validation criteria for animal models of human mental disorders: Learned helplessness as a paradigm case. *Progress in Neuro-Psychopharmacology & Biological Psychiatry*, *10*(6), 677–690. https://doi.org/10.1016/0278-5846(86)90051-5.

Wilsnack, R. W., Wilsnack, S. C., Kristjanson, A. F., Vogeltanz-Holm, N. D., & Gmel, G. (2009). Gender and alcohol consumption: Patterns from the multinational GENACIS project. *Addiction*, *104*(9), 1487–1500. https://doi.org/10.1111/j.1360-0443.2009.02696.x.

Wolitzky-Taylor, K., Sewart, A., Vrshek-Schallhorn, S., Zinbarg, R., Mineka, S., Hammen, C., et al. (2017). The effects of childhood and adolescent adversity on substance use disorders and poor health in early adulthood. *Journal of Youth and Adolescence*, *46*(1), 15–27. https://doi.org/10.1007/s10964-016-0566-3.

World Health Organization. (2018). *International classification of diseases for mortality and morbidity statistics-11*. World Health Orgnaization.

Wrase, J., Schlagenhauf, F., Kienast, T., Wustenberg, T., Bermpohl, F., Kahnt, T., et al. (2007). Dysfunction of reward processing correlates with alcohol craving in detoxified alcoholics. *NeuroImage*, *35*(2), 787–794. https://doi.org/10.1016/j.neuroimage.2006.11.043.

Yehuda, R., Hoge, C. W., McFarlane, A. C., Vermetten, E., Lanius, R. A., Nievergelt, C. M., et al. (2015). Post-traumatic stress disorder. *Nature Reviews. Disease Primers*, *1*. https://doi.org/10.1038/nrdp.2015.57.

Yorgason, J. T., Calipari, E. S., Ferris, M. J., Karkhanis, A. N., Fordahl, S. C., Weiner, J. L., et al. (2016). Social isolation rearing increases dopamine uptake and psychostimulant potency in the striatum. *Neuropharmacology*, *101*, 471–479. https://doi.org/10.1016/j.neuropharm.2015.10.025.

Yorgason, J. T., Espana, R. A., Konstantopoulos, J. K., Weiner, J. L., & Jones, S. R. (2013). Enduring increases in anxiety-like behavior and rapid nucleus accumbens dopamine signaling in socially isolated rats. *The European Journal of Neuroscience*, *37*(6), 1022–1031. https://doi.org/10.1111/ejn.12113.

Young, K. D., Zotev, V., Phillips, R., Misaki, M., Drevets, W. C., & Bodurka, J. (2018). Amygdala real-time functional magnetic resonance imaging neurofeedback for major depressive disorder: A review. *Psychiatry and Clinical Neurosciences*, *72*(7), 466–481. https://doi.org/10.1111/pcn.12665.

Yuan, M., Pantazatos, S. P., Zhu, H., Li, Y., Miller, J. M., Rubin-Falcone, H., et al. (2019). Altered amygdala subregion-related circuits in treatment-naive post-traumatic stress disorder comorbid with major depressive disorder. *European Neuropsychopharmacology*, *29*(10), 1092–1101. https://doi.org/10.1016/j.euroneuro.2019.07.238.

Zakiniaeiz, Y., Scheinost, D., Seo, D., Sinha, R., & Constable, R. T. (2017). Cingulate cortex functional connectivity predicts future relapse in alcohol dependent individuals. *NeuroImage: Clinical, 13*, 181–187. https://doi.org/10.1016/j.nicl.2016.10.019.

Zamorski, M. A., Bennett, R. E., Rusu, C., Weeks, M., Boulos, D., & Garber, B. G. (2016). Prevalence of past-year mental disorders in the canadian armed forces, 2002–2013. *Canadian Journal of Psychiatry, 61*(1 Suppl), 26S–35S. https://doi.org/10.1177/0706743716628854.

Zhang, X., Ge, T. T., Yin, G., Cui, R., Zhao, G., & Yang, W. (2018). Stress-induced functional alterations in amygdala: Implications for neuropsychiatric diseases. *Frontiers in Neuroscience, 12*, 367. https://doi.org/10.3389/fnins.2018.00367.

Zhang, R., & Volkow, N. D. (2019). Brain default-mode network dysfunction in addiction. *NeuroImage, 200*, 313–331. https://doi.org/10.1016/j.neuroimage.2019.06.036.

Zhang, X., Kim, J., & Tonegawa, S. (2020). Amygdala reward neurons form and store fear extinction memory. *Neuron, 105*, 1077–1093. https://doi.org/10.1016/j.neuron.2019.12.025.

CHAPTER FOUR

Cannabis use and posttraumatic stress disorder comorbidity: Epidemiology, biology and the potential for novel treatment approaches

Veronika Kondev[a], Nathan Winters[a,c], and Sachin Patel[a,b,c,d,*]
[a]Department of Psychiatry and Behavioral Sciences, Vanderbilt University Medical Center, Nashville, TN, United States
[b]Department of Molecular Physiology & Biophysics, Vanderbilt University, Nashville, TN, United States
[c]Department of Pharmacology, Vanderbilt University, Nashville, TN, United States
[d]Vanderbilt Center for Addiction Research, Vanderbilt University, Nashville, TN, United States
*Corresponding author: e-mail address: sachin.patel@vanderbilt.edu

Contents

Abstract

Cannabis use is increasing among some demographics in the United States and is tightly linked to anxiety, trauma, and stress reactivity at the epidemiological and biological level. Stress-coping motives are highly cited reasons for cannabis use. However, with increased cannabis use comes the increased susceptibility for cannabis use disorder (CUD). Indeed, CUD is highly comorbid with posttraumatic stress disorder (PTSD). Importantly, endogenous cannabinoid signaling systems play a key role in the regulation of stress reactivity and anxiety regulation, and preclinical data suggest deficiencies in this signaling system could contribute to the development of stress-related psychopathology. Furthermore, endocannabinoid deficiency states, either pre-existing or

International Review of Neurobiology, Volume 157
ISSN 0074-7742
https://doi.org/10.1016/bs.irn.2020.09.007

induced by trauma exposure, could provide explanatory insights into the high rates of comorbid cannabis use in patients with PTSD. Here we review clinical and preclinical literature related to the cannabis use-PTSD comorbidity, the role of endocannabinoids in the regulation of stress reactivity, and potential therapeutic implications of recent work in this area.

1. Introduction

Cannabis (*Cannabis sativa*) is among the most widely used substances in the United States and globally and its ceremonial use by humans can be dated to 500 BC (Ren et al., 2019). Indeed, approximately 8.9% of the United States population age $\geq$12 years reported past-month cannabis use in 2016 (Ahrnsbrak, Bose, Hedden, Lipari, & Park-Lee, 2019). Cannabis contains a plethora of phytocannabinoid compounds including two major constituents, Δ^9-tetrahydrocannabinol (THC) and cannabidiol (CBD), as well as a number of minor constituents (Izzo, Borrelli, Capasso, Di Marzo, & Mechoulam, 2009). THC is the primary psychoactive component of cannabis, responsible for the reported "high" experienced by users and exerts its actions primarily through activation of cannabinoid receptors (Banister, Arnold, Connor, Glass, & McGregor, 2019). In contrast, CBD is non-psychoactive and does not activate cannabinoid receptors, exhibiting much more complex polypharmacological actions on a variety of neuronal signaling systems (Ibeas Bih et al., 2015). Cannabis users self-report a variety of reasons, ranging from recreational use to self-medication of a host of conditions, such as anxiety, epilepsy, insomnia, posttraumatic stress disorder (PTSD), irritable bowel syndrome, and pain (National Academies of Sciences and Medicine et al., 2017). Within the larger population of cannabis users, it is estimated that a subset of approximately 19.5% will meet lifetime diagnostic criteria for Cannabis Use Disorder (CUD) according to the guidelines set by *The Diagnostic and Statistical Manual of Mental Disorders - 5th Edition* (American Psychiatric Association, 2013; Hasin, 2018). Diagnostic criteria address several domains of problematic cannabis use, including interference with work or personal life activities, unwanted escalations or persistent use despite negative consequences, and manifestations of cravings and withdrawal symptoms during abstinence. Of note, craving and withdrawal symptoms were added as an expansion to diagnostic criteria in the DSM—V. Clinical diagnosis of CUD ranges from mild to severe based on how many of the 11 outlined diagnostic criteria are met within a 12-month period; mild CUD requires meeting 2–3 criteria, while severe CUD requires at least six (American Psychiatric Association, 2013). Cannabis use in the United States is increasing (Ahrnsbrak et al., 2019; Hasin, 2018), likely due to more positive public attitudes towards the use of cannabis products, a reduction in perceived harms, and the rapidly expanding availability due to changes in legal status, with over 30 states having legalized cannabis use for medical or recreational use. It is important to note that the high rates of cannabis use and CUD are attributed to these factors, as opposed to an inherently high addiction liability of cannabis (Nutt, King, & Phillips, 2010).

While THC activates cannabinoid receptors to exert subjective and physiological effects, endogenous ligands for cannabinoid receptors, as well as their metabolic pathways, have been discovered in recent decades revealing the critical importance of endogenous cannabinoid systems in the regulation of a variety of physiological processes ranging from pain (Barrie & Manolios, 2017) and anxiety (Lutz, Marsicano, Maldonado, & Hillard, 2015) to metabolic regulation (Carr, Jesch, & Brown, 2008). The endocannabinoid (eCB) system is comprised of the two cannabinoid receptors (CB_1 and CB_2) (Matsuda, Lolait, Brownstein, Young, & Bonner, 1990; Munro, Thomas, & Abu-Shaar, 1993), eCB ligands (2-arachidonoylglycerol [2-AG] and anandamide [AEA]) (Devane et al., 1992; Sugiura et al., 1995), and the enzymes responsible for the synthesis and degradation of the eCBs (Blankman & Cravatt, 2013). This system is widely distributed across several tissues and organ systems, including (but not limited to) the nervous system, immune system, and gastrointestinal tract. Within the brain, the eCB system serves as a neuromodulatory system wherein the lipid-derived eCB ligands act as retrograde messengers to reduce presynaptic neurotransmitter release. In brief, eCBs are synthesized from lipid precursors within the postsynaptic compartment of neurons and diffuse retrogradely to activate presynaptic CB_1 receptors that, in turn, reduce neurotransmitter release through a number of defined biochemical mechanisms. The CB_1 receptor is a Class A G protein-coupled receptor (GPCR) and the most abundant GPCR in the brain (CB_2 expression in the brain is low) (Glass, Dragunow, & Faull, 1997; Herkenham et al., 1990). CB_1 couples to canonical $G_{i/o}$ signaling mechanisms and reduces neurotransmitter release by activating presynaptic potassium channels, inhibiting presynaptic voltage-gated calcium channels, reducing adenylate-cyclase and protein kinase A activity (Howlett, 2002). Collectively, the eCB system serves as a negative feedback loop at the synapse, and can be recruited in either a tonic or phasic (i.e., constitutive or on-demand) manner to regulate the efficacy of synaptic transmission (For reviews, see Kano, Ohno-shosaku, Hashimotodani, Uchigashima, & Watanabe, 2009 and Katona & Freund, 2012).

The eCB system has been implicated in modulating a variety of physiological and behavioral processes, including emotional regulation and the neuroadaptive responses to stress exposure. Indeed, eCB levels are altered by exposure to stress in laboratory animals, with decreases in AEA levels and early elevations in levels of 2-AG being most consistently observed (Bedse, Hill, & Patel, 2020; Hill, Campolongo, Yehuda, & Patel, 2018; Patel, Roelke, Rademacher, Cullinan, & Hillard, 2004; Rademacher et al., 2008). Delayed reductions in brain 2-AG have also been observed in preclinical studies (Hill et al., 2005). Furthermore, genetic and pharmacological disruptions of eCB signaling produce innate anxiety-like phenotypes and enhanced stress susceptibility in mice (Bluett et al., 2017; Shonesy et al., 2014). Correspondingly, some stress-susceptible or anxious phenotypes in rodents are associated with deficient eCB regulation of limbic brain circuits (Bluett et al., 2017; Marcus et al., 2020), and pharmacological augmentation of eCB signaling exhibits a robust anxiolytic effect in rodents exposed to stress (Bedse et al., 2017; Patel, Hill, Cheer, Wotjak, & Holmes, 2017). Exogenous cannabinoid receptor agonists (e.g., THC) exhibit similar anxiolytic properties, albeit with a bi-phasic dose-response relationship with higher doses actually increasing anxiety-like behaviors (Rey, Purrio, Viveros, & Lutz, 2012). In line with this, anxiety and stress relief are major reasons for self-reported cannabis use (Temple, Driver, & Brown, 2014) and paradoxical anxiogenic and panic responses are often seen after high dose cannabis use and cited as reasons for discontinuation of use (Crippa et al., 2009). Collectively, there is substantial evidence to support a critical role for eCB signaling in the regulation of anxiety and stress responsivity, which has led to a growing interest in cannabinoid-based therapies for stress-related neuropsychiatric disorders and may account, at least in part, for the degree of comorbidity between cannabis use and CUD, and mental health disorders (Hasin et al., 2016).

Given the plethora of reported uses for cannabis and correspondingly high usage rates, it follows that CUD exhibits comorbidities with several other conditions, most notably mood and anxiety disorders (Hasin et al., 2016). Of particular interest is PTSD, given the high comorbidity rate (9.4% rate of CUD in PTSD patients, relative to 2.2% in patients without PTSD) (Bilevicius, Sommer, Asmundson, & El-Gabalawy, 2019), and potential for treating PTSD with cannabinoid-based therapies (see Hill et al., 2018 for review). However, empirical studies assessing the connection between cannabinoids and PTSD and the efficacy of cannabinoid-based therapies for this disorder remain critically lacking. Several small clinical pilot studies have addressed this (Cameron, Watson, & Robinson, 2014; Fraser, 2009; Jetly, Heber, Fraser, & Boisvert, 2015; Roitman, Mechoulam, Cooper-Kazaz, & Shalev, 2014) and will be reviewed in detail below. Nonetheless, these studies were conducted with relatively small sample sizes or did not utilize fully comprehensive study designs (i.e., double-blinded, placebo-controlled), highlighting the need for larger-scale well-designed studies to address these questions. The following sections will address the physiology and functions of the eCB system and cannabinoids, the putative pathophysiology of PTSD, and discuss the parallels between these topics that may contribute to the clinical comorbidity between cannabis use and CUD, and PTSD.

2. PTSD

PTSD is the fourth most common psychiatric disorder in the United States (Breslau, Davis, Andreski, & Peterson, 1991; Kessler, Sonnega, Bromet, Hughes, & Nelson, 1995) and results from World Health Organization (WHO) World Mental Health (WMH) Surveys suggest lifetime prevalence of PTSD to be 4–9% (Breslau et al., 1991; Kessler et al., 1995; Kilpatrick et al., 2013; Koenen et al., 2017). While PTSD was once considered an anxiety or fear-based disorder (American Psychiatric Association, 2000), it is now categorized as a trauma or stress-related disorder (American Psychiatric Association, 2013; Pai, Suris, & North, 2017). PTSD is characterized by symptoms in four distinct clusters: intrusion, avoidance, arousal/reactivity, and negative cognition/mood (American Psychiatric Association, 2013; Yehuda, Lehrner, & Rosenbaum, 2015) (Table 1).

Table 1 Symptoms of PTSD.

Intrusion Re-experiencing symptoms of aversive events Flashbacks Nightmares Intrusive memories	Avoidance Avoid triggers or reminders of events Emotional avoidance: thoughts or feelings Behavioral avoidance: places, people, sounds, smells, etc.
Arousal/reactivity Hyperarousal and hypervigilance Irritability Increase in anger outbreaks	Negative cognition/mood Impaired concentration Depressed mood or anxiety Guilt, shame, or anger

These four core features represent consequences of an overwhelmed fear response, resulting in maladaptive coping behaviors to avoid generating a negative affective state (Huether, 1996; Huether, Doering, Rüger, Rüther, & Schüssler, 1999). For diagnosis, these symptoms must persist for at least 1 month and cause significant impairment in daily life (American Psychiatric Association, 2013).

It is posited that in susceptible individuals, exposure to trauma results in the establishment of a prolonged and sustained trauma memory. The DSM-V has defined trauma as "actual or threatened death, serious injury, or sexual violence." This trauma memory includes not only the negative emotional and physical consequences of the trauma exposure, but also the environmental cues associated with it, including the place, people, sounds, or smells. This associative learning process and the clinically significant persistence of this trauma memory over long periods of time has led to the conceptualization of PTSD as a disorder of learning and memory (Careaga, Girardi, & Suchecki, 2016; Myers & Davis, 2007; Ross et al., 2017; Smith, Doran, Sippel, & Harpaz-Rotem, 2017), where dysfunctions in fear acquisition, fear reconsolidation, fear extinction, and generalization may cause the functional impairments characteristic of this disease (Careaga et al., 2016; Hill et al., 2018).

One key question is what factors make certain individuals susceptible to developing PTSD, given that not all people who have a traumatic experience develop PTSD (Breslau et al., 1991; Dursa, Reinhard, Barth, & Schneiderman, 2014; Hill et al., 2018; Liu et al., 2017). Several risk factors and general trends have been identified, including gender, type of trauma, and preexisting health conditions (Breslau et al., 1991). Risk factors for developing PTSD after a trauma involve early separation from parents, preexisting anxiety or depressive disorders, and family history of anxiety-related disorders (Breslau et al., 1991). PTSD is also a sexually dimorphic condition, with women being twice as likely to develop PTSD as men (Breslau et al., 1991; Kessler et al., 1995; Kilpatrick et al., 2013), but one study found contradicting results (Haskell et al., 2010). Furthermore, people exposed to certain traumas, particularly those involving violence, including combat exposure or interpersonal violence, are at higher risk for developing PTSD (Dursa et al., 2014; Graham-Bermann, Castor, Miller, & Howell, 2012; Kessler et al., 2017, 1995; Liu et al., 2017). Consequently, veterans are disproportionately affected by this disease. Vietnam veterans' estimated lifetime prevalence of PTSD is ~20% (Kulka et al., 1990; Magruder et al., 2015), while Iraq and Afghanistan veterans who sought healthcare from the Veterans Health Administration (VHA) is ~23% (Vaughan, Schell, Tanielian, Jaycox, & Marshall, 2014). Ongoing research suggests that whether a person develops PTSD after a trauma depends on a combination of biological and social factors that predispose an individual to PTSD: susceptibility factors that are detectable before the trauma, and sequelae factors that drive post-trauma processes during or shortly after exposure (Alexander, Nalloor, Bunting, & Vazdarjanova, 2020).

Despite progress being made toward understanding the molecular and neural mechanism of susceptibility, there remains a lack of efficacious treatment for PTSD patients. Currently, the first-line of treatment is trauma-focused cognitive behavioral psychotherapy (Graham, Callaghan, & Richardson, 2014; Ney, Matthews, Bruno, & Felmingham, 2019; Rauch, Eftekhari, & Ruzek, 2012) that includes prolonged exposure (PE) therapy and cognitive processing therapy (CPT), where patients gradually confront reminders of their trauma in order to regulate fear of the no-longer threatening situation (Graham et al., 2014). However, the use of these psychotherapies is met with varying degrees of limited empirical support (Benish, Imel, & Wampold, 2008; Forman-Hoffman et al., 2018; Steenkamp, Litz, Hoge, & Marmar, 2015). Evidence suggests patients completing these psychotherapies have heterogeneous outcomes (Bisson, Roberts, Andrew, Cooper, & Lewis, 2013). Study results are variable and have shown that 31–70% (Eftekhari et al., 2013; Litz et al., 2019; Steenkamp et al., 2015; Steenkamp, Litz, & Marmar, 2020) of patients show clinically significant improvement in their PTSD symptom severity. Despite this, most patients that complete treatment retain their PTSD diagnosis (i.e., continue to meet diagnostic criteria) (Litz et al., 2019) and impairment and distress remain significant for years (Schnurr, Lunney, Sengupta, & Waelde, 2003; Suris, Link-Malcolm, Chard, Ahn, & North, 2013). One meta-analysis of nine PE/CPT randomized clinical trials found that 60–72% of veteran PTSD patients retain their diagnosis after treatment (Steenkamp et al., 2015). The efficacy of these treatments is further brought into question by lack of long-term follow-up; limited data suggests that patients often do not show sustained recovery or drop-out before they receive a sufficient amount of care (Berke et al., 2019; Bisson et al., 2013; Shin, Greenbaum, Jain, & Rosen, 2014; Spoont, Murdoch, Hodges, & Nugent, 2010; Suris et al., 2013; Watts et al., 2014). Furthermore, many PTSD patients do not ever initially engage in these first-line psychotherapies (Shin et al., 2014; Suris et al., 2013; Watts et al., 2014), either due to social stigmas or logistical constraints. The lack of longitudinal, empirical data for PE/CPT and the subsequent heterogeneity of responding to treatment highlights issues of tolerability, efficacy, and access to these psychotherapies. Similarly, pharmacological interventions have also been shown to be relatively ineffective and produce considerable side effects (Berger et al., 2009; Bernardy & Friedman, 2015; Sartori & Singewald, 2019; Shin et al., 2014). First-line PTSD pharmacotherapies are selective serotonin reuptake inhibitors (SSRIs), specifically paroxetine and sertraline (Akikia & Abdallah, 2019; Hoskins et al., 2015; Ipser & Stein, 2012; Sartori & Singewald, 2019; Steckler & Risbrough, 2012). However, almost half of patients do not achieve meaningful clinical response and only 20–30% show remission (Berger et al., 2009; Lee et al., 2016; Stein, Ipser, Seedat, Sager, & Amos, 2006; Zohar et al., 2002). Finally, just as with psychotherapies, few patients begin, or continue, to take medication (Schnurr et al., 2003; Shin et al., 2014; Suris et al., 2013). This chronicity and resistance to treatment, combined with significant side effects make current pharmacotherapeutics suboptimal and analyses have concluded potentially high bias and unclear clinical significance (Hoskins et al., 2015; Lee et al., 2016). Given that current psychological and pharmacological treatments are met with low initiation and adherence rates and few PTSD patients show protracted remission, it is clear that novel, tolerable, efficacious treatments are needed (Sartori & Singewald, 2019).

3. PTSD—Cannabis use comorbidity

The lack of responding to current treatment options, high drop-out rates, and discontinuous recovery have led patients to turn to other avenues for symptom relief (Bonn-Miller & Rousseau, n.d.; Bonn-Miller, Vujanovic, Feldner, Bernstein, & Zvolensky, 2007; Kilpatrick et al., 2000). One of the most common coping methods to alleviate PTSD symptoms is cannabis use. The association between PTSD and cannabis use has been well documented (Bonn-Miller & Rousseau, n.d.; Bonn-Miller et al., 2007; Bonn-Miller, Babson, & Vandrey, 2014; Bonn-Miller, Vujanovic, & Drescher, 2011; Cougle, Bonn-Miller, Vujanovic, Zvolensky, & Hawkins, 2011; Kilpatrick et al., 2000; Lipschitz et al., 2003; Potter, Vujanovic, Marshall-Berenz, Bernstein, & Bonn-Miller, 2011; Rohrbach, Grana, Vernberg, Sussman, & Ping, 2009; Vlahov et al., 2002) and surveys have found that some veterans believe cannabis will relieve PTSD symptoms (Earleywine & Bolles, 2014). Clinical studies have consistently identified this link between cannabis use and coping with the emotional and physical symptoms of PTSD. For example, it has been reported that PTSD patients use cannabis to improve their sleep (Bonn-Miller et al., 2014; Bonn-Miller, Babson, Vujanovic, & Feldner, 2010), alleviate hyperarousal (Bremner, Southwick, Darnell, & Charney, 1996; Tull, McDermott, & Gratz, 2016), "forget their worries" (Bonn-Miller et al., 2007), manage pain (Baron, Lucas, Eades, & Hogue, 2018; Chan et al., 2017; Drost et al., 2017; Romero-Sandoval, Fincham, Kolano, Sharpe, & Alvarado-Vázquez, 2018), and avoid emotional triggers (Bordieri, Tull, McDermott, & Gratz, 2014; Pineles et al., 2011). Results from post-hoc analyses of military Veteran PTSD patients ($n=432$) further show that less change in PTSD symptom severity during treatment is significantly predictive of greater frequency of cannabis use 4 months later (Bonn-Miller et al., 2011), again suggesting cannabis is being used by many patients to treat residual symptoms. Indeed, cannabis use for self-medication was corroborated by another study examining veterans with a history of cannabis use ($n=1971$); they found that veterans who reported greater combat exposure and PTSD symptoms were more likely to use cannabis and identify as medicinal cannabis users (Loflin, Earleywine, & Bonn-Miller, 2017). However, self-medicating with cannabis can have medical and psychiatric consequences that are still not fully understood.

One such consequence is the cascade of cannabis use for self-medication to an overreliance on the drug, the development of tolerance and withdrawal, and ultimately the transition to CUD and dependence in some patients. Findings show significant correlations between PTSD, daily cannabis use, dependence, and CUD (Bonn-Miller, Harris, & Trafton, 2012; Bordieri et al., 2014; Cougle et al., 2011; Kevorkian et al., 2015). In support of the symptom coping hypothesis for cannabis use, Bordieri et al. (2014) found that high PTSD symptom severity was associated with a greater likelihood of cannabis dependence (Bordieri et al., 2014). A path analysis of teenagers and young adults further demonstrated that PTSD diagnosis was directly associated with the presence of CUD (Cornelius et al., 2010). This finding was reproduced using data from the National Epidemiologic Survey on Alcohol and Related Conditions (NESARC; $n=34{,}396$) that also showed PTSD was significantly associated with CUD (Kevorkian et al., 2015). A reciprocal relationship has further been demonstrated where veterans that are diagnosed with CUD show higher rates of PTSD than patients diagnosed with other drug use disorders (Bonn-Miller et al., 2012).

From a public health perspective, understanding how CUD and PTSD are related is crucial. Since 1996, over 30 states and the District of Columbia have passed laws legalizing the use of cannabis for medical or recreational purposes (Hasin, 2018). This change in legalization complements increased public perception that cannabis is safe and may ultimately contribute to the increased rates of cannabis use and subsequent CUD (Hasin, 2018). For example, daily use of marijuana has increased 60% in the past decade (Cougle et al., 2011), and CUD has increased 25% in states that have legalized cannabis (Cerdá et al., 2020). Bonn-Miller et al. (2012) found that the prevalence of CUD diagnosis within the Veterans Affairs (VA) healthcare system increased 115% from 2002 to 2009. Importantly, PTSD is often listed as a qualified medical condition for the use of medicinal cannabis at the state level. As of 2018, 21 states approved cannabis as a treatment of PTSD, and it is expected that this number will continue to grow with the expanding legalization of cannabis (Hasin, 2018). At a state level, PTSD has become one of the most common conditions treated by patients who use medical cannabis; a state report from Illinois shows that the overall use of medical cannabis has risen 83% in 1 year as of 2018 (McCoppin, 2018). These data indicate cannabis use by patients with PTSD is common and growing, despite unclear evidence for its efficacy or understanding the potentially adverse medical or psychiatric effects in this population.

These legal changes and the shift in public perception regarding the safety and efficacy of cannabis use is estimated to have potentially profound consequences for the treatment of PTSD, especially in the incidence of PTSD-CUD comorbidity. The percentage of U.S. Veterans accessing VA healthcare with PTSD and substance use disorders who were diagnosed with CUD increased almost 70% from 2002 to 2014 (Bonn-Miller & Rousseau, n.d.). This growing comorbidity is associated with poorer treatment outcomes (Bonn-Miller et al., 2012; Cougle et al., 2011; Orsolini et al., 2019; Wilkinson, Stefanovics, & Rosenheck, 2018). For example, military veteran PTSD patients ($n=260$) that obtained a CUD diagnosis were more likely to experience lower levels of change in PTSD symptoms over time (Bonn-Miller, Boden, Vujanovic, & Drescher, 2013). These quantitative analyses linking PTSD and CUD are comparable to patient reports; patients perceive these two disorders as being functionally related, so the worsening of one disorder can exacerbate the other (Brown, Stout, & Gannon-Rowley, 1998). This comorbidity further complicates the treatment of this heterogenous, complex psychiatric disorder.

However, insight can be gained from the association between PTSD and CUD, particularly whether these disorders share overlapping etiology. Clinical and preclinical evidence suggest that the eCB system plays a crucial role in the pathophysiology of PTSD, and thus may serve to explain the high use of cannabis in PTSD patients (Hill et al., 2018; Patel et al., 2017).

PTSD has been associated with reduced peripheral levels of AEA in some studies (Hill et al., 2013; Neumeister et al., 2013; for review see: Morena, Patel, Bains, & Hill, 2016). Based on these and other preclinical data, it has been proposed that deficits in eCB signaling may underlie susceptibility and promote the long-term impairment and distress characteristic of PTSD. Self-medication with cannabis may thus alleviate these eCB deficits (via direct activation of cannabinoid receptors) and ultimately relieve symptoms. However, the efficacy of cannabis to treat PTSD remains to be demonstrated and is complicated by increasing rates of CUD and other adverse effects of chronic cannabis use. Here, we review why such a significant link between PTSD and CUD seems to occur, focusing on the eCB system. Below, we address (1) how the eCB system could play a role in the development and maintenance of PTSD, (2) how cannabis use may counteract deficits in eCB signaling to alleviate symptoms, and (3) current clinical evidence that may be used to guide future therapeutics to more efficiently target components of the eCB system to potentially treat comorbid PTSD-CUD.

4. eCB deficiency hypothesis of PTSD susceptibility

As discussed above, the eCB system consists of the cannabinoid receptors (CB1R and CB2R), the ligands that bind them, and the biosynthetic and hydrolytic enzymes that synthesize and degrade these ligands (Kano et al., 2009). CB1R are the most abundant GPCR in the central nervous system (Glass et al., 1997; Herkenham et al., 1990) and are found in many areas that regulate stress reactivity and responsivity, emotion modulation circuits, and fear learning and memory processes (Akirav & Maroun, 2007; Hill et al., 2018; Jankord & Herman, 2008; Patel et al., 2017; Taber & Hurley, 2009). Translational preclinical evidence and human studies have led to the hypothesis that the eCB system could represent a causal factor in the susceptibility and pathophysiology of PTSD (Neumeister, Seidel, Ragen, & Pietrzak, 2015). Specifically, that eCB deficits, whether preexisting or stress-induced, may enhance susceptibility for the negative effects of stress or promote PTSD development and maintenance.

The first line of evidence supporting the PTSD-eCB hypothesis is that global reduction in eCB signaling causes increases in anxiety, impairs fear extinction, and is associated with a stress susceptibility phenotype. Genetic or pharmacological disruption of CB1R signaling produces an anxiety-like phenotype in male mice (Haller, Bakos, Szirmay, Ledent, &

Freund, 2002; Haller, Varga, Ledent, Barna, & Freund, 2004; Haller, Varga, Ledent, & Freund, 2004; Hill, Hillard, & McEwen, 2011; Martin, Ledent, Parmentier, Maldonado, & Valverde, 2002; Mikics, Vas, Aliczki, Halasz, & Haller, 2009; Urigüen, Pérez-Rial, Ledent, Palomo, & Manzanares, 2004). Importantly, rodent students have shown that this anxiety-like phenotype, caused genetically in CB1R knock-out mice (Haller, Varga, Ledent, Barna, & Freund, 2004) or pharmacologically using CB1R antagonists (Gamble-George et al., 2013), is exacerbated under high stress conditions, demonstrating that the eCB system is especially important for regulating stress-induced affective behaviors. Furthermore, a role for AEA and 2-AG in regulating emotional behavior has also been extensively studied and evidence clearly supports the notion that eCBs play a key role in regulating anxiety and stress responsivity (Bedse et al., 2017; reviewed: Morena et al., 2016). Rodent studies have begun to identify key emotional processing circuits, such as the basolateral amygdala (BLA) to medial prefrontal cortex (mPFC) circuit, that are highly regulated by eCBs. Importantly, impairment of eCB signaling within these brain regions can mimic the anxiogenic phenotype characteristic of the global CB1R knock-out, providing sites of action where eCBs may exert their effects (Marcus et al., 2020). CB1R knock-out mice also show impaired extinction of fear memories, a finding highly relevant to the pathophysiology of PTSD (Chhatwal, Davis, Maguschak, & Ressler, 2005; Kamprath et al., 2006; Marsicano et al., 2002). Finally, CB1R blockade and 2-AG depletion have also been shown to increase susceptibility to the adverse effects of acute traumatic stress in both male and female mice, making the eCB system a possible factor that determines disease susceptibility (Bluett et al., 2017). Overall, these data indicate that impaired eCB signaling represents a key neurobiological mechanism for affective and stress-related phenotypes, paralleling aspects of PTSD symptomatology.

The second line of evidence that the eCB system presents a functional locus for PTSD pathophysiology is that stress exposure, which is required for the development of PTSD, decreases eCB/CB1R signaling. For example, exposure to stress can reduce levels of both eCBs, AEA and 2-AG (Hill et al., 2005, 2018; Patel, Roelke, Rademacher, & Hillard, 2005; Qin et al., 2015; Rademacher et al., 2008; Zhong et al., 2014). However, it is important to note that these stress-induced effects are regionally specific, temporally dynamic, and dependent on the duration and type of stressor (Patel & Hillard, 2008). For example, acute or repeated homotypic stress in male mice decreases AEA and increases 2-AG levels in the amygdala (AMY), a key emotional processing region (Bedse et al., 2020; Patel et al., 2004, 2005). Over consecutive stress exposures, this short-lived rise in 2-AG levels is sensitized, producing greater enhancements of AMY 2-AG levels during, and immediately after stress (see Hill et al., 2018). In contrast, following some forms of stress, especially unpredictable forms of stress, this initial increase in 2-AG is followed by a significant reduction of 2-AG signaling in key emotional processing areas (Bedse et al., 2020; Hill et al., 2005; Marcus et al., 2020; Qin et al., 2015; Rademacher et al., 2008; Zhong et al., 2014). Due to the slow nature of this reduction in 2-AG levels, this effect has been deemed a delayed onset of 2-AG deficiency. This decrease in eCB signaling is paralleled by rodent studies consistently demonstrating that chronic stress also significantly decreases AEA levels in the mPFC and AMY (Patel et al., 2005; Rademacher et al., 2008). Finally, stress-induced deficiency in eCB signaling is corroborated by preclinical evidence showing that chronic stress, including chronic unpredictable stress and chronic administration of corticosterone, downregulates CB1R expression in a brain region-dependent manner (Hill et al., 2005; Hill, Carrier, Vanessa Ho, et al., 2008; Hu, Zhang, Czéh, Zhang, & Flügge, 2011; Lee & Hill, 2013; Patel & Hillard, 2008; Reich, Taylor, & McCarthy, 2009; Rossi et al., 2008, 2010; Taber & Hurley, 2009; Wamsteeker, Kuzmiski, & Bains, 2010; Wang et al., 2010; for review see: Morena et al., 2016). Taken together, these data indicate that eCB deficiency states, either pre-existing or stress-induced, could contribute to a PTSD susceptibility endophenotype or symptom expression, respectively. Specifically, since eCBs function to reduce anxiety-like behaviors and stress reactivity, deficiencies in this system, if associated with the disease phenotype, would be expected to contribute to disease pathophysiology.

While rodent studies consistently show this relationship between eCBs and the regulation of various processes relevant for developing or maintaining PTSD, the picture is complicated by the inherent heterogeneity of PTSD. Rodent studies have focused almost solely on males, despite women being twice as likely to develop PTSD in the general population (Breslau et al., 1991; Kessler et al., 1995; Kilpatrick et al., 2013). Therefore, evidence investigating the role of eCB signaling and PTSD pathophysiology in female preclinical models is lacking. Furthermore, not everyone that is exposed to a trauma develops PTSD. Animal studies observing stress effects are often based on models that examine main group differences between stress and unstressed controls, limiting investigation of individual differences in stress responsivity. Despite these limitations, these preclinical data ultimately provide a reciprocal model for the PTSD-eCB deficiency hypothesis. On one hand, reductions in eCB signaling may underlie susceptibility and various clinical endophenotypes for PTSD. On the other hand, stress, itself, dysregulates the eCB system towards a state of deficiency. In support of this reduction of eCB signaling, PTSD patients have been found to have reduced circulating levels of AEA (Neumeister et al., 2013) and 2-AG (Hill et al., 2013), although one group found elevations in eCB levels (Hauer et al., 2013). Interestingly, it has been reported that PTSD patients with reduced AEA levels showed compensatory increases in CB1R density, possibly due to decreased AEA availability (Neumeister et al., 2013). These changes in AEA and CB1R levels were specific to PTSD, as trauma-exposed controls did not show these alterations (Neumeister et al., 2013). While these eCB deficits nicely parallel the preclinical rodent data, one question that remains is the timing for causality: do PTSD patients exhibit baseline eCB deficits before being exposed to trauma that increase their susceptibility for developing PTSD, or do PTSD patients develop a larger stress-induced reduction in eCB signaling compared to individuals that are also exposed to trauma but do not develop PTSD? This and other important questions will need to be addressed in a future comprehensive research agenda on PTSD-cannabinoid interactions.

Overall, this eCB deficiency hypothesis, either preexisting or induced by trauma exposure, forms a compelling model for why many PTSD patients self-medicate with cannabis—if these individuals have lower levels of eCBs that account for susceptibility or symptomatology, then administering exogenous cannabinoids, such as THC, may alleviate those deficits. In support of this hypothesis, preclinical rodent models have shown that pharmacologically enhancing AEA and 2-AG levels by inhibiting eCB degradation enzymes may represent a novel avenue to harness eCB augmentation in PTSD therapeutics. First, inhibiting the AEA degradation enzyme, fatty acid amide hydrolase (FAAH), could increase AEA levels and restore the low levels of AEA characteristic of PTSD (Hill et al., 2013; Neumeister et al., 2013; Pietrzak et al., 2014). Kathuria et al. first found that pharmacological FAAH inhibition reduces anxiety in rats (Kathuria et al., 2003). This anxiolytic effect was replicated by many groups (Bedse et al., 2018, 2017; Busquets-Garcia et al., 2011; for review see: Hill et al., 2018) and genetic deletion of FAAH has been shown to reduce anxiety-like behaviors in mice, as well (Moreira, Kaiser, Monory, & Lutz, 2008). Importantly, FAAH inhibition has been shown to be more effective at reducing anxiety during periods of high stress or after trauma exposure (Bluett et al., 2014;

Haller et al., 2009; Naidu et al., 2007; Rossi et al., 2010), illustrating that pharmacologically elevating AEA levels not only decreases basal anxiety states, but also reduces stress-induced effects on affective behaviors. Pharmacological inhibition of FAAH in male mice has also been shown to enhance fear extinction, providing another neurobiological avenue by which increasing AEA levels could alleviate PTSD symptomatology (Gunduz-Cinar et al., 2013). To further substantiate the therapeutic potential of FAAH inhibition, viral-mediated over-expression of FAAH can reduce AEA levels and increase anxiety-like behaviors in rats (Rubino et al., 2008). Finally, a functional polymorphism in the *FAAH* gene (C385A) that results in elevated AEA levels via reduced AEA hydrolysis (Chiang, Gerber, Sipe, & Cravatt, 2004) leads to reduced anxiety and accelerated fear extinction in both rodents and humans (Dincheva et al., 2015; Gee et al., 2016). Individuals with this polymorphism also exhibit lower stress reactivity to threat (Gunduz-Cinar, Hill, McEwen, & Holmes, 2013; Gunduz-Cinar, MacPherson, et al., 2013). Interestingly, one study compared PTSD patients with and without the C385A *FAAH* allele; it was demonstrated that C385A carriers showed decreased subjective anxiety, diminished stress reactivity, and lower arousal (Spagnolo et al., 2016). These data ultimately support the hypothesis that enhancing AEA levels may treat the underlying pathophysiology of PTSD and emerge as a promising target for future stress-related disorders.

Similar to AEA, enhancing 2-AG levels by targeting the 2-AG degradation enzyme, monoacylglycerol lipase (MAGL), has also been investigated as it relates to PTSD-related neural processes. Treatment with MAGL inhibitors has been shown to reduce basal anxiety in some studies using rodent models (Aliczki, Balogh, Tulogdi, & Haller, 2012; Aliczki et al., 2013; Busquets-Garcia et al., 2011; Kinsey, O'Neal, Long, Cravatt, & Lichtman, 2011; Sciolino, Zhou, & Hohmann, 2011). Interestingly, some groups have shown that the therapeutic benefit of MAGL inhibition is more effective following stress or under conditions of high environmental aversiveness (Sciolino et al., 2011). For example, Bedse et al. (2017) showed that MAGL inhibition has a mild basal anxiolytic effect and, critically, prevents the development of stress-induced anxiety phenotypes in mice (Bedse et al., 2017). These data complement evidence implicating FAAH inhibition as therapeutic under not only basal states, but more importantly, stress-induced pathological states. To parallel these studies, genetic overexpression of MAGL in the hippocampus of adult, male mice, which decreases 2-AG levels, increases anxiety (Guggenhuber et al., 2016). The development of novel pharmaceuticals and transgenic mice has also allowed groups to study the contribution of 2-AG synthesis, by targeting its synthetic enzyme, diacylglycerol lipase (DAGL), on PTSD-related behavioral phenotypes. For example, pharmacological inhibition of DAGL in rodents produces anxiogenic effects (Bedse et al., 2017; review: Hill et al., 2018). Similarly, *DAGLα* knock-out mice also exhibit increased anxiety-like behaviors (Jenniches et al., 2016; Shonesy et al., 2014). Finally, pharmacological inhibition and genetic deletion of DAGL have been shown to disrupt fear extinction in mice (Cavener et al., 2018; Jenniches et al., 2016), further demonstrating that targeting 2-AG signaling may also present a novel therapeutic avenue in treating PTSD pathology.

Ultimately, these data implicating CB1R, AEA, and 2-AG in the pathophysiology of PTSD provide a conceptual framework for the eCB deficiency hypothesis. Novel pharmacological and genetic tools continue to reveal different avenues to harness eCB-mediated therapeutics, such as FAAH or MAGL inhibitors. However, future placebo-controlled, clinical studies will be required to determine the efficacy of these compounds in treating PTSD. Until then, there remains a need to better understand how eCBs regulate neurobiological processes relevant to PTSD, and how cannabis affects these stress, learning, and memory systems.

5. Cannabinoid modulation of neural processes underlying PTSD

The data above provide promising evidence illustrating the link between a deficiency in eCB signaling and PTSD susceptibility and/or the development and maintenance of PTSD symptomology; however, the biobehavioral processes affected by eCB signaling are diverse and remain even less well understood. These include: regulation of stress responsivity, modulation of key emotional processing circuits, and fear learning and memory processes. While the role eCBs play in regulating these three processes has been thoroughly reviewed elsewhere (Hill et al., 2018), here we will discuss how eCB regulation of these processes relates to PTSD and how cannabis may serve to counteract eCB signaling deficits contributing to dysregulation within these processes.

5.1 Role of eCBs in regulation of HPA axis and dysregulation in PTSD

As described above, stress decreases AEA levels, CB1R density, and exerts temporally-distinct effects on 2-AG levels in a brain region-specific manner.

The canonical stress response relies on activation of the hypothalamic-pituitary gland-adrenal gland (HPA) axis (Akirav, 2013; Hill & McEwen, 2010; Hill & Tasker, 2012; Jankord & Herman, 2008; Morena et al., 2016; Riebe & Wotjak, 2011; Segev & Akirav, 2016; Steiner & Wotjak, 2008). Activation of the HPA axis constitutes an orchestrated release of hormones to regulate physiological and behavioral sequalae following stress exposure. This involves the release of corticotropin-releasing hormone (CRH) from the paraventricular nucleus of the hypothalamus (PVN), the subsequent release of adrenocorticotropic hormone (ACTH) from the pituitary gland, which stimulates the release of glucocorticoid hormones, such as cortisol (corticosterone in rodents), from the adrenal cortex into the systemic circulation. These glucocorticoids negatively feedback on glucocorticoid receptors to prevent further release of both CRH and ACTH. Some of the stress-induced changes in eCB signaling described above have been shown to be heavily tied to glucocorticoid levels; translational approaches have revealed that administration of corticosterone, the rodent orthologue of cortisol, can be anxiogenic and mimic stress-induced alterations in eCB signaling in rodents (Hill, McLaughlin, et al., 2010; Hill, Patel, et al., 2010; Hill, Carrier, McLaughlin, et al., 2008; Hill, Carrier, Vanessa Ho, et al., 2008; Hong et al., 2011; Morgan et al., 2019; Rossi et al., 2008). For example, chronic corticosterone administration was shown to reduce CB1R density in the hippocampus, paralleling effects from a chronic unpredictable stress model (Hill et al., 2005; Hill, Carrier, McLaughlin, et al., 2008; Hill, Carrier, Vanessa Ho, et al., 2008). Gray et al. showed that the stress hormone, CRH, mediates AEA hydrolysis in the AMY and consequent anxiety-like behaviors, further demonstrating how the neuroendocrine response alters eCB levels and PTSD-related anxiogenic states (Gray et al., 2015). Finally, this link between glucocorticoid and eCB levels has also been found in the periphery where chronic exposure to glucocorticoids increases circulating levels of AEA in rodents (Bowles et al., 2015); similar associations were shown in humans where cortisol levels were positively correlated with circulating levels of AEA (Hill et al., 2013).

However, translational approaches have revealed that eCBs and stress exhibit a reciprocal relationship: while HPA axis activity alters eCB signaling, eCBs also regulate the stress neuroendocrine response. CB1R are highly expressed in both the PVN, as well as upstream regulatory structures such as the amygdala, hippocampus, and prefrontal cortex (Herkenham et al., 1990; Hill, Patel, et al., 2010; Malcher-Lopes et al., 2006). It has also been shown that both AEA and 2-AG are synthesized locally within the PVN

(Malcher-Lopes et al., 2006). In support of the eCB system's ability to regulate this neuroendocrine stress response, genetic and pharmacological disruption of CB1R signaling increases HPA axis activity (Cota et al., 2007; Patel et al., 2004; Steiner & Wotjak, 2008; Wade, Degroot, & Nomikos, 2006). Overall, the eCB system has been consistently shown to be an important negative modulator of the activation and termination of HPA axis function (Dow-Edwards & Silva, 2017; Hill et al., 2009, 2013; Hill & McEwen, 2010; Hill, Miller, Carrier, Gorzalka, & Hillard, 2009; Jennings, Okine, Olango, Roche, & Finn, 2016; Morena et al., 2016; Patel et al., 2005; Rademacher et al., 2008).

With respect to PTSD, studies revealed that PTSD patients exhibit elevated basal levels of CRH (Baker et al., 1999; Bremner et al., 1997; Sautter et al., 2003), but lower levels of cortisol upon awakening (de Kloet et al., 2007; Gill, Vythilingam, & Page, 2008; Labonté, Azoulay, Yerko, Turecki, & Brunet, 2014; McFarlane, Barton, Yehuda, & Wittert, 2011; Wahbeh & Oken, 2013; Yehuda, 2009; Yehuda, Golier, & Kaufman, 2005; Yehuda et al., 1995). This dysregulation of HPA axis function could be relevant to observed stress-induced reductions in eCB signaling, since exogenous intracerebroventricular CRH administration in rodents has been shown to reduce AEA levels in the AMY (see above); in support of this link, 2-AG levels in the PFC have been found to be positively modulated by corticosterone (Hill, McLaughlin, et al., 2011). Therefore, increased levels of CRH may serve to promote decreased AEA levels, while decreased cortisol levels could facilitate the delayed onset of 2-AG deficiency. Importantly, some of these glucocorticoid associations were absent in traumatized individuals without PTSD, suggesting that these hormonal effects are not just a stress response to trauma, but may represent a fundamental aspect of PTSD pathophysiology (Gill et al., 2008; Wahbeh & Oken, 2013; Yehuda et al., 1995). Lastly, when PTSD patients are exposed to personalized reminders of abuse, they exhibit larger increases in cortisol compared to trauma-exposed controls (Elzinga, Schmahl, Vermetten, van Dyck, & Bremner, 2003; Stoppelbein, Greening, & Fite, 2012). It has been suggested that PTSD may reflect an enhanced negative feedback inhibition state (Daskalakis, Lehrner, & Yehuda, 2013; de Kloet et al., 2007; McFarlane et al., 2011; Yehuda, 2009; Yehuda, Halligan, Golier, Grossman, & Bierer, 2004; Yehuda et al., 1995) and could rely on changes in glucocorticoid receptor sensitivity (Binder, 2009). Furthermore, this dysfunctional regulatory feedback in the HPA axis stress response has also been associated with increased catecholamine secretion that can promote symptoms characteristic of PTSD, such as hyperarousal and distress (Daskalakis et al., 2013; Krystal & Neumeister, 2009). These data ultimately illustrate HPA axis dysregulation in PTSD, and tie in how deficits in eCB and neuroendocrine signaling may coordinate to ultimately underlie PTSD pathophysiology.

The reciprocal relationship between the neuroendocrine and eCB system suggests that cannabis should be able to alter HPA axis activity in a CB1R-dependent manner to potentially alleviate PTSD symptoms (for review see: Sbarski & Akirav, 2020). CB1R agonists, like THC, have been shown to dose-dependently activate or inhibit HPA axis activity at high or low doses, respectively. At high doses, exogenous cannabinoids can augment stress-induced corticosterone levels (Jacobs, Dellarca, Manfredi, & Harclerode, 1979; Patel et al., 2004; Sano et al., 2009), while at lower doses they inhibit HPA axis activity (Ganon-Elazar & Akirav, 2012; Mayer, Matar, Kaplan, Zohar, & Cohen, 2014; Patel et al., 2004). Early rodent studies also demonstrated that THC elevates ACTH and corticosterone levels (Johnson, Dewy, Ritter, & Beckner, 1978; Pertwee, 1974; Puder et al., 1982; Weidenfeld, Feldman, & Mechoulam, 1994). While the exact mechanisms by which cannabis use (acute or chronic) could affect HPA axis function and ultimately aspects of PTSD symptomology remain unknown, the key role eCBs and CB1Rs play in HPA axis regulation suggest this warrants further investigation.

5.2 Role of eCBs in regulating amygdala and prefrontal cortex activity in PTSD

As described above, the eCB system is heavily tied to anxiety phenotypes and stress-dependent changes in affective behaviors. Site-specific studies have identified several key brain regions that mediate eCB effects on anxiety and stress reactivity. Of particular importance in relation to PTSD symptomatology are the PFC and the AMY. The PFC plays a role in processing trauma-related information. CB1R and eCB synthesis enzymes have been found in the medial PFC (mPFC) of rodents (Hill, Hillard, & McEwen, 2011; Marcus et al., 2020). As such, eCB-mediated synaptic plasticity has been demonstrated within the PFC, and activation of CB1R, specifically, can increase the activity of principal neurons (Chiu, Puente, Grandes, & Castillo, 2010; Hill, Hillard, & McEwen, 2011; Hill, McLaughlin, et al., 2011; Ji & Neugebauer, 2014; Kiritoshi, Ji, & Neugebauer, 2016; Pistis, Muntoni, Pillolla, & Gessa, 2002; Wędzony & Chocyk, 2009). The AMY processes emotionally salient information (particularly anxiety- and threat-inducing stimuli), regulates fear memory consolidation and extinction, and activates the sympathetic nervous system to induce a "fight or flight" response. A subdivision of the AMY, called the basolateral AMY (BLA) also contains CB1R on both glutamatergic and GABAergic terminals, and exhibits short-term eCB-mediated synaptic plasticity (Katona et al., 2001; Kodirov et al., 2010; Ramikie et al., 2014; Zhu & Lovinger, 2005). Ultimately, these data implicate eCB signaling as critical to regulating both PFC and AMY activity, and suggest that the eCB system may be a neural substrate mediating changes in activity at these emotional processing loci.

In terms of PTSD, both AMY and ventromedial PFC (vmPFC) activity have been shown to be altered, compared to healthy controls. Neuroimaging studies revealed that the AMY is hyperactive in PTSD (Diamond & Zoladz, 2016; Hughes & Shin, 2011; Sheynin & Liberzon, 2017; Shin, Rauch, & Pitman, 2006), while vmPFC activity is reduced (Hughes & Shin, 2011; Shin et al., 2004). For example, Veterans with PTSD exhibited deficits in PFC neural activity when individuals were instructed to reduce negative affect of images, compared to non-PTSD combat Veterans (Rabinak et al., 2014). It is thought that this decrease in activity relates to reduced cognitive control and regulation of affect (Loeffler et al., 2018). Pharmacological activation of CB1R may help reverse PTSD-related changes in neural activity due to the ability of eCBs to regulate both AMY and PFC activity (Chiu et al., 2010; Hill, McLaughlin, et al., 2011; Ji & Neugebauer, 2014; Kiritoshi et al., 2016; Perra, Pillolla, Luchicchi, & Pistis, 2008; Pistis et al., 2004; Pistis et al., 2002; Wędzony & Chocyk, 2009). For example, genetically-induced elevation of AEA has been shown to blunt threat-induced amygdala reactivity in humans (Demers, Drabant Conley, Bogdan, & Hariri, 2016; Gunduz-Cinar, Hill, et al., 2013; Gunduz-Cinar, MacPherson, et al., 2013; Hariri et al., 2009) and in mice (Dincheva et al., 2015).

In support of the hypothesis that THC may help alleviate PTSD pathophysiology, it has been shown that THC can reduce AMY reactivity to threatening images (Gruber, Rogowska, & Yurgelun-Todd, 2009; Phan et al., 2008), suggesting that THC may reverse PTSD-related increases in AMY activity. Similarly, THC induces potentially therapeutic changes in vmPFC activity. For example, THC has also been shown to increase vmPFC activity in healthy controls, compared to placebo (Rabinak, MacNamara, et al., 2014). Importantly, the effects of THC are dose-dependent, as groups have shown that lower doses of THC can increase PFC metabolic activity, while higher doses actually decrease PFC activity, which may correlate with decreasing cognitive control associated with intoxication (review: Egerton, Allison, Brett, & Pratt, 2006). To support THC's therapeutic effects, THC was also shown to increase vmPFC activation and lower threat-related AMY reactivity selectively in trauma-exposed individuals with PTSD, but not trauma-exposed or healthy controls (Rabinak et al., 2020). These data provide further evidence that cannabis may serve to reverse deficits in AMY and vmPFC activity characteristic of PTSD. Finally, it has been shown that PTSD patients demonstrate impaired coupling between vmPFC and AMY (Harris & Gordon, 2015; Kim et al., 2011; Likhtik & Paz, 2015). THC has been shown to further increase the functional coupling of the AMY and the vmPFC (Gorka, Fitzgerald, De Wit, & Phan, 2015; Rabinak et al., 2020), providing another avenue by which cannabis may exert therapeutic effects for PTSD. These studies provide promising preliminary evidence that administration of THC may modulate prefrontal-limbic circuits that ultimately underlie affective behaviors related to PTSD. As PSTD patients typically exhibit elevated AMY activity and decreased PFC activity, the effects of cannabis serve as an encouraging model for innovative pharmacological strategies to treat PTSD and its underlying pathophysiology.

5.3 Role of eCBs in regulating fear extinction

As discussed earlier, PTSD has been heavily suggested to rely on dysfunctional fear learning processes (Careaga et al., 2016; Myers & Davis, 2007; Rabinak, Mori, Lyons, Milad, & Luan Phan, 2017; Ross et al., 2017; Smith et al., 2017). eCBs are critical modulators of the processes underlying fear memory acquisition, consolidation, recall, and extinction (for detailed discussion see: De Bitencourt, Pamplona, & Takahashi, 2013; Hill et al., 2018; Morena & Campolongo, 2014). For PTSD, of paramount importance are fear extinction processes, given that patients show sustained trauma memory that underlie the core features of this disease (Davis, Myers, Chhatwal, & Ressler, 2006; Johansen, Cain, Ostroff, & Ledoux, 2011; Jurkus et al., 2016; Milad, Rauch, Pitman, & Quirk, 2006; Myers & Davis, 2007; Smith et al., 2017). Clinical studies directly demonstrate that PTSD patients exhibit impaired fear extinction (Blechert, Michael, Vriends, Margraf, & Wilhelm, 2007; Jovanovic et al., 2009; Orr et al., 2000; Smith et al., 2017; Wessa & Flor, 2007; Wicking et al., 2016). Using preclinical rodent models of classical associative fear conditioning and extinction paradigms, it has been revealed that the eCB system plays a critical role in modulating fear extinction (Aisenberg, Serova, Sabban, & Akirav, 2017; Cavener et al., 2018; Hartley et al., 2016; Hermanson et al., 2013; Kamprath et al., 2006; Laviolette & Grace, 2006a, 2006b; Lutz et al., 2015; Morena et al., 2018; Ney et al., 2019; Sbarski & Akirav, 2020). For example, mice lacking the CB1R do not extinguish fear behaviors as readily as wild-type mice and show prolonged fear expression (Marsicano et al., 2002). Similarly, CB1R antagonists can impair auditory (Bowers & Ressler, 2015; Chhatwal et al., 2005; Gunduz-Cinar, Hill, et al., 2013; Llorente-Berzal et al., 2015; Marsicano et al., 2002; Pickens & Theberge, 2014; Reich, Mohammadi, & Alger, 2008) and contextual (Marsicano et al., 2002; Pamplona, Bitencourt, & Takahashi, 2008; Pamplona, Prediger, Pandolfo, & Takahashi, 2006; Suzuki et al., 2004) fear conditioning, further illustrating the importance of CB1R signaling in fear extinction.

AEA and 2-AG signaling have also been shown to be critical modulators of fear extinction learning and memory. Marsicano and colleagues reported increased levels of both AEA and 2-AG in the AMY after extinction training, suggesting that enhanced eCB levels may contribute to fear extinction processes (Marsicano et al., 2002). In support of this theory, genetic and pharmacological rodent studies have revealed a causal role for 2-AG and AEA in mediating extinction of fear. For example, genetic studies in mice bearing the FAAH C385A polymorphism show accelerated fear extinction relative to wild-type littermates (Dincheva et al., 2015), demonstrating the importance of AEA signaling in mediating fear extinction. This is paralleled by human data showing that this same polymorphism is associated with more rapid fear extinction learning curves (Dincheva et al., 2015). Pharmacological increases in AEA levels with FAAH inhibitors or eCB uptake inhibitors also enhance contextual fear extinction in rodents (Bowers & Ressler, 2015; Gunduz-Cinar, Hill, et al., 2013; Gunduz-Cinar, MacPherson, et al., 2013; Laricchiuta, Centonze, & Petrosini, 2013; Pamplona et al., 2008, 2006; review: Hill et al., 2018). For example, rats and mice administered FAAH inhibitors showed enhanced extinction retrieval (Godlewski et al., 2010; Lin, Mao, Su, & Gean, 2009). Gunduz-Cinar, et al. recapitulated these results and further showed that infusing the FAAH inhibitor directly into the BLA decreased fear during extinction retrieval test (Gunduz-Cinar, MacPherson, et al., 2013).

To parallel these findings, enhancing 2-AG levels with MAGL inhibitors have also been shown to enhance fear extinction (Morena et al., 2018), but data are mixed (Hartley et al., 2016). These divergent effects may be due to different dosing or timing of MAGL inhibitor administration, ultimately suggesting that optimal levels of 2-AG are required for proper processing of fear responses (Lutz et al., 2015; for review: Gil-Ordóñez, Martín-Fontecha, Ortega-Gutiérrez, & López-Rodríguez, 2018). More consistent results show that decreasing total 2-AG levels by targeting the 2-AG synthesis enzyme, DAGL, can impair fear extinction in mice (see Bedse et al., 2020; Jenniches et al., 2016). For example, reducing 2-AG levels via pharmacological DAGL inhibition impairs extinction learning (Cavener et al., 2018). Jenniches, et al. also demonstrated this fear extinction deficit in DAGL knock-out mice (Jenniches et al., 2016). These data clearly demonstrate that proper eCB system functioning is critical for fear extinction processes.

Based on this evidence, it can be inferred that other CB1R agonists, such as THC, may also serve to promote fear extinction. Preclinical and clinical studies have supported this hypothesis (Berardi, Schelling, & Campolongo, 2016). Administration of CB1R agonists, including a synthetic form of THC, dronabinol, can facilitate extinction in rodents (Abush & Akirav, 2010; Berardi et al., 2016; de Oliveira Alvares, Pasqualini Genro, Diehl, Molina, & Quillfeldt, 2008; Micale, Di Marzo, Sulcova, Wotjak, & Drago, 2013; Pamplona et al., 2008, 2006; Ruhl, Zeymer, & von der Emde, 2017). However, results regarding THC, specifically, in preclinical models are inconclusive. The human data, on the other hand, is a bit more consistent. THC has been shown to increase extinction of fear memory by several groups (Hammoud et al., 2019; Klumpers et al., 2012; Rabinak et al., 2014, 2013). For example, Rabinak et al. (2013) showed, in a double-blind, placebo-controlled, randomized study, that THC (vs. placebo) enhanced extinction in a Pavlovian fear learning task in PTSD patients (Rabinak et al., 2013). Overall, more evidence is necessary to assess the potentially beneficial effects of THC on fear extinction processes.

5.4 Conclusions and limitations

Significant evidence demonstrates that eCBs are critical modulators of several neural processes contributing to PTSD pathophysiology. However, how THC, as a partial agonist of CB1Rs, serves to regulate these brain regions, circuits, and systems remains to be elucidated. Preliminarily, THC could alleviate deficits in glucocorticoid signaling, reverse dysfunctional AMY and vmPFC activity, and potentially enhance fear extinction. However, the therapeutic potential of THC in regulating these different neural processes is notably variable across different species, as well as between research groups. Differences in responding may be due to differences in timing of THC administration (acute vs chronic), environmental factors, pre-exposure to stress, and inconsistencies in age, sex, and dosing (Hammoud et al., 2019; Klumpers et al., 2012; Lin, Mao, Chen, & Gean, 2008; Papini et al., 2017;

Schramm-Sapyta et al., 2007). Similar to other CB1R agonists, THC exerts biphasic dose-dependent effects, with lower doses potentially accounting for the therapeutic alleviation of PTSD pathophysiology, while higher doses may exacerbate symptoms. Particularly important to THC's variability is the differential effects of THC on the adolescent/young adult brain in comparison to the adult brain. Rodent studies have consistently shown that adolescence represents a vulnerable period for the adverse consequences of THC exposure (Hurd, Michaelides, Miller, & Jutras-Aswad, 2014; Lee, 2019; Leishman, Murphy, Mackie, & Bradshaw, 2018; Moretti et al., 2015; Prini, Penna, Sciuccati, Alberio, & Rubino, 2017; Schramm-Sapyta et al., 2007). This is particularly important given that both PTSD and CUD are generally more prevalent in younger generations (Lee, 2019), and may serve to explain variability in THC responsivity, as well as provide potential guidelines for age restrictions when distributing possible THC-based pharmaceuticals.

To further complicate analysis, these neurobiological stress, emotional processing, and fear learning processes are also highly interdependent. Glucocorticoids can modulate activity in the AMY and the PFC, as well as affect fear memory processes (McEwen, Nasca, & Gray, 2016; van Ast, Cornelisse, Meeter, Joëls, & Kindt, 2013). Similarly, AMY and PFC activity have been shown to be critical for proper fear acquisition, consolidation, and extinction and both AMY and PFC are upstream regulators of the HPA axis (Berardi et al., 2016; Lisboa et al., 2017; Milad et al., 2007; Sigurdsson & Duvarci, 2016). Associative fear conditioning models also produce changes in glucocorticoid levels and AMY and PFC activity. As such, pharmacotherapies need to be designed to properly synchronize this interconnected network of processes. THC, as described above, can simultaneously modulate all three neural processes; for example, healthy individuals that receive THC, compared to placebo, showed increased vmPFC activation specifically to a previously extinguished conditioned stimulus during extinction memory recall (Rabinak, Angstadt, et al., 2014), illustrating that the neuroendocrine stress response, cortico-limbic circuitry, and fear extinction are all interconnected. Similarly, Hammoud et al. (2019) illustrated that THC induced acute increases in vmPFC activity during early extinction learning; however, when extinction memory was tested, THC-treated participants exhibited enhanced extinction, but with an absence of changes in neural activity (Hammoud et al., 2019). Further studies carefully defining the timing for THC administration in regulating these processes and subsequent PTSD symptomatology will also need to be conducted to fully capitalize on THC's beneficial effects.

Taken together, these data provide intriguing insight into the effects of THC on biobehavioral processes highly relevant to PTSD and suggest biologically plausible mechanisms by which cannabis use may relieve PTSD symptoms. Importantly, the directionality of these results is consistent with previously reviewed epidemiological data indicating strong associations between cannabis use, PTSD and symptom coping motives, and with the notion that deficient eCB signaling could contribute to the pathophysiology of PTSD. We next review data on the clinical efficacy of cannabis for the treatment of PTSD.

6. Cannabis as treatment for PTSD

High quality evidence directly demonstrating the efficacy of cannabis or cannabinoids in treating PTSD is lacking and requires further investigation, although small studies have been conducted and can provide some early insight into the question of potential efficacy. An open-label, uncontrolled pilot study found that 5 mg oral Δ^9-THC as an add-on therapy to existing stable medication resulted in a significant improvement in sleep quality, nightmare frequency, hyperarousal, and overall symptom severity in a small ($n=10$) cohort of PTSD patients (Roitman et al., 2014). A retrospective chart review analysis found that cannabis resulted in a 75% reduction in Clinician Administered Posttraumatic Scale for DSM-IV (CAPS) symptom scores (Greer, Grob, & Halberstadt, 2014). Observational studies qualitatively describe that cannabis may decrease self-reported patient symptom severity. For example, qualitative data taken from a focus group discussion among military veterans ($n=7$) with chronic PTSD found that medical cannabis exerted a wide range of therapeutic effects, most frequently increasing the quality of sleep (Krediet, Janssen, Heerdink, Egberts, & Vermetten, 2020). Another case report of severe sexual-abuse related PTSD reported that cannabis can decrease anxiety, flash-backs, stress, and self-harm (Passie, Emrich, Karst, Brandt, & Halpern, 2012). These self-reported improvements in symptoms were replicated in PTSD patients using medical cannabis in Canada; patients reported significant improvement in commonly reported symptoms such as depression, anxiety, sleep and pain at the 4 month follow-up, compared to baseline (Drost et al., 2017). A voluntary online survey concluded from PTSD patient-reported outcomes using medical cannabis in Canada ($n=588$) that cannabis use for 4 months resulted in significant improvement in patient's ability to cope with pain, overall quality of life, and general mood, sleep, and concentration (Chan et al., 2017). Another study tracked 404 medical cannabis users who self-identify as having PTSD for 31 months and found that PTSD symptoms were reduced by more than 50% immediately following cannabis use (LaFrance, Glodosky, Bonn-Miller, & Cuttler, 2020). Bremner et al. reported association between marijuana use and alleviation of PTSD symptoms in documented Vietnam veterans diagnosed with PTSD (Bremner et al., 1996). Finally, an unpublished open-label pilot study further demonstrated that medicinal cannabis improved PTSD symptoms in male Israeli combat veterans with PTSD (Mashiah, 2012). Several other clinical trials investigating cannabinoid therapies for PTSD are ongoing or recently completed, although no results from these studies have yet been reported (see clinicaltrials.gov identifiers: NCT02069366; NCT04080427; NCT03008005; NCT00965809; NCT02517424; NCT02759185; NCT03944447).

A synthetic derivative of THC, Nabilone, has also been investigated as a therapeutic intervention for PTSD. A retrospective chart review of Canadian incarcerated males ($N=104$) showed that daily nabilone significantly improved PTSD symptoms; nabilone increased sleep duration, decreased nightmare quantity, and improved overall PTSD symptomology (Cameron et al., 2014). Similarly, a small, randomized placebo-controlled cross-over study with male military personnel ($n=10$) showed that nabilone significantly reduced nightmare occurrence and increased overall well-being, compared to placebo (Jetly et al., 2015). Another uncontrolled, open-label study found that Nabilone significantly reduced the frequency of nightmares in 34/47 (72%) of PTSD patients in the study (Fraser, 2009). Importantly, these effects were shown to be dependent on nabilone administration, as reinstatement of nightmare occurrence coincided with cessation of nabilone treatment, and ceased again with the re-initiation of nabilone.

Another important potential therapeutic compound not yet discussed is the non-psychoactive constituent of cannabis, CBD. In line with the basic psychological and physiological etiology of PTSD, CBD has been shown to reduce expression of learned fear (Do Monte, Souza, Bitencourt, Kroon, & Takahashi, 2013; Gomes et al., 2012; Lemos, Resstel, & Guimarães, 2010; Resstel, Joca, Moreira, Corrêa, & Guimarães, 2006), facilitate fear extinction (Das et al., 2013; Song, Stevenson, Guimaraes, & Lee, 2016), and exhibit anxiolytic properties (Campos, Ferreira, & Guimarães, 2012; Schiavon, Bonato, Milani, Guimarães, & Weffort de Oliveira, 2016) in laboratory animals. For further review on CBD effects on fear and anxiety, see Jurkus et al. (2016) and Lee, Bertoglio, Guimarães, and Stevenson (2017). These properties along with the lack of psychoactive side effects comprise a favorable biobehavioral profile for a PTSD therapeutic, although current evidence for the utility of CBD in treating PTSD remains critically lacking. One case report explored CBD as a therapeutic avenue for a 10-year-old girl with PTSD and found that CBD oil decreased anxiety and improved both quality and quantity of sleep (Shannon & Opila-Lehman, 2016). Another case series ($n=11$) with an open-label, flexible dosing regimen found that CBD significantly decreased PTSD symptoms severity in 91% of patients (Elms, Shannon, Hughes, & Lewis, 2019). However, unlike THC and other cannabinoids, the mechanism of action is less clear—CBD has widespread pharmacology and data are conflicting, with some studies suggesting actions on CB1R, and others suggesting these effects are mediated through the serotonin 5HT1A receptor (Do Monte et al., 2013; Sartim, Guimarães, & Joca, 2016). Taken together, CBD appears to be a promising potential intervention for PTSD; however, further research, including full-scale controlled clinical studies, as well as further insight into the mechanism of action are needed to directly assess its efficacy in treating this disorder.

Despite the amount of promising data suggesting the therapeutic utility of cannabis, its constituents, and related compounds, evidence for the plausibility and efficacy of cannabis for PTSD is variable (Bonn-Miller et al., 2014; Johnson et al., 2016; Wilkinson et al., 2018). A systemic literature review published in 2019 concluded that cannabis use is linked to PTSD onset and poorer clinical course (Botsford, Yang, & George, 2020). Another review of medical cannabis for the treatment of PTSD further reported that out of the five studies identified, only three studies concluded benefit, while two discouraged cannabis use (Shishko, Oliveira, Moore, & Almeida, 2018). It is important to note that critical evidence regarding the therapeutic potential of cannabis and cannabinoids in treating PTSD remains limited – there remains a lack of longitudinal data, and most studies exhibited high bias (O'Neil et al., 2017), small sample sizes, or limited assessment of symptoms. Most observational studies also lack randomization and placebo control groups, making interpretation more difficult.

Another caveat for the therapeutic potential of THC is complicated by the variability in cannabis products, which can contain different strains, strengths, and compositions of cannabinoid drugs. Medical marijuana is typically derived from two subspecies of cannabis, *Cannabis sativa* and *Cannabis indica*, which can produce differential subjective and physiological effects, likely due to different profiles of physiologically active compounds, including differing ratios of THC and CBD. Furthermore, cannabis obtained outside the scope of legal dispensaries (i.e., "from the streets") lacks information on THC content, and its constituents are not controlled or biochemically validated. This variability and possible contamination could heavily impact studies tracking self-reported cannabis users. Finally, while clinical evidence may directly assess the effects of cannabis as a whole, laboratory studies often examine just the effects of THC. Differences in the additional compounds found in cannabis, overall dosage of THC, and characteristics of cannabis subspecies could all produce confounding effects. As cannabis use continues to grow with positive public opinion and the subsequent legalization, there remains a lack of evidence concerning the safety, efficacy, and long-term effects of cannabis. This raises the question of whether increasing rates of cannabis use are doing more harm than good, especially with respect to PTSD. Ultimately, more conclusive, controlled clinical trials will be needed to elucidate the true potential of cannabis in treating PTSD.

7. Conclusion

As cannabis becomes legalized in a growing number of states and municipalities, an increased incidence of CUD is a concern, especially for PTSD patients. Self-medication with cannabis is common, and without substantial empirical evidence for clinical benefit, development of CUD among PTSD patients remains a concern while therapeutic utility remains unclear. Placebo-controlled, blinded, randomized clinical trials will be needed to fully elucidate the therapeutic benefit of cannabis in treating PTSD. However, insight can be gleaned from understanding why this comorbidity may be so prevalent—evidence highlighting the complementary biology between the eCB system and the putative pathophysiology of PTSD suggests that cannabis and its constituents may be useful tools for alleviating symptoms of PTSD mediated via eCB deficient states. Accordingly, an approach for treating PTSD could be the use of FAAH inhibitors that enhance levels of the eCB, AEA. FAAH inhibition has been shown to be anxiolytic (Bedse et al., 2018, 2017; Busquets-Garcia et al., 2011; Kathuria et al., 2003 for review see: Hill et al., 2018), especially during high-stress situations or after trauma exposure (Bluett et al., 2014; Haller et al., 2009; Naidu et al., 2007; Rossi et al., 2010). The C385A *FAAH* allele, which is associated with elevated AEA levels, is associated with reduced anxiety (Dincheva et al., 2015; Gee et al., 2016), decreased stress reactivity (Gunduz-Cinar, Hill, et al., 2013; Gunduz-Cinar, MacPherson, et al., 2013), and enhanced fear extinction in mice and humans (Dincheva et al., 2015). Perhaps even more importantly, a recent double-blind, placebo-controlled phase 2 trial found that FAAH inhibition may even be an effective and safe approach to treat

CUD (D'Souza et al., 2019). As such, these targeted developments into drugs that ultimately enhance eCB tone (in contrast to direct CB1R agonists) to augment deficient eCB signaling characteristic of PTSD pathophysiology, could represent novel approaches in treating PTSD, CUD, and PTSD-CUD comorbidity.

References

Abush, H., & Akirav, I. (2010). Cannabinoids modulate hippocampal memory and plasticity. *Hippocampus, 20*(10), 1126–1138. https://doi.org/10.1002/hipo.20711.

Ahrnsbrak, R. A., Bose, J., Hedden, S. L., Lipari, R. N., & Park-Lee, E. (2019). *Key substance use and mental health indicators in the United States: Results from the 2018 National Survey on Drug Use and Health (HHS Publication No.SMA17-5044, NSDUH Series H-52).* Rockville, MD: Center for Behavioral Health Statistics and Quality. Substance Abuse and Mental Health Services Administration (SAMHSA). Retrieved from https://www.samhsa.gov/data/.

Aisenberg, N., Serova, L., Sabban, E. L., & Akirav, I. (2017). The effects of enhancing endocannabinoid signaling and blocking corticotrophin releasing factor receptor in the amygdala and hippocampus on the consolidation of a stressful event. *European Neuropsychopharmacology, 27*(9), 913–927. https://doi.org/10.1016/j.euroneuro.2017.06.006.

Akikia, T. J., & Abdallah, C. G. (2019). Are there effective psychopharmacologic treatments for PTSD. *The Journal of Clinical Psychiatry, 80*(3). https://doi.org/10.1016/j.physbeh.2017.03.040.

Akirav, I. (2013). Cannabinoids and glucocorticoids modulate emotional memory after stress. *Neuroscience and Biobehavioral Reviews, 37*(10), 2554–2563. https://doi.org/10.1016/j.neubiorev.2013.08.002.

Akirav, I., & Maroun, M. (2007). The role of the medial prefrontal cortex-amygdala circuit in stress effects on the extinction of fear. *Neural Plasticity, 2007*. https://doi.org/10.1155/2007/30873.

Alexander, K. S., Nalloor, R., Bunting, K. M., & Vazdarjanova, A. (2020). Investigating individual pre-trauma susceptibility to a PTSD-like phenotype in animals. *Frontiers in Systems Neuroscience, 13*, 85. https://doi.org/10.3389/fnsys.2019.00085.

Aliczki, M., Balogh, Z., Tulogdi, A., & Haller, J. (2012). The temporal dynamics of the effects of monoacylglycerol lipase blockade on locomotion, anxiety, and body temperature. *Behavioral Pharmacology, 23*(4), 348–357. https://doi.org/10.1097/FBP.0b013e3283564dfa.

Aliczki, M., Zelena, D., Mikics, E., Varga, Z. K., Pinter, O., Bakos, N. V., et al. (2013). Monoacylglycerol lipase inhibition-induced changes in plasma corticosterone levels, anxiety and locomotor activity in male CD1 mice. *Hormones and Behavior, 63*(5), 752–758. https://doi.org/10.1016/j.yhbeh.2013.03.017.

American Psychiatric Association: Washington, DC. (2000). *APA (American Psychiatric Association) Diagnostic and Statistical Manual of Mental Disorders* (4th ed.).

American Psychiatric Association: Washington, DC. (2013). *APA (American Psychiatric Association) Diagnostic and Statistical Manual of Mental Disorders* (5th ed.).

Baker, D. G., West, S. A., Nicholson, W. E., Ekhator, N. N., Kasckow, J. W., Hill, K. K., et al. (1999). Serial CSF corticotropin-releasing hormone levels and adrenocortical activity in combat veterans with posttraumatic stress disorder. *American Journal of Psychiatry, 156*(4), 585–588. https://doi.org/10.1176/ajp.156.4.585.

Banister, S. D., Arnold, J. C., Connor, M., Glass, M., & McGregor, I. S. (2019). Dark classics in chemical neuroscience: Δ9-tetrahydrocannabinol. *ACS Chemical Neuroscience, 10*(5), 2160–2175. https://doi.org/10.1021/acschemneuro.8b00651.

Baron, E. P., Lucas, P., Eades, J., & Hogue, O. (2018). Patterns of medicinal cannabis use, strain analysis, and substitution effect among patients with migraine, headache, arthritis, and chronic pain in a medicinal cannabis cohort. *Journal of Headache and Pain, 19*, 37. https://doi.org/10.1186/s10194-018-0862-2.

Barrie, N., & Manolios, N. (2017). The endocannabinoid system in pain and inflammation: Its relevance to rheumatic disease. *European Journal of Rheumatology, 4*, 210–218. https://doi.org/10.5152/eurjrheum.2017.17025.

Bedse, G., Bluett, R. J., Patrick, T. A., Romness, N. K., Gaulden, A., Kingsley, P. J., et al. (2018). Therapeutic endocannabinoid augmentation for mood and anxiety disorders: Comparative profiling of FAAH, MAGL and dual inhibitors. *Translational Psychiatry, 8*, 92. https://doi.org/10.1038/s41398-018-0141-7.

Bedse, G., Hartley, N. D., Neale, E., Gaulden, A., Patrick, T. A., Kingsley, P. J., et al. (2017). Functional redundancy between canonical endocannabinoid signaling systems in the modulation of anxiety. *Biological Psychiatry, 82*, 488–499. https://doi.org/10.1016/j.biopsych.2017.03.002.

Bedse, G., Hill, M. N., & Patel, S. (2020). 2-Arachidonoylglycerol modulation of anxiety and stress adaptation: from grass roots to novel therapeutics. *Biological Psychiatry, 88*(7), 520–530. https://doi.org/10.1016/j.biopsych.2020.01.015.

Benish, S. G., Imel, Z. E., & Wampold, B. E. (2008). The relative efficacy of bona fide psychotherapies for treating post-traumatic stress disorder: A meta-analysis of direct comparisons. *Clinical Psychology Review, 28*, 746–758. https://doi.org/10.1016/j.cpr.2007.10.005.

Berardi, A., Schelling, G., & Campolongo, P. (2016). The endocannabinoid system and Post Traumatic Stress Disorder (PTSD): From preclinical findings to innovative therapeutic approaches in clinical settings. *Pharmacological Research, 111*, 668–678. https://doi.org/10.1016/j.phrs.2016.07.024.

Berger, W., Mendlowicz, M. V., Marques-Portella, C., Kinrys, G., Fontenelle, L. F., Marmar, C. R., et al. (2009). Pharmacologic alternatives to antidepressants in post-traumatic stress disorder: A systematic review. *Progress in Neuro-Psychopharmacology and Biological Psychiatry, 33*(2), 169–180. https://doi.org/10.1016/j.pnpbp.2008.12.004.

Berke, D. S., Kline, N. K., Wachen, J. S., McLean, C. P., Yarvis, J. S., Mintz, J., et al. (2019). Predictors of attendance and dropout in three randomized controlled trials of PTSD treatment for active duty service members. *Behavior Research and Therapy, 118*, 7–17. https://doi.org/10.1016/j.brat.2019.03.003.

Bernardy, N. C., & Friedman, M. J. (2015). Psychopharmacological strategies in the management of Posttraumatic Stress Disorder (PTSD): What have we learned? *Current Psychiatry Reports, 17*, 20. https://doi.org/10.1007/s11920-015-0564-2.

Bilevicius, E., Sommer, J. L., Asmundson, G. J., & El-Gabalawy, R. (2019). Associations of PTSD, chronic pain, and their comorbidity on cannabis use disorder: Results from an American nationally representative study. *Depression and Anxiety, 36*, 1036–1046. https://doi.org/10.1002/da.22947.

Binder, E. B. (2009). The role of FKBP5, a co-chaperone of the glucocorticoid receptor in the pathogenesis and therapy of affective and anxiety disorders. *Psychoneuroendocrinology, 34S*, S186–S195. https://doi.org/10.1016/j.psyneuen.2009.05.021.

Bisson, J. I., Roberts, N. P., Andrew, M., Cooper, R., & Lewis, C. (2013). Psychological therapies for chronic post-traumatic stress disorder (PTSD) in adults. *Cochrane Database of Systematic Reviews, 12*. https://doi.org/10.1002/14651858.CD003388.pub4.

Blankman, J. L., & Cravatt, B. F. (2013). Chemical probes of endocannabinoid metabolism. *Pharmacological Reviews, 65*, 849–871. https://doi.org/10.1124/pr.112.006387.

Blechert, J., Michael, T., Vriends, N., Margraf, J., & Wilhelm, F. H. (2007). Fear conditioning in posttraumatic stress disorder: Evidence for delayed extinction of autonomic, experiential, and behavioral responses. *Behavior Research and Therapy*, *45*(9), 2019–2033. https://doi.org/10.1016/j.brat.2007.02.012.

Bluett, R. J., Báldi, R., Haymer, A., Gaulden, A. D., Hartley, N. D., Parrish, W. P., et al. (2017). Endocannabinoid signaling modulates susceptibility to traumatic stress exposure. *Nature Communications*, *8*(14), 782. https://doi.org/10.1038/ncomms14782.

Bluett, R. J., Gamble-George, J. C., Hermanson, D. J., Hartley, N. D., Marnett, L. J., & Patel, S. (2014). Central anandamide deficiency predicts stress-induced anxiety: Behavioral reversal through endocannabinoid augmentation. *Translational Psychiatry*, *e408*, 4. https://doi.org/10.1038/tp.2014.53.

Bonn-Miller, M. O., Babson, K. A., & Vandrey, R. (2014). Using cannabis to help you sleep: Heightened frequency of medical cannabis use among those with PTSD. *Drug and Alcohol Dependence*, *136*, 162–165. https://doi.org/10.1016/j.drugalcdep.2013.12.008.

Bonn-Miller, M. O., Babson, K., Vujanovic, A. A., & Feldner, M. T. (2010). Sleep problems and PTSD symptoms interact to predict marijuana use coping motives: A preliminary investigation. *Journal of Dual Diagnosis*, *6*(2), 111–122. https://doi.org/10.1080/15504261003751887.

Bonn-Miller, M. O., Boden, M. T., Vujanovic, A. A., & Drescher, K. D. (2013). Prospective investigation of the impact of cannabis use disorders on posttraumatic stress disorder symptoms among veterans in residential treatment. *Psychological Trauma Theory Research Practice and Policy*, *5*(2), 193–200. https://doi.org/10.1037/a0026621.

Bonn-Miller, M. O., Harris, A. H., & Trafton, J. A. (2012). Prevalence of cannabis use disorder diagnoses among veterans in 2002, 2008, and 2009. *Psychological Services*, *9*(4), 404–416. https://doi.org/10.1037/a0027622.

Bonn-Miller M.O & Rousseau G.S. (n.d.). Marijuana Use and PTSD among veterans. In National Center for PTSD, VA Mental Health. https://doi.org/10.5860/choice.43sup-0732.

Bonn-Miller, M. O., Vujanovic, A. A., & Drescher, K. D. (2011). Cannabis use among military veterans after residential treatment for posttraumatic stress disorder. *Psychology of Addictive Behaviors*, *25*(3), 485–491. https://doi.org/10.1037/a0021945.

Bonn-Miller, M. O., Vujanovic, A. A., Feldner, M. T., Bernstein, A., & Zvolensky, M. J. (2007). Posttraumatic stress symptom severity predicts marijuana use coping motives among traumatic event-exposed marijuana users. *Journal of Traumatic Stress*, *20*(4), 577–586. https://doi.org/10.1002/jts.

Bordieri, M. J., Tull, M. T., McDermott, M. J., & Gratz, K. L. (2014). The moderating role of experiential avoidance in the relationship between posttraumatic stress disorder symptom severity and cannabis dependence. *Journal of Contextual Behavioral Science*, *3*, 273–278. https://doi.org/10.1016/j.jcbs.2014.08.005.

Botsford, S. L., Yang, S., & George, T. P. (2020). Cannabis and cannabinoids in mood and anxiety disorders: impact on illness onset and course, and assessment of therapeutic potential. *American Journal on Addictions*, *29*, 9–26. https://doi.org/10.1111/ajad.12963.

Bowers, M. E., & Ressler, K. J. (2015). Interaction between the cholecystokinin and endogenous cannabinoid systems in cued fear expression and extinction retention. *Neuropsychopharmacology*, *40*, 688–700. https://doi.org/10.1038/npp.2014.225.

Bowles, N. P., Karatsoreos, I. N., Li, X., Vemuri, V. K., Wood, J., Li, Z., et al. (2015). A peripheral endocannabinoid mechanism contributes to glucocorticoid-mediated metabolic syndrome. *Proceedings of the National Academy of Sciences of the United States of America*, *112*(1), 285–290. https://doi.org/10.1073/pnas.1421420112.

Bremner, J. D., Licinio, J., Darnell, A., Krystal, J. H., Owens, M. J., Southwick, S. M., et al. (1997). Elevated CSF corticotropin-releasing factor concentrations in posttraumatic stress disorder. *American Journal of Psychiatry*, *154*(5), 624–629. https://doi.org/10.1176/ajp.154.5.624.

Bremner, J. D., Southwick, S. M., Darnell, A., & Charney, D. S. (1996). Chronic PTSD in Vietnam combat veterans: Course of illness and substance abuse. *The American Journal of Psychiatry*, *153*(3), 369–375. https://doi.org/10.1176/ajp.153.3.369.

Breslau, N., Davis, G. C., Andreski, P., & Peterson, E. (1991). Traumatic events and Posttraumatic Stress Disorder in an urban population of young adults. *Archives of General Psychiatry*, *48*, 216–222. https://doi.org/10.1001/archpsyc.1991.01810270028003.

Brown, P. J., Stout, R. L., & Gannon-Rowley, J. (1998). Substance use disorder-PTSD comorbidity: Patients' perceptions of symptom interplay and treatment issues. *Journal of Substance Abuse Treatment*, *15*(5), 445–448. https://doi.org/10.1016/S0740-5472(97)00286-9.

Busquets-Garcia, A., Puighermanal, E., Pastor, A., de la Torre, R., Maldonado, R., & Ozaita, A. (2011). Differential role of anandamide and 2-arachidonoylglycerol in memory and anxiety-like responses. *Biological Psychiatry*, *70*, 479–486. https://doi.org/10.1016/j.biopsych.2011.04.022.

Cameron, C., Watson, D., & Robinson, J. (2014). Use of a synthetic cannabinoid in a correctional population for posttraumatic stress disorder-related insomnia and nightmares, chronic pain, harm reduction, and other indications: A retrospective evaluation. *Journal of Clinical Psychopharmacology*, *34*(5), 559–564. https://doi.org/10.1097/JCP.0000000000000180.

Campos, A., Ferreira, F. R., & Guimarães, F. S. (2012). Cannabidiol blocks long-lasting behavioral consequences of predator threat stress: Possible involvement of 5HT1A receptors. *Journal of Psychiatric Research*, *46*, 1501–1510. https://doi.org/10.1016/j.jpsychires.2012.08.012.

Careaga, M. D. L., Girardi, C. E. N., & Suchecki, D. (2016). Understanding posttraumatic stress disorder through fear conditioning, extinction and reconsolidation. *Neuroscience and Biobehavioral Reviews*, *71*, 48–57. https://doi.org/10.1016/j.neubiorev.2016.08.023.

Carr, T. P., Jesch, E., & Brown, A. (2008). Endocannabinoids, metabolic regulation, and the role of diet. *Nutrition Research*. https://doi.org/10.1016/j.nutres.2008.08.003.

Cavener, V. S., Gaulden, A., Pennipede, D., Jagasia, P., Uddin, J., Marnett, L. J., et al. (2018). Inhibition of diacylglycerol lipase impairs fear extinction in mice. *Frontiers in Neuroscience*, *12*, 479. https://doi.org/10.3389/fnins.2018.00479.

Cerdá, M., Mauro, C., Hamilton, A., Levy, N. S., Santaella-Tenorio, J., Hasin, D., et al. (2020). Association between recreational marijuana legalization in the United States and changes in marijuana use and cannabis use disorder from 2008 to 2016. *JAMA Psychiatry*, 77(2), 165–171. https://doi.org/10.1001/jamapsychiatry.2019.3254.

Chan, S., Blake, A., Wolt, A., Wan, B. A., Zaki, P., Zhang, L., et al. (2017). Medical cannabis use for patients with post-traumatic stress disorder (PTSD). *Journal of Pain Management*, *10*(4), 385–396.

Chhatwal, J., Davis, M., Maguschak, K. A., & Ressler, K. J. (2005). Enhancing cannabinoid neurotransmission augments the extinction of conditioned fear. *Neuropsychopharmacology*, *30*, 516–524. https://doi.org/10.1038/sj.npp.1300655.

Chiang, K. P., Gerber, A. L., Sipe, J. C., & Cravatt, B. F. (2004). Reduced cellular expression and activity of the P129T mutant of human fatty acid amide hydrolase: Evidence for a link between defects in the endocannabinoid system and problem drug use. *Human Molecular Genetics*, *13*(18), 2113–2119. https://doi.org/10.1093/hmg/ddh216.

Chiu, C. Q., Puente, N., Grandes, P., & Castillo, P. E. (2010). Dopaminergic modulation of endocannabinoid-mediated plasticity at GABAergic synapses in the prefrontal cortex. *Journal of Neuroscience*, *30*(21), 7236–7248. https://doi.org/10.1523/JNEUROSCI.0736-10.2010.

Cornelius, J. R., Kirisci, L., Reynolds, M., Clark, D. B., Hayes, J., & Tarter, R. (2010). PTSD contributes to teen and young adult cannabis use disorders. *Addictive Behaviors*, *35*, 91–94. https://doi.org/10.1016/j.addbeh.2009.09.007.

Cota, D., Steiner, M., Marsicano, G., Cervino, C., Herman, J. P., Grübler, Y., et al. (2007). Requirement of cannabinoid receptor type 1 for the basal modulation of hypothalamic-pituitary-adrenal axis function. *Endocrinology, 148*(4), 1574–1581. https://doi.org/10.1210/en.2005-1649.

Cougle, J. R., Bonn-Miller, M. O., Vujanovic, A. A., Zvolensky, M. J., & Hawkins, K. A. (2011). Posttraumatic stress disorder and cannabis use in a nationally representative sample. *Psychology of Addictive Behaviors, 25*(3), 554–558. https://doi.org/10.1037/a0023076.

Crippa, J., Zuardi, A., Martín-Santos, R., Bhattacharyya, S., Atakan, Z., McGuire, P., et al. (2009). Cannabis and anxiety: A critical review of the evidence. *Human Psychopharmacology, 24*(7), 515–523. https://doi.org/10.1002/hup.1048.

Das, R. K., Kamboj, S. K., Ramadas, M., Yogan, K., Gupta, V., Redman, E., et al. (2013). Cannabidiol enhances consolidation of explicit fear extinction in humans. *Psychopharmacology, 226*, 781–792. https://doi.org/10.1007/s00213-012-2955-y.

Daskalakis, N. P., Lehrner, A., & Yehuda, R. (2013). Endocrine aspects of post-traumatic stress disorder and implications for diagnosis and treatment. *Endocrinology and Metabolism Clinics of North America, 42*, 503–513. https://doi.org/10.1016/j.ecl.2013.05.004.

Davis, M., Myers, K. M., Chhatwal, J., & Ressler, K. J. (2006). Pharmacological treatments that facilitate extinction of fear: Relevance to psychotherapy. *NeuroRx The American Society for Experimental Neurotherapeutics, 3*, 82–96. https://doi.org/10.1016/j.nurx.2005.12.008.

De Bitencourt, R. M., Pamplona, F. A., & Takahashi, R. N. (2013). A current overview of cannabinoids and glucocorticoids in facilitating extinction of aversive memories: Potential extinction enhancers. *Neuropharmacology, 64*, 389–395. https://doi.org/10.1016/j.neuropharm.2012.05.039.

de Kloet, C. S., Vermetten, E., Heijnen, C. J., Geuze, E., Lentjes, E. G. W. M., & Westenberg, H. G. M. (2007). Enhanced cortisol suppression in response to dexamethasone administration in traumatized veterans with and without posttraumatic stress disorder. *Psychoneuroendocrinology, 32*, 215–226. https://doi.org/10.1016/j.psyneuen.2006.12.009.

de Oliveira Alvares, L., Pasqualini Genro, B., Diehl, F., Molina, V. A., & Quillfeldt, J. A. (2008). Opposite action of hippocampal CB1 receptors in memory reconsolidation and extinction. *Neuroscience, 154*, 1648–1655. https://doi.org/10.1016/j.neuroscience.2008.05.005.

Demers, C. H., Drabant Conley, E., Bogdan, R., & Hariri, A. R. (2016). Interactions between anandamide and corticotropin-releasing factor signaling modulate human amygdala function and risk for anxiety disorders: An imaging genetics strategy for modeling molecular interactions. *Biological Psychiatry, 80*, 356–362. https://doi.org/10.1016/j.biopsych.2015.12.021.

Devane, W. A., Hanuš, L., Breuer, A., Pertwee, R. G., Stevenson, L. A., Griffin, G., et al. (1992). Isolation and structure of a brain constituent that binds to the cannabinoid receptor. *Science, 258*(5090), 1946–1949. https://doi.org/10.1126/science.1470919.

Diamond, D. M., & Zoladz, P. R. (2016). Dysfunctional or hyperfunctional? The amygdala in posttraumatic stress disorder is the bull in the evolutionary China shop. *Journal of Neuroscience Research, 94*, 437–444. https://doi.org/10.1002/jnr.23684.

Dincheva, I., Drysdale, A. T., Hartley, C. A., Johnson, D. C., Jing, D., King, E. C., et al. (2015). FAAH genetic variation enhances fronto-amygdala function in mouse and human. *Nature Communications, 6*, 6395. https://doi.org/10.1038/ncomms7395.

Do Monte, F. H., Souza, R. R., Bitencourt, R. M., Kroon, J. A., & Takahashi, R. N. (2013). Infusion of cannabidiol into infralimbic cortex facilitates fear extinction via CB1 receptors. *Behavioral Brain Research, 250*, 23–27. https://doi.org/10.1016/j.bbr.2013.04.045.

Dow-Edwards, D., & Silva, L. (2017). Endocannabinoids in brain plasticity: cortical maturation, HPA axis function and behavior. *Brain Research, 1654*, 157–164. https://doi.org/10.1016/j.brainres.2016.08.037.

Drost, L., Wan, B. A., Blake, A., Chan, S., Wolt, A., Ganesh, V., et al. (2017). Efficacy of different varieties of medical cannabis in relieving symptoms in post-traumatic stress disorder (PTSD) patients. *Journal of Pain Management, 10*(4), 415–422.

D'Souza, D. C., Cortes-Briones, J., Creatura, G., Bluez, G., Thurnauer, H., Deaso, E., et al. (2019). Efficacy and safety of a fatty acid amide hydrolase inhibitor (PF-04457845) in the treatment of cannabis withdrawal and dependence in men: A double-blind, placebo-controlled, parallel group, phase 2a single-site randomized controlled trial. *The Lancet Psychiatry, 6*, 35–45. https://doi.org/10.1016/S2215-0366(18)30427-9.

Dursa, E. K., Reinhard, M. J., Barth, S. K., & Schneiderman, A. I. (2014). Prevalence of a positive screen for PTSD among OEF/OIF and OEF/OIF-era veterans in a large population-based cohort. *Journal of Traumatic Stress, 27*, 542–549. https://doi.org/10.1002/jts.

Earleywine, M., & Bolles, J. R. (2014). Marijuana, expectancies, and post-traumatic stress symptoms: A preliminary investigation. *Journal of Psychoactive Drugs, 46*(3), 171–177. https://doi.org/10.1080/02791072.2014.920118.

Eftekhari, A., Ruzek, J. I., Crowley, J. J., Rosen, C. S., Greenbaum, M. A., & Karlin, B. E. (2013). Effectiveness of national implementation of prolonged exposure therapy in veterans affairs care. *JAMA Psychiatry, 70*(9), 949–955. https://doi.org/10.1001/jamapsychiatry.2013.36.

Egerton, A., Allison, C., Brett, R. R., & Pratt, J. A. (2006). Cannabinoids and prefrontal cortical function: insights from preclinical studies. *Neuroscience and Biobehavioral Reviews, 30*, 680–695. https://doi.org/10.1016/j.neubiorev.2005.12.002.

Elms, L., Shannon, S., Hughes, S., & Lewis, N. (2019). Cannabidiol in the treatment of post-traumatic stress disorder: A case series. *Journal of Alternative and Complementary Medicine, 25*(4), 392–397. https://doi.org/10.1089/acm.2018.0437.

Elzinga, B. M., Schmahl, C. G., Vermetten, E., van Dyck, R., & Bremner, J. D. (2003). Higher cortisol levels following exposure to traumatic reminders in abuse-related PTSD. *Neuropsychopharmacology, 28*, 1656–1665. https://doi.org/10.1038/sj.npp.1300226.

Forman-Hoffman, V., Cook Middleton, J., Feltner, C., Gaynes, B. N., Palmieri Weber, R., Bann, C., et al. (2018). *Psychological and pharmacological treatments for adults with post-traumatic stress disorder: a systematic review update. (Prepared by the RTI International-University of North Carolina at Chapel Hill Evidence-based Practice Center under Contract No. 290–2015-00011-I for AHRQ and PCORI.) AHRQ Publication No. 18-EHC011-EF. PCORI Publication No. 2018-SR-01*. Rockville, MD: Agency for Healthcare Research and Quality. May 2018. Posted final reports are located on the Effective Health Care Program search page https://doi.org/10.23970/AHRQEPCCER207.

Fraser, G. A. (2009). The use of a synthetic cannabinoid in the management of treatment-resistant nightmares in posttraumatic stress disorder (PTSD). *CNS Neuroscience and Therapeutics, 15*, 84–88. https://doi.org/10.1111/j.1755-5949.2008.00071.x.

Gamble-George, J. C., Conger, J. R., Hartley, N. D., Gupta, P., Sumislawski, J. J., & Patel, S. (2013). Dissociable effects of CB1 receptor blockade on anxiety-like and consummatory behaviors in the novelty-induced hypophagia test in mice. *Psychopharmacology, 228*(3), 401–409. https://doi.org/10.1007/s00213-013-3042-8.Dissociable.

Ganon-Elazar, E., & Akirav, I. (2012). Cannabinoids prevent the development of behavioral and endocrine alterations in a rat model of intense stress. *Neuropsychopharmacology, 37*, 456–466. https://doi.org/10.1038/npp.2011.204.

Gee, D. G., Fetcho, R. N., Jing, D., Li, A., Glatt, C. E., Drysdale, A. T., et al. (2016). Individual differences in frontolimbic circuitry and anxiety emerge with adolescent changes in endocannabinoid signaling across species. *Proceedings of the National Academy of Sciences of the United States of America, 113*(16), 4500–4505. https://doi.org/10.1073/pnas.1600013113.

Gill, J., Vythilingam, M., & Page, G. G. (2008). Low cortisol, high DHEA, and high levels of stimulated TNF-alpha and IL-6 in women with PTSD. *Journal of Traumatic Stress, 21*(6), 530–539. https://doi.org/10.1002/jts.

Gil-Ordóñez, A., Martín-Fontecha, M., Ortega-Gutiérrez, S., & López-Rodríguez, M. L. (2018). Monoacylglycerol lipase (MAGL) as a promising therapeutic target. *Biochemical Pharmacology, 157*, 18–32. https://doi.org/10.1016/j.bcp.2018.07.036.

Glass, M., Dragunow, M., & Faull, R. L. M. (1997). Cannabinoid receptors in the humsan brain: A detailed anatomical and quantitative autoradiographic study in the fetal, neonatal and adult human brain. *Neuroscience*, 77(2), 299–318. https://doi.org/10.1016/S0306-4522(96)00428-9.

Godlewski, G., Alapafuja, S. O., Bátkai, S., Nikas, S. P., Cinar, R., Offertáler, L., et al. (2010). Inhibitor of fatty acid amide hydrolase normalizes cardiovascular function in hypertension without adverse metabolic effects. *Chemistry and Biology, 17*, 1256–1266. https://doi.org/10.1016/j.chembiol.2010.08.013.

Gomes, F. V., Reis, D. G., Alves, F. H. F., Corrêa, F. M. A., Guimarães, F. S., & Resstel, L. B. M. (2012). Cannabidiol injected into the bed nucleus of the stria terminalis reduces the expression of contextual fear conditioning via 5-HT 1A receptors. *Journal of Psychopharmacology, 26*(1), 104–113. https://doi.org/10.1177/0269881110389095.

Gorka, S. M., Fitzgerald, D. A., De Wit, H., & Phan, K. L. (2015). Cannabinoid modulation of amygdala subregion functional connectivity to social signals of threat. *International Journal of Neuropsychopharmacology, 18*(3), 1–6. https://doi.org/10.1093/ijnp/pyu104.

Graham, B. M., Callaghan, B. L., & Richardson, R. (2014). Bridging the gap: Lessons we have learnt from the merging of psychology and psychiatry for the optimisation of treatments for emotional disorders. *Behavior Research and Therapy, 62*, 3–16. https://doi.org/10.1016/j.brat.2014.07.012.

Graham-Bermann, S. A., Castor, L. E., Miller, L. E., & Howell, K. H. (2012). The impact of intimate partner violence and additional traumatic events on trauma symptoms and PTSD in preschool-aged children. *Journal of Trauma, 25*, 393–400. https://doi.org/10.1002/jts.

Gray, J. M., Vecchiarelli, H. A., Morena, M., Lee, T. T. Y., Hermanson, D. J., Kim, A. B., et al. (2015). Corticotropin-releasing hormone drives anandamide hydrolysis in the amygdala to promote anxiety. *Journal of Neuroscience, 35*(9), 3879–3892. https://doi.org/10.1523/JNEUROSCI.2737-14.2015.

Greer, G. R., Grob, C. S., & Halberstadt, A. L. (2014). PTSD symptom reports of patients evaluated for the New Mexico medical cannabis program. *Journal of Psychoactive Drugs, 46*(1), 73–77. https://doi.org/10.1080/02791072.2013.873843.

Gruber, S. A., Rogowska, J., & Yurgelun-Todd, D. A. (2009). Altered affective response in marijuana smokers: An FMRI study. *Drug and Alcohol Dependence, 105*, 139–153. https://doi.org/10.1016/j.drugalcdep.2009.06.019.

Guggenhuber, S., Romo-Parra, H., Bindila, L., Leschik, J., Lomazzo, E., Remmers, F., et al. (2016). Impaired 2-AG signaling in hippocampal glutamatergic neurons: Aggravation of anxiety-like behavior and unaltered seizure susceptibility. *International Journal of Neuropsychopharmacology, 19*(2), 1–13. https://doi.org/10.1093/ijnp/pyv091.

Gunduz-Cinar, O., Hill, M. N., McEwen, B. S., & Holmes, A. (2013). Amygdala FAAH and anandamide: mediating protection and recovery from stress. *Trends in Pharmacological Sciences, 34*(11), 637–644. https://doi.org/10.1016/j.tips.2013.08.008.

Gunduz-Cinar, O., MacPherson, K. P., Cinar, R., Gamble-George, J., Sugden, K., Williams, B., et al. (2013). Convergent translational evidence of a role for anandamide

in amygdala-mediated fear extinction, threat processing and stress-reactivity. *Molecular Psychiatry*, *18*, 813–823. https://doi.org/10.1038/mp.2012.72.
Haller, J., Bakos, N., Szirmay, M., Ledent, C., & Freund, T. F. (2002). The effects of genetic and pharmacological blockade of the CB1 cannabinoid receptor on anxiety. *European Journal of Neuroscience*, *16*, 1395–1398. https://doi.org/10.1046/j.1460-9568.2002.02192.x.
Haller, J., Barna, I., Barsvari, B., Gyimesi Pelczer, K., Yasar, S., Panlilio, L. V., et al. (2009). Interactions between environmental aversiveness and the anxiolytic effects of enhanced cannabinoid signaling by FAAH inhibition in rats. *Psychopharmacology*, *204*, 607–616. https://doi.org/10.1007/s00213-009-1494-7.
Haller, J., Varga, B., Ledent, C., Barna, I., & Freund, T. F. (2004). Context-dependent effects of CB1 cannabinoid gene disruption on anxiety-like and social behavior in mice. *European Journal of Neuroscience*, *19*, 1906–1912. https://doi.org/10.1111/j.1460-9568.2004.03293.x.
Haller, J., Varga, B., Ledent, C., & Freund, T. F. (2004). CB1 cannabinoid receptors mediate anxiolytic effects: convergent genetic and pharmacological evidence with CB1-specific agents. *Behavioral Pharmacology*, *15*(4), 299–304. https://doi.org/10.1097/01.fbp.0000135704.56422.40.
Hammoud, M. Z., Peters, C., Hatfield, J. R. B., Gorka, S. M., Phan, K. L., Milad, M. R., et al. (2019). Influence of Δ9-tetrahydrocannabinol on long-term neural correlates of threat extinction memory retention in humans. *Neuropsychopharmacology*, *44*, 1769–1777. https://doi.org/10.1038/s41386-019-0416-6.
Hariri, A. R., Gorka, A., Hyde, L. W., Kimak, M., Halder, I., Ducci, F., et al. (2009). Divergent effects of genetic variation in endocannabinoid signaling on human threat- and reward-related brain function. *Biological Psychiatry*, *66*, 9–16. https://doi.org/10.1016/j.biopsych.2008.10.047.
Harris, A. Z., & Gordon, J. A. (2015). Long-range neural synchrony in behavior. *Annual Review of Neuroscience*, *38*, 171–194. https://doi.org/10.1146/annurev-neuro-071714-034111.
Hartley, N. D., Gunduz-Cinar, O., Halladay, L., Bukalo, O., Holmes, A., & Patel, S. (2016). 2-Arachidonoylglycerol signaling impairs short-term fear extinction. *Translational Psychiatry*, *6*, e749. https://doi.org/10.1038/tp.2016.26.
Hasin, D. S. (2018). US epidemiology of cannabis use and associated problems. *Neuropsychopharmacology*, *43*, 195–212. https://doi.org/10.1038/npp.2017.198.
Hasin, D. S., Kerridge, B. T., Saha, T. D., Huang, B., Pickering, R., Smith, S. M., et al. (2016). Prevalence and correlates of DSM-5 cannabis use disorder, 2012–2013: Findings from the national epidemiologic survey on alcohol and related conditions-III. *American Journal of Psychiatry*, *173*(6), 588–599. https://doi.org/10.1176/appi.ajp.2015.15070907.
Haskell, S. G., Gordon, K. S., Mattocks, K., Duggal, M., Erdos, J., Justice, A., et al. (2010). Gender differences in rates of depression, PTSD, pain, obesity, and military sexual trauma among Connecticut war Veterans of Iraq and Afghanistan. *Journal of Women's Health*, *19*(2), 267–271.
Hauer, D., Schelling, G., Gola, H., Campolongo, P., Morath, J., Roozendaal, B., et al. (2013). Plasma concentrations of endocannabinoids and related primary fatty acid amides in patients with post-traumatic stress disorder. *PLoS One*, *8*(5), e62741. https://doi.org/10.1371/journal.pone.0062741.
Herkenham, M., Lynn, A. B., Little, M. D., Johnson, M. R., Melvin, L. S., de Costa, B. R., et al. (1990). Cannabinoid receptor localization in brain. *Proceedings of the National Academy of Sciences of the United States of America*, *87*, 1932–1936. https://doi.org/10.1073/pnas.87.5.1932.
Hermanson, D. J., Hartley, N. D., Gamble-George, J., Brown, N., Shonesy, B. C., Kingsley, P. J., et al. (2013). Substrate-selective COX-2 inhibition decreases anxiety via endocannabinoid activation. *Nature Neuroscience*, *16*(9), 1291–1299. https://doi.org/10.1038/nn.3480.

Hill, M. N., Bierer, L. M., Makotkine, I., Golier, J. A., Galea, S., McEwen, B. S., et al. (2013). Reductions in circulating endocannabinoid levels in individuals with post-traumatic stress disorder following exposure to the world trade center attacks. *Psychoneuroendocrinology, 38*, 2952–2961. https://doi.org/10.1016/j.psyneuen.2013.08.004.

Hill, M. N., Campolongo, P., Yehuda, R., & Patel, S. (2018). Integrating endocannabinoid signaling and cannabinoids into the biology and treatment of Posttraumatic Stress Disorder. *Neuropsychopharmacology, 43*, 80–102. https://doi.org/10.1038/npp.2017.162.

Hill, M. N., Carrier, E. J., McLaughlin, R. J., Morrish, A. C., Meier, S. E., Hillard, C. J., et al. (2008). Regional alterations in the endocannabinoid system in an animal model of depression: Effects of concurrent antidepressant treatment. *Journal of Neurochemistry, 106*, 2322–2336. https://doi.org/10.1111/j.1471-4159.2008.05567.x.

Hill, M. N., Carrier, E. J., Vanessa Ho, W.-S., Shi, L., Patel, S., Gorzalka, B. B., et al. (2008). Prolonged glucocorticoid treatment decreases cannabinoid CB1 receptor density in the hippocampus. *Hippocampus, 18*, 221–226. https://doi.org/10.1002/hipo.

Hill, M. N., Hillard, C. J., & McEwen, B. S. (2011). Alterations in corticolimbic dendritic morphology and emotional behavior in cannabinoid CB 1 receptor-deficient mice parallel the effects of chronic stress. *Cerebral Cortex, 21*, 2056–2064. https://doi.org/10.1093/cercor/bhq280.

Hill, M. N., & McEwen, B. S. (2010). Involvement of the endocannabinoid system in the neurobehavioral effects of stress and glucocorticoids. *Progress in Neuro-Psychopharmacology and Biological Psychiatry, 34*, 791–797. https://doi.org/10.1016/j.pnpbp.2009.11.001.

Hill, M. N., McLaughlin, R. J., Bingham, B., Shrestha, L., Lee, T. Y., Gray, J. M., et al. (2010). Endogenous cannabinoid signaling is essential for stress adaptation. *Proceedings of the National Academy of Sciences of the United States of America, 107*(20), 9406–9411. https://doi.org/10.1073/pnas.0914661107.

Hill, M. N., McLaughlin, R. J., Morrish, A. C., Viau, V., Floresco, S. B., Hillard, C. J., et al. (2009). Suppression of amygdalar endocannabinoid signaling by stress contributes to activation of the hypothalamic-pituitary-adrenal axis. *Neuropsychopharmacology, 34*, 2733–2745. https://doi.org/10.1038/npp.2009.114.

Hill, M. N., McLaughlin, R. J., Pan, B., Fitzgerald, M. L., Roberts, C. J., Lee, T. T.-Y., et al. (2011). Recruitment of prefrontal cortical endocannabinoid signaling by glucocorticoids contributes to termination of the stress response. *Journal of Neuroscience, 31*(29), 10506–10515. https://doi.org/10.1523/JNEUROSCI.0496-11.2011.

Hill, M. N., Miller, G. E., Carrier, E. J., Gorzalka, B. B., & Hillard, C. J. (2009). Circulating endocannabinoids and N-acyl ethanolamines are differentially regulated in major depression and following exposure to social stress. *Psychoneuroendocrinology, 34*, 1257–1262. https://doi.org/10.1016/j.psyneuen.2009.03.013.

Hill, M. N., Patel, S., Campolongo, P., Tasker, J. G., Wotjak, C. T., & Bains, J. S. (2010). Functional interactions between stress and the endocannabinoid system: from synaptic signaling to behavioral output. *Journal of Neuroscience, 30*(45), 14980–14986. https://doi.org/10.1523/JNEUROSCI.4283-10.2010.

Hill, M. N., Patel, S., Carrier, E. J., Rademacher, D. J., Ormerod, B. K., Hillard, C. J., et al. (2005). Downregulation of endocannabinoid signaling in the hippocampus following chronic unpredictable stress. *Neuropsychopharmacology, 30*, 508–515. https://doi.org/10.1038/sj.npp.1300601.

Hill, M. N., & Tasker, J. G. (2012). Endocannabinoid signaling, glucocorticoid-mediated negative feedback, and regulation of the hypothalamic-pituitary-adrenal axis. *Neuroscience, 204*, 5–16. https://doi.org/10.1016/j.neuroscience.2011.12.030.

Hong, S., Zheng, G., Wu, X., Snider, N. T., Owyang, C., & Wiley, J. W. (2011). Corticosterone mediates reciprocal changes in CB 1 and TRPV1 receptors in primary sensory neurons in the chronically stressed rat. *Gastroenterology, 140*, 627–637. https://doi.org/10.1053/j.gastro.2010.11.003.

Hoskins, M., Pearce, J., Bethell, A., Dankova, L., Barbui, C., Tol, W. A., et al. (2015). Pharmacotherapy for post-traumatic stress disorder: Systematic review and meta-analysis. *British Journal of Psychiatry*, *206*, 93–100. https://doi.org/10.1192/bjp.bp.114.148551.

Howlett, A. C. (2002). The cannabinoid receptors. *Prostaglandins and Other Lipid Mediators*, *68–69*, 619–631. https://doi.org/10.1016/S0090-6980(02)00060-6.

Hu, W., Zhang, M., Czéh, B., Zhang, W., & Flügge, G. (2011). Chronic restraint stress impairs endocannabinoid mediated suppression of GABAergic signaling in the hippocampus of adult male rats. *Brain Research Bulletin*, *85*, 374–379. https://doi.org/10.1016/j.brainresbull.2011.04.005.

Huether, G. (1996). The central adaptation syndrome: psychosocial stress as a trigger for adaptive modification of brain structure and brain function. *Progress in Neurobiology*, *48*, 569–612. https://doi.org/10.1016/0301-0082(96)00003-2.

Huether, G., Doering, S., Rüger, U., Rüther, E., & Schüssler, G. (1999). The stress-reaction process and the adaptive modification and reorganization of neuronal networks. *Psychiatry Research*, *87*, 83–95. https://doi.org/10.1016/S0165-1781(99)00044-X.

Hughes, K. C., & Shin, L. M. (2011). Functional neuroimaging studies of post-traumatic stress disorder. *Expert Review of Neurotherapeutics*, *11*(2), 275–285. https://doi.org/10.1586/ern.10.198.

Hurd, Y. L., Michaelides, M., Miller, M. L., & Jutras-Aswad, D. (2014). Trajectory of adolescent cannabis use on addiction vulnerability. *Neuropharmacology*, *76*, 416–424. https://doi.org/10.1016/j.neuropharm.2013.07.028.

Ibeas Bih, C., Chen, T., Nunn, A. V. W., Bazelot, M., Dallas, M., & Whalley, B. J. (2015). Molecular targets of cannabidiol in neurological disorders. *Neurotherapeutics*, *12*, 699–730. https://doi.org/10.1007/s13311-015-0377-3.

Ipser, J. C., & Stein, D. J. (2012). Evidence-based pharmacotherapy of post-traumatic stress disorder (PTSD). *International Journal of Neuropsychopharmacology*, *15*, 825–840. https://doi.org/10.1017/S1461145711001209.

Izzo, A. A., Borrelli, F., Capasso, R., Di Marzo, V., & Mechoulam, R. (2009). Non-psychotropic plant cannabinoids: new therapeutic opportunities from an ancient herb. *Trends in Pharmacological Sciences*, *30*, 515–527. https://doi.org/10.1016/j.tips.2009.10.007.

Jacobs, J. A., Dellarca, A. J., Manfredi, R. A., & Harclerode. (1979). The effects of delta 9-tetrahydrocannabinol, cannabidiol, and shock on plasma corticosterone concentrations in rats. *The Journal of Pharmacy and Pharmacology*, *31*(5), 341–342. https://doi.org/10.1111/j.2042-7158.1979.tb13516.x.

Jankord, R., & Herman, J. P. (2008). Limbic regulation of hypothalamo-pituitary-adrenocortical function during acute and chronic stress. *Stress, Neurotransmitters, and Hormones: Annals of the New York Academy of Sciences*, *1148*, 64–73. https://doi.org/10.1196/annals.1410.012.

Jenniches, I., Ternes, S., Albayram, O., Otte, D. M., Bach, K., Bindila, L., et al. (2016). Anxiety, stress, and fear response in mice with reduced endocannabinoid levels. *Biological Psychiatry*, *79*, 858–868. https://doi.org/10.1016/j.biopsych.2015.03.033.

Jennings, E. M., Okine, B. N., Olango, W. M., Roche, M., & Finn, D. P. (2016). Repeated forced swim stress differentially affects formalin-evoked nociceptive behavior and the endocannabinoid system in stress normo-responsive and stress hyper-responsive rat strains. *Progress in Neuro-Psychopharmacology and Biological Psychiatry*, *64*, 181–189. https://doi.org/10.1016/j.pnpbp.2015.05.008.

Jetly, R., Heber, A., Fraser, G., & Boisvert, D. (2015). The efficacy of nabilone, a synthetic cannabinoid, in the treatment of PTSD-associated nightmares: a preliminary randomized, double-blind, placebo-controlled cross-over design study. *Psychoneuroendocrinology*, *51*, 585–588. https://doi.org/10.1016/j.psyneuen.2014.11.002.

Ji, G., & Neugebauer, V. (2014). CB1 augments mGluR5 function in medial prefrontal cortical neurons to inhibit amygdala hyperactivity in an arthritis pain model. *European Journal of Neuroscience*, *39*, 455–466. https://doi.org/10.1111/ejn.12432.

Johansen, J. P., Cain, C. K., Ostroff, L. E., & Ledoux, J. E. (2011). Molecular mechanisms of fear learning and memory. *Cell*, *147*, 509–524. https://doi.org/10.1016/j.cell.2011.10.009.

Johnson, K. M., Dewy, W. L., Ritter, K., & Beckner, J. S. (1978). Cannabinoid effects on plasma corticosterone and uptake of 3H-corticosterone by mouse brain. *European Journal of Pharmacology*, *47*, 303–310. https://doi.org/10.1016/0014-2999(78)90238-8.

Johnson, M. J., Pierce, J. D., Mavandadi, S., Klaus, J., Defelice, D., Ingram, E., et al. (2016). Mental health symptom severity in cannabis using and non-using Veterans with probable PTSD. *Journal of Affective Disorders*, *190*, 439–442. https://doi.org/10.1016/j.jad.2015.10.048.

Jovanovic, T., Norrholm, S. D., Fennell, J. E., Keyes, M., Fiallos, A. M., Myers, K. M., et al. (2009). Posttraumatic stress disorder may be associated with impaired fear inhibition: Relation to symptom severity. *Psychiatry Research*, *167*, 151–160. https://doi.org/10.1016/j.psychres.2007.12.014.

Jurkus, R., Day, H. L. L., Guimarães, F. S., Lee, J. L. C., Bertoglio, L. J., & Stevenson, C. W. (2016). Cannabidiol regulation of learned fear: Implications for treating anxiety-related disorders. *Frontiers in Pharmacology*, *7*, 454. https://doi.org/10.3389/fphar.2016.00454.

Kamprath, K., Marsicano, G., Tang, J., Monory, K., Bisogno, T., Di Marzo, V., et al. (2006). Cannabinoid CB1 receptor mediates fear extinction via habituation-like processes. *Journal of Neuroscience*, *26*(25), 6677–6686. https://doi.org/10.1523/JNEUROSCI.0153-06.2006.

Kano, M., Ohno-shosaku, T., Hashimotodani, Y., Uchigashima, M., & Watanabe, M. (2009). Endocannabinoid-mediated control of synaptic transmission. *Physiological Reviews*, *80*, 309–380. https://doi.org/10.1152/physrev.00019.2008.

Kathuria, S., Gaetani, S., Fegley, D., Valiño, F., Duranti, A., Tontini, A., et al. (2003). Modulation of anxiety through blockade of anandamide hydrolysis. *Nature Medicine*, *9*(1), 76–81. https://doi.org/10.1038/nm803.

Katona, I., & Freund, T. F. (2012). Multiple functions of endocannabinoid signaling in the brain. *Annual Review of Neuroscience*, *35*, 529–558. https://doi.org/10.1146/annurev-neuro-062111-150420.

Katona, I., Rancz, E. A., Acsády, L., Ledent, C., Mackie, K., Hájos, N., et al. (2001). Distribution of CB1 cannabinoid receptors in the amygdala and their role in the control of GABAergic transmission. *Journal of Neuroscience*, *21*(23), 9506–9518. https://doi.org/10.1523/jneurosci.21-23-09506.2001.

Kessler, R. C., Aguilar-Gaxiola, S., Alonso, J., Benjet, C., Bromet, E. J., Cardoso, G., et al. (2017). Trauma and PTSD in the WHO world mental health surveys. *European Journal of Psychotraumatology*, *8*, 1353383. https://doi.org/10.1080/20008198.2017.1353383.

Kessler, R. C., Sonnega, A., Bromet, E., Hughes, M., & Nelson, C. B. (1995). Posttraumatic stress disorder in the national comorbidity survey. *Archives of General Psychiatry*, *52*, 1048–1060. https://doi.org/10.1017/CBO9781107415324.004.

Kevorkian, S., Bonn-Miller, M. O., Belendiuk, K., Carney, D. M., Roberson-Nay, R., & Berenz, E. C. (2015). Associations among trauma, posttraumatic stress disorder, cannabis use, and cannabis use disorder in a nationally representative epidemiologic sample. *Psychology of Addictive Behaviors*, *29*(3), 633–638. https://doi.org/10.1037/adb0000110.

Kilpatrick, D. G., Acierno, R., Saunders, B., Resnick, H. S., Best, C. L., & Schnurr, P. P. (2000). Risk factors for adolescent substance abuse and dependence: data from a national sample. *Journal of Consulting and Clinical Psychology*, *68*(1), 19–30. https://doi.org/10.1037/0022-006X.68.1.19.

Kilpatrick, D. G., Resnick, H. S., Milanak, M. E., Miller, M. W., Keyes, K. M., & Friedman, M. J. (2013). National estimates of exposure to traumatic events and PTSD prevalence using DSM-IV and DSM-5 criteria. *Journal of Traumatic Stress, 26*(5), 537–547. https://doi.org/10.1002/jts.21848.

Kim, M. J., Loucks, R. A., Palmer, A. L., Brown, A. C., Solomon, K. M., Marchante, A. N., et al. (2011). The structural and functional connectivity of the amygdala: from normal emotion to pathological anxiety. *Behavioral Brain Research, 223*, 403–410. https://doi.org/10.1016/j.bbr.2011.04.025.

Kinsey, S. G., O'Neal, S. T., Long, J. Z., Cravatt, B. F., & Lichtman, A. H. (2011). Inhibition of endocannabinoid catabolic enzymes elicits anxiolytic-like effects in the marble burying assay. *Pharmacology Biochemistry and Behavior, 98*, 21–27. https://doi.org/10.1016/j.pbb.2010.12.002.

Kiritoshi, T., Ji, G., & Neugebauer, V. (2016). Rescue of impaired mGluR5-driven endocannabinoid signaling restores prefrontal cortical output to inhibit pain in arthritic rats. *Journal of Neuroscience, 36*(3), 837–850. https://doi.org/10.1523/JNEUROSCI.4047-15.2016.

Klumpers, F., Denys, D., Kenemans, J. L., Grillon, C., van der Aart, J., & Baas, J. M. P. (2012). Testing the effects of Δ9-THC and D-cycloserine on extinction of conditioned fear in humans. *Journal of Psychopharmacology, 26*(4), 471–478. https://doi.org/10.1177/0269881111431624.

Kodirov, S. A., Jasiewicz, J., Amirmahani, P., Psyrakis, D., Bonni, K., Wehrmeister, M., et al. (2010). Endogenous cannabinoids trigger the depolarization-induced suppression of excitation in the lateral amygdala. *Learning and Memory, 17*, 43–49. https://doi.org/10.1101/lm.1663410.

Koenen, K. C., Ratanatharathorn, A., Ng, L., McLaughlin, K. A., Bromet, E. J., Stein, D. J., et al. (2017). Posttraumatic stress disorder in the World Mental Health Surveys. *Psychological Medicine, 47*, 2260–2274. https://doi.org/10.1017/S0033291717000708.

Krediet, E., Janssen, D. G. A., Heerdink, E. R., Egberts, T. C. G., & Vermetten, E. (2020). Experiences with medical cannabis in the treatment of veterans with PTSD: results from a focus group discussion. *European Neuropsychopharmacology, 36*, 244–254. https://doi.org/10.1016/j.euroneuro.2020.04.009.

Krystal, J. H., & Neumeister, A. (2009). Noradrenergic and serotonergic mechanisms in the neurobiology of posttraumatic stress disorder and resilience. *Brain Research, 1293*, 13–23. https://doi.org/10.1016/j.brainres.2009.03.044.

Kulka, R. A., Schlenger, W. E., Fairbank, J. A., Hough, R. L., Jordan, B. K., Marmar, C. R., et al. (1990). Trauma and the Vietnam war generation: report of findings from the National Vietnam Veterans Readjustment Study. In *Trauma and the Vietnam war generation: Report of findings from the National Vietnam Veterans Readjustment Study Brunner/Mazel Psychosocial Stress Series, No. 18*.

Labonté, B., Azoulay, N., Yerko, V., Turecki, G., & Brunet, A. (2014). Epigenetic modulation of glucocorticoid receptors in posttraumatic stress disorder. *Translational Psychiatry, 4*, e368. https://doi.org/10.1038/tp.2014.3.

LaFrance, E. M., Glodosky, N. C., Bonn-Miller, M., & Cuttler, C. (2020). Short and long-term effects of cannabis on symptoms of Post-Traumatic Stress Disorder. *Journal of Affective Disorders, 274*, 298–304. https://doi.org/10.1016/j.jad.2020.05.132.

Laricchiuta, D., Centonze, D., & Petrosini, L. (2013). Effects of endocannabinoid and endovanilloid systems on aversive memory extinction. *Behavioral Brain Research, 256*, 101–107. https://doi.org/10.1016/j.bbr.2013.08.010.

Laviolette, S. R., & Grace, A. A. (2006a). Cannabinoids potentiate emotional learning plasticity in neurons of the medial prefrontal cortex through basolateral amygdala inputs. *Journal of Neuroscience, 26*(24), 6458–6468. https://doi.org/10.1523/JNEUROSCI.0707-06.2006.

Laviolette, S. R., & Grace, A. A. (2006b). The roles of cannabinoid and dopamine receptor systems in neural emotional learning circuits: Implications for schizophrenia and addiction. *Cellular and Molecular Life Sciences*, *63*, 1597–1613. https://doi.org/10.1007/s00018-006-6027-5.

Lee, L. (2019). PTSD and aging. *PTSD Research Quarterly*, *30*(4). www.ptsd.va.gov.

Lee, J. L. C., Bertoglio, L. J., Guimarães, F. S., & Stevenson, C. W. (2017). Cannabidiol regulation of emotion and emotional memory processing: relevance for treating anxiety-related and substance abuse disorders. *British Journal of Pharmacology*, *174*, 3242–3256. https://doi.org/10.1111/bph.13724.

Lee, T. T. Y., & Hill, M. N. (2013). Age of stress exposure modulates the immediate and sustained effects of repeated stress on corticolimbic cannabinoid CB1 receptor binding in male rats. *Neuroscience*, *249*, 106–114. https://doi.org/10.1016/j.neuroscience.2012.11.017.

Lee, D. J., Schnitzlein, C. W., Wolf, J. P., Vythilingam, M., Rasmusson, A. M., & Hoge, C. W. (2016). Psychotherapy versus pharmacotherapy for Posttraumatic Stress Disorder: Systemic review and meta-analyses to determine first-line treatments. *Depression and Anxiety*, *33*, 792–806. https://doi.org/10.1002/da.22511.

Leishman, E., Murphy, M., Mackie, K., & Bradshaw, H. B. (2018). Δ9-Tetrahydrocannabinol changes the brain lipidome and transcriptome differentially in the adolescent and the adult. *Molecular and Cell Biology of Lipids*, *1863*, 479–492. https://doi.org/10.1016/j.bbalip.2018.02.001.

Lemos, J. I., Resstel, L. B., & Guimarães, F. S. (2010). Involvement of the prelimbic prefrontal cortex on cannabidiol-induced attenuation of contextual conditioned fear in rats. *Behavioral Brain Research*, *207*, 105–111. https://doi.org/10.1016/j.bbr.2009.09.045.

Likhtik, E., & Paz, R. (2015). Amygdala-prefrontal interactions in (mal)adaptive learning. *Trends in Neurosciences*, *38*(3), 158–166. https://doi.org/10.1016/j.tins.2014.12.007.

Lin, H., Mao, S., Chen, P., & Gean, P. (2008). Chronic cannabinoid administration in vivo compromises extinction of fear memory. *Learning and Memory*, *15*, 876–884. https://doi.org/10.1101/lm.1081908.

Lin, H., Mao, C., Su, C., & Gean, P. (2009). The role of prefrontal cortex CB1 receptors in the modulation of fear memory. *Cerebral Cortex*, *19*, 165–175. https://doi.org/10.1093/cercor/bhn075.

Lipschitz, D. S., Rasmusson, A. M., Anyan, W., Gueorguieva, R., Billingslea, E. M., Cromwell, P. F., et al. (2003). Posttraumatic stress disorder and substance use in inner-city adolescent girls. *Journal of Nervous and Mental Disease*, *191*(11), 714–721. https://doi.org/10.1097/01.nmd.0000095123.68088.da.

Lisboa, S. F., Gomes, F. V., Terzian, A. L. B., Aguiar, D. C., Moreira, F. A., Resstel, L. B. M., et al. (2017). The endocannabinoid system and anxiety. *Vitamins and Hormones*, *103*, 193–279. https://doi.org/10.1016/bs.vh.2016.09.006.

Litz, B. T., Berke, D. S., Kline, N. K., Grimm, K., Rusowicz-Orazem, L., Resick, P. A. F. E. B., et al. (2019). Patterns and predictors of change in trauma-focused treatments for war-related posttraumatic stress disorder. *Journal of Consulting and Clinical Psychology*, 1019–1029. https://doi.org/10.1037/ccp0000426.

Liu, H., Petukhova, M. V., Sampson, N. A., Aguilar-Gaxiola, S., Alonso, J., Andrade, L. H., et al. (2017). Association of DSM-IV posttraumatic stress disorder with traumatic experience type and history in the World Health Organization World Mental Health surveys. *JAMA Psychiatry*, *74*(3), 270–281. https://doi.org/10.1001/jamapsychiatry.2016.3783.

Llorente-Berzal, A., Terzian, A. L. B., di Marzo, V., Micale, V., Viveros, M. P., & Wotjak, C. T. (2015). 2-AG promotes the expression of conditioned fear via cannabinoid receptor type 1 on GABAergic neurons. *Psychopharmacology*, *232*, 2811–2825. https://doi.org/10.1007/s00213-015-3917-y.

Loeffler, L. A. K, Radke, S, Habel, U, Ciric, R, Satterthwaite, T. D, Schneider, F, et al. (2018). The regulation of positive and negative emotions through instructed causal attributions in lifetime depression – A functional magnetic resonance imaging study. *NeuroImage: Clinical*, 1233–1245. https://doi.org/10.1016/j.nicl.2018.10.025.

Loflin, M., Earleywine, M., & Bonn-Miller, M. (2017). Medicinal versus recreational cannabis use: Patterns of cannabis use, alcohol use, and cued-arousal among veterans who screen positive for PTSD. *Addictive Behaviors*, *68*, 18–23. https://doi.org/10.1016/j.addbeh.2017.01.008.

Lutz, B., Marsicano, G., Maldonado, R., & Hillard, C. J. (2015). The endocannabinoid system in guarding against fear, anxiety and stress. *Nature Reviews Neuroscience*, *16*, 705–718. https://doi.org/10.1038/nrn4036.

Magruder, K., Yeager, D., Goldberg, J., Forsberg, C., Litz, B., Vaccarino, V., et al. (2015). Diagnostic performance of the PTSD checklist and the Vietnam Era Twin Registry PTSD scale. *Epidemiology and Psychiatric Sciences*, *24*(5), 415–422. https://doi.org/10.1016/j.physbeh.2017.03.040.

Malcher-Lopes, R., Di, S., Marcheselli, V. S., Weng, F., Stuart, C. T., Bazan, N. G., et al. (2006). Opposing crosstalk between leptin and glucocorticoids rapidly modulates synaptic excitation via endocannabinoid release. *Journal of Neuroscience*, *26*(24), 6643–6650. https://doi.org/10.1523/JNEUROSCI.5126-05.2006.

Marcus, D. M., Bedse, G., Gaulden, A. D., Ryan, J. D., Kondev, V., Winters, N., et al. (2020). Endocannabinoid signaling collapse mediates stress-induced amygdalo-cortical strengthening. *Neuron*, *105*, 1062–1076.e6. https://doi.org/10.1016/j.neuron.2019.12.024.

Marsicano, G., Wotjak, C. T., Azad, S. C., Bisogno, T., Rammes, G., Cascioll, M. G., et al. (2002). The endogenous cannabinoid system controls extinction of aversive memories. *Nature*, *418*, 530–534. https://doi.org/10.1038/nature00839.

Martin, M., Ledent, C., Parmentier, M., Maldonado, R., & Valverde, O. (2002). Involvement of CB1 cannabinoid receptors in emotional behavior. *Psychopharmacology*, *159*, 379–387. https://doi.org/10.1007/s00213-001-0946-5.

Mashiah, M. (2012). Medical cannabis as treatment for chronic combat PTSD: Promising results in an open pilot study. In *Patients Out of Time Conference*.

Matsuda, L. A., Lolait, S. J., Brownstein, M. J., Young, A. C., & Bonner, T. I. (1990). Structure of a cannabinoid receptor and functional expression of the cloned cDNA. *Nature*, *346*(346), 561–564. https://doi.org/10.1038/346561a0.

Mayer, T. A., Matar, M. A., Kaplan, Z., Zohar, J., & Cohen, H. (2014). Blunting of the HPA-axis underlies the lack of preventive efficacy of early post-stressor single-dose delta-9-tetrahydrocannabinol (THC). *Pharmacology Biochemistry and Behavior*, *122*, 307–318. https://doi.org/10.1016/j.pbb.2014.04.014.

McCoppin. (2018). Medical marijuana use up 83% in state; PTSD most common condition treated. *Chicago Tribune*. https://herald-review.com/news/state-and-regional/medical-marijuana-use-up-83-in-state-ptsd-most-common-condition-treated/article_75ae7e95-eda7–5153-9e6c-163ca275c1eb.html.

McEwen, B. S., Nasca, C., & Gray, J. D. (2016). Stress effects on neuronal structure: hippocampus, amygdala, and prefrontal cortex. *Neuropsychopharmacology Reviews*, *41*, 3–23. https://doi.org/10.1038/npp.2015.171.

McFarlane, A. C., Barton, C. A., Yehuda, R., & Wittert, G. (2011). Cortisol response to acute trauma and risk of posttraumatic stress disorder. *Psychoneuroendocrinology*, *36*, 720–727. https://doi.org/10.1016/j.psyneuen.2010.10.007.

Micale, V., Di Marzo, V., Sulcova, A., Wotjak, C. T., & Drago, F. (2013). Endocannabinoid system and mood disorders: Priming a target for new therapies. *Pharmacology and Therapeutics*, *138*, 18–37. https://doi.org/10.1016/j.pharmthera.2012.12.002.

Mikics, E., Vas, J., Aliczki, M., Halasz, J., & Haller, J. (2009). Interactions between the anxiogenic effects of CB1 gene disruption and 5-HT3 neurotransmission. *Behavioral Pharmacology, 20,* 265–272. https://doi.org/10.1097/FBP.0b013e32832c70b1.

Milad, M. R., Rauch, S. L., Pitman, R. K., & Quirk, G. J. (2006). Fear extinction in rats: Implications for human brain imaging and anxiety disorders. *Biological Psychology, 73,* 61–71. https://doi.org/10.1016/j.biopsycho.2006.01.008.

Milad, M. R., Wright, C. I., Orr, S. P., Pitman, R. K., Quirk, G. J., & Rauch, S. L. (2007). Recall of fear extinction in humans activates the ventromedial prefrontal cortex and hippocampus in concert. *Biological Psychiatry, 62,* 446–454. https://doi.org/10.1016/j.biopsych.2006.10.011.

Moreira, F. A., Kaiser, N., Monory, K., & Lutz, B. (2008). Reduced anxiety-like behavior induced by genetic and pharmacological inhibition of the endocannabinoid-degrading enzyme fatty acid amide hydrolase (FAAH) is mediated by CB1 receptors. *Neuropharmacology, 54,* 141–150. https://doi.org/10.1016/j.neuropharm.2007.07.005.

Morena, M., Berardi, A., Colucci, P., Palmery, M., Trezza, V., Hill, M. N., et al. (2018). Enhancing endocannabinoid neurotransmission augments the efficacy of extinction training and ameliorates traumatic stress-induced behavioral alterations in rats. *Neuropsychopharmacology, 43,* 1284–1296. https://doi.org/10.1038/npp.2017.305.

Morena, M., & Campolongo, P. (2014). The endocannabinoid system: an emotional buffer in the modulation of memory function. *Neurobiology of Learning and Memory, 112,* 30–43. https://doi.org/10.1016/j.nlm.2013.12.010.

Morena, M., Patel, S., Bains, J. S., & Hill, M. N. (2016). Neurobiological interactions between stress and the endocannabinoid system. *Neuropsychopharmacology Reviews, 41,* 80–102. https://doi.org/10.1038/npp.2015.166.

Moretti, S., Franchi, S., Castelli, M., Amodeo, G., Somaini, L., Panerai, A., et al. (2015). Exposure of adolescent mice to delta-9-tetrahydrocannabinol induces long-lasting modulation of pro- and anti-inflammatory cytokines in hypothalamus and hippocampus similar to that observed for peripheral macrophages. *Journal of Neuroimmune Pharmacology, 10,* 371–379. https://doi.org/10.1007/s11481-015-9592-2.

Morgan, A., Kondev, V., Bedse, G., Baldi, B., Marcus, D., & Patel, S. (2019). Cyclooxygenase-2 inhibition reduces anxiety-like behavior and normalizes enhanced amygdala glutamatergic transmission following chronic oral corticosterone treatment. *Neurobiology of Stress, 11,* 100190. https://doi.org/10.1016/j.ynstr.2019.100190.

Munro, S., Thomas, K. L., & Abu-Shaar, M. (1993). Molecular characterization of a peripheral receptor for cannabinoids. *Nature, 365,* 61–65. https://doi.org/10.1038/365061a0.

Myers, K. M., & Davis, M. (2007). Mechanisms of fear extinction. *Molecular Psychiatry, 12,* 120–150. https://doi.org/10.1038/sj.mp.4001939.

Naidu, P. S., Varvel, S. A., Ahn, K., Cravatt, B. F., Martin, B. R., & Lichtman, A. H. (2007). Evaluation of fatty acid amide hydrolase inhibition in murine models of emotionality. *Psychopharmacology, 192,* 61–70. https://doi.org/10.1007/s00213-006-0689-4.

National Academies of Sciences and Medicine, Division, Practice, & Agenda. (2017). *The health effects of cannabis and cannabinoids: The current state of evidence and recommendations for research.* National Academies Press.

Neumeister, A., Normandin, M. D., Pietrzak, R. H., Piomelli, D., Zheng, M. Q., Gujarro-Anton, A., et al. (2013). Elevated brain cannabinoid CB 1 receptor availability in post-traumatic stress disorder: A positron emission tomography study. *Molecular Psychiatry, 18,* 1034–1040. https://doi.org/10.1038/mp.2013.61.

Neumeister, A., Seidel, J., Ragen, B. J., & Pietrzak, R. H. (2015). Translational evidence for a role of endocannabinoids in the etiology and treatment of posttraumatic stress disorder. *Psychoneuroendocrinology, 51,* 577–584. https://doi.org/10.1016/j.psyneuen.2014.10.012.

Ney, L. J., Matthews, A., Bruno, R., & Felmingham, K. L. (2019). Cannabinoid interventions for PTSD: Where to next? *Progress in Neuro-Psychopharmacology and Biological Psychiatry*, *93*, 124–140. https://doi.org/10.1016/j.pnpbp.2019.03.017.

Nutt, D. J., King, L. A., & Phillips, L. D. (2010). Drug harms in the UK: A multicriteria decision analysis. *The Lancet*, *376*, 1558–1565. https://doi.org/10.1016/S0140-6736(10)61462-6.

O'Neil, M. E., Nugent, S. M., Morasco, B. J., Freeman, M., Low, A., Kondo, K., et al. (2017). Benefits and harms of plant-based Cannabis for posttraumatic stress disorder a systematic review. *Annals of Internal Medicine*, *167*, 332–340. https://doi.org/10.7326/M17-0477.

Orr, S. P., Metzger, L. J., Lasko, N. B., Macklin, M. L., Peri, T., & Pitman, R. K. (2000). De novo conditioning in trauma-exposed individuals with and without posttraumatic stress disorder. *Journal of Abnormal Psychology*, *109*(2), 290–298. https://doi.org/10.1037/0021-843X.109.2.290.

Orsolini, L., Chiappini, S., Volpe, U., De Berardis, D., Latini, R., Papanti, G. D., et al. (2019). Use of medicinal cannabis and synthetic cannabinoids in post-traumatic stress disorder (PTSD): A systematic review. *Medicina (Lithuania)*, *55*, 525. https://doi.org/10.3390/medicina55090525.

Pai, A., Suris, A. M., & North, C. S. (2017). Posttraumatic stress disorder in the DSM-5: Controversy, change, and conceptual considerations. *Behavioral Science*, 7, 7. https://doi.org/10.3390/bs7010007.

Pamplona, F. A., Bitencourt, R. M., & Takahashi, R. N. (2008). Short- and long-term effects of cannabinoids on the extinction of contextual fear memory in rats. *Neurobiology of Learning and Memory*, *90*, 290–293. https://doi.org/10.1016/j.nlm.2008.04.003.

Pamplona, F. A., Prediger, R. D. S., Pandolfo, P., & Takahashi, R. N. (2006). The cannabinoid receptor agonist WIN 55,212–2 facilitates the extinction of contextual fear memory and spatial memory in rats. *Psychopharmacology*, *188*, 641–649. https://doi.org/10.1007/s00213-006-0514-0.

Papini, S., Ruglass, L. M., Lopez-Castro, T., Powers, M. B., Smits, J. A. J., & Hien, D. A. (2017). Chronic cannabis use is associated with impaired fear extinction in humans. *Journal of Abnormal Psychology*, *126*(1), 117–124. https://doi.org/10.1037/abn0000224.

Passie, T., Emrich, H. M., Karst, M., Brandt, S. D., & Halpern, J. H. (2012). Mitigation of post-traumatic stress symptoms by Cannabis resin: A review of the clinical and neurobiological evidence. *Drug Testing and Analysis*, *4*, 649–659. https://doi.org/10.1002/dta.1377.

Patel, S., Hill, M. N., Cheer, J. F., Wotjak, C. T., & Holmes, A. (2017). The endocannabinoid system as a target for novel anxiolytic drugs. *Neuroscience and Biobehavioral Reviews*, *76*, 56–66. https://doi.org/10.1016/j.neubiorev.2016.12.033.

Patel, S., & Hillard, C. J. (2008). Adaptations in endocannabinoid signaling in response to repeated homotypic stress: A novel mechanism for stress habituation. *European Journal of Neuroscience*, *27*, 2821–2829. https://doi.org/10.1111/j.1460-9568.2008.06266.x.

Patel, S., Roelke, C. T., Rademacher, D. J., Cullinan, W. E., & Hillard, C. J. (2004). Endocannabinoid signaling negatively modulates stress-induced activation of the hypothalamic-pituitary-adrenal axis. *Endocrinology*, *145*(12), 5431–5438. https://doi.org/10.1210/en.2004-0638.

Patel, S., Roelke, C. T., Rademacher, D. J., & Hillard, C. J. (2005). Inhibition of restraint stress-induced neural and behavioral activation by endogenous cannabinoid signaling. *European Journal of Neuroscience*, *21*(4), 1057–1069. https://doi.org/10.1111/j.1460-9568.2005.03916.x.

Perra, S., Pillolla, G., Luchicchi, A., & Pistis, A. (2008). Alcohol inhibits spontaneous activity of basolateral amygdala projection neurons in the rat: Involvement of the endocannabinoid system. *Alcoholism: Clinical and Experimental Research*, *32*(3), 443–449. https://doi.org/10.1111/j.1530-0277.2007.00588.x.

Pertwee, R. G. (1974). Tolerance to the effect of Δ1-tetrahydrocannabinol on corticosterone levels in mouse plasma produced by repeated administration of cannabis extract or Δ1-tetrahydrocannabinol. *British Journal of Pharmacology*, *51*, 391–397. https://doi.org/10.1111/j.1476-5381.1974.tb10674.x.

Phan, K. L., Angstadt, M., Golden, J., Onyewuenyi, I., Popovska, A., & De Wit, H. (2008). Cannabinoid modulation of amygdala reactivity to social signals of threat in humans. *Journal of Neuroscience*, *28*(10), 2313–2319. https://doi.org/10.1523/JNEUROSCI.5603-07.2008.

Pickens, C. L., & Theberge, F. R. (2014). Blockade of CB1 receptors prevents retention of extinction but does not increase low pre-incubated conditioned fear in the fear incubation procedure. *Behavioural Pharmacology*, *25*(1), 23–31. https://doi.org/10.1097/FBP.0000000000000020.Blockade.

Pietrzak, R. H., Huang, Y., Corsi-Travali, S., Zheng, M., Lin, S., Henry, S., et al. (2014). Cannabinoid type 1 receptor availability in the amygdala mediates threat processing in trauma survivors. *Neuropsychopharmacology*, *39*, 2519–2528. https://doi.org/10.1038/npp.2014.110.

Pineles, S., Mostoufi, S. M., Ready, C. B., Street, A. E., Griffin, M. G., & Resick, P. A. (2011). Trauma reactivity, avoidant coping, and PTSD symptoms: a moderating relationship? *Journal of Abnormal Psychology*, *120*(1), 240–246. https://doi.org/10.1037/a0022123.Trauma.

Pistis, M., Ferraro, L., Pira, L., Flore, G., Tanganelli, S., Gessa, G. L., et al. (2002). Δ9-Tetrahydrocannabinol decreases extracellular GABA and increases extracellular glutamate and dopamine levels in the rat prefrontal cortex: An in vivo microdialysis study. *Brain Research*, *948*, 155–158. https://doi.org/10.1016/S0006-8993(02)03055-X.

Pistis, M., Muntoni, A. L., Pillolla, G., & Gessa, G. L. (2002). Cannabinoids inhibit excitatory inputs to neurons in the shell of the nucleus accumbens: An in vivo electrophysiological study. *European Journal of Neuroscience*, *15*, 1795–1802. https://doi.org/10.1046/j.1460-9568.2002.02019.x.

Pistis, M., Perra, S., Pillolla, G., Melis, M., Gessa, G. L., & Muntoni, A. L. (2004). Cannabinoids modulate neuronal firing in the rat basolateral amygdala: Evidence for CB1- and non-CB1-mediated actions. *Neuropharmacology*, *46*, 115–125. https://doi.org/10.1016/j.neuropharm.2003.08.003.

Potter, C. M., Vujanovic, A. A., Marshall-Berenz, E. C., Bernstein, A., & Bonn-Miller, M. O. (2011). Posttraumatic stress and marijuana use coping motives: The mediating role of distress tolerance. *Journal of Anxiety Disorders*, *25*, 437–443. https://doi.org/10.1016/j.janxdis.2010.11.007.

Prini, P., Penna, F., Sciuccati, E., Alberio, T., & Rubino, T. (2017). Chronic Δ9-THC exposure differently affects histone modifications in the adolescent and adult rat brain. *International Journal of Molecular Sciences*, *18*, 2094. https://doi.org/10.3390/ijms18102094.

Puder, M., Weidenfeld, J., Chowers, I., Nir, I., Conforti, N., & Siegel, R. A. (1982). Corticotrophin and corticosterone secretion following Δ1-tetrahydrocannabinol, in intact and in hypothalamic deafferentated male rats. *Experimental Brain Research*, *46*, 85–88. https://doi.org/10.1007/BF00238101.

Qin, Z., Zhou, X., Pandey, N. R., Vecchiarelli, H. A., Stewart, C. A., Zhang, X., et al. (2015). Chronic stress induces anxiety via an amygdalar intracellular cascade that impairs endocannabinoid signaling. *Neuron*, *85*, 1319–1331. https://doi.org/10.1016/j.neuron.2015.02.015.

Rabinak, C. A., Angstadt, M., Lyons, M., Mori, S., Milad, M. R., Liberzon, I., et al. (2014). Cannabinoid modulation of prefrontal-limbic activation during fear extinction learning and recall in humans. *Neurobiology of Learning and Memory*, *113*, 125–134. https://doi.org/10.1016/j.nlm.2013.09.009.

Rabinak, C. A., Angstadt, M., Sripada, C. S., Abelson, J. L., Liberzon, I., Milad, M. R., et al. (2013). Cannabinoid facilitation of fear extinction memory recall in humans. *Neuropharmacology*, *64*, 396–402. https://doi.org/10.1016/j.neuropharm.2012.06.063.

Rabinak, C. A., Blanchette, A., Zabik, N. L., Peters, C., Marusak, H. A., Iadipaolo, A., et al. (2020). Cannabinoid modulation of corticolimbic activation to threat in trauma-exposed adults: A preliminary study. *Psychopharmacology*, *237*, 1813–1826. https://doi.org/10.1007/s00213-020-05499-8.

Rabinak, C. A., MacNamara, A., Kennedy, A. E., Angstadt, M., Stein, M. B., Liberzon, I., et al. (2014). Focal and aberrant prefrontal engagement during emotion regulation in veterans with posttraumatic stress disorder. *Depression and Anxiety*, *31*, 851–861. https://doi.org/10.1002/da.22243.

Rabinak, C. A., Mori, S., Lyons, M., Milad, M. R., & Luan Phan, K. (2017). Acquisition of CS-US contingencies during Pavlovian fear conditioning and extinction in social anxiety disorder and posttraumatic stress disorder. *Journal of Affective Disorders*, *207*, 76–85. https://doi.org/10.1016/j.jad.2016.09.018.

Rademacher, D. J., Meier, S. E., Shi, L., Vanessa Ho, W.-S., Jarrahian, A., & Hillard, C. J. (2008). Effects of acute and repeated restraint stress on endocannabinoid content in the amygdala, ventral striatum, and medial prefrontal cortex in mice. *Neuropharmacology*, *54*, 108–116. https://doi.org/10.1016/j.neuropharm.2007.06.012.

Ramikie, T. S., Nyilas, R., Bluett, R. J., Gamble-George, J. C., Hartley, N. D., Mackie, K., et al. (2014). Multiple mechanistically distinct modes of endocannabinoid mobilization at central amygdala glutamatergic synapses. *Neuron*, *81*, 1111–1125. https://doi.org/10.1016/j.neuron.2014.01.012.

Rauch, S. A. M., Eftekhari, A., & Ruzek, J. I. (2012). Review of exposure therapy: A gold standard for PTSD treatment. *Journal of Rehabilitation Research and Development*, *49*(5), 679–688. https://doi.org/10.1682/JRRD.2011.08.0152.

Reich, C. G., Mohammadi, M. H., & Alger, B. E. (2008). Endocannabinoid modulation of fear responses: Learning and state-dependent performance effects. *Journal of Psychopharmacology*, *22*(7), 769–777. https://doi.org/10.1177/0269881107083999.

Reich, C. G., Taylor, M. E., & McCarthy, M. M. (2009). Differential effects of chronic unpredictable stress on hippocampal CB1 receptors in male and female rats. *Behavioral Brain Research*, *203*, 264–269. https://doi.org/10.1016/j.bbr.2009.05.013.

Ren, M., Tang, Z., Wu, X., Spengler, R., Jiang, H., Yang, Y., et al. (2019). The origins of cannabis smoking: Chemical residue evidence from the first millennium BCE in the Pamirs. *Science Advances*, *5*, eaaw1391. https://doi.org/10.1126/sciadv.aaw1391.

Resstel, L. B. M., Joca, S. R. L., Moreira, F. A., Corrêa, F. M. A., & Guimarães, F. S. (2006). Effects of cannabidiol and diazepam on behavioral and cardiovascular responses induced by contextual conditioned fear in rats. *Behavioral Brain Research*, *172*, 294–298. https://doi.org/10.1016/j.bbr.2006.05.016.

Rey, A. A., Purrio, M., Viveros, M., & Lutz, B. (2012). Biphasic effects of cannabinoids in anxiety responses: CB1 and GABA B receptors in the balance of GABAergic and glutamatergic neurotransmission. *Neuropsychopharmacology*, *37*, 2624–2634. https://doi.org/10.1038/npp.2012.123.

Riebe, C. J., & Wotjak, C. T. (2011). Endocannabinoids and stress. *Stress*, *14*(4), 384–397. https://doi.org/10.3109/10253890.2011.586753.

Rohrbach, L. A., Grana, R., Vernberg, E., Sussman, S., & Ping, S. (2009). Impact of Hurricane Rita on adolescent substance use. *Psychiatry*, *72*(3), 222–237. https://doi.org/10.1521/psyc.2009.72.3.222.

Roitman, P., Mechoulam, R., Cooper-Kazaz, R., & Shalev, A. (2014). Preliminary, open-label, pilot study of add-on oral Δ9-tetrahydrocannabinol in chronic posttraumatic stress disorder. *Clinical Drug Investigation*, *34*, 587–591. https://doi.org/10.1007/s40261-014-0212-3.

Romero-Sandoval, E. A., Fincham, J. E., Kolano, A. L., Sharpe, B. N., & Alvarado-Vázquez, P. A. (2018). Cannabis for chronic pain: challenges and considerations. *Pharmacotherapy*, *38*(6), 651–662. https://doi.org/10.1002/phar.2115.

Ross, D. A., Arbuckle, M. R., Travis, M. J., Dwyer, J. B., van Schalkwyk, G. I., & Ressler, K. J. (2017). An integrated neuroscience perspective on formulation and treatment planning for posttraumatic stress disorder: An educational review. *JAMA Psychiatry*, *74*(4), 407–415. https://doi.org/10.1001/jamapsychiatry.2016.3325.

Rossi, S., De Chiara, V., Musella, A., Kusayanagi, H., Mataluni, G., Bernardi, G., et al. (2008). Chronic psychoemotional stress impairs cannabinoid-receptor-mediated control of GABA transmission in the striatum. *Journal of Neuroscience*, *28*(29), 7284–7292. https://doi.org/10.1523/JNEUROSCI.5346-07.2008.

Rossi, S., De Chiara, V., Musella, A., Sacchetti, L., Cantarella, C., Castelli, M., et al. (2010). Preservation of striatal cannabinoid CB1 receptor function correlates with the antianxiety effects of fatty acid amide hydrolase inhibition. *Molecular Pharmacology*, *78*(2), 260–268. https://doi.org/10.1124/mol.110.064196.

Rubino, T., Realini, N., Castiglioni, C., Guidali, C., Viganó, D., Marras, E., et al. (2008). Role in anxiety behavior of the endocannabinoid system in the prefrontal cortex. *Cerebral Cortex*, *18*, 1292–1301. https://doi.org/10.1093/cercor/bhm161.

Ruhl, T., Zeymer, M., & von der Emde, G. (2017). Cannabinoid modulation of zebrafish fear learning and its functional analysis investigated by c-fos expression. *Pharmacology Biochemistry and Behavior*, *153*, 18–31. https://doi.org/10.1016/j.pbb.2016.12.005.

Sano, K., Koushi, E., Irie, K., Higuchi, S., Tsuchihashi, R., Kinjo, J., et al. (2009). Delta(9)-tetrahydrocannabinol enhances an increase of plasma corticosterone levels induced by forced swim-stress. *Biological and Pharmaceutical Bulletin*, *32*(12), 2065–2067. https://doi.org/10.1248/bpb.32.2065.

Sartim, A. G., Guimarães, F. S., & Joca, S. R. L. (2016). Antidepressant-like effect of cannabidiol injection into the ventral medial prefrontal cortex-possible involvement of 5-HT1A and CB1 receptors. *Behavioral Brain Research*, *303*, 218–227. https://doi.org/10.1016/j.bbr.2016.01.033.

Sartori, S. B., & Singewald, N. (2019). Novel pharmacological targets in drug development for the treatment of anxiety and anxiety-related disorders. *Pharmacology and Therapeutics*, *204*, 107402. https://doi.org/10.1016/j.pharmthera.2019.107402.

Sautter, F. J., Bissette, G., Wiley, J., Manguno-Mire, G., Schoenbachler, B., Myers, L., et al. (2003). Corticotropin-releasing factor in posttraumatic stress disorder (PTSD) with secondary psychotic symptoms, nonpsychotic PTSD, and healthy control subjects. *Biological Psychiatry*, *54*, 1382–1388. https://doi.org/10.1016/S0006-3223(03)00571-7.

Sbarski, B., & Akirav, I. (2020). Cannabinoids as therapeutics for PTSD. *Pharmacology and Therapeutics*, *211*(107), 551. https://doi.org/10.1016/j.pharmthera.2020.107551.

Schiavon, A. P., Bonato, J. M., Milani, H., Guimarães, F. S., & Weffort de Oliveira, R. M. (2016). Influence of single and repeated cannabidiol administration on emotional behavior and markers of cell proliferation and neurogenesis in non-stressed mice. *Progress in Neuro-Psychopharmacology and Biological Psychiatry*, *64*, 27–34. https://doi.org/10.1016/j.pnpbp.2015.06.017.

Schnurr, P. P., Lunney, C. A., Sengupta, A., & Waelde, L. C. (2003). A descriptive analysis of PTSD chronicity in Vietnam veterans. *Journal of Traumatic Stress*, *16*(6), 545–553. https://doi.org/10.1023/B:JOTS.0000004077.22408.cf.

Schramm-Sapyta, N. L., Cha, Y. M., Chaudhry, S., Wilson, W. A., Swartzwelder, H. S., & Kuhn, C. M. (2007). Differential anxiogenic, aversive, and locomotor effects of THC in adolescent and adult rats. *Psychopharmacology*, *191*, 867–877. https://doi.org/10.1007/s00213-006-0676-9.

Sciolino, N. R., Zhou, W., & Hohmann, A. G. (2011). Enhancement of endocannabinoid signaling with JZL184, an inhibitor of the 2-arachidonoylglycerol hydrolyzing enzyme monoacylglycerol lipase, produces anxiolytic effects under conditions of high environmental aversiveness in rats. *Pharmacological Research*, *64*, 226–234. https://doi.org/10.1016/j.phrs.2011.04.010.

Segev, A., & Akirav, I. (2016). Cannabinoids and glucocorticoids in the basolateral amygdala modulate hippocampal-accumbens plasticity after stress. *Neuropsychopharmacology*, *41*, 1066–1079. https://doi.org/10.1038/npp.2015.238.

Shannon, S., & Opila-Lehman, J. (2016). Effectiveness of cannabidiol oil for pediatric anxiety and insomnia as part of posttraumatic stress disorder: A case report. *The Permanente Journal*, *20*(4). 16-005 https://doi.org/10.7812/TPP/16-005.

Sheynin, J., & Liberzon, I. (2017). Circuit dysregulation and circuit-based treatments in posttraumatic stress disorder. *Neuroscience Letters*, *649*, 133–138. https://doi.org/10.1016/j.neulet.2016.11.014.

Shin, H. J., Greenbaum, M. A., Jain, S., & Rosen, C. S. (2014). Associations of psychotherapy dose and SSRI or SNRI refills with mental health outcomes among veterans with PTSD. *Psychiatric Services*, *65*(10), 1244–1248. https://doi.org/10.1176/appi.ps.201300234.

Shin, L. M., Orr, S. P., Carson, M. A., Rauch, S. L., Macklin, M. L., Lasko, N. B., et al. (2004). Regional cerebral blood flow in the amygdala and medial prefrontal cortex during traumatic imagery in male and female Vietnam Veterans with PTSD. *Archives of General Psychiatry*, *61*, 168–176. https://doi.org/10.1001/archpsyc.61.2.168.

Shin, L. M., Rauch, S. L., & Pitman, R. K. (2006). Amygdala, medial prefrontal cortex, and hippocampal function in PTSD. *Annals of the New York Academy of Sciences*, *1071*, 67–79. https://doi.org/10.1196/annals.1364.007.

Shishko, I., Oliveira, R., Moore, T. A., & Almeida, K. (2018). A review of medical marijuana for the treatment of posttraumatic stress disorder: Real symptom re-leaf or just high hopes? *Mental Health Clinician*, *8*(2), 86–94. https://doi.org/10.9740/mhc.2018.03.086.

Shonesy, B. C., Bluett, R. J., Ramikie, T. S., Báldi, R., Hermanson, D. J., Kingsley, P. J., et al. (2014). Genetic disruption of 2-arachidonoylglycerol synthesis reveals a key role for endocannabinoid signaling in anxiety modulation. *Cell Reports*, *9*, 1644–1653. https://doi.org/10.1016/j.celrep.2014.11.001.

Sigurdsson, T., & Duvarci, S. (2016). Hippocampal-prefrontal interactions in cognition, behavior and psychiatric disease. *Frontiers in Systems Neuroscience*, *9*, 190. https://doi.org/10.3389/fnsys.2015.00190.

Smith, N. B., Doran, J. M., Sippel, L. M., & Harpaz-Rotem, I. (2017). Fear extinction and memory reconsolidation as critical components in behavioral treatment for posttraumatic stress disorder and potential augmentation of these processes. *Neuroscience Letters*, *649*, 170–175. https://doi.org/10.1016/j.neulet.2017.01.006.

Song, C., Stevenson, C. W., Guimaraes, F. S., & Lee, J. L. C. (2016). Bidirectional effects of cannabidiol on contextual fear memory extinction. *Frontiers in Pharmacology*, *7*, 493. https://doi.org/10.3389/fphar.2016.00493.

Spagnolo, P. A., Ramchandani, V. A., Schwandt, M. L., Kwako, L. E., George, D. T., Mayo, L. M., et al. (2016). FAAH gene variation moderates stress response and symptom severity in patients with posttraumatic stress disorder and comorbid alcohol dependence. *Alcoholism: Clinical and Experimental Research*, *40*(11), 2426–2434. https://doi.org/10.1111/acer.13210.

Spoont, M. R., Murdoch, M., Hodges, J., & Nugent, S. (2010). Treatment receipt by veterans after a PTSD diagnosis in PTSD, mental health, or general medical clinics. *Psychiatric Services*, *61*(1), 58–63.

Steckler, T., & Risbrough, V. (2012). Pharmacological treatment of PTSD—Established and new approaches. *Neuropharmacology*, *62*, 617–627. https://doi.org/10.1016/j.neuropharm.2011.06.012.

Steenkamp, M. M., Litz, B. T., Hoge, C. W., & Marmar, C. R. (2015). Psychotherapy for military-related PTSD: a review of randomized clinical trials. *JAMA: The Journal of the American Medical Association, 314*(5), 489–500. https://doi.org/10.1001/jama.2015.8370.

Steenkamp, M. M., Litz, B. T., & Marmar, C. R. (2020). First-line psychotherapies for military-related PTSD. *JAMA, 323*(7), 656–657. https://doi.org/10.1037/ccp0000426.

Stein, D. J., Ipser, J. C., Seedat, S., Sager, C., & Amos, T. (2006). Pharmacotherapy for post traumatic stress disorder (PTSD). *Cochrane Database of Systematic Reviews*, (Issue 1). Art. No.:CD002795 https://doi.org/10.1002/14651858.cd002795.pub2.

Steiner, M. A., & Wotjak, C. T. (2008). Role of the endocannabinoid system in regulation of the hypothalamic-pituitary-adrenocortical axis. *Progress in Brain Research, 170*, 397–432. https://doi.org/10.1016/S0079-6123(08)00433-0.

Stoppelbein, L., Greening, L., & Fite, P. (2012). The role of cortisol in PTSD among women exposed to a trauma-related stressor. *Journal of Anxiety Disorders, 26*, 352–358. https://doi.org/10.1016/j.janxdis.2011.12.004.

Sugiura, T., Kondo, S. S., Nakane, S., Shinoda, A., Itoh, K., Yamashita, A., et al. (1995). 2-arachidonoylglycerol: A possible endogenous cannabinoid receptor ligand in brain. *Biochemical and Biophysical Research Communications, 215*(1), 89–97. https://doi.org/10.1006/bbrc.1995.2437.

Suris, A., Link-Malcolm, J., Chard, K., Ahn, C., & North, C. (2013). A randomized clinical trial of cognitive processing therapy for veterans with PTSD related to military sexual trauma. *Journal of Traumatic Stress, 26*, 28–37. https://doi.org/10.1002/jts.

Suzuki, A., Josselyn, S. A., Frankland, P. W., Masushige, S., Silva, A. J., & Kida, S. (2004). Memory reconsolidation and extinction have distinct temporal and biochemical signatures. *Journal of Neuroscience, 24*(20), 4787–4795. https://doi.org/10.1523/JNEUROSCI.5491-03.2004.

Taber, K. H., & Hurley, R. A. (2009). Endocannabinoids: stress, anxiety, and fear. *Journal of Neuropsychiatry and Clinical Neurosciences, 21*(2), 108–113. https://doi.org/10.1176/jnp.2009.21.2.iv.

Temple, E. C., Driver, M., & Brown, R. F. (2014). Cannabis use and anxiety: Is stress the missing piece of the puzzle? *Frontiers in Psychiatry, 5*, 168. https://doi.org/10.3389/fpsyt.2014.00168.

Tull, M. T., McDermott, M. J., & Gratz, K. L. (2016). Marijuana dependence moderates the effect of posttraumatic stress disorder on trauma cue reactivity in substance dependent patients. *Drug and Alcohol Dependence, 159*, 219–226. https://doi.org/10.1016/j.drugalcdep.2015.12.014.

Urigüen, L., Pérez-Rial, S., Ledent, C., Palomo, T., & Manzanares, J. (2004). Impaired action of anxiolytic drugs in mice deficient in cannabinoid CB1 receptors. *Neuropharmacology, 46*, 966–973. https://doi.org/10.1016/j.neuropharm.2004.01.003.

van Ast, V. A., Cornelisse, S., Meeter, M., Joëls, M., & Kindt, M. (2013). Time-dependent effects of cortisol on the contextualization of emotional memories. *Biological Psychiatry, 74*, 809–816. https://doi.org/10.1016/j.biopsych.2013.06.022.

Vaughan, C. A., Schell, T. L., Tanielian, T., Jaycox, L. H., & Marshall, G. N. (2014). Prevalence of mental health problems among Iraq and Afghanistan veterans who have and have not received VA services. *Psychiatric Services, 65*(6), 833–835. https://doi.org/10.1176/appi.ps.201300111.

Vlahov, D., Galea, S., Resnick, H., Ahern, J., Boscarino, J. A., Bucuvalas, M., et al. (2002). Increased use of cigarettes, alcohol, and marijuana among Manhattan, New York, residents after the September 11th terrorist attacks. *American Journal of Epidemiology, 155*(11), 988–996. https://doi.org/10.1093/aje/155.11.988.

Wade, M. R., Degroot, A., & Nomikos, G. G. (2006). Cannabinoid CB1 receptor antagonism modulates plasma corticosterone in rodents. *European Journal of Pharmacology, 551*, 162–167. https://doi.org/10.1016/j.ejphar.2006.08.083.

Wahbeh, H., & Oken, B. S. (2013). Salivary cortisol lower in posttraumatic stress disorder. *Journal of Traumatic Stress*, *26*, 241–248. https://doi.org/10.1002/jts.

Wamsteeker, J. I., Kuzmiski, B., & Bains, J. S. (2010). Repeated stress impairs endocannabinoid signaling in the paraventricular nucleus of the hypothalamus. *Journal of Neuroscience*, *30*(33), 11188–11196. https://doi.org/10.1523/JNEUROSCI.1046-10.2010.

Wang, W., Sun, D., Pan, B., Roberts, C. J., Sun, X., Hillard, C. J., et al. (2010). Deficiency in endocannabinoid signaling in the nucleus accumbens induced by chronic unpredictable stress. *Neuropsychopharmacology*, *35*, 2249–2261. https://doi.org/10.1038/npp.2010.99.

Watts, B. V., Shiner, B., Zubkoff, L., Carpenter-Song, E., Ronconi, J. M., & Coldwell, C. M. (2014). Implementation of evidence-based psychotherapies for posttraumatic stress disorder in VA specialty clinics. *Psychiatric Services*, *65*(5), 648–653. https://doi.org/10.1176/appi.ps.201300176.

Wędzony, K., & Chocyk, A. (2009). Cannabinoid CB1 receptors in rat medial prefrontal cortex are colocalized with calbindin- but not parvalbumin- and calretinin-positive GABA-ergic neurons. *Pharmacological Reports*, *61*, 1000–1007. https://doi.org/10.1016/S1734-1140(09)70161-6.

Weidenfeld, J., Feldman, S., & Mechoulam, R. (1994). Effect of the brain constituent anandamide, a cannabinoid receptor agonist, on the hypothalamo-pituitary-adrenal axis in the rat. *Neuroendocrinology*, *59*, 110–112. https://doi.org/10.1159/000126646.

Wessa, M., & Flor, H. (2007). Failure of extinction of fear responses in posttraumatic stress disorder: evidence from second-order conditioning. *American Journal of Psychiatry*, *164*, 1684–1692. https://doi.org/10.1176/appi.ajp.2007.07030525.

Wicking, M., Steiger, F., Nees, F., Diener, S. J., Grimm, O., Ruttorf, M., et al. (2016). Deficient fear extinction memory in posttraumatic stress disorder. *Neurobiology of Learning and Memory*, *136*, 116–126. https://doi.org/10.1016/j.nlm.2016.09.016.

Wilkinson, S. T., Stefanovics, E., & Rosenheck, R. A. (2018). Marijuana use is associated with worse outcomes in symptom severity and violent behavior in patients with PTSD. *The Journal of Clinical Psychiatry*, *76*(9), 1174–1180. https://doi.org/10.4088/JCP.14m09475.Marijuana.

Yehuda, R. (2009). Status of glucocorticoid alterations in post-traumatic stress disorder. *Glucocorticoids and Mood: Annals of the New York Academy of Sciences*, *1179*, 56–69. https://doi.org/10.1111/j.1749-6632.2009.04979.x.

Yehuda, R., Golier, J. A., & Kaufman, S. (2005). Circadian rhythm of salivary cortisol in Holocaust survivors with and without PTSD. *American Journal of Psychiatry*, *162*, 998–1000. https://doi.org/10.1176/appi.ajp.162.5.998.

Yehuda, R., Halligan, S. L., Golier, J. A., Grossman, R., & Bierer, L. M. (2004). Effects of trauma exposure on the cortisol response to dexamethasone administration in PTSD and major depressive disorder. *Psychoneuroendocrinology*, *29*, 389–404. https://doi.org/10.1016/S0306-4530(03)00052-0.

Yehuda, R., Kahana, B., Binder-Brynes, K., Southwick, S. M., Mason, J. W., & Giller, E. L. (1995). Low urinary cortisol excretion in Holocaust survivors with posttraumatic survivors. *American Journal of Psychiatry*, *152*(7), 982–986.

Yehuda, R., Lehrner, A., & Rosenbaum, T. Y. (2015). PTSD and sexual dysfunction in men and women. *The Journal of Sexual Medicine*, *12*, 1107–1119. https://doi.org/10.1111/jsm.12856.

Zhong, P., Wang, W., Pan, B., Liu, X., Zhang, Z., Long, J. Z., et al. (2014). Monoacylglycerol lipase inhibition blocks chronic stress-induced depressive-like behaviors via activation of mTOR signaling. *Neuropsychopharmacology*, *39*, 1763–1776. https://doi.org/10.1038/npp.2014.24.

Zhu, P. J., & Lovinger, D. M. (2005). Retrograde endocannabinoid signaling in a postsynaptic neuron/synaptic bouton preparation from basolateral amygdala. *Journal of Neuroscience, 25*(26), 6199–6207. https://doi.org/10.1523/JNEUROSCI.1148-05.2005.

Zohar, J., Amital, D., Miodownik, C., Kotler, M., Bleich, A., Lane, R. M., et al. (2002). Double-blind placebo-controlled pilot study of sertraline in military veterans with post-traumatic stress disorder. *Journal of Clinical Psychopharmacology, 22*(2), 190–195. https://doi.org/10.1097/00004714-200204000-00013.

CHAPTER FIVE

Traumatic brain injury and the misuse of alcohol, opioids, and cannabis

Alejandra Jacotte-Simancas[a,b], Elizabeth A. Fucich[a,b], Zachary F. Stielper[a,b], and Patricia E. Molina[a,b,*]
[a]Department of Physiology, Louisiana State University Health Sciences Center, New Orleans, LA, United States
[b]Alcohol and Drug Abuse Center of Excellence, Louisiana State University Health Sciences Center, New Orleans, LA, United States
*Corresponding author: e-mail address: pmolin@lsuhsc.edu

Contents

Abstract

Traumatic brain injury (TBI), most often classified as concussion, is caused by biomechanical forces to the brain resulting in short- or long-term impairment in brain function. TBI resulting from military combat, sports, violence, falls, and vehicular accidents is a major cause of long-term physical, cognitive, and psychiatric dysfunction. Psychiatric disorders associated with TBI include depression, anxiety, and substance use disorder,

International Review of Neurobiology, Volume 157
ISSN 0074-7742
https://doi.org/10.1016/bs.irn.2020.09.003

all having significant implications for post-TBI recovery and rehabilitation. This chapter reviews the current preclinical and clinical literature describing the bidirectional relationship between TBI and misuse of three commonly abused drugs: alcohol, opioids, and cannabis. We highlight the influence of each of these drugs on the incidence of TBI, as well as trends in their use after TBI. Furthermore, we discuss factors that may underlie post-injury substance use. Understanding the complex relationship between TBI and substance misuse will enhance the clinical treatment of individuals suffering from these two highly comorbid conditions.

1. Traumatic brain injury (TBI)

Traumatic brain injury (TBI) is a condition defined by impairment in normal brain function brought about by physical trauma to the head. There were approximately 2.87 million TBI-related emergency department visits, hospitalizations, and deaths in the United States in 2014 (Centers for Diseases Control and Prevention, 2019). Estimates from data collected in 2003 suggest that nearly 43% of patients hospitalized for TBI develop long-term disability (Selassie et al., 2008), and it is predicted that up to 5.3 million Americans suffer from TBI-related disability (Centers for Disease Control and Prevention, 2015). The incidence of TBI and prevalence of TBI-related disability are likely underestimates, however, as they do not reflect nontreatment-seeking individuals or those who received care outside of a hospital setting, and a national surveillance system that tracks TBI-related disability over time is only in early phases. Furthermore, TBI accounts for approximately 30% of injury-related deaths (Faul, Xu, Wald, & Coronado, 2010). Estimates of the annual economic burden of TBI range from $98 to $302 million in direct costs depending on the study (Humphreys, Wood, Phillips, & Macey, 2013). Although TBI-related hospitalizations and deaths have declined between 2006 and 2014, in the same timeframe, rates of TBI-related emergency department visits have more than doubled (Centers for Disease Control and Prevention, 2019).

Among the nearly 3 million reported annual TBI cases, the three most common causes are falls, motor vehicle accidents, and impacts with objects (Faul et al., 2010). Children below age four account for the highest numbers of TBI emergency room visits, followed by adolescents and adults over the age of 65, whereas the highest rates of TBI-related hospitalizations and deaths are seen in adults over 75 (Faul et al., 2010). Overall, males have a higher incidence of TBI than females, with TBI 1.4 times more prevalent in males than females (Faul et al., 2010); however, certain subpopulations of women (e.g., female athletes) may be more at risk for TBI (Covassin, Moran, & Elbin, 2016). TBI is also a common condition within the military population. Given the high incidence of TBI in military personnel returning from recent conflicts, it has been called the "signature injury" of the Iraq and Afghanistan wars (Taylor, 2012). It has been estimated that 12–23% of returning service members and Veterans suffer from TBI, with mild TBIs being the most common (MacGregor et al., 2010; O'Neil et al., 2013). In combat, most TBIs occur as a consequence of blast exposure accounting for approximately 60% of military related TBI (Bell et al., 2009).

TBI presents as a heterogeneous condition: differences in the injury mechanism, severity, and location result in varied clinical presentation (Saatman et al., 2008). The physical forces that produce TBI can be blunt force or penetrating objects; acceleration, deceleration, or rotational forces; or shockwave-like blasts; and these external forces can result in focal or diffuse injury to the brain (Fig. 1). TBIs are classified as mild, moderate, or severe based on criteria that can include posttraumatic amnesia, loss of consciousness, structural imaging, and the Glasgow coma score (Centers for Disease Control and Prevention, 2019). The majority (75–80%) of TBIs are considered mild TBIs, or concussions. Although moderate and severe TBI can be diagnosed using these criteria in combination with neurological exam findings, diagnosing mild TBI is often complicated by the lack of immediate post-concussive symptomatology and absence of imaging findings. Post-injury symptoms are myriad and vary from case to case, but can include deficits in autonomic (Takahashi, Hinson, & Baguley, 2015), motor (Kozlowski, Leasure, & Schallert, 2013), sensory (Callahan & Lim, 2018), and cognitive functions (Arciniegas, Held, & Wagner, 2002), as well as behavioral or emotional problems (Koponen, Taiminen, Hiekkanen, & Tenovuo, 2011). There are currently no therapeutic strategies to treat TBI itself; rather, TBI patients are treated based on the individual symptoms with which they present.

1.1 Mechanisms underlying brain injury

Both preclinical and clinical studies have contributed to our understanding of the pathophysiology of brain injury. Common methods of experimental TBI in laboratory animals include controlled cortical impact, fluid percussion injury, weight drop, and blast TBI (as reviewed in Dai, Ma, Le, Cao, & Wang, 2018; Xiong, Mahmood, & Chopp, 2013). These models have provided insight into the molecular and cellular changes that occur in the brain

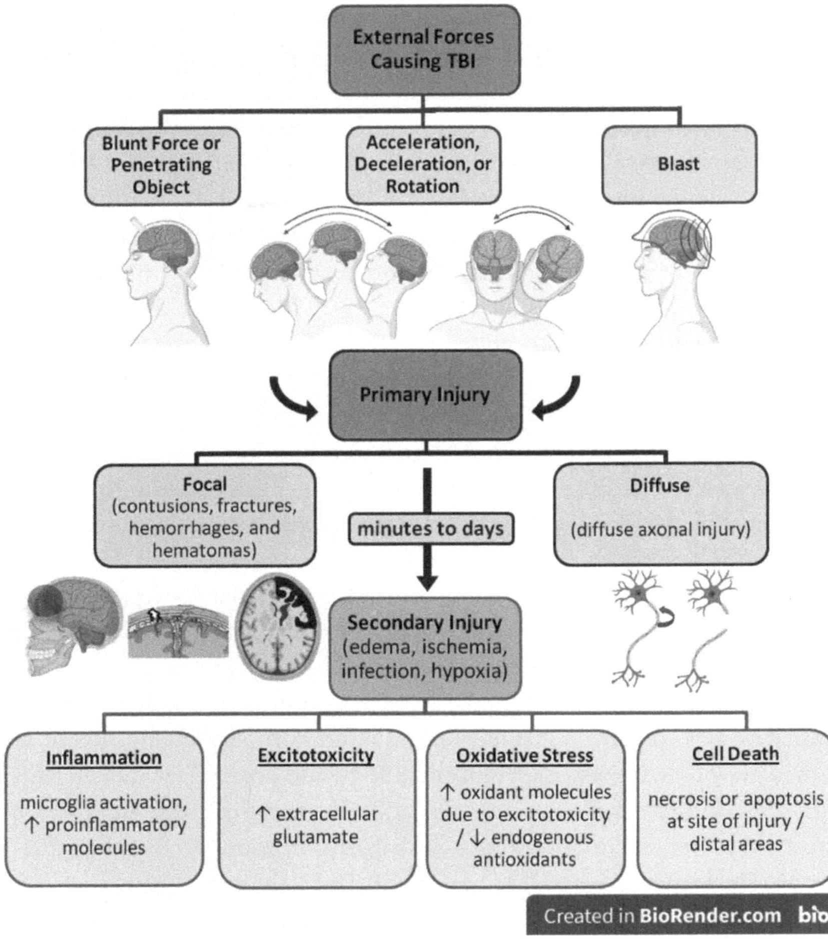

Fig. 1 Mechanisms of brain injury. Traumatic brain injury (TBI) is caused when external physical forces impair normal brain function. These external forces include blunt force to the head or objects that penetrate the skull; acceleration, deceleration, or rotational forces to the head; or shockwave-like blast forces that move through the brain. These external forces cause the immediate primary injury, which may be focal (such as contusions, fractures, hemorrhages, or hematomas) or diffuse (including diffuse axonal injury). Further damage occurs minutes to days after the primary insult via complications such as edema, ischemia, infection, or hypoxia. This secondary damage can result in brain cell damage, death, and dysfunction through mechanisms such as inflammation, excitotoxicity, oxidative stress, and cell death pathways. *Portions of this figure were created with BioRender.com.*

following TBI. Additionally, techniques such as magnetic resonance imaging (MRI), diffusion tensor imaging, functional MRI, and magnetic resonance spectroscopy have allowed for the evaluation of post-TBI structural and molecular changes without sacrificing the animal (Barkhoudarian, Hovda, & Giza, 2016).

Brain damage following TBI results from direct (primary) and indirect (secondary) injury (see Fig. 1). Primary injury is produced immediately by the traumatic external force and can be focal or diffuse. Focal injury may involve cortical or subcortical contusions, fractures, hemorrhages, and hematomas, while diffuse injury most frequently involves diffuse axonal injury (DAI; Rao & Lyketsos, 2002). DAI is produced because of acceleration–deceleration forces and angular movements of the head that lead to shearing and eventual tearing of axonal fibers (Gennarelli & Graham, 1998). Secondary brain injury occurs minutes to days after the primary injury and involves processes such as edema, ischemia, infection, and hypoxia, which can trigger molecular cascades leading to neuroinflammation, excitotoxicity, oxidative stress, and ultimately cell death in the brain.

Neuroinflammation is one of the principal underlying mechanisms of neuronal injury (Werner & Engelhard, 2007). Processes like ischemia and infection secondary to TBI can result in release of immune mediators (e.g., cytokines and chemokines) within the central nervous system. Although the early inflammatory response plays an important role in recovery from injury (Lenzlinger, Morganti-Kossmann, Laurer, & McIntosh, 2001), sustained neuroinflammation contributes both to acute pathology following TBI, such as cerebral edema, and to longer-term neuronal damage and cognitive impairment (Lloyd, Somera-Molina, Van Eldik, Watterson, & Wainwright, 2008). Thus, modulation of neuroinflammation is an attractive therapeutic target to mitigate negative consequences yet allow for reparative functions of glia.

Another process associated with secondary damage following TBI is excitotoxicity (Werner & Engelhard, 2007). Neuronal membranes become damaged as a consequence of mechanical insult, which results in excessive glutamate release. This excessive extracellular glutamate affects both neurons and astrocytes by overstimulating ionotropic and metabotropic glutamate receptors, leading to the opening of ion channels and consequential influx of sodium, calcium, and potassium (Werner & Engelhard, 2007). The overwhelming calcium influx causes indiscriminate activation of cellular signaling cascades and enzymes that can culminate in neuronal death (Wang et al., 2016).

Secondary damage can also occur from oxidative stress (Abdul-Muneer, Chandra, & Haorah, 2015). Free radicals are produced as a result of excitotoxicity and exhaustion of endogenous antioxidative mechanisms after TBI, causing DNA oxidation, lipid peroxidation, and protein carbonylation (Ladak, Enam, & Ibrahim, 2019; Werner & Engelhard, 2007). Reactive oxygen species (ROS) and reactive nitrogen species (RNS) are the main sources of oxidative stress after brain injury (Abdul-Muneer et al., 2015). Superoxide (O_2^-) is the most common ROS and is produced by mitochondria (Abdul-Muneer et al., 2015). The enzyme NADPH oxidase also plays a role in the cellular production of O_2^- after TBI. For example, an increase in NADPH was observed in mice 1h and 24–96h after controlled cortical impact, and that increase contributed to elevation of O_2^- after injury (Zhang et al., 2012). Increases in hydroxyl radicals (HO^-) have also been reported after experimental TBI. Marklund, Clausen, Lewander, and Hillered (2001) found a 250% increase in HO^- in rats after severe TBI, and that increase remained above baseline for 90min post-TBI. Nitric oxide (NO) is an RNS that is also elevated post-TBI (Clark et al., 1996; Wada, Chatzipanteli, Kraydieh, Busto, & Dietrich, 1998). Importantly, NO produces peroxynitrite when coupled with O_2^-, which can produce DNA damage, mitochondrial dysfunction, and apoptosis (Gahm, Holmin, Wiklund, Brundin, & Mathiesen, 2006).

Cell death can be the result of primary or secondary damage after TBI (Allen, 2016). Primary cell loss occurs at the site of injury immediately after trauma because of the impact, and secondary cell loss is the result of biochemical changes, which can occur proximally or distally to the site of injury minutes to days later (Stoica & Faden, 2010). Cell death may be produced via several cellular death paradigms, including apoptosis and necrosis (Allen, 2016). Apoptosis is an energy-dependent process that requires synthesis of proteins and is characterized by condensation and fragmentation of the cytoplasm and nucleus (Ladak et al., 2019). Other cells die via necrosis as a result of severe damage caused by hypoxia and ischemia (Johnston et al., 2009; Miñambres et al., 2008). Necrosis is a passive process that results from ionic imbalance within the cell that leads to swelling and rupture (Allen, 2016; Stoica & Faden, 2010). Neuronal loss after TBI has been observed in both humans and animals. For example, light microscopy evaluation revealed hippocampal cell loss in 86% of samples from human subjects that suffered fatal head injury (Kotapka, Graham, Adams, & Gennarelli, 1994). Studies in various experimental models of TBI have identified cell loss in the hippocampus and in the cortex (Henninger et al., 2005; Nakajima et al., 2010;

Tsuda et al., 2016). Such cell death is thought to contribute to cognitive dysfunction after TBI (Tsuda et al., 2016). Palacios et al. (2013) observed atrophy involving the frontal and parietal lobes in patients with severe and diffuse TBI, which correlated with deficits in declarative memory. Findings using animal models echo these results; for example, Park et al. (2014) found that animals with TBI during early life showed deficits in memory that were sustained into adulthood, and these deficits were attributed to apoptosis.

1.2 TBI and psychiatric disorders

TBI often produces neurobehavioral symptoms, including psychiatric, cognitive, and/or behavioral dysfunction (Rao & Lyketsos, 2002). Although some of this dysfunction resolves within days after the TBI, in many cases it is persistent or emerges months or years later. Those with TBI are challenged by deficits that affect professional, social, and personal areas of their lives, which may also influence the development of psychiatric disorders (Jorge, 2015).

Mood disorders are common following TBI, with prevalence rates higher than in the average population (Deb, Lyons, Koutzoukis, Ali, & McCarthy, 1999). The development of mood disorders post-BI is thought to reflect the interaction of preinjury risk factors, TBI parameters (e.g., severity, mechanism), and post-injury factors that occur in the context of cognitive and emotional dysfunction produced by TBI (Jorge, 2015). Among mood disorders, depression is one of the most common following TBI. A recent study found the prevalence of depression to be 56.3% 10 weeks post injury (Singh, Mason, Lecky, & Dawson, 2018), almost fivefold higher than the 9.4% reported in the general population (Haarasilta, Marttunen, Kaprio, & Aro, 2001). Several factors have been identified that increase the risk for developing depression after TBI, such as the severity of the injury, being female, having a history of psychiatric conditions, and intoxication at the time of injury (Singh et al., 2018). With respect to injury severity as a risk factor, this same study reported depression in 83.9% of patients with severe TBI. However, other studies (Bombardier et al., 2010) have found no association between the severity of TBI and the risk of depression. While the risk of developing depression after TBI is higher in the first year, it remains increased for decades after injury. A longitudinal study examined the severity and the trajectory of depressive disorders 2 years post injury in a sample of 1089 subjects enrolled in the

Traumatic Brain Injury Model Systems database and found that 26% of the individuals who did not have depression in the first year post injury developed depressive disorders during the second year after TBI (Hart et al., 2012). Moreover, three-fourths of individuals who were depressive during the first year after injury continued showing depressive symptomatology 2 years post-TBI (Hart et al., 2012).

TBI is also associated with anxiety and stress-related disorders, including generalized anxiety, panic disorder, phobia, posttraumatic stress, and obsessive-compulsive disorders. The most frequent anxiety disorder in TBI patients is generalized anxiety disorder (Whelan-Goodinson, Ponsford, Johnston, & Grant, 2009), which is defined as an uncontrollable and extreme worry or distress about different areas of life such as work, relationships, school, etc., accompanied by somatic symptoms such as irritability, sleep disturbances, and fatigue (Mallya, Sutherland, Pongracic, Mainland, & Ornstein, 2015). Posttraumatic stress disorder (PTSD) also commonly emerges following TBI (Whelan-Goodinson et al., 2009). TBI is often produced by car accidents, assault, gunshots, or other psychologically traumatic events, which may precipitate PTSD. One characteristic of PTSD involves re-experiencing the traumatic event via nightmares or intrusive thoughts about the trauma (Moore, Terryberry-Spohr, & Hope, 2006). Because TBI can result in loss of consciousness and posttraumatic amnesia, it has been proposed that PTSD may develop after mild TBI where the loss of consciousness and amnesia are less severe. In fact, Bryant and Harvey (1998) found that 82% of patients with mild TBI who were diagnosed with acute stress disorder 1 month following TBI satisfied the criteria for PTSD 6 months later.

Substance use disorder (SUD) is reported as the most common psychiatric disorder preinjury in TBI patients and the third most common psychiatric disorder diagnosed post-TBI (Whelan-Goodinson et al., 2009), with substance intoxication being a major risk factor for head injury. Alcohol is the most frequently used substance after TBI (Merkel et al., 2017). Following alcohol, cannabis is the most commonly used illicit substance before and after TBI (Beaulieu-Bonneau et al., 2018; Kolakowsky-Hayner et al., 2002; Whelan-Goodinson et al., 2009). In addition to using a variety of non-prescription substances, many TBI patients are prescribed opioids following injury (Hammond et al., 2015; Seal et al., 2018), which could contribute to developing an opioid use disorder (OUD). As with other psychiatric disorders, several factors may increase the risk for developing SUD after TBI, such as pre-existing SUD, age at injury, previous TBI, and sex (see Fig. 2). Age is an independent risk factor for SUD, and young

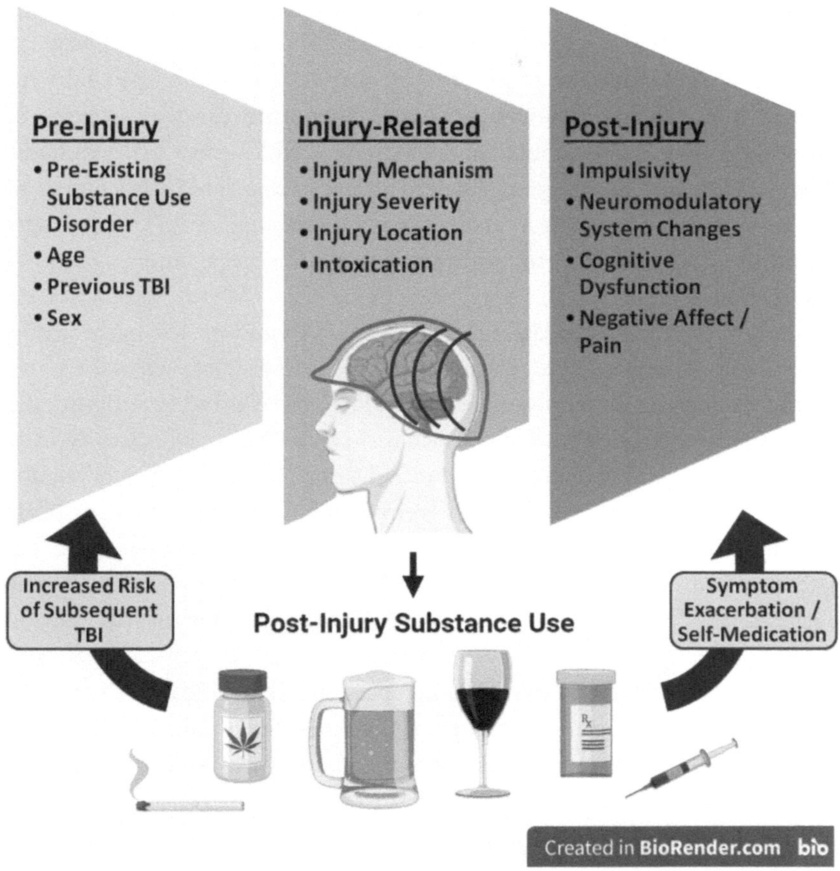

Fig. 2 Potential factors mediating post-TBI substance use. Post-TBI substance use results from the interaction of pre-existing factors prior to injury, injury-related factors, and post-injury factors. Pre-existing diagnoses of substance use disorder or TBI increase the risk of not only incurring a TBI but also subsequent substance use, as does young age and male gender. The mechanism, severity, and location of brain injury all modify the risk of substance use following injury, and intoxication at time of injury predicts post-TBI substance use. Finally, factors in recovery, such as increases in impulsivity or cognitive dysfunction and the emergence of negative affective symptoms and pain, all of which may be driven by disruptions in neuromodulatory system alterations, can contribute to post-TBI use of substances such as alcohol, opioids, and cannabis. Post-TBI substance use may acutely ameliorate TBI symptomatology but also poses the risk of exacerbating many of these symptoms or impairing TBI recovery. Furthermore, substance use post-injury increases the risk of incurring a subsequent injury, potentially perpetuating a cycle. *Portions of this figure were created with BioRender.com.*

people with a history of TBI are more likely to smoke cigarettes, binge drink alcohol, and use illicit and prescription drugs compared to young people without a history of TBI (Ilie et al., 2015). Cannella, McGary, and Ramirez (2019) postulate that a TBI occurring under the age of 19 may result in alterations in ongoing neurodevelopmental processes or neuronal circuitry, which may contribute to developing a SUD post-TBI. Sustaining multiple TBIs or TBIs of greater severity are also associated with higher risk of developing SUD (Corrigan et al., 2013; Merkel et al., 2017). Importantly, comorbid SUD also has implications for physiological and psychosocial outcomes after TBI.

Here, we review the relationship between TBI and the use of three common substances of abuse: alcohol, opioids, and cannabis, as well as the potential impact of using these substances on TBI-related morbidity and mortality. Among substances of abuse, alcohol has received the most scrutiny in the context of TBI, and studies have identified a potential bidirectional relationship between alcohol and TBI, whereby alcohol misuse represents both a risk factor for TBI and a detrimental factor in outcomes following TBI. Cannabis and opioids are the most commonly used illicit drugs of abuse in the U.S. (SAMHSA & Center for Behavioral Health Statistics and Quality, 2018). In the face of increasing medical and recreational legalization of cannabis use and the ongoing opioid epidemic, it is critical that we understand how these substances influence TBI incidence and outcomes and whether TBI represents a risk factor for cannabis or opioid misuse.

2. Alcohol use and TBI

TBI and alcohol use are highly comorbid. It is well-known that alcohol use is a predominant risk factor for traumatic injuries, with up to half of all TBIs being alcohol related (Honkanen & Visuri, 1976; Savola, Niemelä, & Hillbom, 2005). Rogers and Read (2007) concluded that the reciprocal relationship, TBI as a risk factor for subsequent substance abuse, was not supported by clinical evidence at the time, but noted the importance of identifying problem drinking in TBI patients, as alcohol impedes recovery. More recently, the question of TBI as a risk factor for alcohol use has been revisited, with Bjork and Grant (2009) calling for more in-depth clinical and preclinical work to address whether TBI can increase hazardous drinking in the absence of preinjury risk factors, as well as whether TBI could trigger relapse or escalate alcohol use in patients with a history of alcohol use disorder (AUD). Weil, Corrigan, and Karelina (2016) and

Weil, Karelina, Gaier, Corrigan, and Corrigan (2016) have since summarized preclinical studies that begin to address these questions and reiterate the need for studying and treating post-TBI problem drinking.

The underlying causes of alcohol use after TBI are debated, considering the link between TBI and alcohol in addition to several other TBI comorbidities that interact with alcohol use (see Weil, Corrigan, & Karelina, 2018). Regardless of the exact role of TBI in subsequent alcohol drinking, post-TBI alcohol use can have long-term neurobiological and psychosocial consequences, exacerbating damage to or impairing recovery of the brain following injury as well as negatively impacting a patient's quality of life for years after TBI. Treating problem drinking in patients with a history of TBI is therefore imperative and may require special considerations for this unique population.

2.1 Post-TBI alcohol use

Several clinical studies suggest a common pattern of alcohol use following TBI, in which patients will decrease their alcohol use in the initial months after injury, potentially due to hospitalization, disability, or adherence to clinical guidelines to abstain from alcohol, with drinking returning to preinjury levels in a subset of patients by 2 years post-TBI (Bombardier, Temkin, Machamer, & Dikmen, 2003; Kreutzer, Witol, & Marwitz, 1996; Ponsford, Whelan-Goodinson, & Bahar-Fuchs, 2007). Another common finding across studies is that individuals who drink most heavily post injury are those who drank most heavily preinjury (Horner et al., 2005; Kreutzer et al., 1996; Ponsford et al., 2007). There is mixed clinical evidence regarding whether a TBI increases the risk for developing AUD, with some studies showing no increased risk in TBI patients (Heltemes, Dougherty, MacGregor, & Galarneau, 2011; Miles, Graham, & Teng, 2015), and other studies demonstrating a higher prevalence of AUD in subjects with a history of TBI (Johnson, Eick-Cost, Jeffries, Russell, & Otto, 2015; Miller et al., 2013). Because of the strong interrelationship between TBI and alcohol, it is difficult to determine whether the latter finding is causal (i.e., whether TBI directly leads to the development of AUD), or whether individuals with higher risk for AUD are the same individuals that have a higher risk for TBI (see Bjork & Grant, 2009). Preclinical evidence is similarly mixed (see Weil, Corrigan, & Karelina, 2016), with some studies showing that experimental TBI did not significantly increase post-injury alcohol drinking (Lim et al., 2015; Lowing et al., 2014; Poznanski, Lesniak, Korostynski, &

Sacharczuk, 2020; Smith, Moore, & Klein, 2019) but others, including studies from our own lab, showing increased alcohol sensitivity, preference, consumption, or motivation following TBI (Fucich et al., 2019; Karelina, Nicholson, & Weil, 2018; Lowing et al., 2014; Mayeux, Teng, Katz, Gilpin, & Molina, 2015; Weil, Karelina, et al., 2016). Some preclinical work recapitulates the clinical findings that alcohol consumption may decrease initially after injury (Lowing et al., 2014; Poznanski et al., 2020), and, as in our study, that subjects with high preinjury drinking will consume higher amounts of alcohol post-TBI (Lim et al., 2015; Mayeux et al., 2015).

Several factors have been examined to determine whether they affect post-TBI alcohol use, including injury severity, number of injuries, age and sex. Some studies have found that patients with mild TBI will return to alcohol use earlier than patients with moderate or severe TBI, perhaps due to an earlier recovery of physical functioning or lack of treatment that would restrict or advise against drinking (Beaulieu-Bonneau et al., 2018), though other studies have not found a correlation between post-TBI drinking patterns and TBI severity (Horner et al., 2005). Corrigan, Bogner, and Holloman (2012) found in a secondary analysis of two cohorts that persons who had a severe TBI in adulthood, a mild TBI in adolescence, or multiple mild TBIs had greater AUD severity, and in a separate study (Corrigan et al., 2013) found that incurring a prior TBI, particularly before age six, significantly increased the likelihood of substance abuse. Indeed, other studies have shown that early life TBI increases risk for later alcohol misuse (as reviewed in Cannella et al., 2019), and preclinical work in juvenile rodents supports these clinical findings (Karelina et al., 2018; Weil, Karelina, et al., 2016). Human studies also suggest that males have a higher risk of alcohol misuse after TBI; for example, Ponsford et al. (2007) found that male gender was a significant predictor of heavy drinking 2 years post-TBI. Only recently have preclinical studies examined post-TBI alcohol consumption in females, with findings suggesting that females are not at a high risk of increasing alcohol consumption if injured in adulthood (Stielper et al., 2020), though females injured during adolescence escalate alcohol consumption in adulthood beyond that of males (Weil, Karelina, et al., 2016). Yet, it remains difficult to disentangle these risk factors for post-TBI alcohol use, as they are independent risk factors for alcohol use or TBI in themselves. Most post-TBI alcohol users are young and male; however, being young and male are also independent risk factors for using alcohol or incurring TBI (Ponsford et al., 2007). Incurring one TBI, particularly if alcohol related, increases the risk for incurring another TBI (Winqvist et al., 2008). Finally, although childhood

TBIs would precede hazardous alcohol drinking, they arise predominantly from falls and disproportionately from acts of violence (parental abuse), the latter of which is also a risk factor for developing alcohol problems in adulthood and often stems from problematic alcohol drinking in the parent (Duhaime et al., 1992; Hawkins, Catalano, & Miller, 1992; Leventhal, Martin, & Asnes, 2010).

2.2 Factors related to post-TBI alcohol use

Several potential mechanisms have been postulated to contribute to post-TBI alcohol use, including preinjury risk factors (such as those described above), post-injury psychosocial factors (incentive-motivation or self-medication), or potentially TBI neuropathology (e.g., neuroinflammation, dopamine dysregulation, or cognitive deficits) (Bjork & Grant, 2009; Rogers & Read, 2007; Weil, Corrigan, et al., 2016; see Fig. 2). Here we highlight some common TBI comorbidities that may affect, or be affected by, post-TBI alcohol use, including pre-existing alcohol use disorder, post-TBI emotional dysregulation, and pain.

2.2.1 Pre-TBI alcohol use disorder

The prevalence of a pre-TBI history of AUD has been found to range from 37% to 66% across studies (Corrigan, 1995; Parry-Jones, Vaughan, & Miles Cox, 2006). Perhaps not surprisingly, a history of problem drinking predicted post-TBI alcohol-related problems assessed 1 year after injury (Bombardier et al., 2003). Though the relationship between pre-TBI AUD and post-TBI alcohol misuse is strong, the picture is more complex. Not all patients with an AUD at the time of injury resume their problem drinking or to the same extent post-TBI, not all patients who develop problem drinking after TBI have a history of pre-TBI AUD, and not all patients with an AUD history will relapse following a brain injury (Bombardier et al., 2003; see Herrold et al., 2015). Future studies in patients with a pre-TBI AUD should investigate post-TBI drinking across different TBI severities or at extended time points to better characterize the relationship between pre-TBI AUD and post-TBI drinking and elucidate the influence of time since injury or injury-related hospitalization or treatment on drinking in this prevalent patient population.

While consumption of alcohol post-TBI worsens TBI outcomes, a history of alcohol misuse has also been found to worsen outcomes after TBI, including increasing the risk for suffering more severe TBI and developing post-TBI mood disorders, increasing hospitalization lengths, worsening vocational outcomes, and reducing neurocognitive test scores and frontal gray matter volume (Bigler et al., 1996; Jorge et al., 2005; Kaplan & Corrigan, 1992; Tate, Freed, Bombardier, Harter, & Brinkman, 1999). Further investigations are needed to delineate whether these worsened TBI outcomes contribute to post-TBI problem drinking or if post-TBI drinking is solely a function of pre-TBI AUD in these patients. Regardless, the fact that both pre- and post-TBI alcohol misuse worsens TBI outcomes emphasizes the need to develop treatments for comorbid TBI and AUD to serve this large patient population.

2.2.2 Post-TBI emotional dysregulation

Emotional dysregulation is a common outcome following TBI, and a strong association has been found between TBI and the development of mood disorders, anxiety disorders, and PTSD (Hayes, 2019; Hibbard, Uysal, Kepler, Bogdany, & Silver, 1998). As mentioned above, alcohol misuse can increase the risk for developing emotional dysregulation post-TBI, but there is also evidence for the inverse relationship, that post-TBI emotional dysregulation can lead to increased alcohol use. For example, a study of over 4000 college-aged subjects found that alcohol use in subjects with a concussion history was potentially mediated by depression or panic disorder (Newman, Grantz, Brooks, Gutierrez, & Kawata, 2020). Furthermore, post-TBI development of mood disorders was found to increase the risk of alcohol abuse relapse in patients with closed-head injury (Jorge et al., 2005).

The mechanism of post-TBI emotional dysregulation leading to subsequent alcohol use requires further study. Because the orbital frontal cortex is particularly susceptible to physical trauma, due to its proximity to bony ridges on the orbital surface of the skull, Bjork and Grant (2009) postulate that orbitofrontal damage and resultant cognitive deficits could underlie both emotional dysregulation (e.g., organic personality disorder) and impact reward-directed decision making and uniquely contribute to substance misuse in a post-TBI condition. More recent work has shown changes in OFC connectivity in TBI patients that were associated with increased aggression (McGlade, Rogowska, & Yurgelun-Todd, 2015), but more work is necessary to explore a specific connection between TBI-related OFC dysfunction and post-injury substance use. Weil, Corrigan, et al. (2016) and Weil, Karelina, et al. (2016) discuss a possible self-medication mechanism, by which patients may seek out alcohol to reduce negative affect or anxiety that develops post-TBI. Indeed, the idea that emotional dysregulation drives alcohol use via negative reinforcement (i.e., drinking to relieve negative emotions) is a prevailing one in the current understanding of AUD, as discussed in chapters "Pain + Alcohol" by Edwards et al. and "Traumatic Stress + Alcohol" by Weiner et al., in this volume. Therefore overall, treating emotional dysregulation in TBI patients may effectively reduce alcohol use and improve TBI outcomes generally.

2.2.3 Post-TBI pain

Pain, ranging from headaches to chronic pain, is frequently reported following TBI, and post-TBI pain has been found to be both independently associated with the TBI as well as mediated by a psychiatric comorbidity such as PTSD (Nampiaparampil, 2008; Tyrer & Lievesley, 2003). Post-TBI pain is described in greater detail in Section 3.2. Chronic alcohol use can lead to diffuse peripheral neuropathy, which could complicate a patient's post-TBI pain (Ivanhoe & Hartman, 2004), and conversely, many speculate post-TBI pain could drive alcohol use (Grossbard et al., 2017; Nampiaparampil, 2008), in line with the belief that alcohol analgesia and hyperalgesia can promote alcohol misuse in general (Egli, Koob, & Edwards, 2012; see chapter "Pain + Alcohol" by Edwards et al., in this volume). A history of AUD is an important consideration when managing a TBI patient's pain and may limit their pharmacological treatment options for severe chronic pain (e.g., opioids) due to abuse potential (Tyrer & Lievesley, 2003).

The role of post-TBI pain in alcohol use after injury is complex. Post-TBI pain can be influenced by the physical brain injury and subsequent neuropathology, and also by the emotional context of the injury and subsequent disability (Tyrer & Lievesley, 2003). There is some speculation that post-TBI pain could drive alcohol use as a form of self-medication to mitigate the experience of pain itself, or alternatively as a coping strategy for the stress caused by pain (Grossbard et al., 2017; Nampiaparampil, 2008). Further investigation into the relationship between post-TBI pain and alcohol use is warranted to explore these hypotheses, with particular consideration given to psychiatric comorbidities such as those described above that may drive both pain and alcohol use.

2.3 Potential impact of alcohol on TBI outcomes

It is accepted that both pre- and post-TBI alcohol use worsens TBI outcomes, ranging from physiological to psychosocial, and this topic has been reviewed previously (Bjork & Grant, 2009; Molina et al., 2015;

Weil, Corrigan, et al., 2016, 2018). Briefly, alcohol use is associated with poorer vocational outcomes, poorer rehabilitation trajectory, and increased risk for mood disorders, seizures, and subsequent brain injury in TBI patients (Corrigan, 1995; Ellerd & Moore, 1992; Jorge et al., 2005; Vaaramo, Puljula, Tetri, Juvela, & Hillbom, 2014a, 2014b). Preclinical work has shown that post-TBI alcohol exposure exacerbates neurological impairment and neuroinflammation in rats (Teng et al., 2015) as well as neuronal damage, neuroinflammation, and learning and memory impairments in mice (Karelina et al., 2017), though intoxication at the time of injury may be neuroprotective (e.g., Christensen et al., 2020). Further study of the effects of post-TBI alcohol exposure is important for identifying potential therapies to ameliorate negative consequences of alcohol on outcomes of and recovery from TBI or to reduce post-TBI drinking itself.

Because of the many negative consequences of alcohol on TBI recovery and comorbidities, reducing or abstaining from alcohol use after TBI could aid physical rehabilitation, mitigate psychological dysfunction, and have a lasting positive impact on outcomes following brain injury. Indeed, many clinicians recommend abstaining from alcohol for up to 2 years after TBI, and treatment of substance abuse was recommended for inclusion in TBI rehabilitation programs by the National Institutes of Health in 1999 (Laborde, 2000). The initial decline in drinking in the first month following TBI provides a therapeutic window for treating a patient's problem drinking (Pagulayan, Temkin, Machamer, & Dikmen, 2016). Although, the complex neuropathology and psychiatric comorbidities of TBI may call for special considerations in treating this patient population (see Weil et al., 2018).

It has been suggested that traditional AUD treatments may be less effective in TBI patients due to post-TBI cognitive, behavioral, emotional, or physical deficits (Corrigan, Lamb-Hart, & Rust, 1995). Graham and Cardon (2008) summarized findings regarding the treatment of substance abuse in TBI populations, concluding that a multidisciplinary community-based approach was optimal, including use of skills-based treatments and peer-based support systems. Though tailored treatments for comorbid TBI and AUD are still limited, Herrold et al. (2015) describe several promising options based on more recent clinical data, including behavioral, pharmacological, and neuromodulatory treatments. There are also relatively few preclinical studies investigating treatments to reduce post-TBI drinking or mitigate the negative consequences of post-TBI alcohol use on TBI outcomes. To date, such preclinical work suggests that therapies that treat TBI neuropathology independently of alcohol, including environmental enrichment (Weil, Karelina, et al., 2016), inhibition of microglia activation (Karelina et al., 2018), and endocannabinoid degradation inhibition (Fucich et al., 2019), can also reduce post-TBI drinking and improve other TBI neurobiological and behavioral outcomes in alcohol-drinking rats. Ongoing investigations are key to expanding effective treatment options for comorbid TBI and AUD and to improve outcomes for a large portion of these overlapping patient populations.

3. Opioid use and TBI

Opioids are commonly used to treat acute and chronic pain (Nafziger & Barkin, 2018). Opioids have analgesic properties when binding mu-opioid receptors, and engagement of mu-opioid receptors is also responsible for the rewarding effect of opioids (Volkow, Jones, Einstein, & Wargo, 2019). The endogenous opioid system is comprised of three types of homologous G-protein coupled receptors (mu-, delta-, and kappa-opioid receptors; Darcq & Kieffer, 2018) as well as endogenous ligands of these receptors, including dynorphin, beta-endorphin, and enkephalin (Stein, 2016), which are derived from specific precursor hormones: preprodynorphin (Pdyn), proopiomelanocortin (POMC), and preproenkephalin (Penk), respectively (Le Merrer, Becker, Befort, & Kieffer, 2009). These endogenous opioids have different functional effects and binding characteristics (Froehlich, 1997), and they function as neurotransmitters or neuromodulators in the central nervous system, playing a role in nociception, analgesia, mood, locomotion, thermoregulation, stress, sexual behavior, drinking, eating, and immune function (Le Merrer et al., 2009; Oswald & Wand, 2004; for a comprehensive review of the behavioral effects of the endogenous opioid system, see Bodnar, 2016). The analgesic effects of endogenous opioids are primarily via receptors within the central nervous system (see Corder, Castro, Bruchas, & Scherrer, 2018).

The use of opioids for pain has increased in recent years, contributing to the over-prescription of opioids and the current global opioid crisis (Makary, Overton, & Wang, 2017). While opioids are highly effective in reducing acute pain, they are not equally effective for chronic pain (Koob, 2020). Some research has also found that opioids are not effective for all kinds of chronic pain (Arnér & Meyerson, 1988). Furthermore, although preclinical studies seem to be more consistent, showing that females require higher doses of opioids to relieve pain compared to males (Ji, Murphy, & Traub, 2006; Loyd, Wang, & Murphy, 2008; Posillico,

Terasaki, Bilbo, & Schwarz, 2015), studies in humans regarding sex differences in opioid analgesia remain controversial; while some studies have shown no differences (Fillingim et al., 2005), others have reported a decreased analgesic efficacy in women (Cepeda & Carr, 2003).

Moreover, opioids can produce numerous adverse effects such as gastrointestinal and immunological dysfunction, hormonal changes, sleep disturbances, sedation, and hyperalgesia after chronic use (Benyamin et al., 2008). Continued administration of opioids can enhance pain sensitivity (paradoxical hyperalgesia), driving increases in opioid doses and further exacerbation of the pain (Volkow, Benveniste, & McLellan, 2018). Additionally, long-term opioid use can lead to physical dependence, which manifests when opioid use is disrupted, and is characterized by the appearance of withdrawal symptoms including insomnia, nausea, diarrhea, and body aches, among others (Volkow et al., 2019). These withdrawal symptoms can contribute to continued opioid use or relapse via negative reinforcement (Volkow et al., 2019).

Despite their abuse potential, opioids are still commonly prescribed to patients with a TBI history, even though this population may be particularly vulnerable to substance misuse and substance use disorders. Some studies have included post-TBI opioid use in their measures, and we postulate here the factors that contribute to this use as well as the implications for opioid use on TBI outcomes. Overall, better post-TBI pain management and substance abuse treatment could improve several aspects of patient health and prevent negative consequences like poor quality of life, exacerbated brain damage, or opioid overdose.

3.1 Post-TBI opioid use

After TBI, patients may use prescribed opioids to treat pain, misuse prescription opioids or use illicit opioids, or meet criteria for OUD. Misuse refers to the use of opioids other than as prescribed (Volkow et al., 2019). OUD is a substance use disorder in the DSM-5 defined as a problematic pattern of opioid use leading to clinically significant impairment or distress and is characterized by compulsive self-administration of opioids in the absence of or in excess of treating a medical condition (American Psychiatric Association, 2013). Though many studies in TBI patients report substance use rates in general, fewer have reported the prevalence of opioid prescriptions (discussed in Section 3.2) or included separate measures of opioid use specifically. For example, Whelan-Goodinson et al. (2009) found that, in the

21% of individuals with a post-TBI substance use disorder 0.5–5.5 years following injury, opioids were the third most commonly used non-alcohol drug class (15%) after marijuana (45%) and stimulants (20%). Perron and Howard (2008) reported that substance use was overall higher in individuals with TBI compared to individuals without TBI in a sample of 720 youth delinquents residing in rehabilitation facilities, and included measures for heroin (11% of the TBI group were users vs 5% in the non-TBI group), cocaine (36% vs 21%), marijuana (93% vs 85%), and ecstasy (33% vs 17%). Similar results were found in a sample of 6288 Canadian high school students divided by TBI history, where 27.2% of the 1215 students with a TBI history had used opioid pain relievers (e.g., Percocet) without a prescription in the previous year compared with only 12.6% of the 5073 students without a TBI history, reflecting a significantly higher odds ratio (2.6) in students with a TBI history. Only cannabis use (39.5% vs 25.2%; OR = 2.00) and alcohol use (77% vs 63.7%; OR = 2.01) were more prevalent than opioid use of the seven drugs found to have significantly higher odds ratios in the TBI group (Ilie, Mann, et al., 2015). Likewise, in a sample of 1999 Canadian adults, those who reported having a history of TBI (16.8%) had higher odds of using nonmedical opioids (AOR = 2.90) and cannabis (AOR = 2.80; Ilie et al., 2015). A study of a sample of 200 Australian prison entrants, where 82% had a lifetime history of TBI and 43% reported four or more TBIs, revealed that those with a TBI history were more likely to have alcohol problems (46% vs 36%; OR = 1.59) and use illicit drugs daily (51% vs 31%; OR = 2.42). Of the illicit drugs used in the overall sample, 75 individuals used cannabis, 25 used amphetamine, 24 used heroin, 10 used non-prescribed methadone or other opioids, 2 used cocaine, and 1 used volatile solvents daily (Schofield et al., 2006). Though the numbers for specific drugs used were not parsed out based on TBI history, these data suggest that individuals who sustain TBI and those who misuse opioids are overlapping populations.

Though seemingly not as prevalent as comorbid TBI and alcohol use, more studies are needed to better understand the prevalence, pattern, and factors that influence opioid use in TBI patients. Some clinical findings have begun to address these points. For example, while Schofield et al. (2006) found that prisoners with daily illicit drug use were more than twice as likely to have incurred a TBI, neither the number of TBIs, loss of consciousness status, nor recency of TBI had any gradient effects on the odds of substance use described above, which included heroin and nonmedical opioids. However, Darke, McDonald, Kaye, and Torok (2012) found an association between the number of TBIs and early onset of heroin use in a sample of 175 treatment-seeking heroin users. Clearly, more clinical and preclinical work is necessary to investigate the relationship between TBI and opioid use, particularly considering the current opioid epidemic.

3.2 Factors related to post-TBI opioid use

Several pathways to the development of OUD have been self-reported by patients, including inadequately-controlled physical pain, exposure to opioids after acute pain conditions, previous substance use disorders, emotional distress, and recreational opioid use, many of which may play a role in the development of or transition to opioid misuse after TBI (Stumbo, Yarborough, McCarty, Weisner, & Green, 2017). Here we discuss in more detail post-TBI pain, its treatment with opioids, and the consequential implications for post-TBI opioid use. We also review other factors that may promote opioid use in TBI patients: pre-TBI substance use disorders, psychiatric comorbidities, and post-injury cognitive dysfunction (Fig. 2).

3.2.1 Post-TBI pain

Pain is a complex condition with different physiological manifestations (Mehalick & Glueck, 2018). The International Association for the Study of Pain (IASP) defines pain as "an unpleasant sensory and emotional experience associated with or resembling that is associated with actual or potential tissue damage" (International Association for the Study of Pain (IASP), 2017). Acute pain, considered part of an alert response to injury, is provoked by a specific disease or injury and is self-limited, while chronic pain could be considered a disease state given that it outlasts the normal time of healing (Swieboda, Filip, Prystupa, & Drozd, 2013). Chronic pain is a debilitating condition that alters the normal functioning of a person, affecting daily activities and quality of life.

The prevalence of pain after TBI is high; Hoffman, Pagulayan, Zawaideh, and Bell (2005) observed that 72.5% of individuals with TBI reported pain 1 year post-TBI, with 47.2% reporting mild pain and 25.3% reporting moderate to severe pain. The prevalence of pain after TBI is higher than after other neurological disorders (Sherman, Goldberg, & Bell, 2006), and although pain may be more frequent acutely after the injury, in some cases it not only persists but also worsens several years after the trauma (Brickell, Lange, & French, 2014). This co-occurrence could be due to the nature of events that cause TBI, which can include injuries to other parts of the head or body leading to pain, and/or the alteration of brain regions that affect sensory and motor functioning or modify the perception of pain (Sherman et al., 2006). Several pain conditions are very common post-TBI, including headache, neuropathic pain, and others produced by spasticity, venous thrombosis, fractures, etc. (Ivanhoe & Hartman, 2004). Additionally, pain may result from different kinds of injuries associated with TBI, such as limb amputation producing phantom limb syndrome (Rauh et al., 2013). It is noteworthy that different types and locations of pain can co-occur in the same patient and is more common in cases of less severe TBI (Sherman et al., 2006). Headache is the most frequent pain reported after TBI, commonly referred to as post-traumatic headache (PTH), followed by neck, shoulder, back, and upper limb pain (Khoury & Benavides, 2018). Headache can be acute (resolving within 3 months of onset) or persistent (>3 months) and is experienced by patients after all severities of TBI (Ashina et al., 2019). The most common types of PTH are tension, migraine, or a mix of both (Irvine & Clark, 2018). A recent study showed that among patients with new or worse headache after injury, those with mild TBI reported higher rates of tension-type headaches (32%) compared to those with moderate-severe TBI (14%) 12 months post-injury (Hoffman et al., 2020). Additionally, those participants with moderate to severe TBI reported a higher headache frequency (headaches per week) compared to mild TBI (43% vs 27%) (Hoffman et al., 2020).

The transition from acute post-TBI pain to chronic pain involves a complex interaction between neurobiological and psychosocial factors (Khoury & Benavides, 2018). Ofek and Defrin (2007) noted that chronic pain emerges with a relatively late onset (6.6 months) post-TBI. Emergence and persistence of TBI-induced hyperalgesia may also worsen future experiences of TBI, further hindering therapeutic strategies to manage chronic pain. Unfortunately, very few preclinical mechanistic studies have directly compared pain-related neurobiological adaptations following single versus repeated TBI.

In the civilian population, pain prevalence seems to be inversely correlated with TBI severity, with pain being more common in cases of mild TBI. Nampiaparampil (2008) found that the prevalence of chronic pain in patients diagnosed with mild TBI was 75.3% compared with 32.1% in patients with moderate and severe TBI. Although this inverse correlation seems to be paradoxical, one explanation for this relationship may be that people with moderate or severe TBI are treated with additional support (e.g., paralytic agents) or are immobilized to facilitate healing and recovery of damaged muscles and ligaments, unlike people with mild TBI who do not receive this treatment (Martelli, Zasler, Bender, & Nicholson, 2004; Zasler, 1999).

Both pain and TBI are also frequently reported among military personnel. In a sample of 613,391 Veterans who returned from Afghanistan and Iraq and received care from Veterans Health Administration facilities from 2009 to 2011, 40.2% were diagnosed with pain (Cifu et al., 2013). With respect to TBI status; head, back, and neck pain have been found to be higher among Veterans with TBI (48%, 47%, and 16%, respectively) compared to those without TBI (9%, 25%, and 6%, respectively; Taylor, 2012). The constellation of pain-related symptoms extends beyond these areas of the body. In a recent study, two-thirds of Veterans with a history of TBI reported chronic pain in multiple locations (King, Beehler, & Wade, 2015).

Currently there are no medications approved by the U.S. Food and Drug Administration for the treatment of TBI. Frequently, individuals with TBI present to inpatient facilities with a series of cognitive, behavioral, and psychiatric symptoms that require medications. These medications are used for symptom management, while others are used to prevent or reduce comorbidities (Hammond et al., 2015). Individuals with TBI and comorbid pain are commonly treated with prescription opioids. A multi-center investigation found that 72% of individuals with TBI receiving inpatient rehabilitation were prescribed opioids during their hospital stay (Hammond et al., 2015). A similar situation occurs in military populations. Although the clinical practice guidelines of the U.S. Department of Veterans Affairs (VA) and the Department of Defense (DoD) advise against the use of opioid medications for individuals with a diagnosis of TBI (Management of Concussion/mTBI Working Group, 2009), opioids continue to be prescribed to this population. Hudson et al. (2018) found that 21% of a sample of over 35,000 Veterans diagnosed with TBI and treated in the VA health system, initiated opioids within a year following injury, and 23% of new initiators used opioids chronically. People between ages 36–45 had 8.3% greater odds of receiving opioids compared with those in the 26–35 age group, and while women had higher odds of initiating opioid use compared to men, they were less likely to use them chronically. Additionally, those initiating opioid use were more likely to have at least one pain diagnosis compared to those who did not initiate opioid medications (Hudson et al., 2018). Seal et al. (2018) found similar numbers, where 22% of a sample of over 53,000 Veterans received opioid prescriptions within a year of their TBI. They also found that Veterans with moderate to severe TBI are more likely to be prescribed long-term opioid therapy for the treatment of pain compared to those with mild TBI. More studies on the administration of opioids during the acute and late states of TBI are necessary. Controlling for severity of injury in such studies will provide information about patterns of opioid administration in mild and moderate to severe TBI. It is important to highlight that there is not rigorous evidence for the efficacy of many psychoactive drugs that are being used in the TBI population (Whyte, 2010), which demands new investigations to guide clinics in the treatment of these patients to avoid future complications. Of note, long-term opioid use, leading to opioid misuse and OUD, is commonly initiated by the prescription of opioid analgesics by a physician (Paulozzi, Jones, Mack, & Rudd, 2011). Indeed, Golub and Bennett (2013) found that opioid misuse in Veterans is initiated during deployment with medical use of opioids for pain. All these data are concerning, considering not only the risk for transitioning to opioid misuse or OUD after receiving prescription opioids in general, but also the risk associated with multiple other factors present in the TBI population discussed below.

3.2.2 Pre-TBI substance use disorder

One factor associated with the susceptibility of people with brain injuries to opioid misuse is the high prevalence of preinjury SUD in this population. As mentioned previously, rates of SUD, including alcohol and opioid use disorders, are high in those with a TBI history, and many TBIs are substance related (Bjork & Grant, 2009; Savola et al., 2005). Evidence suggests prior substance abuse is a risk factor for developing an OUD after chronic exposure to opioid analgesics, with very few patients who developed OUD after chronic opioid therapy lacking a substance abuse history (Frishbain, Cole, Lewis, Rosomoff, & Rosomoff, 2008). Additionally, few new onsets of substance misuse have been reported in patients post-TBI compared with pre-TBI incidence (Whelan-Goodinson et al., 2009), which may support the notion that post-TBI opioid misuse is strongly linked to prior SUD. However, more studies addressing the prevalence of OUD or opioid misuse in individuals with TBI are necessary.

3.2.3 Post-TBI psychiatric disorders

As discussed in Section 1.2, psychiatric disorders commonly develop in patients with a history of TBI, and these illnesses may contribute to post-TBI substance misuse. Studies in the U.S. military population have provided insight into the relationship between TBI-related psychiatric illnesses—particularly PTSD, depression, and anxiety—and post-injury opioid use.

The most common psychiatric illness associated with recent military conflicts is PTSD. A recent study found that of Veterans with deployment

TBI, 64% were diagnosed with PTSD, 28% had depressive disorders, 21% had anxiety, and 16% had a SUD (McGlinchey, Milberg, Fonda, & Fortier, 2017). In Veterans with TBI, PTSD can develop because of head injury and psychological trauma produced by blasts, gunshot wounds, or motor vehicle accidents during combat. Individuals with PTSD have a high rate of SUD compared with people without PTSD, and substance use could be a way of self-medicating to reduce PTSD symptoms (Jacobsen, Southwick, & Kosten, 2001). The risk of opioid misuse among individuals with PTSD is, in part, due to stress-induced increases in both the reinforcing properties of opioids and the aversive effects of opioid withdrawal (Logrip, Zorrilla, & Koob, 2012). In one study, Seal et al. (2012) found higher rates of opioid-related accidents and overdoses in Veterans with TBI and PTSD compared to those with no psychiatric condition or those with psychiatric conditions other than PTSD. Fonda et al. (2020) also found that PTSD has a stronger impact on the association between TBI and opioid overdose compared to other psychiatric disorders.

Studies conducted in Veterans have also provided insight into comorbid pain and emotional dysregulation following TBI, which could further exacerbate post-TBI opioid misuse. In the sample of Veterans with a TBI history studied by King et al. (2015) in which 79% of individuals reported chronic pain, more than half reported concurrent emotional health concerns. TBI and chronic pain each increase the risk of developing anxiety and depression, and anxiety and depression can also increase opioid misuse (Fonda et al., 2020). Thus, it is possible that comorbid chronic pain and TBI could enhance the risk of developing anxiety and depression and, subsequently, opioid misuse. Veterans also often experience what has been termed the "polytrauma clinical triad": chronic pain, post-concussion syndrome, and PTSD. For instance, Lew et al. (2009) found that 66.8% of Veterans sampled had post-concussion symptoms, 81.5% had chronic pain, and 68.2% met the criteria to be diagnosed with PTSD. In the same study, it was reported that 42.1% of the Veterans sample were diagnosed with all three conditions simultaneously. This triple comorbidity could similarly exacerbate post-TBI opioid misuse beyond that of either chronic pain or PTSD in isolation. In fact, a recent study found that 27.6% of 16,590 Veterans presented with the polytrauma clinical triad, and the prescription and utilization of greater than 30 days of opioids was most common in these patients (36.8% and 20.2%, respectively), compared to patients with PTSD only (9.5% and 2.9%), pain only (26.3% and 11,2%), or TBI only (8.9% and 1.8%)

(Adams, Larson, Meerwijk, Williams, & Harris, 2019). As mentioned previously, more research is necessary to elucidate the effectiveness of treatments in these patients.

Overall, individuals with mental health problems are among those at the highest risk for receiving chronic opioid prescriptions (Dobscha, Morasco, Duckart, MacEy, & Deyo, 2013). Indeed, Seal et al. (2018) reported that mental health problems co-occurring with TBI in their Veteran sample, such as PTSD and depression, increased the likelihood of receiving long- and short-term opioid prescriptions. Therefore, the high prevalence of psychiatric disorders following TBI may increase the risk of opioid misuse and overdose.

3.3 Potential impact of opioids on TBI outcomes

Opioids are commonly used to treat pain, and despite clinical guidelines to the contrary, individuals with TBI are often prescribed opioid analgesics. Although opioids have demonstrated clinical efficacy for the treatment of various pain conditions, regular exposure to opioids causes analgesic tolerance, opioid-induced hyperalgesia (a hypersensitivity to painful stimuli that develops following chronic opioid administration and generates an exacerbated pain sensation (Roeckel, Le Coz, Gavériaux-Ruff, & Simonin, 2016)), and may produce dependence, even under the care of a physician (Schirmer & Seale, 2018). Moreover, opioids also produce euphoria, which increases the likelihood of misuse, abuse, and overdose.

Opioid misuse may exacerbate negative outcomes of TBI. It has been speculated that OUD increases the risk of future brain injury (Corrigan & Adams, 2019). Additionally, opioid overdose reduces respiration and consequently decreases oxygen supply to the brain, which could precipitate hypoxic injury and thereby increase neuronal death, inflammation, and oxidative stress (Schirmer & Seale, 2018; Zhao, Yang, & Cui, 2017), and potentially alter several cognitive domains including visuo-motor skills, attention, and executive functions (Schirmer & Seale, 2018), ultimately affecting the daily activities of patients with TBI.

Moving forward, appropriate management of pain after TBI could help reduce opioid use, misuse, and the associated negative consequences. Some authors have suggested that a major barrier to effective management of pain following TBI is the lack of its accurate assessment, specifically in those people with a moderate to severe injury whose physical condition together with cognitive alterations may complicate the use of standard methods to determine pain (Bosco, Murphy, & Clark, 2013; Clark, Scholten, Walker, & Gironda, 2009). Additionally, the myriad of psychiatric disorders associated with TBI can alter the experience of pain and complicate the accurate measurement of pain. Determining the etiology of pain could improve post-TBI pain management. Fonda et al. (2020) suggest that treating TBI and psychiatric conditions with behavioral therapy may be effective in reducing the sensation of pain, and this will lead to a reduction in opioid use in those individuals who are at risk for misusing opioids.

4. Cannabis use and TBI

Cannabis is one of the most commonly used substances of abuse. In 2018, approximately 10% of Americans over 12 years of age reported cannabis use in the prior month, and 45% reported cannabis use in the past year (SAMHSA & Center for Behavioral Health Statistics and Quality, 2018). Cannabis contains a rich diversity of phytocannabinoids, but the psychoactive effects of cannabis are mediated primarily by Δ^9-tetrahydrocannabinol (THC). THC exerts its effects through modulation of the endocannabinoid (eCB) system, particularly by acting as a partial agonist at the cannabinoid receptor 1 (CB1R). The eCB system includes the lipid-derived endocannabinoids, 2-arachidonoylglycerol (2-AG) and *N*-arachidonoylethanolamine (anandamide, AEA); enzymes associated with eCB synthesis and degradation; and the receptors to which the eCBs bind—namely CB1R and cannabinoid receptor 2 (CB2R). The eCB system is widely distributed throughout the body; CB1R is highly expressed in the brain and is also expressed in the gastrointestinal tract, adipose tissue, reproductive organs, and liver, while CB2R is expressed by immunologic and hematologic cells of the body (Hillard, 2018; Pacher, Bátkai, & Kunos, 2006). CB2Rs are present at much lower levels than CB1R in the brain and CB2R expression is largely restricted to microglia (Stella, 2010). Neuronal CB2Rs have been reported (e.g., Zhang et al., 2014), but CB2R expression in neurons remains contentious within the cannabinoid field (Atwood & Mackie, 2010). The expression of CB2R in microglia seems to be dependent on activation state. Little to no CB2R expression is expected in ramified (surveying) microglia; however, the receptor is upregulated following activation (Stella, 2010). The endocannabinoids have been measured in almost every tissue and fluid in the body examined thus far (Hillard, 2018). Within the central nervous system, eCB signaling is active across several regions of the brain and spinal cord. Neuronal eCB signaling regulates synaptic transmission through retrograde suppression of neurotransmission, whereby neuronal activity stimulates production of eCBs that activate presynaptic CB1R and inhibit further presynaptic neurotransmitter release (Castillo, Younts, Chávez, & Hashimotodani, 2012). In addition to modulating synaptic transmission, brain eCBs have also been implicated in the regulation of neuroinflammation (Walter & Stella, 2004). Anti-inflammatory properties of the eCB system are primarily attributed to the expression of CB2R in immune cells. Microglial CB2R activity modulates neuroinflammation in part through modulating microglial polarization (Mecha, Carrillo-Salinas, Feliú, Mestre, & Guaza, 2016). 2-AG acts as a full agonist at CB2R, while AEA is a weak partial agonist with low affinity, and these two eCBs may differentially influence microglial polarization (Mecha et al., 2016).

To date, very few studies have examined the relationship between TBI and cannabis use. Unlike alcohol use, which has clearly been implicated as a major risk factor for TBI, studies have not examined cannabis use as a risk factor for TBI. A report by the National Academies of Sciences, Engineering, and Medicine (2017) concluded that there is "substantial evidence of a statistical association between cannabis use and an increased risk of motor vehicle crashes." Motor vehicle accidents (MVA) represent the third most common mechanism of TBI, accounting for 13.2% of all TBI-related emergency department visits (Centers for Disease Control and Prevention, 2019), and so it is therefore possible that cannabis use could increase the risk of incurring TBI via increases in MVA risk. Studies explicitly designed to examine this issue are warranted and necessary.

4.1 Post-TBI cannabis use

Cannabis is the most commonly used illicit drug in drug-using individuals with TBI (Beaulieu-Bonneau et al., 2018; Whelan-Goodinson et al., 2009). Currently available data suggest that individuals with TBI are more likely to use cannabis than individuals without TBI. A cross-sectional retrospective study found that adults with a lifetime history of TBI had an enhanced likelihood (AOR = 2.8) for consuming cannabis (Ilie, Adlaf, et al., 2015). Another study identified that patients with a positive lifetime history of TBI associated with loss of consciousness had a higher average number of months using cannabis relative to those without TBI, independent of the number of months using alcohol or other substances (Walker, Cole, Logan, & Corrigan, 2007). A limitation of these studies is that the frequency and dosing of cannabis use was not examined or reported. However, an

Australian study did show that, when including patients unable to identify their age at the time of TBI or with TBI more than 4 years prior to the study, TBI patients were at increased risk of developing problematic cannabis use (Tait, Anstey, & Butterworth, 2010). In this study, being male was associated with an increased risk of developing problematic cannabis use in individuals with TBI. Problematic cannabis use was identified in this study by asking patients if they had ever used more cannabis than intended or if they ever desired to decrease their cannabis use. To date, there are no studies that have specifically evaluated the development of cannabis use disorder following TBI.

Existing data suggest that few individuals will initiate cannabis use de novo following TBI, and of those with a history of regular pre-TBI cannabis use, many will discontinue or decrease use acutely post-TBI. Beaulieu-Bonneau et al. (2018) found that substance use decreased within 4 months after TBI but approached preinjury levels by 1 year following injury. Only one of 150 individuals who did not use drugs prior to TBI was diagnosed with a SUD post injury. A prospective cohort study of 307 TBI patients determined that 44% of TBI patients who regularly consumed cannabis prior to injury discontinued use during the acute post-concussion period; only about 2.3% of patients who did not consume cannabis prior to injury reported cannabis use within 1 month post-TBI (Lawrence et al., 2020). Similarly, a study of TBI patients in Colorado showed that only 3% of participants reported using cannabis post-TBI without a history of use prior to the injury (Hawley, Ketchum, Morey, Collins, & Charlifue, 2018).

Childhood and adolescence may represent a window of time in which individuals are particularly vulnerable to problematic cannabis use following TBI. A cross-sectional school survey of adolescents reported that individuals with a history of TBI are approximately twice as likely to use cannabis relative to peers without a TBI history (Ilie et al., 2014). The relationship between early life TBI and problematic cannabis use was examined using data collected from the Avon Longitudinal Study of Parents and Children (ALSPAC). Problematic cannabis use was identified using the Cannabis Abuse Screening Test (CAST), which assesses patterns of use, negative consequences of use, and inability to stop use, and data showed that individuals with TBI were at increased risk of problematic cannabis use relative to individuals with an orthopedic injury and uninjured peers after adjusting for pre-birth and childhood confounding factors. Furthermore, data from this study demonstrated that TBIs occurring both in childhood (birth-age 11) and adolescence (ages 12–16) were associated with problematic cannabis use at age 17 (Kennedy, Heron, & Munafò, 2017).

4.2 Factors related to post-TBI cannabis use

The factors that drive post-TBI cannabis use have not been well studied but may include neurological changes induced by the injury as well as psychosocial factors that influence its use (see Fig. 2). The wave of legalization of medical marijuana in the U.S. has promoted the use of cannabis for the treatment of myriad health conditions, although controlled studies providing scientific evidence supporting the efficacy of cannabis in the treatment of many of these conditions are still needed. There are many symptoms that can emerge post-TBI that have been proposed to be ameliorated by cannabis, including anxiety, pain, irritability, insomnia, and spasticity. Currently, TBI is a qualifying condition for medical marijuana in four states (Illinois Department of Public Health, 2019; New Hampshire Department of Health and Human Services, 2017; Ohio Medical Marijuana Control Program, n.d.; Washington State Department of Health, n.d.), and many conditions (e.g., pain, PTSD, seizures) that can arise secondary to TBI qualify individuals for medical marijuana in many more states.

In addition to THC, cannabis contains numerous other phytocannabinoids, including cannabidiol (CBD). CBD has been proposed as a treatment for a wide array of medical conditions, many of which can arise secondary to TBI, including anxiety, insomnia, and pain (Crippa, Guimarães, Campos, & Zuardi, 2018). Unlike THC, CBD is not intoxicating and lacks abuse liability. CBD acts as a negative allosteric modulator at the CB1 receptor (Laprairie, Bagher, Kelly, & Denovan-Wright, 2015), and it is also an agonist at the 5-HT_{1A} receptor (Russo, Burnett, Hall, & Parker, 2005) and the vanilloid receptor 1 (TRPV1; Costa, Giagnoni, Franke, Trovato, & Colleoni, 2004). The relative concentrations of THC, CBD, minor phytocannabinoids like cannabinol (CBN) and cannabichromene (CBC), and other compounds (e.g., terpenes) are thought to account for differences in the effects of different strains of marijuana. The combination of cannabinoids found in cannabis is thought to confer entourage effects—synergy in the effects of cannabis conferred by the variety of cannabinoids (Russo, 2019). Growing commercialization of the cannabis industry has facilitated the advent of cannabis vaping products that utilize oil-based concentrates (Cranford, Bohnert, Perron, Bourque, & Ilgen, 2016). These concentrates can be designed to contain specific ratios of major and minor cannabinoids and marketed as beneficial for varying maladies. The marketing of medical marijuana and its derivatives, such as CBD, alone may entice patients with TBI to try cannabis as a therapeutic.

Assessing motives for cannabis use among cannabis-using TBI patients revealed that these individuals most commonly report using marijuana recreationally; however, more than half of these patients also reported its use for stress/anxiety relief and for improving sleep (Hawley et al., 2018). Therefore, any relief of TBI sequelae by cannabis use may drive further cannabis use through negative reinforcement.

4.2.1 Post-TBI endocannabinoid dysregulation

Dysregulation of eCB signaling in the brain represents a potential molecular mechanism that contributes to post-TBI cannabis use. Alterations in eCB signaling have been implicated in AUD (Basavarajappa, Joshi, Shivakumar, & Subbanna, 2019) and SUD (Manzanares et al., 2018). Changes in eCB tone have the potential to affect neuronal circuits regulating reward, learning and memory, habit formation, and negative affect, all of which may contribute to problematic substance use. Evidence suggests that TBI causes alterations in eCB signaling. Work from the laboratory of Raphael Mechoulam has shown in a murine closed-head injury model that TBI induces the release of the predominant brain eCB 2-arachidonylglycerol (2-AG) in the ipsilateral hemisphere, and 2-AG levels remain elevated for up to 24 h post injury (Panikashvili et al., 2001). A study of adolescent male mice injured using the weight drop model of TBI showed that CB1R mRNA and protein expression are decreased 1 and 3 days post-TBI in ipsilateral brain tissue and that CB1R expression was negatively associated with brain edema. Data also showed a progressive increase in CB2R expression over 2 weeks post-TBI, with a significant increase in CB2R protein expression at 24 h post injury (Lopez-Rodriguez, Acaz-Fonseca, Viveros, & Garcia-Segura, 2015). Work from our lab using the lateral fluid percussion injury model in adult female rats shows that TBI reduces 2-AG levels in the basolateral nucleus of the amygdala 11 days following TBI, and 2-AG levels are negatively associated with injury pressure severity (Stielper et al., 2020). Together, these data suggest that TBI alters eCB signaling in tissue proximal to the injury as well as in subcortical sites. Furthermore, inhibition of 2-AG degradation post-TBI ameliorates enhanced motivation to consume alcohol (measured by a progressive ratio alcohol self-administration paradigm) following injury, indicating that eCBs modulate the injury response and regulate alcohol consumption after TBI (Fucich et al., 2019). Whether TBI-induced changes in eCB signaling mediate changes in post-injury cannabis use or other substance use remains to be determined.

4.3 Potential impact of cannabis on TBI outcomes

The consequences of cannabis use following TBI are currently unclear, as no prospective, randomized, placebo-controlled clinical trials have evaluated the impact of cannabis use on TBI outcomes, nor have there been studies evaluating the long-term consequences of chronic cannabis use in TBI patients. In 2017, the National Academies of Sciences, Engineering, and Medicine published a review of the health effects of cannabis use and concluded that "there is limited evidence of a statistical association between cannabinoids and better outcomes (i.e., mortality, disability) after traumatic brain injury." This review did not describe any findings associating cannabinoids and worse outcomes. The limited data on the subject does suggest a positive effect of cannabis on TBI-related outcomes. Two retrospective observational studies have examined TBI outcomes in patients who do and do not test positive for urine THC at the time of emergency department admission (Nguyen et al., 2014; Wu et al., 2018). In these observational studies, the incidence of TBI patients screening positive for urine THC was 8.9% and 18.4%, and both studies report that positive THC screens were associated with improved survival. However, one of these studies also reported a lack of effect of a positive THC screen on the duration of hospital stay (Nguyen et al., 2014; Wu et al., 2018). Similarly, a prospective study of 307 patients with TBI showed that cannabis use in the acute post-injury period was associated with lower unadjusted symptom severity scores at 3- and 4-weeks post injury but was not associated with recovery of cognitive and physical activity (Lawrence et al., 2020). A retrospective analysis of 66 patient charts at the Dent Neurologic Institute in New York revealed a higher subjective quality of life in TBI patients who used medicinal cannabis (McVige, Bargnes, Shukri, & Mechtler, 2018).

While not fully reflective of the rich variety of phytocannabinoids contained in cannabis, clinical trials have evaluated the use of synthetic THC (dexanabinol) in severe TBI patients. A phase III, randomized, placebo-controlled clinical trial evaluated the safety and efficacy of a single dose of dexanabinol delivered within 6h of the injury on the extended Glasgow outcome scale in patients with severe TBI (Maas et al., 2006). Results from this study indicate that dexanabinol was safe for human administration, but there was no effect of dexanabinol on the extended Glasgow outcome scale at 6 months post injury relative to placebo. In a randomized, double-blind, placebo-controlled phase II trial, 97 comatose TBI patients were treated with a 24h infusion of a dual CB1R/CB2R agonist or placebo beginning no later than 4.5 h following TBI (Firsching et al., 2012). Results show enhanced survival in the treatment groups compared with placebo groups at 1 month post injury, but there was no effect on 6-month mortality rates. This study also found that critical intracranial pressure and cranial perfusion pressure occurred less frequently and were less extreme in the treatment groups. It is important to note that these clinical trials were conducted in patients who experienced severe TBI, so how these drugs may affect outcomes following mild or moderate TBI remains to be tested. Additionally, both of these studies evaluated the efficacy of a single dose of cannabinoid given acutely post-TBI. It remains possible that repeated dosing in early recovery may yield greater clinical effects. There is extensive preclinical literature supporting the therapeutic efficacy of cannabinoids or endocannabinoid modulators in TBI animal models (see review by Schurman & Lichtman, 2017). These studies have identified positive effects of endocannabinoid or cannabinoid receptor signaling modulation on many aspects of TBI pathophysiology, including neuroinflammation, oxidative stress, blood-brain barrier permeability, and excitotoxicity. While these findings are promising, it remains unclear how post-TBI cannabis use will affect patient outcomes.

Altogether, data currently suggests that TBI may increase cannabis use post injury, and this effect is strongest when TBI occurs in childhood or adolescence. This may be driven by the emergence of symptomatology thought to be ameliorated by cannabis, or personality changes that increase risky behaviors and substance use in general. Dysregulation of endocannabinoid signaling may also contribute to post-TBI substance use. There is a clear need for more data on the relationship between TBI and cannabis, including whether cannabis use is a risk factor for TBI, how TBI affects cannabis use or the development of cannabis use disorder, and what the consequences of post-TBI cannabis use are on health outcomes.

5. Post-TBI cognitive dysfunction and substance misuse

Cognitive impairment is one of the main consequences of TBI. It has been reported that up to 65% of individuals with moderate to severe TBI and approximately 15% of those with mild TBI have some kind of cognitive dysfunction (Rabinowitz & Levin, 2014). Adams, Corrigan, and Dams-O'Connor (2019) postulate that many neurological, behavioral, and cognitive consequences of TBI may contribute to increased risk for medication mismanagement, SUD, and substance misuse. Alterations in executive function are very common following TBI and are independent of injury severity (McDonald, Flashman, & Saykin, 2002). They are mediated by the pre-frontal cortex, an area in the brain that is especially vulnerable to the mechanical forces implicated in TBI (Corrigan & Cole, 2008). Executive functions are higher-order cognitive processes like working memory, inhibitory control, reasoning, cognitive flexibility, and problem solving (Cristofori, Cohen-Zimerman, & Grafman, 2019) that allow for regulation of thoughts and actions during goal-directed behavior (Friedman & Miyake, 2017). For example, inhibitory control involves being able to control thoughts, behaviors, and/or emotions (Diamond, 2013). Alterations in inhibitory control can increase impulsivity and habitual behaviors (Diamond, 2013). Increased impulsivity is a common and debilitating consequence of TBI (James, Strom, & Leskela, 2014) and includes increases in the proclivity to engage in certain behaviors; therefore, alterations in this cognitive function may contribute to substance misuse following TBI. Cannabis use has also been associated with executive dysfunction, specifically increased impulsivity (Ansell, Laws, Roche, & Sinha, 2015). Structural imaging studies of regular or chronic cannabis users have revealed abnormalities in brain regions associated with executive control, such as reductions in pre-frontal cortical volume (Wrege et al., 2014). The temporal relationship between impulsivity and cannabis use has not been fully established; however, synthesizing data from studies of addiction and other drugs of abuse suggests that impulsivity is both a risk factor for and consequence of chronic substance use (Jentsch et al., 2014). TBI-induced increased impulsivity may then represent a factor that mediates post-concussive substance misuse. However, studies addressing the mechanisms by which post-TBI cognitive impairment leads to misuse or overdose are needed.

6. Summary

Traumatic brain injury incidence is likely vastly underreported but nevertheless on the rise, globally affecting populations vulnerable to substance misuse. As discussed above, TBI patients demonstrate a bimodal age distribution, with children and adolescents as well as the elderly being primarily affected (Faul et al., 2010). While TBI in the elderly population has greater mortality, TBI in the younger population could have implications for increased morbidity considering the brain is still developing. TBI is more prevalent in males, the same demographic more likely to misuse substances including alcohol, cannabis, and opioids (Faul et al., 2010; Ponsford et al., 2007). Moreover, the arguable majority of TBIs are either substance-related or occur in patients with a history of substance misuse (Bjork & Grant, 2009; Savola et al., 2005). In addition to the demographics of individuals who incur TBI already being at risk for substance misuse, the heterogeneity of post-TBI symptoms complicates both the study and treatment of post-injury substance misuse. Physical disability, pain, cognitive dysfunction, and behavioral/emotional disruptions following brain injury could all potentially contribute to, be affected by, or complicate treatment for post-TBI substance misuse (Bjork & Grant, 2009). Thus, there is a clear need for treating comorbid TBI and substance use disorders as well as other TBI comorbidities to improve long-term outcomes of TBI patients.

Though decades of clinical and preclinical research have been conducted to understand the neurobiological mechanisms underlying brain injury, TBI constitutes a global health problem with no existing treatments beyond rehabilitation and symptom management. Only recently have animal models been employed to study the impact of TBI on the response to substances of abuse and, conversely, the impact of substances of abuse on TBI outcomes. Neurobiological processes at the intersection of TBI and SUD may underlie the bidirectional relationship between these common comorbidities, ranging from neuroinflammation and oxidative stress to neurochemical alterations in dopamine, endocannabinoid, opioid, or other neuromodulatory systems. Moving forward, further research is critical to identify effective treatments for TBI and comorbid SUD.

Acknowledgment

This work was supported by the National Institutes of Alcohol Abuse and Alcoholism of the National Institutes of Health under award numbers R01AA025792 (Molina) and F32AA026779 (Fucich).

References

Abdul-Muneer, P. M., Chandra, N., & Haorah, J. (2015). Interactions of oxidative stress and neurovascular inflammation in the pathogenesis of traumatic brain injury. *Molecular Neurobiology*, *51*(3), 966–979. https://doi.org/10.1007/s12035-014-8752-3.

Adams, R. S., Corrigan, J. D., & Dams-O'Connor, K. (2019). Opioid use among individuals with traumatic brain injury: A perfect storm? *Journal of Neurotrauma*, *216*, 211–216. https://doi.org/10.1089/neu.2019.6451.

Adams, R. S., Larson, M. J., Meerwijk, E. L., Williams, T. V., & Harris, A. (2019). Post deployment polytrauma diagnoses among soldiers and Veterans using the Veterans

Health Affairs polytrauma system of care and receipt of opioids, nonpharmacologic, and mental health treatments. *The Journal of Head Trauma Rehabilitation, 34*(3), 167–175. https://doi.org/10.1097/HTR.0000000000000481.

Allen, K. A. (2016). Pathophysiology and treatment of severe traumatic brain injuries in children. *Journal of Neuroscience Nursing, 48*(1), 15–27. https://doi.org/10.1097/JNN.0000000000000176.

American Psychiatric Association. (2013). *Diagnostic and statistical manual of mental disorders* (5th ed.). Arlington: American Psychiatric Publishing.

Ansell, E. B., Laws, H. B., Roche, M. J., & Sinha, R. (2015). Effects of marijuana use on impulsivity and hostility in daily life. *Drug and Alcohol Dependence, 148*, 136–142. https://doi.org/10.1016/j.drugalcdep.2014.12.029.

Arciniegas, D. B., Held, K., & Wagner, P. (2002). Cognitive impairment following traumatic brain injury. *Current Treatment Options in Neurology, 4*(1), 43–57. https://doi.org/10.1007/s11940-002-0004-6.

Arnér, S., & Meyerson, B. A. (1988). Lack of analgesic effect of opioids on neuropathic and idiopathic forms of pain. *Pain, 33*(1), 11–23. https://doi.org/10.1016/0304-3959(88)90198-4.

Ashina, H., Porreca, F., Anderson, T., Mohammad Amin, F., Ashina, M., Winther Schytz, H., et al. (2019). Post-traumatic headache: Epidemiology and pathophysiological insights. *Nature Reviews Neurology, 15*(10), 607–617. https://doi.org/10.1038/s41582-019-0243-8.

Atwood, B. K., & Mackie, K. (2010). CB2: A cannabinoid receptor with an identity crisis. *British Journal of Pharmacology, 160*(3), 467–479. https://doi.org/10.1111/j.1476-5381.2010.00729.x.

Barkhoudarian, G., Hovda, D. A., & Giza, C. C. (2016). The molecular pathophysiology of concussive brain injury—An update. *Physical Medicine and Rehabilitation Clinics of North America, 27*(2), 373–393. https://doi.org/10.1016/j.pmr.2016.01.003.

Basavarajappa, B. S., Joshi, V., Shivakumar, M., & Subbanna, S. (2019). Distinct functions of endogenous cannabinoid system in alcohol abuse disorders. *British Journal of Pharmacology, 176*(17), 3085–3109. https://doi.org/10.1111/bph.14780.

Beaulieu-Bonneau, S., St-Onge, F., Blackburn, M. C., Banville, A., Paradis-Giroux, A. A., & Ouellet, M. C. (2018). Alcohol and drug use before and during the first year after traumatic brain injury. *The Journal of Head Trauma Rehabilitation, 33*(3), E51–E60. https://doi.org/10.1097/HTR.0000000000000341.

Bell, R. S., Vo, A. H., Neal, C. J., Tigno, J., Roberts, R., Mossop, C., et al. (2009). Military traumatic brain and spinal column injury: A 5-year study of the impact blast and other military grade weaponry on the central nervous system. *Journal of Trauma and Acute Care Surgery, 66*(4), S104. https://doi.org/10.1097/TA.0b013e31819d88c8.

Benyamin, R., Trescot, A. M., Datta, S., Buenaventura, R., Adlaka, R., Sehgal, N., et al. (2008). Opioids complications and side effects. *Pain Physician, 11*(Spec. Iss. 2), 105–120. https://doi.org/10.1007/978-3-540-35423-9_14.

Bigler, E. D., Blatter, D. D., Johnson, S. C., Anderson, C. V., Russo, A. A., Gale, S. D., et al. (1996). Traumatic brain injury, alcohol and quantitative neuroimaging: Preliminary findings. *Brain Injury, 10*(3), 197–206. https://doi.org/10.1080/026990596124511.

Bjork, J. M., & Grant, S. J. (2009). Does traumatic brain injury increase risk for substance abuse? *Journal of Neurotrauma, 26*(7), 1077–1082. https://doi.org/10.1089/neu.2008.0849.

Bodnar, R. J. (2016). Endogenous opiates and behavior: 2014. *Peptides, 75*, 18–70. https://doi.org/10.1016/j.peptides.2015.10.009.

Bombardier, C. H., Fann, J. R., Temkin, N. R., Esselman, P. C., Barber, J., & Dikmen, S. S. (2010). Rates of major depressive disorder and clinical outcomes following traumatic brain injury. *JAMA, 303*(19), 1938–1945. https://doi.org/10.1001/jama.2010.599.

Bombardier, C. H., Temkin, N. R., Machamer, J., & Dikmen, S. S. (2003). The natural history of drinking and alcohol-related problems after traumatic brain injury. *Archives of Physical Medicine and Rehabilitation*, *84*(2), 185–191. https://doi.org/10.1053/apmr.2003.50002.

Bosco, M. A., Murphy, J. L., & Clark, M. E. (2013). Chronic pain and traumatic brain injury in OEF/OIF service members and Veterans. *Headache*, *53*(9), 1518–1522. https://doi.org/10.1111/head.12172.

Brickell, T. A., Lange, R. T., & French, L. M. (2014). Health-related quality of life within the first 5 years following military-related concurrent mild traumatic brain injury and polytrauma. *Military Medicine*, *179*(8), 827–838. https://doi.org/10.7205/milmed-d-13-00506.

Bryant, R. A., & Harvey, A. G. (1998). Relationship between acute stress disorder and post-traumatic stress disorder following mild traumatic brain injury. *The American Journal of Psychiatry*, *155*(5), 625–629. https://doi.org/10.1176/ajp.155.5.625.

Callahan, M. L., & Lim, M. M. (2018). Sensory sensitivity in TBI: Implications for chronic disability. *Current Neurology and Neuroscience Reports*, *18*(9), 56. https://doi.org/10.1007/s11910-018-0867-x.

Cannella, L. A., McGary, H., & Ramirez, S. H. (2019). Brain interrupted: Early life traumatic brain injury and addiction vulnerability. *Experimental Neurology*, *317*(February), 191–201. https://doi.org/10.1016/j.expneurol.2019.03.003.

Castillo, P. E., Younts, T. J., Chávez, A. E., & Hashimotodani, Y. (2012). Endocannabinoid signaling and synaptic function. *Neuron*, *76*(1), 70–81. https://doi.org/10.1016/j.neuron.2012.09.020.

Centers for Disease Control and Prevention. (2015). *Report to congress on traumatic brain injury in the United States: Epidemiology and rehabilitation*. Atlanta GA: National Center for Injury Prevention and Control; Division of Unintentional Injury Prevention and Control.

Centers for Disease Control and Prevention. (2019). *Surveillance report of traumatic brain injury-related emergency department visits, hospitalizations and deaths—United States, 2014*. Centers for Disease Control and Prevention, U.S. Department of Health and Human Services.

Cepeda, M. S., & Carr, D. B. (2003). Women experience more pain and require more morphine than men to achieve a similar degree of analgesia. *Anesthesia & Analgesia*, *97*(5), 1464–1468. https://doi.org/10.1213/01.ane.0000080153.36643.83.

Christensen, J., Eyolfson, E., Salberg, S., Bhatt, D., Weerawardhena, H., Tabor, J., et al. (2020). When two wrongs make a right: The effect of acute and chronic binge drinking on traumatic brain injury outcomes in young adult female rats. *Journal of Neurotrauma*, *37*(2), 273–285. https://doi.org/10.1089/neu.2019.6656.

Cifu, D. X., Taylor, B. C., Carne, W. F., Bidelspach, D., Sayer, N. A., Scholten, J., et al. (2013). Traumatic brain injury, posttraumatic stress disorder, and pain diagnoses in OIF/OEF/OND Veterans. *Journal of Rehabilitation Research and Development*, *50*(9), 1169–1176. https://doi.org/10.1682/JRRD.2013.01.0006.

Clark, R. S., Kochanek, P. M., Schwarz, M. A., Schiding, J. K., Turner, D. S., Chen, M., et al. (1996). Inducible nitric oxide synthase expression in cerebrovascular smooth muscle and neutrophils after traumatic brain injury in immature rats. *Pediatric Research*, *39*(5), 784–790. https://doi.org/10.1203/00006450-199605000-00007.

Clark, M. E., Scholten, J. D., Walker, R. L., & Gironda, R. J. (2009). Assessment and treatment of pain associated with combat-related polytrauma. *Pain Medicine*, *10*(3), 456–469. https://doi.org/10.1111/j.1526-4637.2009.00589.x.

Corder, G., Castro, D. C., Bruchas, M. R., & Scherrer, G. (2018). Endogenous and exogenous opioids in pain. *Annual Review of Neuroscience*, *41*, 453–473. https://doi.org/10.1146/annurev-neuro-080317-061522.

Corrigan, J. D. (1995). Substance abuse as a mediating factor in outcome from traumatic brain injury. *Archives of Physical Medicine and Rehabilitation, 76*(4), 302–309. https://doi.org/10.1016/s0003-9993(95)80654-7.

Corrigan, J. D., & Adams, R. S. (2019). The intersection of lifetime history of traumatic brain injury and the opioid epidemic. *Addictive Behaviors, 90*, 143–145 (October 2018) https://doi.org/10.1016/j.addbeh.2018.10.030.

Corrigan, J. D., Bogner, J., & Holloman, C. (2012). Lifetime history of traumatic brain injury among persons with substance use disorders. *Brain Injury, 26*(2), 139–150. https://doi.org/10.3109/02699052.2011.648705.

Corrigan, J. D., Bogner, J., Mellick, D., Bushnik, T., Dams-O'Connor, K., Hammond, F. M., et al. (2013). Prior history of traumatic brain injury among persons in the traumatic brain injury model systems national database. *Archives of Physical Medicine and Rehabilitation, 94*(10), 1940–1950. https://doi.org/10.1016/j.apmr.2013.05.018.

Corrigan, J. D., & Cole, T. B. (2008). Substance use disorders and clinical management of traumatic brain injury and posttraumatic stress disorder. *JAMA: The Journal of the American Medical Association, 300*(6), 720–721. https://doi.org/10.1001/jama.300.6.720.

Corrigan, J. D., Lamb-Hart, G. L., & Rust, E. (1995). A programme of intervention for substance abuse following traumatic brain injury. *Brain Injury, 9*(3), 221–236. https://doi.org/10.3109/02699059509008195.

Costa, B., Giagnoni, G., Franke, C., Trovato, A. E., & Colleoni, M. (2004). Vanilloid TRPV1 receptor mediates the antihyperalgesic effect of the nonpsychoactive cannabinoid, cannabidiol, in a rat model of acute inflammation. *British Journal of Pharmacology, 143*(2), 247–250. https://doi.org/10.1038/sj.bjp.0705920.

Covassin, T., Moran, R., & Elbin, R. J. (2016). Sex differences in reported concussion injury rates and time loss from participation: An update of the national collegiate Athletic association injury surveillance program from 2004–2005 through 2008–2009. *Journal of Athletic Training, 51*(3), 189–194. https://doi.org/10.4085/1062-6050-51.3.05.

Cranford, J. A., Bohnert, K. M., Perron, B. E., Bourque, C., & Ilgen, M. (2016). Prevalence and correlates of "Vaping" as a route of cannabis administration in medical cannabis patients. *Drug and Alcohol Dependence, 169*, 41–47. https://doi.org/10.1016/j.drugalcdep.2016.10.008.

Crippa, J. A., Guimarães, F. S., Campos, A. C., & Zuardi, A. W. (2018). Translational investigation of the therapeutic potential of cannabidiol (CBD): Toward a new age. *Frontiers in Immunology, 9*, 2009. https://doi.org/10.3389/fimmu.2018.02009.

Cristofori, I., Cohen-Zimerman, S., & Grafman, J. (2019). Executive functions. *Handbook of Clinical Neurology, 163*, 197–219. https://doi.org/10.1016/B978-0-12-804281-6.00011-2.

Dai, J.-X., Ma, Y.-B., Le, N.-Y., Cao, J., & Wang, Y. (2018). Large animal models of traumatic brain injury. *International Journal of Neuroscience, 128*(3), 243–254. https://doi.org/10.1080/00207454.2017.1380008.

Darcq, E., & Kieffer, B. L. (2018). Opioid receptors: Drivers to addiction? *Nature Reviews. Neuroscience, 19*(8), 499–514. https://doi.org/10.1038/s41583-018-0028-x.

Darke, S., McDonald, S., Kaye, S., & Torok, M. (2012). Prevalence and correlates of traumatic brain injury amongst heroin users. *Addiction Research & Theory, 20*(6), 522–528. https://doi.org/10.3109/16066359.2012.672600.

Deb, S., Lyons, I., Koutzoukis, C., Ali, I., & McCarthy, G. (1999). Rate of psychiatric illness 1 year after traumatic brain injury. *American Journal of Psychiatry, 156*(3), 374–378. https://doi.org/10.1176/ajp.156.3.374.

Diamond, A. (2013). Executive functions. *Annual Review of Psychology, 64*, 135–168. https://doi.org/10.1146/annurev-psych-113011-143750.

Dobscha, S. K., Morasco, B. J., Duckart, J. P., MacEy, T., & Deyo, R. A. (2013). Correlates of prescription opioid initiation and long-term opioid use in Veterans with persistent pain. *Clinical Journal of Pain, 29*(2), 102–108. https://doi.org/10.1097/AJP.0b013e3182490bdb.

Duhaime, A. C., Alario, A. J., Lewander, W. J., Schut, L., Sutton, L. N., Seidl, T. S., et al. (1992). Head injury in very young children: Mechanisms, injury types, and ophthalmologic findings in 100 hospitalized patients younger than 2 years of age. *Pediatrics, 90*(2), 179–185. https://doi.org/10.1016/s0196-0644(05)80277-x.

Egli, M., Koob, G. F., & Edwards, S. (2012). Alcohol dependence as a chronic pain disorder. *Neuroscience & Biobehavioral Reviews, 36*(10), 2179–2192. https://doi.org/10.1016/j.neubiorev.2012.07.010.

Ellerd, D. A., & Moore, S. C. (1992). Follow-up at twelve and thirty months of persons with traumatic brain injury engaged in supported employment placements. *Journal of Applied Rehabilitation Counseling, 23*(3), 48–50. https://doi.org/10.1891/0047-2220.23.3.48.

Faul, M., Xu, L., Wald, M. M., & Coronado, V. G. (2010). *Traumatic brain injury in the United States: Emergency department visits, hospitalizations and deaths 2002–2006*. Atlanta (GA): Centers for Disease Control and Prevention, National Center for Injury Prevention and Control. https://www.cdc.gov/traumaticbraininjury/pdf/blue_book.pdf.

Fillingim, R. B., Ness, T. J., Glover, T. L., Campbell, C. M., Hastie, B. A., Price, D. D., et al. (2005). Morphine responses and experimental pain: Sex differences in side effects and cardiovascular responses but not analgesia. *The Journal of Pain, 6*(2), 116–124. https://doi.org/10.1016/j.jpain.2004.11.005.

Firsching, R., Piek, J., Skalej, M., Rohde, V., Schmidt, U., & Striggow, F. (2012). Early survival of comatose patients after severe traumatic brain injury with the dual cannabinoid CB1/CB2 receptor agonist KN38-7271: A randomized, double-blind, placebo-controlled phase II trial. *Journal of Neurological Surgery, Part A: Central European Neurosurgery, 73*(4), 204–216. https://doi.org/10.1055/s-0032-1304815.

Fonda, J. R., Gradus, J. L., Brogly, S. B., Mcglinchey, R. E., Milberg, W. P., & Fredman, L. (2020). Traumatic brain injury and opioid overdose among post-9/11 Veterans with long-term opioid treatment of chronic pain. *The Journal of Head Trauma Rehabilitation, 35*(3), 209–217. https://doi.org/10.1097/HTR.0000000000000546.

Friedman, N. P., & Miyake, A. (2017). Unity and diversity of executive functions: Individual differences as a window on cognitive structure. *Cortex, 86*, 186–204. https://doi.org/10.1016/j.cortex.2016.04.023.

Frishbain, D. A., Cole, B., Lewis, J., Rosomoff, H. L., & Rosomoff, R. S. (2008). What percentage of chronic nonmalignant pain patients exposed to chronic opioid analgesic therapy develop abuse/addiction and/or aberrant drug-related behaviors? A structured evidence-based review. *Pain Medicine, 9*(4), 444–459. https://doi.org/10.1111/j.1526-4637.2007.00370.x.

Froehlich, J. C. (1997). Opioid peptides. *Alcohol Health and Research World, 21*(2), 132–136. https://pubs.niaaa.nih.gov/publications/toc20-4.htm.

Fucich, E. A., Mayeux, J. P., McGinn, M. A., Gilpin, N. W., Edwards, S., & Molina, P. E. (2019). A novel role for the endocannabinoid system in ameliorating motivation for alcohol drinking and negative behavioral affect after traumatic brain injury in rats. *Journal of Neurotrauma, 36*(11), 1847–1855. https://doi.org/10.1089/neu.2018.5854.

Gahm, C., Holmin, S., Wiklund, P. N., Brundin, L., & Mathiesen, T. (2006). Neuroprotection by selective inhibition of inducible nitric oxide synthase after experimental brain contusion. *Journal of Neurotrauma, 23*(9), 1343–1354. https://doi.org/10.1089/neu.2006.23.1343.

Gennarelli, T. A., & Graham, D. I. (1998). Neuropathology of the head injuries. *Seminars in Clinical Neuropsychiatry, 3*(3), 160–175.

Golub, A., & Bennett, A. S. (2013). Prescription opioid initiation, correlates, and consequences among a sample of OEF/OIF military personnel. *Substance Use & Misuse, 48*(10), 811–820. https://doi.org/10.3109/10826084.2013.796988.

Graham, D. P., & Cardon, A. L. (2008). An update on substance use and treatment following traumatic brain injury. *Annals of the New York Academy of Sciences, 1141*, 148–162. https://doi.org/10.1196/annals.1441.029.

Grossbard, J., Malte, C. A., Lapham, G., Pagulayan, K., Turner, A. P., Rubinsky, A. D., et al. (2017). Prevalence of alcohol misuse and follow-up care in a national sample of OEF/OIF VA patients with and without TBI. *Psychiatric Services, 68*(1), 48–55. https://doi.org/10.1176/appi.ps.201500290.

Haarasilta, L., Marttunen, M., Kaprio, J., & Aro, H. (2001). The 12-month prevalence and characteristics of major depressive episode in a representative nationwide sample of adolescents and young adults. *Psychological Medicine, 31*(7), 1169–1179. https://doi.org/10.1017/S0033291701004573.

Hammond, F. M., Barrett, R. S., Shea, T., Seel, R. T., McAlister, T. W., Kaelin, D., et al. (2015). Psychotropic medication use during inpatient rehabilitation for traumatic brain injury. *Archives of Physical Medicine and Rehabilitation, 96*(8), S256–S273.e14. https://doi.org/10.1016/j.apmr.2015.01.025.

Hart, T., Hoffman, J. M., Pretz, C., Kennedy, R., Clark, A. N., & Brenner, L. A. (2012). A longitudinal study of major and minor depression following traumatic brain injury. *Archives of Physical Medicine and Rehabilitation, 93*(8), 1343–1349. https://doi.org/10.1016/j.apmr.2012.03.036.

Hawkins, J. D., Catalano, R. F., & Miller, J. Y. (1992). Risk and protective factors for alcohol and other drug problems in adolescence and early adulthood: Implications for substance abuse prevention. *Psychological Bulletin, 112*(1), 64–105. https://doi.org/10.1037/0033-2909.112.1.64.

Hawley, L. A., Ketchum, J. M., Morey, C., Collins, K., & Charlifue, S. (2018). Cannabis use in individuals with spinal cord injury or moderate to severe traumatic brain injury in Colorado. *Archives of Physical Medicine and Rehabilitation, 99*(8), 1584–1590. https://doi.org/10.1016/j.apmr.2018.02.003.

Hayes, J. P. (2019). PTSD and TBI comorbidity. *PTSD Research Quarterly, 30*(2), 1–18. https://www.ptsd.va.gov/publications/ptsd_rq.asp.

Heltemes, K. J., Dougherty, A. L., MacGregor, A. J., & Galarneau, M. R. (2011). Alcohol abuse disorders among U.S. service members with mild traumatic brain injury. *Military Medicine, 176*(2), 147–150. https://doi.org/10.7205/milmed-d-10-00191.

Henninger, N., Dützmann, S., Sicard, K. M., Kollmar, R., Bardutzky, J., & Schwab, S. (2005). Impaired spatial learning in a novel rat model of mild cerebral concussion injury. *Experimental Neurology, 195*(2), 447–457. https://doi.org/10.1016/j.expneurol.2005.06.013.

Herrold, A. A., Sander, A. M., Wilson, K. V., Scimeca, L. M., Cobia, D. J., & Breiter, H. C. (2015). Dual diagnosis of traumatic brain injury and alcohol use disorder: Characterizing clinical and neurobiological underpinnings. *Current Addiction Reports, 2*(4), 273–284. https://doi.org/10.1007/s40429-015-0078-3.

Hibbard, M. R., Uysal, S., Kepler, K., Bogdany, J., & Silver, J. (1998). Axis I psychopathology in individuals with traumatic brain injury. *The Journal of Head Trauma Rehabilitation, 13*(4), 24–39. https://doi.org/10.1097/00001199-199808000-00003.

Hillard, C. J. (2018). Circulating endocannabinoids: From whence do they come and where are they going? *Neuropsychopharmacology, 43*(1), 155–172. https://doi.org/10.1038/npp.2017.130.

Hoffman, J., Lucas, S., Dikmen, S., & Temkin, N. (2020). Clinical perspectives on headache after traumatic brain injury. *PM & R: The Journal of Injury, Function, and Rehabilitation*, 1–8. https://doi.org/10.1002/pmrj.12338.

Hoffman, J., Pagulayan, K., Zawaideh, N., & Bell, K. (2005). Poster 46. Pain after traumatic brain injury: Predictors and correlates. *Archives of Physical Medicine and Rehabilitation, 86*(10), e17. https://doi.org/10.1016/j.apmr.2005.08.069.

Honkanen, R., & Visuri, T. (1976). Blood alcohol levels in a series of injured patients with special reference to accident and type of injury. *Annales Chirurgiae et Gynaecologiae, 65*(4), 287–294.

Horner, M. D., Ferguson, P. L., Selassie, A. W., Labbate, L. A., Kniele, K., & Corrigan, J. D. (2005). Patterns of alcohol use 1 year after traumatic brain injury: A population-based, epidemiological study. *Journal of the International Neuropsychological Society, 11*(3), 322–330. https://doi.org/10.1017/S135561770505037X.

Hudson, T. J., Painter, J. T., Gressler, L. E., Lu, L., Williams, J. S., Booth, B. M., et al. (2018). Factors associated with opioid initiation in OEF/OIF/OND Veterans with traumatic brain injury. *Pain Medicine, 19*(4), 774–787. https://doi.org/10.1093/pm/pnx208.

Humphreys, I., Wood, R. L., Phillips, C. J., & Macey, S. (2013). The costs of traumatic brain injury: A literature review. *ClinicoEconomics and Outcomes Research, 5*, 281–287. https://doi.org/10.2147/CEOR.S44625.

Ilie, G., Adlaf, E. M., Mann, R. E., Boak, A., Hamilton, H., Asbridge, M., et al. (2014). The moderating effects of sex and age on the association between traumatic brain injury and harmful psychological correlates among adolescents. *PLoS One, 9*(9), e108167. https://doi.org/10.1371/journal.pone.0108167.

Ilie, G., Adlaf, E. M., Mann, R. E., Ialomiteanu, A., Hamilton, H., Rehm, J., et al. (2015). Associations between a history of traumatic brain injuries and current cigarette smoking, substance use, and elevated psychological distress in a population sample of Canadian adults. *Journal of Neurotrauma, 32*(14), 1130–1134. https://doi.org/10.1089/neu.2014.3619.

Ilie, G., Mann, R. E., Hamilton, H., Adlaf, E. M., Boak, A., Asbridge, M., et al. (2015). Substance use and related harms among adolescents with and without traumatic brain injury. *The Journal of Head Trauma Rehabilitation, 30*(5), 293–301. https://doi.org/10.1097/HTR.0000000000000101.

Illinois Department of Public Health. (2019). *Debilitating conditions*. http://www.dph.illinois.gov/topics-services/prevention-wellness/medical-cannabis/debilitating-conditions.

International Association for the Study of Pain (IASP). (2017). *IASP terminology*. Washington, D.C.: International Association for the Study of Pain. https://www.iasp-pain.org/terminology?navItemNumber=576#Pain.

Irvine, J. A., & Clark, J. D. (2018). Chronic pain after traumatic brain injury: Pathophysiology and pain mechanisms. *Pain Medicine, 19*(7), 1315–1333. https://doig.org/10.1093/pm/pnx153.

Ivanhoe, C. B., & Hartman, E. T. (2004). Clinical caveats on medical assessment and treatment of pain after TBI. *The Journal of Head Trauma Rehabilitation, 19*(1), 29–39. https://doi.org/10.1097/00001199-200401000-00004.

Jacobsen, L. K., Southwick, S. M., & Kosten, T. R. (2001). Substance use disorders in patients with posttraumatic stress disorder: A review of the literature. *The American Journal of Psychiatry, 158*(8), 1184–1190. https://doi.org/10.1176/appi.ajp.158.8.1184.

James, L. M., Strom, T. Q., & Leskela, J. (2014). Risk-taking behaviors and impulsivity among Veterans with and without PTSD and mild TBI. *Military Medicine, 179*(4), 357–363. https://doi.org/10.7205/milmed-d-13-00241.

Jentsch, J. D., Ashenhurst, J. R., Cervantes, M. C., Groman, S. M., James, A. S., & Pennington, Z. T. (2014). Dissecting impulsivity and its relationships to drug addictions. *Annals of the New York Academy of Sciences, 1327*(1), 1–26. https://doi.org/10.1111/nyas.12388.

Ji, Y., Murphy, A. Z., & Traub, R. J. (2006). Sex differences in morphine-induced analgesia of visceral pain are supraspinally and peripherally mediated. *American Journal of Physiology. Regulatory, Integrative and Comparative Physiology, 291*(2), R307–R314. https://doi.org/10.1152/ajpregu.00824.2005.

Johnson, L. A., Eick-Cost, A., Jeffries, V., Russell, K., & Otto, J. L. (2015). Risk of alcohol use disorder or other drug use disorder among U.S. service members following traumatic brain injury, 2008–2011. *Military Medicine, 180*(2), 208–215. https://doi.org/10.7205/milmed-d-14-00268.

Johnston, M. V., Ishida, A., Ishida, W. N., Matsushita, H. B., Nishimura, A., & Tsuji, M. (2009). Plasticity and injury in the developing brain. *Brain & Development, 31*(1), 1–10. https://10.10.1016/j.braindev.2008.03.014.

Jorge, R. E. (2015). *Mood disorders. Vol. 128* (1st ed.). Elsevier Ltd. https://doi.org/10.1016/B978-0-444-63521-1.00038-8.

Jorge, R. E., Starkstein, S. E., Arndt, S., Moser, D., Crespo-Facorro, B., & Robinson, R. G. (2005). Alcohol misuse and mood disorders following traumatic brain injury. *Archives of General Psychiatry, 62*(7), 742–749. https://doi.org/10.1001/archpsyc.62.7.742.

Kaplan, C. P., & Corrigan, J. D. (1992). Effect of blood alcohol level on recovery from severe closed head injury. *Brain Injury, 6*(4), 337–349. https://doi.org/10.3109/02699059209034948.

Karelina, K., Gaier, K. R., Prabhu, M., Wenger, V., Corrigan, T. E. D., & Weil, Z. M. (2017). Binge ethanol in adulthood exacerbates negative outcomes following juvenile traumatic brain injury. *Brain, Behavior, and Immunity, 60*, 304–311. https://doi.org/10.1016/j.bbi.2016.11.009.

Karelina, K., Nicholson, S., & Weil, Z. M. (2018). Minocycline blocks traumatic brain injury-induced alcohol consumption and nucleus accumbens inflammation in adolescent male mice. *Brain, Behavior, and Immunity, 69*, 532–539. https://doi.org/10.1016/j.bbi.2018.01.012.

Kennedy, E., Heron, J., & Munafò, M. (2017). Substance use, criminal behavior and psychiatric symptoms following childhood traumatic brain injury: Findings from the ALSPAC cohort. *European Child and Adolescent Psychiatry, 26*(10), 1197–1206. https://doi.org/10.1007/s00787-017-0975-1.

Khoury, S., & Benavides, R. (2018). Pain with traumatic brain injury and psychological disorders. *Progress in Neuro-Psychopharmacology and Biological Psychiatry, 87*(June 2017), 224–233. https://doi.org/10.1016/j.pnpbp.2017.06.007.

King, P. R., Beehler, G. P., & Wade, M. J. (2015). Self-reported pain and pain management strategies among Veterans with traumatic brain injury: A pilot study. *Military Medicine, 180*(8), 863–868. https://doi.org/10.7205/MILMED-D-14-00472.

Kolakowsky-Hayner, S. A., Gourley, E. V., 3rd, Kreutzer, J. S., Marwitz, J. H., Meade, M. A., & Cifu, D. X. (2002). Post-injury substance abuse among persons with brain injury and persons with spinal cord injury. *Brain Injury, 16*(7), 583–592. https://doi.org/10.1080/02699050110119475.

Koob, G. F. (2020). Neurobiology of opioid addiction: Opponent process, hyperkatifeia, and negative reinforcement. *Biological Psychiatry, 87*(1), 44–53. https://doi.org/10.1016/j.biopsych.2019.05.023.

Koponen, S., Taiminen, T., Hiekkanen, H., & Tenovuo, O. (2011). Axis I and II psychiatric disorders in patients with traumatic brain injury: A 12-month follow-up study. *Brain Injury, 25*(11), 1029–1034. https://doi.org/10.3109/02699052.2011.607783.

Kotapka, M. J., Graham, D. I., Adams, J. H., & Gennarelli, T. A. (1994). Hippocampal pathology in fatal human head injury without high intracranial pressure. *Journal of Neurotrauma, 11*(3), 317–324. https://doi.org/10.1089/neu.1994.11.317.

Kozlowski, D. A., Leasure, J. L., & Schallert, T. (2013). The control of movement following traumatic brain injury. *Comprehensive Physiology*, *3*(1), 121–139. https://doi.org/10.1002/cphy.c110005.

Kreutzer, J. S., Witol, A. D., & Marwitz, J. H. (1996). Alcohol and drug use among young persons with traumatic brain injury. *Journal of Learning Disabilities*, *29*(6), 649–651. https://doi.org/10.1177/002221949602900608.

Laborde, A. (2000). NIH consensus development panel on rehabilitation of persons with traumatic brain injury. *The Journal of Head Trauma Rehabilitation*, *15*(1), 761–762.

Ladak, A. A., Enam, S. A., & Ibrahim, M. T. (2019). A review of the molecular mechanisms of traumatic brain injury. *World Neurosurgery*, *131*, 126–132. https://doi.org/10.1016/j.wneu.2019.07.039.

Laprairie, R. B., Bagher, A. M., Kelly, M. E., & Denovan-Wright, E. M. (2015). Cannabidiol is a negative allosteric modulator of the cannabinoid CB1 receptor. *British Journal of Pharmacology*, *172*(20), 4790–4805. https://doi.org/10.1111/bph.13250.

Lawrence, D. W., Foster, E., Comper, P., Langer, L., Hutchison, M. G., Chandra, T., et al. (2020). Cannabis, alcohol and cigarette use during the acute post-concussion period. *Brain Injury*, *34*(1), 42–51. https://doi.org/10.1080/02699052.2019.1679885.

Le Merrer, J., Becker, J. A. J., Befort, K., & Kieffer, B. L. (2009). Reward processing by the opioid system in the brain. *Physiological Reviews*, *89*(4), 1379–1412. https://doi.org/10.1152/physrev.00005.2009.

Lenzlinger, P. M., Morganti-Kossmann, M. C., Laurer, H. L., & McIntosh, T. K. (2001). The duality of the inflammatory response to traumatic brain injury. *Molecular Neurobiology*, *24*(1–3), 169–181. https://doi.org/10.1385/MN:24:1-3:169.

Leventhal, J. M., Martin, K. D., & Asnes, A. G. (2010). Fractures and traumatic brain injuries: Abuse versus accidents in a US database of hospitalized children. *Pediatrics*, *126*(1), e104–e115. https://doi.org/10.1542/peds.2009-1076.

Lew, H. L., Otis, J. D., Tun, C., Kerns, R. D., Clark, M. E., & Cifu, D. X. (2009). Prevalence of chronic pain, posttraumatic stress disorder, and persistent postconcussive symptoms in OIF/OEF Veterans: Polytrauma clinical triad. *Journal of Rehabilitation Research and Development*, *46*(6), 697–702. https://doi.org/10.1682/JRRD.2009.01.0006.

Lim, Y. W., Meyer, N. P., Shah, A. S., Budde, M. D., Stemper, B. D., & Olsen, C. M. (2015). Voluntary alcohol intake following blast exposure in a rat model of mild traumatic brain injury. *PLoS One*, *10*(4), e0125130. https://doi.org/10.1371/journal.pone.0125130.

Lloyd, E., Somera-Molina, K., Van Eldik, L. J., Watterson, D. M., & Wainwright, M. S. (2008). Suppression of acute proinflammatory cytokine and chemokine upregulation by post-injury administration of a novel small molecule improves long-term neurologic outcome in a mouse model of traumatic brain injury. *Journal of Neuroinflammation*, *5*, 28. https://doi.org/10.1186/1742-2094-5-28.

Logrip, M. L., Zorrilla, E. P., & Koob, G. F. (2012). Stress modulation of drug self-administration: Implications for addiction comorbidity with post-traumatic stress disorder. *Neuropharmacology*, *62*(2), 552–564. https://doi.org/10.1016/j.neuropharm.2011.07.007.

Lopez-Rodriguez, A. B., Acaz-Fonseca, E., Viveros, M. P., & Garcia-Segura, L. M. (2015). Changes in cannabinoid receptors, aquaporin 4 and vimentin expression after traumatic brain injury in adolescent male mice. Association with edema and neurological deficit. *PLoS One*, *10*(6), e0128782. https://doi.org/10.1371/journal.pone.0128782.

Lowing, J. L., Susick, L. L., Caruso, J. P., Provenzano, A. M., Raghupathi, R., & Conti, A. C. (2014). Experimental traumatic brain injury alters ethanol consumption and sensitivity. *Journal of Neurotrauma*, *31*(20), 1700–1710. https://doi.org/10.1089/neu.2013.3286.

Loyd, D. R., Wang, X., & Murphy, A. Z. (2008). Sex differences in micro-opioid receptor expression in the rat midbrain periaqueductal gray are essential for eliciting sex differences in morphine analgesia. *The Journal of Neuroscience, 28*(52), 14007–14017. https://doi.org/10.1523/JNEUROSCI.4123-08.2008.

Maas, A. I. R., Murray, G., Henney, H., Kassem, N., Legrand, V., Mangelus, M., et al. (2006). Efficacy and safety of dexanabinol in severe traumatic brain injury: Results of a phase III randomised, placebo-controlled, clinical trial. *Lancet Neurology, 5*(1), 38–45. https://doi.org/10.1016/S1474-4422(05)70253-2.

MacGregor, A. J., Shaffer, R. A., Dougherty, A. L., Galarneau, M. R., Raman, R., Baker, D. G., et al. (2010). Prevalence and psychological correlates of traumatic brain injury in operation Iraqi freedom. *The Journal of Head Trauma Rehabilitation, 25*(1), 1–8. https://doi.org/10.1097/HTR.0b013e3181c2993d.

Makary, M. A., Overton, H. N., & Wang, P. (2017). Overprescribing is major contributor to opioid crisis. *British Medical Journal, 359*, 19–20. https://doi.org/10.1136/bmj.j4792.

Mallya, S., Sutherland, J., Pongracic, S., Mainland, B., & Ornstein, T. J. (2015). The manifestation of anxiety disorders after traumatic brain injury: A review. In *Vol. 32. Journal of neurotrauma* Mary Ann Liebert Inc. https://doi.org/10.1089/neu.2014.3504.

Management of Concussion/mTBI Working Group. (2009). VA/DoD clinical practice guideline for management of concussion/mild traumatic brain injury. *Journal of Rehabilitation Research and Development, 46*(6), CP1–68. https://doi.org/10.1682/jrrd.2009.06.0076.

Manzanares, J., Cabañero, D., Puente, N., García-Gutiérrez, M. S., Grandes, P., & Maldonado, R. (2018). Role of the endocannabinoid system in drug addiction. *Biochemical Pharmacology, 157*, 108–121. https://doi.org/10.1016/j.bcp.2018.09.013.

Marklund, N., Clausen, F., Lewander, T., & Hillered, L. (2001). Monitoring of reactive oxygen species production after traumatic brain injury in rats with microdialysis and the 4-hydroxybenzoic acid trapping method. *Journal of Neurotrauma, 18*(11), 1217–1227. https://doi.org/10.1089/089771501317095250.

Martelli, M. F., Zasler, N. D., Bender, M. C., & Nicholson, K. (2004). Psychological, neuropsychological, and medical considerations in assessment and management of pain. *The Journal of Head Trauma Rehabilitation, 19*(1), 10–28. https://doi.org/10.1097/00001199-200401000-00003.

Mayeux, J. P., Teng, S. X., Katz, P. S., Gilpin, N. W., & Molina, P. E. (2015). Traumatic brain injury induces neuroinflammation and neuronal degeneration that is associated with escalated alcohol self-administration in rats. *Behavioral Brain Research, 279*, 22–30. https://doi.org/10.1016/j.bbr.2014.10.053.

McDonald, B. C., Flashman, L. A., & Saykin, A. J. (2002). Executive dysfunction following traumatic brain injury: Neural substrates and treatment strategies. *NeuroRehabilitation, 17*(4), 333–344. https://doi.org/10.3233/nre-2002-17407.

McGlade, E., Rogowska, J., & Yurgelun-Todd, D. (2015). Sex differences in orbitofrontal connectivity in male and female Veterans with TBI. *Brain Imaging and Behavior, 9*(3), 535–549. https://doi.org/10.1007/s11682-015-9379-3.

McGlinchey, R. E., Milberg, W. P., Fonda, J. R., & Fortier, C. B. (2017). A methodology for assessing deployment trauma and its consequences in OEF/OIF/OND Veterans: The TRACTS longitudinal prospective cohort study. *International Journal of Methods in Psychiatric Research, 26*(3), e1556. https://doi.org/10.1002/mpr.1556.

McVige, J., Bargnes, V. H., Shukri, S., & Mechtler, L. (2018). Cannabis, concussion, and chronic pain: An ongoing retrospective analysis at Dent Neurologic Institute in Buffalo, NY. *Neurology, 91*(23 Suppl. 1), S18.3–S19. https://doi.org/10.1212/01.wnl.0000550692.23055.1f.

Mecha, M., Carrillo-Salinas, F. J., Feliú, A., Mestre, L., & Guaza, C. (2016). Microglia activation states and cannabinoid system: Therapeutic implications. *Pharmacology & Therapeutics*, *166*, 40–55. https://doi.org/10.1016/j.pharmthera.2016.06.011.

Mehalick, M. L., & Glueck, A. C. (2018). Examining the relationship and clinical management between traumatic brain injury and pain in military and civilian populations. *Brain Injury*, *32*(11), 1307–1314. https://doi.org/10.1080/02699052.2018.1495339.

Merkel, S. F., Cannella, L. A., Razmpour, R., Lutton, E., Raghupathi, R., Rawls, S. M., et al. (2017). Factors affecting increased risk for substance use disorders following traumatic brain injury: What we can learn from animal models. *Neuroscience and Biobehavioral Reviews*, 77, 209–218. https://doi.org/10.1016/j.neubiorev.2017.03.015.

Miles, S. R., Graham, D. P., & Teng, E. J. (2015). Examining the influence of mild traumatic brain injury and posttraumatic stress disorder on alcohol use disorder in OEF/OIF Veterans. *Military Medicine*, *180*(1), 45–52. https://doi.org/10.7205/milmed-d-14-00187.

Miller, S. C., Baktash, S. H., Webb, T. S., Whitehead, C. R., Maynard, C., Wells, T. S., et al. (2013). Risk for addiction-related disorders following mild traumatic brain injury in a large cohort of active-duty U.S. airmen. *American Journal of Psychiatry*, *170*(4), 383–390. https://doi.org/10.1176/appi.ajp.2012.12010126.

Miñambres, E., Ballesteros, M. A., Mayorga, M., Marin, M. J., Muñoz, P., Figols, J., et al. (2008). Cerebral apoptosis in severe traumatic brain injury patients: An in vitro, in vivo, and postmortem study. *Journal of Neurotrauma*, *25*(6), 581–591. https://doi.org/10.1089/neu.2007.0398.

Molina, P. E., Katz, P. S., Souza-Smith, F., Ford, S. M., Teng, S. X., Dodd, T. Y., et al. (2015). Alcohol's burden on immunity following burn, hemorrhagic shock, or traumatic brain injury. *Alcohol Research: Current Reviews*, *37*(2), 263–278. https://www.arcr.niaaa.nih.gov/.

Moore, E. L., Terryberry-Spohr, L., & Hope, D. A. (2006). Mild traumatic brain injury and anxiety sequelae: A review of the literature. *Brain Injury*, *20*(2), 117–132. https://doi.org/10.1080/02699050500443558.

Nafziger, A. N., & Barkin, R. L. (2018). Opioid therapy in acute and chronic pain. *Journal of Clinical Pharmacology*, *58*(9), 1111–1122. https://doi.org/10.1002/jcph.1276.

Nakajima, Y., Horiuchi, Y., Kamata, H., Yukawa, M., Kuwabara, M., & Tsubokawa, T. (2010). Distinct time courses of secondary brain damage in the hippocampus following brain concussion and contusion in rats. *The Tohoku Journal of Experimental Medicine*, *221*(3), 229–235. https://doi.org/10.1620/tjem.221.229.

Nampiaparampil, D. E. (2008). Prevalence of chronic pain after traumatic brain injury: A systematic review. *JAMA*, *300*(6), 711–719. https://doi.org/10.1001/jama.300.6.711.

National Academies of Sciences, Engineering, and Medicine. (2017). Health and Medicine Division, Board on Population Health and Public Health Practice, & Committee on the Health Effects of Marijuana: An Evidence Review and Research Agenda. In *The health effects of Cannabis and Cannabinoids: The current state of evidence and recommendations for research* (pp. 14–17). USA: National Academies Press. https://doi.org/10.17226/24625.

New Hampshire Department of Health and Human Services. (2017). *Qualifying medical conditions*. Therapeutic Cannabis Program. https://www.dhhs.nh.gov/oos/tcp/medical-conditions.htm.

Newman, S. D., Grantz, J. G., Brooks, K., Gutierrez, A., & Kawata, K. (2020). Association between history of concussion and substance use is mediated by mood disorders. *Journal of Neurotrauma*, *37*(1), 146–151. https://doi.org/10.1089/neu.2019.6550.

Nguyen, B. M., Kim, D., Bricker, S., Bongard, F., Neville, A., Putnam, B., et al. (2014). Effect of marijuana use on outcomes in traumatic brain injury. *American Surgeon*, *80*(10), 979–983. https://www.americanjournalofsurgery.com/.

O'Neil, M., Carlson, K., Storzbach, D., Brenner, L., Freeman, M., Quinones, A., et al. (2013). *Complications of mild traumatic brain injury in Veterans and military personnel: A systematic review* (pp. 1–162). Department of Veterans Affairs.

Ofek, H., & Defrin, R. (2007). The characteristics of chronic central pain after traumatic brain injury. *Pain, 131*(3), 330–340. https://doi.org/10.1016/j.pain.2007.06.015.

Ohio Medical Marijuana Control Program. (n.d.). Patients and caregivers. https://www.medicalmarijuana.ohio.gov/patients-caregivers. accessed 30 March 2020.

Oswald, L. M., & Wand, G. S. (2004). Opioids and alcoholism. *Physiology & Behavior, 81*(2), 339–358. https://doi.org/10.1016/j.physbeh.2004.02.008.

Pacher, P., Bátkai, S., & Kunos, G. (2006). The endocannabinoid system as an emerging target of pharmacotherapy. *Pharmacological Reviews, 58*(3), 389–462. https://doi.org/10.1124/pr.58.3.2.

Pagulayan, K. F., Temkin, N. R., Machamer, J. E., & Dikmen, S. S. (2016). Patterns of alcohol use after traumatic brain injury. *Journal of Neurotrauma, 33*(14), 1390–1396. https://doi.org/10.1089/neu.2015.4071.

Palacios, E. M., Sala-Llonch, R., Junque, C., Fernandez-Espejo, D., Roig, T., Tormos, J. M., et al. (2013). Long-term declarative memory deficits in diffuse TBI: Correlations with cortical thickness, white matter integrity and hippocampal volume. *Cortex, 49*(3), 646–657. https://doi.org/10.1016/j.cortex.2012.02.011.

Panikashvili, D., Simeonidou, C., Ben-Shabat, S., Hanuš, L., Breuer, A., Mechoulam, R., et al. (2001). An endogenous cannabinoid (2-AG) is neuroprotective after brain injury. *Nature, 413*(6855), 527–531. https://doi.org/10.1038/35097089.

Park, M. S., Oh, H. A., Ko, I. G., Kim, S. E., Kim, S. H., Kim, C. J., et al. (2014). Influence of mild traumatic brain injury during pediatric stage on short-term memory and hippocampal apoptosis in adult rats. *Journal of Exercise Rehabilitation, 10*(3), 148–154. https://doi.org/10.12965/jer.140109.

Parry-Jones, B. L., Vaughan, F. L., & Miles Cox, W. (2006). Traumatic brain injury and substance misuse: A systematic review of prevalence and outcomes research (1994-2004). *Neuropsychological Rehabilitation, 16*(5), 537–560. https://doi.org/10.1080/09602010500231875.

Paulozzi, L. J., Jones, C. M., Mack, K. A., & Rudd, R. A. (2011). Vital signs: Overdoses of prescription opioid pain relievers—United States, 1999–2008. *Morbidity and Mortality Weekly Report, 60*(43), 1487–1492. https://www.cdc.gov/mmwr/preview/mmwrhtml/mm6043a4.htm.

Perron, B. E., & Howard, M. O. (2008). Prevalence and correlates of traumatic brain injury among delinquent youth. *Criminal Behavior and Mental Health, 18*(4), 243–255. https://doi.org/10.1002/cbm.702.

Ponsford, J., Whelan-Goodinson, R., & Bahar-Fuchs, A. (2007). Alcohol and drug use following traumatic brain injury: A prospective study. *Brain Injury, 21*(13–14), 1385–1392. https://doi.org/10.1080/02699050701796960.

Posillico, C. K., Terasaki, L. S., Bilbo, S. D., & Schwarz, J. M. (2015). Examination of sex and minocycline treatment on acute morphine-induced analgesia and inflammatory gene expression along the pain pathway in Sprague-Dawley rats. *Biology of Sex Differences, 6*, 33. https://doi.org/10.1186/s13293-015-0049-3.

Poznanski, P., Lesniak, A., Korostynski, M., & Sacharczuk, M. (2020). Ethanol consumption following mild traumatic brain injury is related to blood-brain barrier permeability. *Addiction Biology, 25*(1), e12683. https://doi.org/10.1111/adb.12683.

Rabinowitz, A. R., & Levin, H. S. (2014). Cognitive sequelae of traumatic brain injury. *Psychiatric Clinics of North America, 37*(1), 1–11. https://doi.org/10.1016/j.psc.2013.11.004.

Rao, V., & Lyketsos, C. G. (2002). Psychiatric aspects of traumatic brain injury. *The Psychiatric Clinics of North America, 25*(1), 43–69. https://doi.org/10.1016/s0193-953x(03)00052-2.

Rauh, M., Aralis, H., Melcer, T., Macera, C., Sessoms, P., Barlett, J., et al. (2013). Effect of traumatic brain injury among U.S. servicemembers with amputation. *Journal of Rehabilitation Research and Development*, *50*(2), 161–172. https://doi.org/10.1682/jrrd.2011.11.0212.

Roeckel, L. A., Le Coz, G. M., Gavériaux-Ruff, C., & Simonin, F. (2016). Opioid-induced hyperalgesia: Cellular and molecular mechanisms. *Neuroscience*, *338*, 160–182. https://doi.org/10.1016/j.neuroscience.2016.06.029.

Rogers, J. M., & Read, C. A. (2007). Psychiatric comorbidity following traumatic brain injury. *Brain Injury*, *21*(13–14), 1321–1333. https://doi.org/10.1080/02699050701765700.

Russo, E. B. (2019). The case for the entourage effect and conventional breeding of clinical cannabis: No"strain,"no gain. *Frontiers in Plant Science*, *9*, 1969. https://doi.org/10.3389/fpls.2018.01969.

Russo, E. B., Burnett, A., Hall, B., & Parker, K. K. (2005). Agonistic properties of cannabidiol at 5-HT1a receptors. *Neurochemical Research*, *30*, 1037–1043. https://doi.org/10.1007/s11064-005-6978-1.

Saatman, K. E., Duhaime, A. C., Bullock, R., Maas, A., Valadka, A., Manley, G., et al. (2008). Classification of traumatic brain injury for targeted therapies. *Journal of Neurotrauma*, *25*(7), 719–738. https://doi.org/10.1089/neu.2008.0586.

SAMHSA, & Center for Behavioral Health Statistics and Quality. (2018). *Detailed tables*. National Survey on Drug Use and Health. https://www.samhsa.gov/data/data-we-collect/nsduh-national-survey-drug-use-and-health.

Savola, O., Niemelä, O., & Hillbom, M. (2005). Alcohol intake and the pattern of trauma in young adults and working aged people admitted after trauma. *Alcohol and Alcoholism*, *40*(4), 269–273. https://doi.org/10.1093/alcalc/agh159.

Schirmer, D., & Seale, G. S. (2018). *Non-lethal opiod overdose and acquired brain injury: A position statement of the Brain Injury Association of America*. www.biausa.org.

Schofield, P. W., Butler, T. G., Hollis, S. J., Smith, N. E., Lee, S. J., & Kelso, W. M. (2006). Neuropsychiatric correlates of traumatic brain injury (TBI) among Australian prison entrants. *Brain Injury*, *20*(13–14), 1409–1418. https://doi.org/10.1080/02699050601130443.

Schurman, L. D., & Lichtman, A. H. (2017). Endocannabinoids: A promising impact for traumatic brain injury. *Frontiers in Pharmacology*, *8*(Feb), 69. https://doi.org/10.3389/fphar.2017.00069.

Seal, K. H., Bertenthal, D., Barnes, D. E., Byers, A. L., Gibson, C. J., Rife, T. L., et al. (2018). Traumatic brain injury and receipt of prescription opioid therapy for chronic pain in Iraq and Afghanistan Veterans: Do clinical practice guidelines matter? *The Journal of Pain*, *19*(8), 931–941. https://doi.org/10.1016/j.jpain.2018.03.005.

Seal, K. H., Shi, Y., Cohen, G., Cohen, B. E., Maguen, S., Krebs, E. E., et al. (2012). Association of mental health disorders with prescription opioids and high-risk opioid use in US Veterans of Iraq and Afghanistan. *JAMA*, *307*(9), 940–947. https://doi.org/10.1001/jama.2012.234.

Selassie, A. W., Zaloshnja, E., Langlois, J. A., Miller, T., Jones, P., & Steiner, C. (2008). Incidence of long-term disability following traumatic brain injury hospitalization, United States, 2003. *The Journal of Head Trauma Rehabilitation*, *23*(2), 123–131. https://doi.org/10.1097/01.HTR.0000314531.30401.39.

Sherman, K. B., Goldberg, M., & Bell, K. R. (2006). Traumatic brain injury and pain. *Physical Medicine and Rehabilitation Clinics of North America*, *17*(2), 473–490. https://doi.org/10.1016/j.pmr.2005.11.007.

Singh, R., Mason, S., Lecky, F., & Dawson, J. (2018). Prevalence of depression after TBI in a prospective cohort: The SHEFBIT study. *Brain Injury*, *32*(1), 84–90. https://doi.org/10.1080/02699052.2017.1376756.

Smith, K., Moore, S. D., & Klein, R. (2019). Voluntary ethanol drinking behavior in an Alzheimer's mouse model following repeated mild traumatic brain injury. In *Neuroscience meeting planner* (p. 189.02). https://doi.org/10.1371/journal.pone.0125130.

Stein, C. (2016). Opioid receptors. *Annual Review of Medicine, 67*(1), 433–451. https://doi.org/10.1146/annurev-med-062613-093100.

Stella, N. (2010). Cannabinoid and cannabinoid-like receptors in microglia, astrocytes, and astrocytomas. *Glia, 58*(9), 1017–1030.

Stielper, Z. F., Fucich, E. A., Middleton, J., Hillard, C., Edwards, S., Molina, P., et al. (2020). Traumatic brain injury (TBI) and alcohol drinking alter basolateral amygdala (BLA) endocannabinoids in female rats. *Journal of Neurotrauma*. https://doi.org/10.1089/neu.2020.7175. Advance Online Publication.

Stoica, B. A., & Faden, A. I. (2010). Cell death mechanisms and modulation in traumatic brain injury. *Neurotherapeutics*, 7(1), 3–12. https://doi.org/10.1016/j.nurt.2009.10.023.

Stumbo, S. P., Yarborough, B. J. H., McCarty, D., Weisner, C., & Green, C. A. (2017). Patient-reported pathways to opioid use disorders and pain-related barriers to treatment engagement. *Journal of Substance Abuse Treatment, 73*, 47–54. https://doi.org/10.1016/j.jsat.2016.11.003.

Swieboda, P., Filip, R., Prystupa, & Drozd, M. (2013). Assessment of pain: Types, mechanism and treatment. *Annals of Agricultural Environmental Medicine, 20*, 2–7. Spec. No. 1.

Tait, R. J., Anstey, K. J., & Butterworth, P. (2010). Incidence of self-reported brain injury and the relationship with substance abuse: Findings from a longitudinal community survey. *BMC Public Health, 10*, 171. https://doi.org/10.1186/1471-2458-10-171.

Takahashi, C., Hinson, H. E., & Baguley, I. J. (2015). Autonomic dysfunction syndromes after acute brain injury. *Handbook of Clinical Neurology, 128*, 539–551. https://doi.org/10.1016/B978-0-444-63521-1.00034-0.

Tate, P. S., Freed, D. M., Bombardier, C. H., Harter, S. L., & Brinkman, S. (1999). Traumatic brain injury: Influence of blood alcohol level on post-acute cognitive function. *Brain Injury, 13*(10), 767–784. https://doi.org/10.1080/026990599121160.

Taylor, B. (2012). *Fiscal year 2012 VA utilization report for Iraq and Afghanistan war Veterans diagnosed with TBI*. www.queri.research.va.gov/ptbri.

Teng, S. X., Katz, P. S., Maxi, J. K., Mayeux, J. P., Gilpin, N. W., & Molina, P. E. (2015). Alcohol exposure after mild focal traumatic brain injury impairs neurological recovery and exacerbates localized neuroinflammation. *Brain, Behavior, and Immunity, 45*, 145–156. https://doi.org/10.1016/j.bbi.2014.11.006.

Tsuda, S., Hou, J., Nelson, R. L., Wilkie, Z. J., Mustafa, G., Sinharoy, A., et al. (2016). Prolonged hippocampal cell death following closed-head traumatic brain injury in rats. *Neuroreport, 27*(10), 724–729. https://doi.org/10.1097/WNR.0000000000000598.

Tyrer, S., & Lievesley, A. (2003). Pain following traumatic brain injury: Assessment and management. *Neuropsychological Rehabilitation, 13*(1–2), 189–210. https://doi.org/10.1080/09602010244000381.

Vaaramo, K., Puljula, J., Tetri, S., Juvela, S., & Hillbom, M. (2014a). Head trauma sustained under the influence of alcohol is a predictor for future traumatic brain injury: A long-term follow-up study. *European Journal of Neurology, 21*(2), 293–298. https://doi.org/10.1111/ene.12302.

Vaaramo, K., Puljula, J., Tetri, S., Juvela, S., & Hillbom, M. (2014b). Predictors of new-onset seizures: A 10-year follow-up of head trauma subjects with and without traumatic brain injury. *Journal of Neurology, Neurosurgery and Psychiatry, 85*(6), 598–602. https://doi.org/10.1136/jnnp-2012-304457.

Volkow, N., Benveniste, H., & McLellan, A. T. (2018). Use and misuse of opioids in chronic pain. *Annual Review of Medicine, 69*, 451–465. https://doi.org/10.1146/annurev-med-011817-044739.

Volkow, N. D., Jones, E. B., Einstein, E. B., & Wargo, E. M. (2019). Prevention and treatment of opioid misuse and addiction: A review. *JAMA Psychiatry*, *76*(2), 208–216. https://doi.org/10.1001/jamapsychiatry.2018.3126.

Wada, K., Chatzipanteli, K., Kraydieh, S., Busto, R., & Dietrich, W. D. (1998). Inducible nitric oxide synthase expression after traumatic brain injury and neuroprotection with aminoguanidine treatment in rats. *Neurosurgery*, *43*(6), 1427–1436. https://doi.org/10.1097/00006123-199812000-00096.

Walker, R., Cole, J. E., Logan, T. K., & Corrigan, J. D. (2007). Screening substance abuse treatment clients for traumatic brain injury: Prevalence and characteristics. *The Journal of Head Trauma Rehabilitation*, *22*(6), 360–367. https://doi.org/10.1097/01.HTR.0000300231.90619.50.

Walter, L., & Stella, N. (2004). Cannabinoids and neuroinflammation. *British Journal of Pharmacology*, *141*(5), 775–785. https://doi.org/10.1038/sj.bjp.0705667.

Wang, K., Cui, D., & Gao, L. (2016). Traumatic brain injury: A review of characteristics, molecular basis and management. *Frontiers in Bioscience (Landmark Edition)*, *21*, 890–899. https://doi.org/10.2741/4426.

Washington State Department of Health. (n.d.). Medical marijuana qualifying conditions. https://www.doh.wa.gov/YouandYourFamily/Marijuana/MedicalMarijuana/PatientInformation/QualifyingConditions. accessed 30 March 2020.

Weil, Z. M., Corrigan, J. D., & Karelina, K. (2016). Alcohol abuse after traumatic brain injury: Experimental and clinical evidence. *Neuroscience and Biobehavioral Reviews*, *62*, 89–99. https://doi.org/10.1016/j.neubiorev.2016.01.005.

Weil, Z. M., Corrigan, J. D., & Karelina, K. (2018). Alcohol use disorder and traumatic brain injury. *Alcohol Research: Current Reviews*, *39*(2), 171–180. https://www.ncbi.nlm.nih.gov/pmc/articles/PMC6561403/.

Weil, Z. M., Karelina, K., Gaier, K. R., Corrigan, T. E. D., & Corrigan, J. D. (2016). Juvenile traumatic brain injury increases alcohol consumption and reward in female mice. *Journal of Neurotrauma*, *33*(9), 895–903. https://doi.org/10.1089/neu.2015.3953.

Werner, C., & Engelhard, K. (2007). Pathophysiology of traumatic brain injury. *British Journal of Anaesthesia*, *99*(1), 4–9. https://doi.org/10.1093/bja/aem131.

Whelan-Goodinson, R., Ponsford, J., Johnston, L., & Grant, F. (2009). Psychiatric disorders following traumatic brain injury: Their nature and frequency. *The Journal of Head Trauma Rehabilitation*, *24*(5), 324–332. https://doi.org/10.1097/HTR.0b013e3181a712aa.

Whyte, J. (2010). Pharmacologic treatment of cognitive and behavioral sequelae of traumatic brain injury: Practicing in the absence of strong evidence. *European Journal of Physical and Rehabilitation Medicine*, *46*(4), 557–562.

Winqvist, S., Luukinen, H., Jokelainen, J., Lehtilahti, M., Näyhä, S., & Hillbom, M. (2008). Recurrent traumatic brain injury is predicted by the index injury occurring under the influence of alcohol. *Brain Injury*, *22*(10), 780–785. https://doi.org/10.1080/02699050802339397.

Wrege, J., Schmidt, A., Walter, A., Smieskova, R., Bendfeldt, K., Radue, E.-W., et al. (2014). Effects of cannabis on impulsivity: A systematic review of neuroimaging findings. *Current Pharmaceutical Design*, *20*(13), 2126–2137. https://doi.org/10.2174/13816128113199990428.

Wu, A., Khan, M. N., Kulvatunyou, N., Hamidi, M., Gries, L. M., Zakaria, E. R., et al. (2018). Cannabinoids use is associated with improved outcomes in traumatic brain injury. *Journal of the American College of Surgeons*, *227*(4), e230. https://doi.org/10.1016/j.jamcollsurg.2018.08.622.

Xiong, Y., Mahmood, A., & Chopp, M. (2013). Animal models of traumatic brain injury. *Nature Reviews. Neuroscience*, *14*(2), 128–142. https://doi.org/10.1038/nrn3407.

Zasler, N. D. (1999). Posttraumatic headache: Caveats and controversies. *The Journal of Head Trauma Rehabilitation*, *14*(1), 1–8. https://doi.org/10.1097/00001199-199902000-00003.

Zhang, Q. G., Laird, M. D., Han, D., Nguyen, K., Scott, E., Dong, Y., et al. (2012). Critical role of NADPH oxidase in neuronal oxidative damage and microglia activation following traumatic brain injury. *PLoS One*, 7(4), e34504. https://doi.org/10.1371/journal.pone.0034504.

Zhang, H. Y., et al. (2014). Cannabinoid CB2 receptors modulate midbrain dopamine neuronal activity and dopamine-related behavior in mice. *Proceedings of the National Academy of Sciences of the United States of America*, *111*(46), E5007–E5015. https://doi.org/10.1073/pnas.1413210111.

Zhao, F., Yang, J., & Cui, R. (2017). Effect of hypoxic injury in mood disorder. *Neural Plasticity*, *2017*, 6986983. https://doi.org/10.1155/2017/6986983.

CHAPTER SIX

Depression and substance use disorders: Clinical comorbidity and shared neurobiology

Cali A. Calarco and Mary Kay Lobo*
Department of Anatomy and Neurobiology, University of Maryland School of Medicine, Baltimore, MD, United States
*Corresponding author: e-mail address: mklobo@som.umaryland.edu

Contents

Abstract

Mood disorders, including major depressive disorder (MDD), are the most prevalent psychiatric illnesses, and pose an incredible burden to society, both in terms of disability and in terms of costs associated with medical care and lost work time. MDD has extremely high rates of comorbidity with substance use disorders (SUD) as many of the same neurobiological circuits and molecular mechanisms regulate the reward pathways disrupted in both conditions. MDD may induce SUDs, SUD may contribute to MDD development, or underlying vulnerabilities and common life experience may confer risk to developing both conditions. In this chapter we explore theories of MDD and SUD

International Review of Neurobiology, Volume 157
ISSN 0074-7742
https://doi.org/10.1016/bs.irn.2020.09.004

comorbidity, the neurobiological underpinnings of depression, overlapping cellular and molecular pathways for both conditions, and current treatment approaches for these comorbid conditions.

1. Introduction

Major depressive disorder (MDD) is among the top three leading causes of disability and lost productivity worldwide (leading cause for ages 15–44 yo) and affects more than 260 million people (James et al., 2018). MDD diagnosis is currently more prevalent in women, and while MDD can be diagnosed at any age, elevated risk for the disorder begins in the early teen years and increases through late adolescence. Combining direct medical costs and lost productivity, the economic burden associated with MDD increased from $173.2 billion to $210.5 billion per year (inflation-adjusted dollars; 21.5%) from 2005 to 2010 (Greenberg, Fournier, Sisitsky, Pike, & Kessler, 2015), and these numbers continue to rise. Interestingly, only 38% of this total is due to directly to MDD itself, while the rest is spent on associated comorbid conditions. The current diagnostic criteria for MDD in the DSM-V, adopted in 2013, dictate that an individual must have five or more of the following symptoms persistently and one of the five must been either depressed mood or loss of interest or pleasure (anhedonia): depressed mood, loss of interest in pleasure, significant weight loss or gain, sleep disturbance, slowed thought and physical activity, loss of energy, feeling worthless or excessive guilt, diminished ability to think or concentrate, or recurrent thoughts of death or suicidality. These symptoms must cause clinically significant distress or impairment and must not be the result of another medical condition (American Psychiatric Association, 2013). Importantly, for a diagnosis of MDD, these symptoms must not be caused by a substance use disorder (SUD), but SUD is listed as a common co-occurring disorder. MDD has a high rate of clinical comorbidity with other psychiatric illnesses, particularly anxiety, in addition to post-traumatic stress disorder (PTSD), and depressive episodes are also part of bipolar disorder (BD). MDD comorbidities with these conditions are reviewed elsewhere (Cerdá, Sagdeo, Johnson, & Galea, 2010; Colston et al., 2016; Flory & Yehuda, 2015; Mcgrath & Miller, 2010), and will not be discussed in this chapter. Here, we focus on the comorbidity of MDD with SUDs, the neurobiological basis for depression, how stress influences drug intake, how drug withdrawal can precipitate depressive symptoms, specific comorbid neurobiology for individual drugs of abuse, and finally treatment options specifically for patients with a dual diagnosis.

Between one quarter and one third of people with any mental illness also have a substance use disorder. Conversely, more than one third of those who abuse alcohol and 40–60% of those who abuse other drugs also have a mental illness (Balhara, Kuppili, & Gupta, 2017; Conway, Compton, Stinson, & Grant, 2006; Substance Abuse and Mental Health Services Administration, 2018; Torrens, Gilchrist, & Domingo-Salvany, 2011). These numbers vary slightly when considering only depression, but have remained largely consistent over many years and rounds of national surveys (Compton, Thomas, Stinson, & Grant, 2007; Grant, 1995; Volkow, 2004). Rates of comorbidity and depression for multiple classes of illicit drugs in adults, as assessed in the National Epidemiologic Survey on Alcohol and Related Conditions (NESARC), are summarized in Table 1 (Conway et al., 2006). For rates of co-occurring major depressive episodes and substance use among adolescents ages 12–17, see Substance Abuse and Mental Health Services Administration (2018). The rates of depression in Table 1 do not specifically include alcohol or nicotine abuse/dependence, nor do they account for multi-drug use/abuse. Both alcohol and nicotine abuse/dependence are highly comorbid with the lifetime abuse of other drugs (Compton et al., 2007; Kessler, Wai, Demler, & Walters, 2005; McCabe & West, 2017) and, of those with a history of depression, ~20% have concurrent alcohol use disorder (AUD) and ~40% are nicotine dependent. Further, those using multiple substances are more likely to have a mood disorder and nicotine dependence (McCabe & West, 2017). In the NESARC sample, major depression was the most prevalent (29.2–59.8%) mood disorder among those with specific drug use disorders. Mood and anxiety disorders overall were greater among women than men, and drug use disorders were more prevalent in men; however, the comorbidity of drug use and mood disorders was more prevalent in women, despite overall lower rate of drug abuse (Conway et al., 2006). While most cases of comorbid mood or anxiety disorders were independent of the drug use, it is more likely for depression to be alcohol-induced in those who abuse alcohol. Further, substance users with comorbid psychiatric conditions reported greater psychosocial and medical problems and had poorer treatment prognoses than substance users without comorbid conditions (Torrens et al., 2011). Individuals with comorbid disorders also tend to have an earlier age of onset for depression and increased risk for suicide. Finally, treatment rates for SUDs in the NESARC sample were about half that for major depression, but those with alcohol or drug dependence were twice as likely to receive treatment for depression (Hasin & Kilcoyne, 2012).

Table 1 Lifetime prevalence of comorbid MDD with substance abuse or dependence by substance class.

Drug use disorder	% with any mood disorder	% with depression
Any drug use disorder	21.6	20.7
Any drug abuse	13.5	13.1
Any drug dependence	8.2	7.7
Sedative use disorder	3.0	3.1
Sedative abuse	2.1	2.2
Sedative dependence	0.9	0.9
Tranquilizer use disorder	2.7	2.7
Tranquilizer abuse	1.9	1.9
Tranquilizer dependence	0.8	0.8
Opioid disorder	4.0	3.9
Opioid abuse	2.7	2.7
Opioid dependence	1.3	1.2
Amphetamine use disorder	5.2	5.0
Amphetamine abuse	3.3	3.2
Amphetamine dependence	2.0	1.9
Hallucinogen use disorder	4.5	4.3
Hallucinogen abuse	3.6	3.5
Hallucinogen dependence	0.9	0.7
Marijuana use disorder	17.1	16.3
Marijuana abuse	13.1	12.6
Marijuana dependence	4.0	3.7
Cocaine use disorder	6.5	6.1
Cocaine abuse	3.3	3.2
Cocaine dependence	3.1	2.8
Inhalant/solvent abuse	0.8	0.8

Adapted from data in Conway, K. P., Compton, W., Stinson, F. S., & Grant, B. F. (2006). Lifetime comorbidity of DSM-IV mood and anxiety disorders and specific drug use disorders: Results from the National Epidemiologic Survey on Alcohol and Related Conditions. *Journal of Clinical Psychiatry*, *67*(2), 247–257. https://doi.org/10.4088/jcp.v67n0211.

Understanding the unique profile of comorbid MDD and SUD compared to each disorder individually is critical for appropriate diagnosis and treatment and there are multiple hypotheses regarding the etiology of comorbidity. First, a common hypothesis is that the mood disorder drives repeated drug use as individuals seek to self-medicate their psychiatric symptoms (Gómez-Coronado et al., 2018; Volkow, 2004). The neurobiological changes associated with the mood disorder might also predispose individuals to be more reactive to the effects of certain drugs (National Institute on Drug Abuse, 2018). Conversely, depressive symptoms can be a side effect both of acute drug withdrawal and of the life impairment associated with a SUD (Quello, Brady, & Sonne, 2005). Further, repeated substance use can also induce neuroplastic changes in the brain that induce vulnerability to a mood disorder (Gómez-Coronado et al., 2018; National Institute on Drug Abuse, 2018). It is possible that these disorders occur completely independently with independent neurobiological mechanisms, but a third hypothesis is that there are shared risk factors that contribute to both disorders. Genetic or developmental predispositions may drive how an individual reacts to drugs of abuse or to stress (Volkow, 2004). Indeed, both trauma and chronic stress, especially during development, shape the way the body responds to stress and are risk factors for both MDD and SUD (Brady & Sinha, 2005; Rao, 2010). The recent theory involving allostatic load, or the accumulated cost to the body of maintaining a new homeostatic baseline after a stressor, may explain how recurrent mood episodes and substance use can amplify vulnerability and even cause sensitization to future mood symptoms and future drug use, creating a vicious cycle (Gómez-Coronado et al., 2018; Koob & Moal, 2001). Understanding the etiology of these comorbid conditions, shared or otherwise, is critical for determining appropriate treatments for both conditions. While specifics vary for individual drug classes, and will be examined later in the chapter, understanding the neurobiology of depression allows for a deeper understanding of the ways in which drug abuse can influence vulnerability to depression.

2. Neurobiology of depression

2.1 Clinical findings

Decades of research on MDD and depressive episodes have led to an increased understanding of both the brain regions involved in producing these symptoms and the underlying neurobiology, although research has been limited by the heterogeneity of the disease. MDD is associated with anatomical differences, but there are also differences at the functional and circuit level. Finally, cellular, molecular, and genetic changes are also implicated in depression etiology.

A combination of post-mortem and neuroimaging analyses have reveled anatomical differences in brains of individuals with a history of depressive episodes and healthy controls. These studies have defined critical nodes for depression, and many of these regions are also critical for reward processing. The frontal cortex, and more specifically the orbitofrontal prefrontal cortex (ofPFC), the dorsolateral prefrontal cortex (dlPFC), and the subgenual anterior cingulate cortex (ACC) show both functional and structural abnormalities in depressed patients (Gong & He, 2015; Pandya, Altinay, Malone, & Anand, 2012; Zhang, Peng, Sweeney, Jia, & Gong, 2018). Orbitofrontal, dorsolateral, and subgenual prefrontal regions show reductions in gray matter, decreased volume, and other structural abnormalities, in addition to reduced metabolic activity (Gold, Machado-Vieira, & Pavlatou, 2015; Pandya et al., 2012; Zhang et al., 2018). Anterior and posterior cingulate cortical regions in addition to parts of parietal cortex have reduced gray matter volumes (Gong & He, 2015; Zhang et al., 2018). Subcortically, a history of major depression is linked to significantly smaller left and right hippocampal volumes with no differences in total cerebral volumes, and the degree of hippocampal volume reduction correlated with total duration of major depression (Sheline, Wang, Gado, Csernansky, & Vannier, 1996). Regions of the amygdala, thalamus and caudate nucleus also show reduced gray matter in individuals with depression (Gong & He, 2015; Pandya et al., 2012; Zhang et al., 2018). Diffusion tensor imaging (DTI) has revealed altered white matter volume in the hippocampus, cingulate, parietal and other cortical regions (Zhang et al., 2018). Many of these regions also show altered function either with respect to baseline metabolic activity, resting state activity, or task-based activity in functional magnetic resonance imaging scans (fMRI). While prefrontal cortical regions often show decreased activity, regions like the amygdala show increased activity, and regions of the striatum have increased, decreased, or no change in activity depending on the study or the task used (Pandya et al., 2012; Russo & Nestler, 2013; Zhang et al., 2018). For a comprehensive review of human imaging findings in subjects with MDD, see Savitz and Drevets (2009).

Since so many interconnected brain regions are implicated in the etiology of depression, and no singular region seems to be directly responsible for symptom severity, the connectivity of these regions and the function of these circuits has been examined. The frontal and parietal lobes as well as the thalamus, putamen, and hippocampus are hubs in the common functional networks identified (Zhang et al., 2018). Abnormal functional activity has been observed in the default mode network (DMN, including posterior cingulate cortex, medial prefrontal cortex, angular gyrus and connected regions), ACC-thalamus, ACC-insula, prefrontal-limbic-thalamus, as well as in the corpus collosum, with a general shift toward more randomized connectivity in depressed groups (Gong & He, 2015). Increased regional connectivity and functional connectivity in both the caudate nucleus and in the DMN correlate with disease duration and severity, in addition to increased nodal connectivity in the putamen correlating with depressive episode number but independent of current symptoms (Gong & He, 2015). Higher regional connectivity is also observed in prefrontal-limbic circuits including the amygdala and the ventromedial PFC, coupled with overall reduced clustering in gray matter networks and, in cases of geriatric depression, disruptions in whole brain DTI, suggesting less specialized organization and generally disrupted topographical organization (Gong & He, 2015). Exposure to early life stress, a risk factor for both depression and substance abuse, can also cause dysregulated functional connectivity in these circuits, independent of an actual depression diagnosis (Gong & He, 2015). With respect to generally dysregulated systems, depression is associated with an overactive glucocorticoid system, including the hypothalamic-pituitary-adrenal axis (HPA), hypercortisolism, and an overactive norepinephrine system, indicating dysregulation of adaptive stress systems (Gold et al., 2015). These anatomical and circuit changes are built from cellular, molecular, and genetic changes, which may be acting across multiple brain regions to cause these widespread functional abnormalities, while changes in specific brain regions may contribute to individual symptoms and disturbances.

While cellular and molecular studies are more challenging in human populations and are often limited by brain tissue availability, anatomical and functional changes have been observed in clinical populations. Individuals with a short (S) allele variant in the promoter region for the serotonin transporter gene have reduced functional connectivity between the amygdala and ACC, and these effects covary with respect to the VAL (compared to the MET) allele in the brain-derived neurotrophic factor (BDNF) gene (Gong & He, 2015). Both the S allele for the serotonin transporter and the VAL allele in the BDNF gene confer risk for depression, and some work has been done attempting to validate changes in DTI or functional connectivity as a biomarker for depression in general, or subtypes of depression with specific symptom clusters or trauma history. Postmortem studies have shown that some of the anatomical volume reductions described above may be due to reduced soma size and glial number in the PFC, increased pyramidal neuron packing in the hippocampus, and reduced synapse number in regions of the dlPFC (Duman, Sanacora, & Krystal, 2019). Further, reduction in markers for certain types of inhibitory interneurons, specifically somatostatin/calbindin neurons, has been observed in the PFC and somatostatin mRNA is decreased in dlPFC, ACC, and the amygdala of depressed subjects (Duman et al., 2019). A recent hypothesis for the broad circuit changes seen in depression is a model including an imbalance between excitatory (glutamatergic) and inhibitory (GABAergic) signaling, with reduced connectivity resulting from atrophy in glutamatergic neurons or stress-induced glutamatergic toxicity, coupled with dysregulation of inhibitory neurons (Duman et al., 2019). For review of the functional, anatomical, and cellular changes observed in humans with depression, see Russo and Nestler (2013). A general limitation of human and clinical studies is often small sample size, or heterogeneity of symptom/treatment history, and direct manipulation of cellular and molecular function, outside of pharmacological treatment, is untenable. Therefore, a huge variety of work has been done to develop models of depression in preclinical systems to explore the circuits and underlying neurobiology in a more controlled manner.

2.2 Modeling depression preclinically

Preclinical studies have been crucial for studying the interaction between reward-associated brain regions in both addiction and depression. While the impacts of exposure to drugs of abuse are straightforward, the complex symptoms that make up human SUDs are harder to model effectively, and various creative protocols have been developed for different drug classes (Deroche-Gamonet & Piazza, 2014; Everitt & Robbins, 2016; Koob, 2013; Koob & Kreek, 2007; Li, Venniro, & Shaham, 2016). Similarly, MDD and its associated complex emotional symptoms are difficult to accurately model in rodent species. Nevertheless, there are useful preclinical models with reasonable face, predictive, and construct validity, and these models are currently the standard by which many antidepressant drugs are screened (Czéh, Fuchs, Wiborg, & Simon, 2016; Muir, Lopez, & Bagot, 2019). In addition to rodents selectively bred or genetically manipulated for "depressive-like" behaviors, various forms of stress are regularly used to induce anhedonia or alter reward-seeking behavior, often measured as a loss of preference for otherwise rewarding stimuli, like sucrose solutions or interacting with a novel conspecific, or reduced time spent struggling against inescapable stressors such as forced swimming or tail suspension. It is important to note that while the latter behaviors (FST and TST) may have predictive validity for antidepressant dosage, they lack face validity because they have no relation to the symptoms of human depression (Bale et al., 2019). Chronic stress models such as chronic predictable or unpredictable stress, learned helplessness, chronic social defeat stress, and various forms of early life stress all reduce hedonic or reward-seeking, thus are rodent models for depression with face validity (Czéh et al., 2016; Fox & Lobo, 2019; Hollis & Kabbaj, 2014; Knowland & Lim, 2018; Muir et al., 2019). Commonly used and well validated models of stress and behavioral outcome tests are listed in Table 2. As noted in the table, some behavioral tests map very effectively onto human symptoms (e.g., anhedonia, change in body weight, etc.), however, other symptom classes like suicidal thoughts and excessive guilt are impossible to model in rodents.

Animal models are incredibly useful and, coupled with recently developed technology like optogenetics (Belzung, Turiault, & Griebel, 2014; Tye & Deisseroth, 2012), have led to abundant insights into the circuit, molecular, and genetic underpinnings of both depressive-like behaviors themselves and susceptibility to developing these behaviors after stress. They act as important intermediate phenotypes that can inform the human condition (Donaldson & Hen, 2014), and in tandem with Research Domain Criteria (RDoC) used in human clinical research, they can help parse out common underlying biological processes and circuitry across a number of symptom classes and behaviors (Deisseroth, Etkin, & Malenka, 2015; Marton & Sohal, 2016; Monteggia, 2016). These animal models move us closer to the eventual goals of clinical biomarkers for depression to improve diagnosis and treatment (Deisseroth et al., 2015).

It is important to note that historically most preclinical experiments involving stress have been conducted exclusively in male animals. Stress paradigms like chronic mild stress are effectively used on both sexes, but other models, like chronic social defeat stress pose challenges for use in females, due to low rates of male-female or female-female aggression in rodents (Czéh et al., 2016; Knowland & Lim, 2018). This is especially critical in the case of depression research as women are nearly twice as likely to receive a depression diagnosis with the current diagnostic criteria, and more frequently have comorbid drug abuse, despite lower overall drug use in women. While this may be due in part to socialized sex differences in emotion expression, and in one study, rates equilibrated after aggressive

Table 2 Depression symptoms and behavioral tests used to model these symptoms in rodents.

DSM 5 criteria	Test	Procedure	Common interpretation	Indicator of susceptibility
Depressed mood	*Coping Tests*			
	Forced Swim	Mouse/rat is place in an inescapable cylinder of water. After initial struggling (swimming, climbing), animals will eventually stop struggling and float immobile. Latency to first become immobile and total time immobile are measured	"Behavioral despair" or passive v. active coping	Increased immobility
	Tail Suspension	Mouse is suspended by tail. After initial struggling, mice stop struggling and hang immobile. Latency to first become immobile and total time immobile are measured	"Behavioral despair" or passive v. active coping	Increased immobility
	Learned Helplessness	Mice are exposed to one or more sessions of repeated inescapable electric shock and later are given the opportunity to escape further signaled shocks	"Behavioral despair" or passive v. active coping	Increased latency or failure to escape from escapable shock
	Splash Test	Mouse/rat is sprayed with a 10% sucrose solution and recorded for 5 min. Latency to groom and time spent grooming are measured	Motivation to groom or self-care	Increased latency to groom, reduced time spent grooming
Markedly diminished interest or pleasure in activities	*Reward Based Tests*			

	Sucrose Preference	Mouse/rat has free access to bottles of water and sucrose solution in home cage or in a discrete test. The volume of each solution consumed in measured and compared to calculate a preference score	Anhedonia	Decreased preference for sucrose
	Female Urine Sniffing Test	Cotton-tipped applicator dipped in urine of a female from the same stain as the mouse/rat is placed in the home cage for a 3 min exposure	Anhedonia	Reduced time spent sniffing urine
	Intracranial Self-Stimulation	Operant paradigm in which rodents self-administer rewarding electrical stimulation through electrodes implanted into the brain (usu. medial forebrain bundle)	Reward seeking	Reduced self-stimulation
	Drug Self-Administration	Operant paradigm where the rodent makes an action (level press, nose poke, wheel turn, etc.) to trigger the delivery of a dose of drug, usually through an intravenous catheter	Reward seeking	Reduced drug administration
Dysfunctional social behavior, frequent negative social interactions	Social Interaction	Mouse is placed in an arena with a mesh enclosure that is empty on an initial "no target" trial and contains a novel mouse on a "target" trial. Time spent in vicinity of the enclosure (interaction zone) while the social target is present vs absent is measured and used to calculate a social interaction ratio	Social withdrawal	Reduced time interacting with social target
Anxiety, agitation or restlessness	*Anxiety Based Tests*			

Continued

Table 2 Depression symptoms and behavioral tests used to model these symptoms in rodents.—cont'd

DSM 5 criteria	Test	Procedure	Common interpretation	Indicator of susceptibility
	Open Field	Mouse/rat is placed in an empty arena and time spent exploring the center vs periphery is measured. Manipulations associated with increased anxiety reduce center exploration	Anxiety	Reduced time in center
	Elevated Plus Maze	Mouse/rat is placed in the center of a plus (+) shaped maze with two walled "closed" arms and two "open" arms. Time spent in open and closed arms is measured. Manipulations associated with increased anxiety reduce time spent in open arms	Anxiety	Reduced time in open arms
	Dark-Light	Mouse/rat is placed in a two-chamber apparatus; one side is brightly lit and open, the other is covered and dark	Anxiety	Reduced time in the bright chamber
	Novelty-Induced Hypophagia	Mildly food-restricted mouse/rat is placed in a novel open area with food in the center for 5 min	Anxiety	Increased time to approach the food and reduced food intake
Fatigue or loss of energy/psychomotor agitation or retardation	*Activity/Compulsivity Tests*			
	Home Cage Activity	Distance, movements, and time spent active are recorded while the mouse/rat is in their home cage	Activity	Reduced distance traveled or bouts
	Treadmill Running	Mouse/rat has access to a running wheel in their home cage	Activity	Reduced distance traveled or bouts

	Marble Burying	Twenty marbles are arrayed on the surface of clean bedding. The number of marbles buried in a 30 min session is scored	Compulsivity	Increased number of buried marbles
	Nestlet Shredding	A nestlet comprised of pulped cotton fiber is preweighed and placed on top of cage bedding and the amount of the nestlet remaining intact after a 30 min test session is determined	Compulsivity	Reduced time spent or incomplete nest building
Diminished ability to think or concentrate, or indecisiveness	*Cognition Based Tests*			
	Morris Water Maze	Mouse/rat is placed in a large circular pool of water and required to escape from water onto a hidden platform over repeated trials	Spatial memory	Increased time or failure to find the platform
	Barnes Maze	Mouse/rat is placed on an open, circular platform with up to 20 holes around the circumference. One hole has an "escape box" which the animal must find based on spatial cues over repeated trials	Spatial memory	Increased time or failure to find the escape box
	Radial Arm Maze	Mouse/rat is placed in an eight-arm maze, with some or all arms baited with a reward	Spatial and working memory	Visiting un-baited arms or reentering empty arms
	T-maze/Y-maze	Mouse/rat is placed at the bottom of a T or Y shaped maze. At the start of a trial animals walk to the choice point and must successfully retrieve rewards from alternately baited arms	Working memory	Increase in failed trials, when the animal enters the previously baited arm rather than the currently baited arm

Continued

Table 2 Depression symptoms and behavioral tests used to model these symptoms in rodents.—cont'd

DSM 5 criteria	Test	Procedure	Common interpretation	Indicator of susceptibility
	Novel Object Exploration	A mouse is presented with two similar objects during the first session, and then one of the two objects is replaced by a new object during a second session	Recognition memory	Reduced time spent interacting with the new object or equal time spent with both objects
Significant weight loss or gain	Body Weight Measurements	Body weight measured	Weight change	Significant weight change
Feelings of worthlessness or excessive or inappropriate guilt	Not Modellable	N/A	N/A	N/A
Recurrent thoughts of death, recurrent suicidal ideation	Not Modellable	N/A	N/A	N/A

Compiled and expanded from Czéh, B., Fuchs, E., Wiborg, O., & Simon, M. (2016). Animal models of major depression and their clinical implications. *Progress in Neuro-Psychopharmacology and Biological Psychiatry*, *64*, 293–310. https://doi.org/10.1016/j.pnpbp.2015.04.004; Muir, J., Lopez, J., & Bagot, R. C. (2019). Wiring the depressed brain: Optogenetic and chemogenetic circuit interrogation in animal models of depression. *Neuropsychopharmacology*, *44*, 1013–1026. https://doi.org/10.1038/s41386-018-0291-6; Nestler, E. J., Barrot, M., Dileone, R. J., Eisch, A. J., Gold, S. J., & Monteggia, L. M. (2002). Neurobiology of depression. *Cell*, *34*, 13–25.

symptoms were used as diagnostic criteria for men (Martin, Neighbors, & Griffith, 2013), it is clear that there are relevant sex difference in the development and expression of depression (Hodes, Walker, Labonté, Nestler, & Russo, 2017). Current research efforts are striving to be more balanced with respect to including female animals, and social stress models are being expanded to include witnessing of social defeat or "psychological" stress (Iñiguez et al., 2018; Warren et al., 2013). Stressors that don't involve physical discomfort/pain also beneficially disentangle confounds related to altered pain processing in depression and depression models, which is also crucial for understanding how opiate abuse relates to pain management.

While the majority of stress models disrupt hedonic or reward-seeking states in young-adult/adult mice, early life stress and maternal separation models are important for understanding windows for reward circuitry development and periods of vulnerability for later life depression or addiction (Bolton, Molet, Ivy, & Baram, 2017; Kim et al., 2017; Korosi et al., 2010; Molet, Maras, Avishai-eliner, & Baram, 2014; Nishi, Horii-hayashi, & Sasagawa, 2014; Peña et al., 2017, 2019). Models specifically using older mice are not represented in the literature, thus failing to capture unique features of senior depression, which makes up a significant portion of clinical cases. Finally, not all humans or animals that endure trauma or stress go on to develop clinical depression, disrupted motivation or anhedonia; they are considered "resilient." Just as it is critical to model symptoms with the goal of developing more effective treatments, it is also worthwhile to understand underlying neurobiology that confers resilience in the face of adversity (Han & Nestler, 2017).

2.3 Systems and circuits

Years of developing and refining preclinical models, in tandem with corroborating clinical studies, have provided insight into numerous systems that function differently after stress in a depressed state. Prominent discoveries have led to multiple theories for the etiology of depression including the biorhythm, neuroimmune, neuroendocrine, monoamine, and neuroplasticity theories (Loonen & Ivanova, 2016b). These theories all attempt to explain altered circuit function that underlies the observable behavioral changes in animals after stress (described in Table 3). Indeed, the number of systems involved in depression is inconsistent with the idea that depression is a single disease, and more consistent with the notion that it is a heterogeneous syndrome comprised of numerous diseases of distinct causes

Table 3 Behavioral stress paradigms used to model depression in rodents.

Paradigm	Duration	Description	Outcomes
Chronic social defeat stress	Chronic (10 d)	Experimental mice experience brief aggressive encounters (5–10 min) in the home cage of an aggressive mouse and are then housed in the same cage separated by a divider to maintain sensory exposure with the aggressor. This is repeated with a new aggressor daily	Social avoidance, reduced center time in open field, immobility in FST/TST, reduced sucrose preference, circadian and metabolic changes
Chronic witness defeat stress	Chronic (10 d)	Experimental procedures are the same as for CSDS, but the target mouse is separated from the defeat by a divider. The target mouse is cohoused with the aggressor mouse, separated by a divider as above	Social avoidance, reduced sucrose preference, reduced body weight, elevated corticosterone
Chronic mild/unpredictable/variable stress	Chronic (6 d–12 wks)	Mice are exposed to daily mild stressors for a variable period. Stressors vary considerably between protocols but can include: Cage tilt, foot shock, tail suspension, restraint, wet cage, social isolation, perturbation of light/dark cycle, cage changing, temperature perturbation, cage shaking, intruder, noise	Reduced sucrose preference, immobility in FST/TST, increased latency to eat in a novel environment, decreased grooming
Learned helplessness	Sub-chronic (2–3 d)	Mice are exposed to one or more sessions of repeated inescapable electric shock and later are given the opportunity to escape further signaled shocks	Failure to escape from escapable, signaled shock
Forced swim/tail suspension	Acute (5 m–1 h)	Mice are suspended from their tail (TST) or placed in an inescapable body of water (FST)	Immobility in FST/TST

Maternal separation	Acute (1 d) or Chronic (up to 21 d)	Pups are removed from the dam and left alone or with their littermates during post-natal development. Timing varies from a single day of 24h separation, to 1h or more/day for repeated days, up to P21 or weaning. "Handling" or removal for 15 min or less/day encourages maternal behavior, lowering depression/anxiety behavior	Results vary based on developmental window of separation; HPA axis disruption, reduced locomotion in open field, increased immobility in FST, increased closed arm time in EPM
Olfactory bulbectomy	Lesion	Chemical or surgical lesions of the olfactory bulb	Hyperlocomotion in open field, increased immobility in FST/TST, reduced sucrose preference, disrupted learning and hippocampal functioning
Fear conditioning	Acute (1 or more days)	Repeated pairing of neutral cues (lights or tones) with aversive stimuli (shock)	Freezing or fear related behavior in response to previously neutral cues like tones or contexts

Adapted and expanded from Muir, J., Lopez, J., & Bagot, R. C. (2019). Wiring the depressed brain: Optogenetic and chemogenetic circuit interrogation in animal models of depression. *Neuropsychopharmacology*, *44*, 1013–1026. https://doi.org/10.1038/s41386-018-0291-6; Nestler, E. J., Barrot, M., Dileone, R. J., Eisch, A. J., Gold, S. J., & Monteggia, L. M. (2002). Neurobiology of depression. *Cell*, *34*, 13–25.

and pathophysiologies (Nestler et al., 2002). As its name suggests, the biorhythm theory focuses on the disrupted circadian rhythms observed in those with MDD and replicable in animal models, specifically disruption in the suprachiasmatic nucleus (SCN), pineal gland, and proper entrainment to diurnal cues. This may be especially relevant to cases of seasonal affective disorder (SAD)-related depressed mood, which can be lessened with bright light therapy (Loonen & Ivanova, 2016b). The neuroimmune theory relies on observed increases in global and neural inflammation after stress and increases in inflammatory cytokines like interleukin-6 (IL-6) and tumor necrosis factor (TNF). Additionally, the resident immune cells in the brain, microglia, are activated by stress and may regulate how inflammation influences brain function and neuronal plasticity (Douma & de Kloet, 2020; Fox & Lobo, 2019; Kreisel et al., 2014; Tynan et al., 2010). Dysregulated inflammation and immune system function are also associated with higher levels of corticotrophin releasing hormone (CRH), the basis for the neuroendocrine hypothesis.

Altered levels of CRH and glucocorticoids (cortisol in humans, corticosterone in rodents) with broadly abnormal HPA axis function is a consistent MDD finding, and many rodent models are validated based on their ability to increase or disrupt corticosterone release. When functioning properly, sensory cues signal the paraventricular nucleus (PVN) of the hypothalamus to secrete CRH, which stimulates the synthesis and release of adrenocorticotropin (ACTH) from the anterior pituitary. ACTH stimulates the adrenal cortex to produce glucocorticoids. This process is regulated by both the hippocampus (inhibitory) and the amygdala (excitatory). CRH and subsequent glucocorticoid release leads to increased heart rate, arousal, and vigilance with decreased appetite and sexual drive. After an acute stressor, glucocorticoid signaling leads to a negative feedback effect stopping the production of CRH in the PVN and ending the stress response in ~60–120 min. In the case of chronic or prolonged stress, brain region specific adaptations, such as the downregulation of glucocorticoid receptors in the PVN or upregulation of the CRH system in the amygdala, lead to persistent activation of the HPA axis with failure to restrain this hyperfunction (Makino, Hashimoto, & Gold, 2002). A hyperfunctioning HPA axis may lead to higher glucocorticoid level spikes in response to acute stress, prolonged glucocorticoid spikes, or might lead to chronically elevated glucocorticoid levels at baseline with a subsequent blunting of glucocorticoid spikes in response to acute stressors. Prolonged stress and elevated glucocorticoid levels lead to atrophy in the hippocampus, which is a cellular explanation for reduced hippocampal volumes seen in patients, however it is unclear if a dysregulated HPA axis is a cause of depression or rather a correlated downstream symptom (Nestler et al., 2002). CRH and glucocorticoid signaling are also fundamental to changes in plasticity seen in the VTA and other relevant brain regions (Douma & de Kloet, 2020).

The monoaminergic hypothesis developed in response to the clinical discovery that monoamine-altering drugs, like selective serotonin reuptake inhibitors (SSRIs) or monoamine oxidase inhibitors (MAO-Is), reduced depressive symptoms; however, these drugs, which are currently still first-line treatments, can take 4–6 weeks or more to reduce symptoms. Understanding the role of monoaminergic signaling, specifically serotonin, norepinephrine, and dopamine, in depression and exactly why increasing their concentrations reduces depression symptoms at such long timescales has led to important findings about neurogenesis and synaptic signaling that are the foundation for the neuroplasticity model of depression. This theory, also called the kindling theory, borrowing terminology from epilepsy models, posits that the accumulation of a number of smaller cellular and circuit function changes reach a tipping point to cause a depressive episode (Loonen & Ivanova, 2016b). Most recently, the theory that depression, and other psychiatric illnesses, are disorders of network function, with clinical symptoms resulting from a combination of molecular and cellular lesions distributed throughout large-scale circuits is growing in popularity (Loonen & Ivanova, 2016a, 2016b; Marton & Sohal, 2016). Within this framework, it is easy to conceptualize how drugs of abuse might contribute to these distributed functional lesions and influence depression etiology. To probe cellular and circuit level changes in neuronal plasticity in relevant anatomical nodes implicated in depressive symptoms, the field has turned to new tools that allow for unprecedented specificity of neuronal control during and after stress exposure.

Optogenetics and chemogenetics, which allow for experimenter controlled neuronal excitation or inhibition, coupled with genetic driver lines, which allow for cellular selection within heterogeneous anatomical regions, have revolutionized the field of preclinical psychiatric research. In recent years, numerous studies have probed both reward circuits broadly and more specifically the circuits and parameters that increase or decrease susceptibility to stress. Indeed, optogenetics has prompted the development of circuit models for a number of psychiatric illnesses including depression, using supra-physiologic circuit stimulation to both establish causal links between neuronal pathways and complex behaviors, and to understand how manipulations in these pathways can inform treatment (Ilango & Lobo, 2015;

Marton & Sohal, 2016; Steinberg, Christoffel, Deisseroth, & Malenka, 2014; Tye & Deisseroth, 2012). Optogenetics experiments have yielded important insights and a more comprehensive understanding of how the brain codes "reward" in general, which has been useful in the study of both addiction and depression to understand reward seeking and anhedonia (Der-Avakian & Markou, 2012; Hu, 2016; Kringelbach & Berridge, 2009; Lüscher, 2016). Indeed, many of the commonly used laboratory paradigms assay motivation, reward, and social interaction, leading early optogenetic studies probing stress outcomes to focus on mesolimbic and mesocortical pathways involving the PFC, VTA, and NAc (Chaudhury et al., 2013; Covington et al., 2010; Francis et al., 2015; Friedman et al., 2014; Marton & Sohal, 2016; Tye et al., 2013; Warden et al., 2012), favoring dopaminergic rather than the serotonergic or adrenergic systems (Belzung et al., 2014). An important difference between types of optogenetic experiments is that some studies strive to experimentally induce plasticity changes on or by synaptic contacts via particular stimulation protocols (Lüscher, 2016). On the other hand, other studies seek to induce or reverse more complex affective states, leading to more conflicting and complex experimental results. Chemogenetics and the prolonged activation of G-protein coupled receptors lends itself readily to examining this type of chronic state change, and is being used to further clarify circuit functions in reward-related behaviors (Marton & Sohal, 2016; Muir et al., 2019). Also, while some experiments investigate circuits after a stress induction protocol, many studies examine circuit function in otherwise "normal" rodents, which may not fully capture the way these circuits function in the depressed brain (Belzung et al., 2014; Marton & Sohal, 2016).

In the following paragraphs, we will summarize the major circuit findings that have come out of optogenetic studies focusing on the well-studied mesolimbic and mesocortical pathways involving the PFC, VTA, and NAc in the context of stress-induced and anhedonic behaviors. The discussed circuits involving these regions and others implicated in depression are summarized in Fig. 1. The details of each study and brain region are beyond the scope of this chapter, but for comprehensive reviews including tables listing studies by brain region, stress paradigm, and finding see Belzung et al. (2014); Douma and de Kloet (2020); Muir et al. (2019). In this discussion, certain caveats for the use of these types of experiments should be kept in mind. Firstly, the insertion of exogenous ion channels may impact endogenous neurotransmitter release or release probability, the illuminated field may not capture an entire anatomical brain region, and usually only one

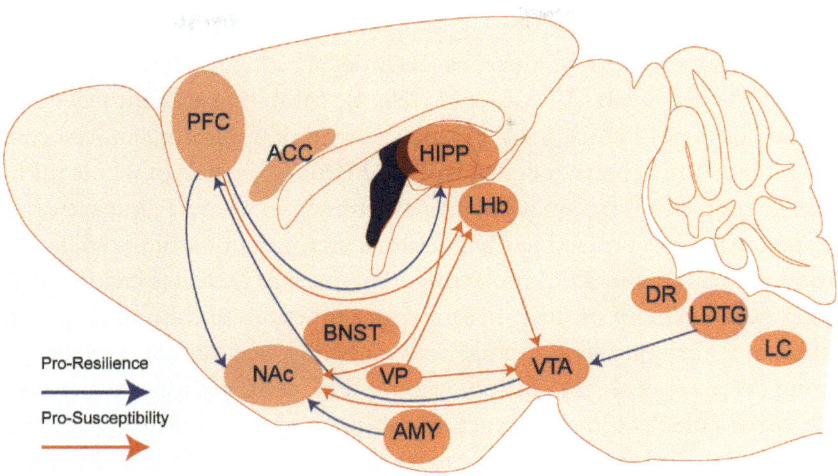

Fig. 1 Schematic of brain regions and circuits involved in depression in the rodent brain. Pathways corresponding to optogenetic or chemogenetic experiments testing depression-like behavior are depicted. Blue arrows represent pro-resilience findings, and orange arrows represent pro-susceptibility findings in tests like social interaction, forced swim, etc. Regions not connected with arrows are also implicated in depression-like behavior and provide important regulatory or neuromodulatory control over behavioral output. Abbreviations: ACC, anterior cingulate cortex; AMY, amygdala; BNST, bed nucleus of the stria terminalis; DR, dorsal raphe; HIPP, hippocampus; LC, locus coeruleus; LDTG, laterodorsal tegmental area; NAc, nucleus accumbens; PFC, prefrontal cortex; VP, ventral pallidum; VTA, ventral tegmental area. *Adapted from Muir, J., Lopez, J., & Bagot, R. C. (2019). Wiring the depressed brain: Optogenetic and chemogenetic circuit interrogation in animal models of depression.* Neuropsychopharmacology, 44, *1013–1026. https://doi.org/10.1038/s41386-018-0291-6.*

cell type of interest is examined or manipulated in each study. Further, stimulation protocols may not mimic or directly compare to the timing of endogenous tonic or stimulus-based circuit activity. Finally, while optogenetic neuronal manipulation is an elegant and effective technique, interpretations are limited by the behavioral paradigms used, which are often simple and have been predominantly focused on the functioning of the dopaminergic system, as noted above (Belzung et al., 2014).

2.4 Prefrontal cortex (PFC)

The PFC is highly interconnected with many other brain areas involved in reward and emotion processing, and reduced PFC activity is associated with depression in humans and depression-like states in rodents. High frequency optogenetic activation of the medial PFC (mPFC) increases resilience after

CSDS in 10 week old male mice (Covington et al., 2010), specifically driven by NAc-projecting mPFC neurons (Bagot et al., 2015), not mPFC projections to the amygdala (Vialou et al., 2014). Inhibiting this mPFC—NAc pathway does not induce social avoidance by itself, indicating a more complex modulation of behavior (Bagot et al., 2015). Activation of whole mPFC in male Long Evans rats does not alter swimming in the FST, but activation of lateral habenula projecting PFC neurons increases immobility while activation of dorsal raphe (DR) projecting PFC neurons increases kicking, indicating heterogeneity of downstream target functions and highlighting the importance for considering exogenous vs endogenous circuit activity and timing (Warden et al., 2012). Inhibition of PFC-DR neurons during social defeat stress blocks the emergence of social avoidance and promotes resilience in 8–12 week old male mice (Challis, Beck, & Berton, 2014; Challis et al., 2013; Steinberg et al., 2014), and chemogenetic silencing of PFC parvalbumin-expressing GABAergic interneurons facilitates learned helplessness acquisition in 5–6 week old male mice (Muir et al., 2019; Perova, Delevich, & Li, 2015). Here, decreasing the activity of a specific PFC pathway or reducing local PFC inhibition was antidepressant. On the other hand, inhibiting dopaminergic projections from the VTA into the mPFC reduces social interaction after a subthreshold defeat stress, inducing susceptibility in 7–9 week old male mice (Chaudhury et al., 2013; Knowland & Lim, 2018). Indeed, activating mPFC pyramidal neurons, as a whole in adult male Sprague-Dawley rats or only specifically those that contain the D1-type dopamine receptor, but not the D2-type, in 8–12 week old male and female mice is sufficient to produce reduced FST immobility and increased EPM open arm time (Fuchikami et al., 2015; Hare et al., 2019). Finally, increasing the excitability of mPFC pyramidal neurons with a stabilized step-function opsin blunts concurrent dopaminergic input into the NAc, which Ferenczi and colleagues elegantly demonstrated with concurrent optogenetic stimulation and MRI BOLD recording in male and female rats (Ferenczi et al., 2016; Robbins, 2016). While these neurons may be hypoactive at baseline, they are also hyperexcitable in response to input, which may in turn provide a top-down inhibitory impact on NAc to produce blunted reward seeking and reduced CPP (Ferenczi et al., 2016). Together these findings indicate both the importance of the PFC in regulating depression-like behaviors, and the importance of context, stimulation timing, and stress history on circuit function.

2.5 Ventral tegmental area (VTA)

The VTA is one of the most well studied brain regions with respect to reward processing, making it a critical hub for both depression and addiction (Hu, 2016). The firing patterns and connectivity of VTA dopamine neurons have captivated the scientific community due to their regulation of reward-related behavior, and their broad connectivity to other hubs of reward processing (Hu, 2016; Koob, 2009; Russo & Nestler, 2013). The VTA is most often noted for its ability to assign prediction and valence to salient stimuli, with positive valence indicating reward, but the VTA also assigns negative valence to fear and uncertainty (Douma & de Kloet, 2020; Keiflin & Janak, 2015; Steinberg et al., 2013). Successful coping with a stressor can take on a positive valence, while chronic and inescapable stress can trigger neuroadaptations that lead to susceptible phenotypes. For a comprehensive review of cellular and molecular plasticity in VTA neurons after stress, see Douma and de Kloet (2020).

Briefly, 7 week old male mice susceptible to social defeat, but not those that are resilient, have elevated firing rates of VTA dopamine neurons (Cao et al., 2010; Krishnan et al., 2007). Optogenetic activation of VTA dopamine neurons can induce susceptibility in resilient mice after CSDS (Chaudhury et al., 2013), but in susceptible mice, the same stimulation induces resilience by normalizing the hyperactivity caused by succumbing to stress (Friedman et al., 2014). Chronic mild stress on the other hand leads to reduced dopamine neuron firing in rats, and optogenetic activation increased resilience in mice (Tye et al., 2013). Many of these effects are recapitulated in studies activating or silencing VTA projections terminating in NAc. Optogenetic stimulation of these terminals induces susceptibility after a subthreshold defeats stress, mediated by both neurotrophic signaling (BDNF) and dopamine signaling via D1 receptors (Koo et al., 2016; Walsh et al., 2014). VTA projections to the PFC mediate a response opposite to VTA projections to NAc, and, as mentioned above, where inhibition induces susceptibility to subthreshold stress, and stimulation can induce resilience in susceptible mice (Chaudhury et al., 2013). The VTA neurons projecting to either NAc or PFC manifest these different outcomes partially via receiving input from different sources, with the NAc-projecting neurons receiving input from the laterodorsal tegmental area (LDTG) and promoting reward-related behaviors, while the PFC-projecting neurons receive input from the lateral habenula (LHb) and promote aversive behaviors (Lammel et al., 2012).

Glutamatergic lateral habenula neurons also synapse onto GABAergic neurons of the VTA (Lammel et al., 2012). While the dopamine neurons are the most studied, the role of GABAergic projection neurons of the VTA, as well as glutamatergic neurons, and neurons that co-release these neurotransmitters are more rigorously being interrogated (Bouarab, Thompson, & Polter, 2019; Buck, Torregrossa, Logan, & Freyberg, 2020; Granger, Wallace, & Sabatini, 2017; Yoo et al., 2016). Together, electrophysiological and optogenetic experiments have established that VTA dopamine neuron firing orchestrates behavioral responses to both acute and chronic stress via plasticity in projection-defined neuronal subpopulations.

2.6 Nucleus accumbens (NAc)

NAc integrates information from both PFC and VTA, turning this input into behavioral output via basal ganglia loops and signaling to the thalamus and cortex. Ninety five percent of neurons in NAc are medium spiny projection neurons (MSN) differentiated by their expression of either D1- or D2-type dopamine receptors. D1-MSNs project to the ventral pallidum (VP), lateral hypothalamus (LH), and the VTA, and are often associated with reward. D2 MSNs mainly project to the VP, with some input to the LH (O'Connor et al., 2015) and are often associated with aversion. While these circuits are treated as independent, there are a proportion of MSNs that do express both D1 and D2 receptors, and the proportion of these mixed MSNs varies by location in NAc, with the shell ranging from 5% to 30% overlap in some reports (Castro & Bruchas, 2019; Kupchik et al., 2015). Further, there is anatomical specificity to NAc, and it is commonly divided into three functional regions: the NAc core and the medial and lateral NAc shell. The NAc core is associated with reward seeking and cognitive reward assignment and is functionally and anatomically similar to the dorsal striatum. Conversely, NAc shell is more associated with "limbic" or emotion function, and avoidance of negative outcomes, with the medial shell specifically having more variation in interneuron populations and more varied input from neuropeptide signals and brain regions, including the lateral hypothalamus and nucleus of the solitary tract (Castro & Bruchas, 2019; Loonen & Ivanova, 2016a, 2016b). Due to the small size of NAc, the distinction between core and shell is not always made in studies using mice.

Within NAc, D1- and D2-MSNs respond differently to chronic stress and show different forms of plasticity. Chronic stress (CSDS) oppositely regulates the frequency of excitatory transmission, with increased intrinsic excitability and reduced frequency of mEPSCs in D1-MSNs, but increased frequencies of mEPSCs in D2-MSNs associated with susceptibility in 8 week old male mice (Francis et al., 2015; Francis & Lobo, 2017). This is consistent with studies demonstrating reduced excitatory input onto D1-MSNs after chronic multimodal or restraint stress in 2–4 month old male mice (LeGates et al., 2018; Lim, Huang, Grueter, Rothwell, & Malenka, 2012). Further, stress resilient animals show increased synaptic strength of D1-MSNs and a down regulation of D2-MSN spine strength, while susceptibility does not significantly alter synaptic strength after CSDS in 8 week old male mice (Khibnik et al., 2016). Spine density, on the other hand, is increased in NAc MSNs in CSDS susceptible 9–11 week old male mice (Christoffel et al., 2011), and this is driven by increases in D2-MSNs (Fox, Figueiredo, Menken, & Lobo, 2020). In a CSDS paradigm, optogenetic stimulation of D1-MSNs rescues the social interaction and sucrose preference of susceptible mice and stimulation of D2-MSNs induces social avoidance after subthreshold stress (Francis et al., 2015). Further, chronic chemogenetic inhibition of D1-MSNs, but not D2-MSNs, induces susceptibility in previously resilient mice. By bidirectionally dictating behavior, D2-MSN plasticity may control the transition to a depressed state, while D1-MSN activity may still retain the ability to override this depression-like behavior (Francis & Lobo, 2017). With respect to input beyond the PFC and VTA, stimulation of terminals from the ventral hippocampus in NAc induce susceptibility and stimulation of terminals from the basolateral amygdala induces resilience following CSDS in male mice (Bagot et al., 2015). Ventral hippocampal synaptic strength is enhanced onto D1-MSNs in anhedonia, and passive coping after a 3-day foot shock stressor is accompanied by D1-MSN hyperexcitability in 2–3-month-old male and female mice (Pignatelli et al., 2020). The latter is similar to hyperexcitability occurring in D1-MSNs in CSDS susceptible conditions (Francis et al., 2015, 2017). However, other models of chronic stress (CMS) reduce synaptic strength and impair LTP at ventral hippocampus-NAc D1-MSN synapses (LeGates et al., 2018). Consistent with CSDS, where D1-MSNs display reduced dendritic complexity resulting in decreased synaptic input but enhanced excitability (Fox, Chandra, et al., 2020; Fox, Figueiredo, et al., 2020; Francis et al., 2015, 2017), the latter is a potentially homeostatic mechanism to maintain equal strength of activity in the face of reduced excitatory transmission.

While the PFC, VTA and NAc are the most well studied circuit hubs with respect to depression and addiction, numerous other brain regions that either provide input or receive output from these regions also play a role in regulating depression-like behavior. First, as an output region for NAc, VP is emerging as important hub of reward integration. Silencing VP projections to lateral habenula (LHb) reduces immobility following defeat, and inhibiting VP projections to the VTA increases social interaction in 8–12 week old male mice (Knowland et al., 2017). Upstream, both the hippocampus and the amygdala play important roles in tracking salient information and cues related to rewards or threats, and the bed nucleus of the stria terminalis (BNST) connects limbic brain regions with the hypothalamus to control hormone secretion (Heshmati & Russo, 2015; Muir et al., 2019). Finally, neuromodulation of these systems has been an important entry point for depression treatment, and cholinergic input from the LDTG into the VTA or from cholinergic interneurons in NAc, serotonergic input from the dorsal raphe, and norepinephrine from the locus coeruleus all influence the plasticity of depression-related brain regions (Belzung et al., 2014; Knowland & Lim, 2018; Muir et al., 2019). The initial optogenetic studies in these circuits provided important information on what these pathways are *capable* of doing with experimenter-driven activity. Having identified these pathways, more naturalistic work moving forward, possibly coupling optogenetics with other techniques like fMRI, fiber photometry, or single-cell calcium imaging, will determine what these circuits and specific cell populations *endogenously* do in the face of stress experience and the expression of depression-like behaviors. Finally, preclinical circuit mechanisms of comorbid models for depression and addiction are lacking. However, we can use this knowledge of stress-induced circuit mechanisms combined with addictive drug-induced circuit mechanisms as an initial foundation for such studies.

2.7 Molecular and transcriptional regulators of depression

Within these brain circuits, several important molecular, transcriptional, and epigenetic changes drive the cellular plasticity underlying observed behaviors after stress. A great deal of work has gone into identifying target molecules and understanding both their function and their role in producing or protecting from depression-like behavior (Nestler, 2015a). Here we will discuss important molecules with respect to their function: signaling molecules, transcription factors, and epigenetic regulators. Signaling molecules include any neurotransmitter, neuropeptide, neurotrophic factor, hormone, or the receptors for any of the above that facilitate signaling between cells, brain regions, or body systems. As previously mentioned, glutamatergic,

GABAergic, and neuromodulatory neurotransmitter signaling play fundamental roles in circuit function. As such, disrupting this signaling by knocking out either individual receptors or transporter molecules alters the response to natural rewards (Der-Avakian & Markou, 2012). Briefly, knocking out tyrosine hydroxylase, D1 receptors, D2 receptors, the serotonin transporter, or either the $\alpha1$, $\alpha5$, or $\gamma2$ subunit of the $GABA_A$ receptor increases anhedonia, while knocking out the dopamine transporter, or the serotonin 1A receptor decreases anhedonia (Der-Avakian & Markou, 2012). At a cell type specific level, reducing expression of the serotonin transporter protein p11 selectively in cholinergic interneurons of NAc induces depression-like behavior in 8–10 week old male mice (Warner-Schmidt et al., 2012).

Neuropeptide signaling also regulates reward processing, and unlike the fast neurotransmitters, both the release and the window of action for these molecules are on longer time scales. Opioid molecules include dynorphin, enriched in D1-MSNs and associated with aversive signaling, and enkephalin, enriched in D2-MSNs and associated with adaptation to stress and pain. Knocking out the μ-opioid receptor, β-endorphin/enkephalin, or preprodynorphin increase anhedonia (Blendov, Walker, Martinez, & Harris, 2006; Hayward, Pintar, & Low, 2002; Papaleo, Kieffer, Tabarin, & Contarino, 2007). Inhibiting dynorphin signaling with κ-opioid receptor antagonists in the NAc produces antidepressant effects (Mague et al., 2003) while activation of κ-opioid receptors promotes aversion in male Sprague-Dawley rats (Newton et al., 2002); agonism of the δ-opioid receptor promotes antidepressant responses (Broom et al., 2002), and indeed, chronic antidepressant treatment increases enkephalin levels in NAc and striatum of male rats (DeFelipe, DeCeballos, Gil, & Fuentes, 1985). In contrast, reduced enkephalins in NAc mediate susceptible behavior and restoring enkephalin levels causes a resilient response to social defeat stress in 7–8 week old male mice (Nam et al., 2019). For a more comprehensive review of opioid and neuropeptide signaling in NAc, see Castro and Bruchas (2019) and for signaling in the VTA, see Douma and de Kloet (2020).

Other types of signaling and stabilizing molecules promote neuronal health. For example, neuroligin (NLGN) is a membrane protein that stabilizes spine formation, and reduced NLGN-2 is associated with depression. CSDS reduces NGLN-2 expression in D1-MSNs, but not D2-MSNs, and knocking NGLN-2 down in D1-MSNs promotes susceptibility while knockdown in D2-MSNs promotes resilience (Heshmati et al., 2018). Finally, one of the most well studied signaling factors with respect to depression is BDNF, a neurotrophin which promotes neurogenesis and spine stability via activation of its receptor tropomyosin receptor kinase B (TrkB). BDNF is released in response to stress and is downstream of glucocorticoid signaling. CSDS increases BDNF in NAc of susceptible mice, but unlike in PFC or hippocampus where BDNF is associated with resilience, BDNF release from the VTA in NAc promotes susceptibility to stress. Indeed, it is BDNF, not dopamine, that drives susceptible behavior to CSDS after optogenetic stimulation of VTA-NAc neurons (Berton et al., 2006; Koo et al., 2016; Krishnan et al., 2007; Walsh et al., 2014). Because of the cell-type-selective, inverse functions of D1- and D2-MSNs with respect to susceptibility, BDNF may support and stabilize plasticity associated with depression-like behaviors via D2-MSNs rather than strengthen top down control, as in PFC, or counteract atrophy, as in the hippocampus, explaining these region-specific effects of the same molecule.

To orchestrate these changes in molecular expression, a series of transcription factor proteins influence susceptibility and resilience to stress. Nuclear factor κB (NF-κB) is a transcription factor activated downstream of inflammatory cytokine and interleukin signaling and is associated with stress-induced impairment. In the hippocampus, NF-κB is required for the expression of stress-induced anhedonia (reduced sucrose preference) in adult male rats (Koo, Russo, Ferguson, Nestler, & Duman, 2010). In NAc CSDS increases NF-κB activity which in turn induced the formation of new immature spines, selectively in susceptible mice (Christoffel et al., 2011). cAMP response element-binding protein (CREB) functions in a neuroprotective manner in the hippocampus, but in NAc is related to susceptibility. Prominent targets upregulated by CREB in reward circuits are BDNF, dynorphin, glutamate receptor subunits, and other neuropeptides, making it part of a feed forward loop to sustain susceptibility (Nestler et al., 2002; Russo & Nestler, 2013). Conversely, other transcription factors, such as ΔFosB and β-catenin, mediate resilience in NAc. ΔFosB, a splice variant of FosB, is induced by chronic social defeat stress in 8 week old male mice age and enhanced in D1-MSNs of resilient mice, but not in D2-MSNs of susceptible mice, and when expressed in NAc can induce resilient responses to social defeat stress in 8–11 week old male mice (Lobo et al., 2013; Vialou et al., 2010). As with the other transcription factors, ΔFosB has an opposite role in the PFC, where it is induced in the prelimbic PFC after CSDS, and overexpression in this region induces susceptibility via signaling through the cholecystokinin pathway in 8–10 week old male mice (Vialou et al., 2014). For review on the role of ΔFosB in depression, see Nestler (2015b). β-catenin, downstream of the WNT signaling pathway, and ΔFosB represent two of the most highly regulated gene hubs in NAc after CSDS in resilient animals based on genome-wide studies in mice, together orchestrating molecular signatures of resilience (Bagot et al., 2016; Han & Nestler, 2017; Nestler, 2015b; Wilkinson et al., 2011). Finally, transcription factor targets have been examined to determine the molecular pathways that underlie stress-induced functional adaptations in neurons. For example, the transcription factor, early growth response 3 (Egr3) is enhanced in NAc D1-MSNs in 8–10 week old male mice that are susceptible to social defeat stress. Egr3 enrichment, on the promoter for the cytoskeleton molecule RhoA is increased in NAc of male susceptible mice (Francis, Gaynor, Chandra, Fox, & Lobo, 2019). RhoA in NAc D1-MSNs, underlies the reduced dendritic complexity and accompanying physiological changes noted in the above section, in susceptible male mice (Fox, Chandra, et al., 2020). Further, enhancement of RhoA in unstressed 8–10 week old male and female mice produces sex specific disruptions in motivated behaviors (Fox, Chandra, et al., 2020).

The sweeping transcriptional changes associated with induced susceptible or resilient states after stress are created not only with transcription factor activation or repression, but also with changes to the chromatin structure of the genome, facilitating or inhibiting transcription via more or less accessible DNA respectively. Chromatin-modifying enzymes including specific histone deacetylases (HDACs), histone methyltransferases (HMTs) and DNA methyltransferases (DNMTs), accomplish these posttranslational modifications of chromatin histones and DNA and play a role in stress susceptibility or resilience. Epigenetic modifications are sufficient to alter gene expression, which has been demonstrated specifically with experimental manipulation of acetylation and methylation at the FosB promoter, and these epigenetic-driven changes in expression were sufficient to alter both drug- and stress-related behavior (Heller et al., 2014). Further, the repressive epigenetic methylation tag, H3K9me2, is enriched at the FosB promoter in NAc of depressed patients relative to controls (Heller et al., 2014). Chromatin-modifying enzymes are themselves enhanced or reduced after stress, and concurrently global regional changes in acetylated or methylated histones in both NAc and hippocampus occur after stress (Russo & Nestler, 2013). Manipulation of HDACs or DNMTs have also been effective at producing antidepressant-like effects in animal models. While stress can influence epigenetic patterns acutely, gene expression regulation via epigenetics is both crucial for proper development, and some modifications can be passed on trans-generationally. Further, there are important sex differences in epigenetic regulation that may both drive sex differences in susceptibility to depression and differentially influence the impact of either maternal or paternal stress on the stress reactivity of offspring (Hodes et al., 2017).

Interestingly, it is not just gene coding regions of DNA that are impacted by depression. In both humans and mice, clinical depression or stress respectively is associated with an increase in mitochondrial DNA and a reduction of telomeric DNA in saliva (Cai, Chang, et al., 2015; Cai, Li, et al., 2015). Follow-up studies indicated this was a tissue specific effect and understanding the role of both mitochondrial gene expression and telomere length in individual brain regions is a new avenue of research. Finally, as with circuit mechanisms noted above, there are limited molecular studies of combined stress and addictive drugs. Below we will discuss overlapping molecular mechanisms that could provide insight into potential mechanisms of depression and addiction comorbidity.

3. Comorbid depression and addiction

Many of the same brain regions and molecular players that mediate depression are involved in regulation of drug intake (Lenz & Lobo, 2013; Lüscher, 2016; Peters, Dunlop, & Downar, 2016; Steinberg et al., 2014). Circuit functions related to drug-induced plasticity, primarily elucidated in the context of cocaine, are briefly summarized in Fig. 2. There is a need for preclinical studies focused on circuit and molecular mechanisms in comorbid models for depression and addiction; however, researchers can use current knowledge of overlapping mechanisms to begin building frameworks for understanding mechanisms of comorbidity. The neuropeptides and transcription factors mentioned above as mediators of depressive behavior have also been studied extensively in the context of addiction, especially the transcriptional regulators CREB, ΔFosB, NFκB and their downstream targets BDNF and CRF (Briand & Blendy, 2009; Nestler, 2012; Robison & Nestler, 2011). Exposure to drugs of abuse alters BDNF expression in the mesolimbic structures discussed above. Cocaine, the drug often used to probe BDNF function in reward circuits, increases BDNF in NAc, PFC, and VTA, and manipulation of BDNF levels in these regions can alter both drug intake behavior as well as vulnerability to relapse of drug seeking (Mccarthy, Brown, & Bhide, 2012; McGinty, Whitfield, & Berglind, 2010; Vargas-Perez et al., 2009). BDNF signaling is critical for drug-induced plasticity of mesolimbic circuits after exposure to drugs of abuse (Barker, Taylor, De Vries, & Peters, 2015; Li & Wolf, 2015), and interestingly, stress can alter

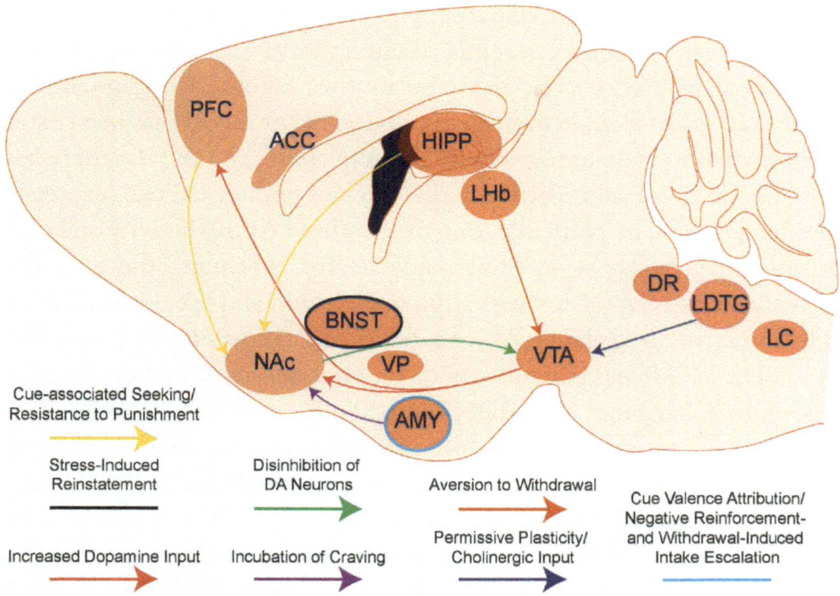

Fig. 2 Schematic of brain regions and circuits involved in drug-evoked synaptic plasticity in a sagittal section of the rodent brain. The circuit functions represented here have been described most thoroughly in the context of cocaine and may vary with respect to other drugs of abuse. The behavioral role of these pathways is noted by color. Regions not connected with arrows are also implicated in addiction and drug-related behaviors and provide important regulatory or neuromodulatory control over behavioral output. Abbreviations: ACC, anterior cingulate cortex; AMY, amygdala; BNST, bed nucleus of the stria terminalis; DR, dorsal raphe; HIPP, hippocampus; LC, locus coeruleus; LDTG, laterodorsal tegmental area; NAc, nucleus accumbens; PFC, prefrontal cortex; VP, ventral pallidum; VTA, ventral tegmental area. *Adapted from Lüscher, C. (2016). The emergence of a circuit model for addiction.* Annual Review of Neuroscience, 39, *257–276. https://doi.org/10.1146/annurev-neuro-070815-013920, with additional information Hsiang, H. L. L., Epp, J. R., van den Oever, M. C., Yan, C., Rashid, A. J., Insel, N., et al. (2014). Manipulating a "cocaine engram" in mice.* Journal of Neuroscience, 34*(42), 14115–14127. https://doi.org/10.1523/JNEUROSCI.3327-14.2014; Koob, G. F. (2003). Neuroadaptive mechanisms of addiction: Studies on the extended amygdala.* European Neuropsychopharmacology, 13*(6), 442–452. https://doi.org/10.1016/j.euroneuro.2003.08.005; Meil, W. M., & See, R. E. (1997). Lesions of the basolateral amygdala abolish the ability of drug associated cues to reinstate responding during withdrawal from self-administered cocaine.* Behavioural Brain Research, 87*(2), 139–148. https://doi.org/10.1016/S0166-4328(96)02270-X.*

the response of BDNF to cocaine in male rats, indicating the intersection of these molecular systems (Fumagalli, Caffino, Racagni, & Riva, 2009). This specific interaction of BDNF and stress has been studied with respect to alcoholism (Moonat & Pandey, 2012), but such comprehensive work has not been done for all drug classes.

As mentioned above, both transcription factors and epigenetic regulation work together to induce sweeping transcriptional changes, and the same processes at play after stress also regulate the response to drugs of abuse and the transition to addiction. Transcription factor ΔFosB has been extensively studied in NAc in response to drugs of abuse, and thus far there have been no drugs of abuse studied that do not induce ΔFosB expression (Nestler, 2013). The particular long time scale of ΔFosB activity and accumulation have made it an ideal candidate for explaining the long term plasticity changes associated with drug sensitization and self-administration/addiction across drug classes. As with depression, ΔFosB induction occurs in D1- vs D2-MSNs and is another node for cross-sensitization between these conditions. Changes in both BDNF and ΔFosB expression, as well as many other genes of interest, are influenced by epigenetic regulation in response to drugs of abuse (Schmidt, Mcginty, West, & Sadri-Vakili, 2014). Cocaine increases H4-acetylation of *FosB*, induces *Dnmt3a* and *Dnmt3b*, and decreases the expression of HDAC5 in mice (Anier, Malinovskaja, Aonurm-Helm, Zharkovsky, & Kalda, 2010; Kumar et al., 2005; LaPlant et al., 2010; Renthal et al., 2007). Opioid use leads to hypermethylation and reduced expression of the μ-opioid receptor, (encoded by the OPRM1 gene), and alcohol can cause demethylation of CpG islands the NMDA NR2B subunit gene (Nielsen, Utrankar, Reyes, Simons, & Kosten, 2012). Epigenetic regulation may be another place where stress and depression collide to alter susceptibility. Indeed, previous cocaine exposure confers susceptibility to future subthreshold social stressors, and this is mediated via a reduction in H3K9 methylation in NAc by the histone methyltransferase G9 in male mice (Covington et al., 2011). Not all epigenetic changes signal susceptibility, and some changes are actually associated with resilience to SUDs as well as depression (Bagot et al., 2016; Cadet, 2016). Therefore, further investigation is warranted, and newly developed tools can manipulate epigenetic signatures allowing for a deeper understanding of the causal relationship between epigenetics and behavioral regulation (Hamilton & Nestler, 2019).

While depression and addiction both involve reward learning and processing, no predictable relationship has been found between the effects of a given protein in the NAc, with respect to promoting specific behaviors (Russo & Nestler, 2013). For example, the same molecule in NAc can promote anhedonia after stress but also enhance drug self-administration. Further, depending on their activation patterns, either population of D1- or D2-MSNs can induce reward or aversion, with different types of stimulation leading to variable downstream pathway activation (Soares-Cunha et al., 2019). The same brain regions can drive drastically different behavior, nevertheless both depression and addiction result from neuroplastic changes at the level of the synapse that create changes at the circuit and activity engram level (Lüthi & Lüscher, 2014; Muir et al., 2019). The high rates of comorbidity for addiction and depression indicates a strong neurobiological interplay between the two disorders, weather in shared etiology or in specific vulnerabilities one disorder causes for the other.

Depression is associated with an increased risk of developing SUD compared with non-depressed populations, especially in adolescents (Rao, Daley, & Hammen, 1998). In a prospective study, adolescents with elevated cortisol levels and stressful life events were more likely to develop SUD over the 5-year follow-up window. Night cortisol levels and stressful life experience sufficiently explained the increase in frequency of SUDs in subjects diagnosed with depression or at risk for depression (Rao, Hammen, & Poland, 2009). Individuals with a SUD were also more likely to develop depression, possibly related to changes in HPA axis function, however the sample size in this study was small. While having one disorder increases risk for developing the other, individuals that have comorbid conditions are neurobiologically unique on several measures compared to those that have only one disorder, or neither. For comprehensive reviews on the specific neurobiological changes in individuals with comorbid substance use and depression, see Balhara et al. (2017) and Gómez-Coronado et al. (2018). Further, comorbidity rates and neurobiological adaptations differ depending on the class of drug considered, and we will highlight specific drug classes here.

3.1 Alcohol

Compared to individuals with AUD who have never been diagnosed with depression, individuals with both AUD and depression have higher daily alcohol intake, higher lifetime rates of diagnosis of other anxiety or affective disorders or abuse of other drugs, are more likely to have AUD in both parents, and more frequently attempt suicide (Roy, Dejong, & Ferraro, 1991; Roy, Dejong, Lamparski, George, & Linnoila, 1991). MDD and AUD have extremely high comorbidity, with ~40% of those with lifetime MDD also having a lifetime diagnosis of AUD, and among those with AUD, about 20% have co-occurring depression (Boschloo et al., 2011; Cranford, Nolen-Hoeksema, & Zucker, 2011; Tseng, Ganz, Mitton, & Tsuang, 2017).

Men are twice as likely as women to be diagnosed with AUD, however, among those with AUDs, women are about twice as likely as men to be diagnosed with co-occurring depression. Further, having an AUD increases the odds of depression about twofold over the general population (Cranford et al., 2011; Tseng et al., 2017). In a small study of those actively experiencing a major depressive episode (MDE), manic symptoms, suicidal ideation, and childhood trauma were more common in those with AUD; however higher anxiety levels and anxious temperament were more common in those without AUD (Hong, Ha, Lee, Oh, & Myung, 2019). The rates of AUD tend to increase with age, but younger adults (18–25) more frequently have co-occurring depression. Indeed, in a longitudinal study of comorbidity, depression preceded AUD, lending support to the hypothesis that depressed individuals may be using alcohol to "self-medicate" their depressive symptoms (Tseng et al., 2017). Of those with a history of alcohol abuse or dependence, a MDE was linked to higher odds of drinking to enhance mood, and, in men, a greater number of days drinking alone (Cranford et al., 2011). Further, of those with AUD, depression predicts an earlier age of onset for their AUD, as well as poorer treatment outcomes for AUD, higher relapse rates, and greater risk of suicide (Cranford et al., 2011). Among adolescents, especially young adolescents, those that have already developed AUD are more than twice as likely also have a mood disorder as a teen compared to those with subthreshold AUD or non-problematic drinking habits (Pedrelli, Shapero, Archibald, & Dale, 2016). Over half of cases of depression have first onset by young adulthood or earlier, and heavy drinking in youth is associated with suicide, motor vehicle accidents, accidental injury, STIs, sexual assault, and impairments in memory and attention, which may also be related to the etiology of depressive symptoms (Pedrelli et al., 2016).

As the balance of neuronal inhibition is disrupted in depression and ethanol directly acts on GABA receptors there are, unsurprisingly, specific variants in the GABAergic system related to comorbidity, and individuals with comorbid depression have higher levels of GABA in their cerebrospinal fluid (Gómez-Coronado et al., 2018; Roy, Dejong, & Ferraro, 1991; Roy, Dejong, Lamparski, et al., 1991). Further, gene variants in the corticotropin-releasing hormone-binding protein (CRHBP) the μ-opioid receptor (OPRM1) and the β1 subunit of GABA A (GABAA) receptors (GABRB1) are all associated with higher risk for depressive symptoms in patients with AUD (Kertes et al., 2011). The short allele variant of the SLC6A4 serotonin transporter gene is represented comparably in groups of depressed individuals with or without comorbid alcohol use disorder, and at higher frequency than those without depression (Nellissery et al., 2003).

In adolescent women, homozygous individuals for the long allele were more likely to have alcohol dependence without a comorbid depression diagnosis, but in another study of adult women homozygosity for the long allele variant is associated with more severe depression symptoms specifically in those with comorbid alcohol dependence (Balhara et al., 2017; Gokturk et al., 2008; Verhagen et al., 2009). Gene variants in the alcohol dehydrogenase gene are specifically associated with comorbid depression and alcohol dependence, but not with either condition alone (Balhara et al., 2017; Gómez-Coronado et al., 2018). Individuals with the Taq1A gene variant in the DRD2 gene show higher novelty seeking and comorbid alcohol abuse with depression/anxiety compared to those with alcohol dependence alone (Gómez-Coronado et al., 2018; Lin et al., 2007). Finally, at the whole genome level, a number of SNPs, including multiple in glutamate receptors, oxytocin, and also some non-gene-coding regions on the X chromosome, have been linked with comorbid alcohol use disorder and depression (Edwards et al., 2012).

At the protein and cellular levels, BDNF signaling, neuroimmune function, and overall cell density are also disrupted in individuals with comorbid depression and alcohol dependence. Lower serum BDNF was related to recent depressive symptoms, however higher serum BDNF was related to a history of more sever alcohol use and binging (Neupane et al., 2015). While the BDNF Val/Met polymorphism was not related to depression in alcohol dependence, the A allele of BDNF rs6265 is overrepresented in alcohol dependent patients with depression compared to those without (Gómez-Coronado et al., 2018; Su et al., 2011). Within patients with alcohol use disorder, depression scores correlate with insulin-like growth factor IL-6, TNF and tryptophan concentrations, but not IL-10 or nerve growth factor, and prolactin levels (Gómez-Coronado et al., 2018; Han et al., 2016; Neupane et al., 2014). Neuroanatomically, those with alcohol use disorder and comorbid depression have smaller glial cell nuclei in the DLPFC compared to those without comorbid depression in a post-mortem brain analysis (Miguel-Hidalgo et al., 2002). Finally, those with alcohol dependence that completed suicide had an increase in glial cell density in ACC, compared to those with alcohol dependence with non-suicide sudden deaths (Balhara et al., 2017; Hercher et al., 2009).

3.2 Tobacco (nicotine)

Nicotine is the most well-studied active ingredient of cigarettes, e-cigarettes and other tobacco products. Around 20% of the global population smoke cigarettes, however, this rate is doubled for individuals that have a history of depression (Kutlu, Parikh, & Gould, 2015; Mathew, Hogarth, Leventhal, Cook, & Hitsman, 2017). Many smokers begin smoking during adolescence, and this carries a fourfold higher risk of developing depressive symptoms (Kutlu et al., 2015). Conversely, adolescents with a history of MDD are twice as likely to experiment with cigarettes and 50% more likely to progress to daily smoking (Audrain-McGovern, Leventhal, & Strong, 2015). Rates of smoking remain higher even when adjusting for comorbid risk factors like low self-esteem (Fergusson, Lynskey, & Horwood, 1996). Nicotine dependence is associated with a vulnerability for depression that is greater with higher rates of smoking, but comorbid depression impacts smoking initiation, maintenance, and cessation (Kutlu et al., 2015). On all measures, smokers with depression tend to consume more nicotine, and experience greater withdrawal symptoms upon abstinence. Individuals with mental illnesses make up around 7% of smokers, yet they consume over 30% of cigarettes sold (Morisano, Bacher, Audrain-McGovern, & George, 2009). Nicotine has the highest rates of comorbidity with other substance use and addictive disorders and plays a huge role in both polysubstance abuse and in persistent substance use. Nearly 70% of those with addictive disorders, including pathological gambling, also smoke, with rates being specifically higher for those with AUD (70–80%), cocaine use (80%), and opioid use (80%) (Morisano et al., 2009). While cannabis use disorder is only present in 0.8% of never smokers, the rate is around 6.5% in smokers. Consumption of alcohol, compared to placebo, increased smoking behavior in a laboratory study (Adermark, Söderpalm, & Burkhardt, 2014; Morisano et al., 2009). Depressive symptoms increase the likelihood of smoking in those on maintenance methadone, and heroin abstinence increased nicotine consumption (Morisano et al., 2009). Further, among cocaine users, those that also smoke tend to have earlier onset of cocaine use, more severe use, and more legal problems (Morisano et al., 2009). Preclinically, nicotine also potentiates cocaine self-administration behavior in male rats (Horger, Giles, & Schenk, 1992; Morisano et al., 2009). Broadly, nicotine has been shown to increase the salience of rewarding stimuli and reward-associated cues, which may underlie the higher prevalence of other SUDs in smokers (Audrain-McGovern et al., 2015).

Like with alcohol, the "self-medication" hypothesis is extremely prevalent to explain elevated nicotine use in depressed individuals. Indeed, an early hypothesis regarding the etiology of depression suggests it results from an imbalance of the cholinergic system, specifically a hypercholinergic state, and consistent use of cigarettes leads to desensitization of nicotinic acetylcholine receptors, effectively reducing cholinergic signaling (Janowsky, Davis, El-Yousef, & Sekerke, 1972; Mineur & Picciotto, 2010; Philip, Carpenter, Tyrka, & Price, 2010). Unfortunately, smoking also leads to an upregulation of nAChRs, which may contribute to increased cholinergic signaling and exacerbated depressed mood upon cessation (Mineur & Picciotto, 2010; Picciotto, Lewis, Van Schalkwyk, & Mineur, 2015). Smoking also is linked to reduced monoamine oxidase (MAO) activity, which is normally elevated in depression, proving another mechanism for self-treatment (Morisano et al., 2009). Nicotine itself can produce antidepressant effects in rodents, possibly through the desensitization of nAChRs, as nicotinic antagonists also produce antidepressant effects (Kutlu et al., 2015; Mineur & Picciotto, 2010; Picciotto et al., 2015). Interestingly, the smoking cessation aid and partial nAChR agonist varenicline has antidepressant-like effects in preclinical studies, but in clinical use it is linked to increased suicidal ideation in smokers attempting cessation (Mineur & Picciotto, 2010). Further, the impacts of cholinergic signaling are brain region specific; decreasing cholinergic tone in the hippocampus is antidepressant (Mineur et al., 2013), but decreasing cholinergic transmission in the striatum can induce depression-like behavior (Picciotto et al., 2015). The high rates of comorbid depression with smoking indicate that nicotine alone is not an effective antidepressant, and while smokers often endorse that smoking improves their mood, consistent elevated mood is achieved more successfully with smoking cessation. Clinically, nicotinic drugs make better antidepressant adjuvants than primary treatments (Mathew, Hogarth, Leventhal, Cook, & Hitsman, 2017; Mineur & Picciotto, 2010; Philip et al., 2010).

Smokers with depression, like those dependent on alcohol, tend to have more significant health burdens than those without comorbid depression. Depressed smokers are more likely to have been hospitalized in the previous month, to have cardiovascular or lung disease, more work related disability, and attempt suicide more frequently (Gómez-Coronado et al., 2018). A primary cause of death for both those with mental illness and those with other SUDs is actually tobacco-related medical illness, and those with mental illnesses die up to 25 years earlier than the general population, in part due to smoking-related chronic conditions (Mathew et al., 2017; Morisano et al., 2009). Smokers with depression have higher levels of inflammatory markers like hs-CRP, IL-6, TNF-α, NOx, AOPP, ESR, and fibrinogen than non-depressed smokers or never smokers (Gómez-Coronado et al., 2018). It is unclear if these differences in inflammatory markers are a result of the comorbid conditions, or if they reflect a shared vulnerability to both disorders.

Multiple studies have found variance in the dopaminergic system in depressed smokers, which is consistent with nicotine directly impacting dopamine release dynamics and may again represent underlying vulnerabilities to comorbidity. PET imaging has revealed that depressed smokers have lower [^{11}C]-raclopride binding than control non-smokers or depressed non-smokers, indicating lower availability of D2 dopamine receptors. Further, smokers with MDD have greater smoking-induced dopamine release, and greater smoking-induced dopamine release is associated with higher depression and anxiety symptoms (Gómez-Coronado et al., 2018). A SNP in the D3 dopamine receptor has a significant association with comorbid MDD and tobacco use disorder, with only a trending association for MDD alone; smokers with two minor alleles (AA) have a sixfold higher risk for MDD compared to those without nicotine dependence and with two major alleles (TT) (Korhonen et al., 2014). In the GABAergic system, there is no difference in overall availability for GABA-A benzodiazepine binding sites between smokers and non-smokers, while cortical receptor availability is negatively correlated with anxiety/depression symptoms in non-smokers but positively correlated in smokers. This may explain subsyndromal anxiety and depression symptoms that are also associated with smoking and cycling, acute withdrawal between cigarettes (Esterlis et al., 2009).

Cessation rates are lower for depressed smokers (~26%) than the general population (~40%), and even lower for those with co-occurring AUD (~17%), despite comparable drive to quit and rate of cessation attempts (Mathew et al., 2017; Morisano et al., 2009). While cross-sectionally daily smokers have twice the rate of depression as non-daily smokers, over time, rates in non-daily smokers increase to be comparable to daily smokers, however non-daily smokers are less likely to attempt quitting (Weinberger et al., 2017). Upon cessation, 40% of smokers with a history of MDD have increased depressive symptoms during early abstinence. These individuals tend to be less attentive to alternate sources of reinforcement, and anhedonia predicts preference for smoking over other rewards (Audrain-McGovern et al., 2015). Depressed smokers have higher increases in negative affect and greater drops in positive affect during withdrawal which precipitate smoking behavior, and endorse greater increases in positive affect following a cigarette (Mathew et al., 2017). Further, depressed smokers have a higher degree of cognitive impairment, reduced attention and executive function in withdrawal, which may discourage cessation. Independent of depression status, nicotine boosts attention in smokers, which may be a contributing factor to continued smoking in depressed smokers (Audrain-McGovern et al., 2015). Both mood management behavioral therapy and antidepressant treatment increase cessation rates, but not significantly more so in depressed smokers. Unfortunately, individuals with a history of MDD are often excluded from smoking cessation studies (Audrain-McGovern et al., 2015). The recent rise of e-cigarettes has been clinically useful in helping chronic smokers reduce nicotine intake and eventually quit successfully. Unfortunately, the prevalence of these devices and their marketing, including the unregulated use of fruity flavorings, has made them exceptionally popular among teens, leading to drastic increases in nicotine use in this population where it had previously been declining (Krishnan-Sarin, Morean, Camenga, Cavallo, & Kong, 2015). As adolescence is a highly vulnerable time for the development of both MDD and nicotine use disorder, this new development in drug use patterns may have yet-to-be-determined profound impacts on the long-term mental health of current adolescent e-cigarette users.

3.3 Cannabis

After alcohol and nicotine, cannabis use disorders are the most prevalent in the general population. Of those with a mood disorder, 17% have a co-occurring cannabis use disorder, again with a higher comorbidity in woman than in men (Conway et al., 2006). In a population of Native American adolescents, cannabis use was more associated with depression symptoms in boys, indicating more complicated factors in underlying sex and individual differences in comorbidity (Gilder & Ehlers, 2012). Like alcohol and tobacco, use of cannabis often begins in adolescence, and cannabis use may change the sensitivity of the endogenous cannabinoid regulation of the HPA axis during development, altering stress reactivity and increasing vulnerability to depression (Gilder & Ehlers, 2012).

Cannabis use and depression comorbidity also results in unique neurobiology, which again could yield underlying risk for both conditions. Among adolescents, those with the $\geq$7R/$\geq$7R genotype in the D4 dopamine receptor have a 2.5 $\times$ greater risk of having comorbid depression and cannabis use (Bobadilla, Vaske, & Asberg, 2013). In young adults (18–25) with depression, amygdala reactivity to threatening stimuli is inversely related to the level of cannabis intake (Cornelius, Aizenstein, & Hariri, 2010). Acute THC reduces amygdala reactivity to threat stimuli in healthy, recreational cannabis users, and this effect may drive increased intake in those with depression. Cannabis use and other hallucinogens are not often associated with self-treatment hypotheses of depression; however, cannabis is often used as a treatment for chronic pain, which is often correlated with depression. Some of the medicinal benefits from cannabis may be from other compounds in the plant, such as cannabidiol (CBD), rather than the high-inducing tetrahydrocannabinol (THC). As legalization of CBD, THC, and cannabis proceeds and as clinical and recreational use increase, it will be interesting to see if rates of cannabis use disorder and by extension cannabis use disorder comorbid with depression also increase. Further, as more products isolating these components of cannabis become widely used, such as CBD containing foods and oils or THC specific vaping solutions, research on the effects of these specific compounds can be conducted and contrasted with that on use of cannabis itself in clinical populations.

3.4 Opiates and other drugs

Overall rates of use of other drugs of abuse are much lower than those for alcohol, nicotine and marijuana, thus the number of individuals with depression who also abuse them is much lower. About 3–5% of those with any mood disorder have a co-occurring cocaine, amphetamine, hallucinogen, opioid, or sedative use disorder (Conway et al., 2006). As such, studies on these drugs specifically have lower participant numbers, and are often conducted in the context of treatment studies, which may reduce the generalizability of results. These populations frequently have a high psychopathology burden, and as mentioned above, are more frequently polysubstance users. Of cocaine users seeking treatment, those with MDD, either independent of their drug use or secondary their substance use, have higher psychological and emotional distress, and reduced psychosocial functioning compared to those with cocaine dependence alone, although those with independent MDD show more overall psychopathology (Leventhal, Mooney, DeLaune, & Schmitz, 2006). In a study of heroin users, after treatment, rates of depression decreased at 12-month follow up, but those who had persistent depression spent more time detoxing than on maintenance treatment and residential rehabilitation. They also reported heavier heroin use, poorer health, and greater psychopathy (Havard et al., 2006).

The most common illicit drug used in the US, after marijuana, which is no longer illegal in some states, is non-prescription use of opiate medications (NPOU). Of those in treatment for NPOU, 27% of individuals also have

MDD, which may be up to one-third more prevalent than comorbidity in the general population including those not seeking treatment (Goldner, Lusted, Roerecke, Rehm, & Fischer, 2014). Lifetime NPOU is associated with higher incidence of MDD and opiate use disorder (OUD) (Martins et al., 2012). Among those with NPOU and experience of a major depressive episode poly-drug use, AUD, unemployment, and clustered burdens of lower socioeconomic status were more pronounced (Fink et al., 2015). Incarcerated, institutionalized, and homeless adults were not included in this study, potentially under-capturing the scope of the comorbidity prevalence. Among adolescents, major depressive episodes often precede NPOU or occur in the same year, and rates are somewhat buffered by having higher family support (Edlund et al., 2015). In longitudinal studies of adults, NPOU predicts development of a mood disorder (Martins et al., 2012). It is unclear from these studies if individuals initiate and use opiates to achieve euphoria or to provide analgesia, and like other drugs, there is evidence to support a self-medication hypothesis of comorbidity, as well as a precipitation model, or a shared vulnerability model (Goldner et al., 2014; Martins et al., 2012). For review of how opiates specifically influence the neurobiology of the brain to promote addiction and influence mood, see Koob (2020). Comparable processes are also at play for other non-prescription abused medications such as stimulants, tranquilizers and sedatives (Schepis & Hakes, 2011; Zullig & Divin, 2012).

4. Stress-induced drug intake

As depression and drug addiction both impact circuits related to reward learning and processing, there has been considerable overlap in research approaches to these disorders, both using techniques across diagnostic lines, as well as exploring shared vulnerabilities, mostly in the form of stress. To further understand the feature of anhedonia, researchers have employed a technique used to study reward in the context of drug abuse and looked at the effects of depression-model-induced stress on internal reward thresholds (Nestler et al., 2002). Intracranial self-stimulation (ICSS), involving operant responding by the animal to activate electrical stimulation in the brain, often in the lateral hypothalamus or the medial forebrain bundle, is an effective method to probe the internal motivational state of an animal (Negus & Miller, 2014). ICSS stimulation thresholds are reduced in conditions that facilitate reward, such as active drug intoxication, and increased in conditions characterized by anhedonia, such as withdrawal from drugs of abuse or after exposure to stress (Scheggi, De Montis, & Gambarana, 2018). CSDS reduces ICSS responding, and this reduction is reversed by overexpression of ΔFosB in the striatum in 8 week old male mice (Donahue, Muschamp, Russo, Nestler, & Carlezon, 2014). Just as depression is associated with increased and more persistent drug abuse, stress also influences drug intake in preclinical models. For review of how timing and type of stress influence cocaine, alcohol, and opioid intake, see Newman, Leonard, Arena, De Almeida, and Miczek (2018). While acute or intermittent stress can augment cocaine intake, more sustained social stress can reduce cocaine intake (Miczek, Nikulina, Shimamoto, & Covington, 2011; Newman, Leonard, et al., 2018; Shimamoto, Holly, Boyson, Debold, & Miczek, 2015; Tidey & Miczek, 1996). On the other hand, CSDS more consistently increases both ethanol and opioid consumption (Cooper et al., 2017; Hwa, Holly, Debold, & Miczek, 2016; Newman, Albrechet-Souza, et al., 2018; Newman, Leonard, et al., 2018).

Stress precipitates many neurological adaptations necessary for these stress-induced changes in reward- and drug-related behaviors. Stress can alter burst firing of VTA dopamine neurons, and increase extracellular dopamine concentrations in NAc and PFC, which could induce plasticity in these regions, priming them for altered responding to drugs of abuse (Newman, Leonard, et al., 2018). Stress also influences multiple hormones and neuropeptides (opioids, glucocorticoids, orexin, oxytocin, etc.), but the most comprehensively studied has been CRF, which may be driving the changes seen in VTA dopamine neuron signaling as well as numerous other changes to neuronal signaling throughout the brain. A CRF-R1 antagonist infused into the VTA blunts alcohol intake after stress, and chronically elevated CRF levels after stress sensitize the VTA-mesolimbic system to increase cocaine intake. CRF and its receptors also modulate responses to both drugs and stress via actions in the basolateral (BLA) and extended (BNST, CeA, etc.) amygdala (Haass-Koffler & Bartlett, 2012; Logrip, Koob, & Zorrilla, 2011). Signaling through CRF-R1 in the extended amygdala is important for regulating negative emotional states that drive alcohol intake (Haass-Koffler & Bartlett, 2012; Hansson, Cippitelli, Sommer, Ciccocioppo, & Heilig, 2007). Finally, with respect to opioids, CRF acts in opposition to the endogenous opioid enkephalin to increase firing rate of adrenergic cells of the locus coeruleus, creating an imbalanced state that may prime increased opioid drug intake (Newman, Leonard, et al., 2018). As mentioned above, increased CRF signaling leads to downstream increases in glucocorticoid levels and signaling. Glucocorticoids impact the dopaminergic system by upregulating tyrosine hydroxylase, the rate-limiting enzyme responsible for dopamine synthesis, downregulating monoamine oxidase (MAO), which breaks down dopamine, and can decrease dopamine reuptake, all leading to increased dopamine signaling (Mukhara, Banks, & Neigh, 2018). These changes may come together to exaggerate glutamatergic plasticity in the VTA-NAc circuit, involving other molecular players influenced by stress including CREB, cAMP, ΔFosB, and CDK5, all leading to increased sensitization and a positive feedback loop for drug addiction susceptibility (Koob & Kreek, 2007; Mukhara et al., 2018; Piazza & Moal, 1998).

Transition to addiction is not the only vulnerability facing individuals dealing with both depression and drug use. Stress can effectively precipitate drug taking or relapse after abstinence, a hallmark of the drug abuse recovery process. Experimentally, there are three well established and effective methods for precipitating renewed drug seeking, both in humans and in preclinical models; exposure to cues previously associated with drug intake, a small "priming" dose of the drug, or exposure to stress. Both cue- and drug-primed reinstatement depend on signaling through the mesolimbic dopamine system. Cue-induced reinstatement involves dopamine and glutamate signaling in the VTA, as well as opioid, serotonin, and endocannabinoid signaling (Crombag, Bossert, Koya, & Shaham, 2008; Liu & Weiss, 2002). The BLA, dorsal medial PFC, dorsal hippocampus, dorsal striatum and lateral hypothalamus also regulate aspects of cue-induced relapse (Crombag et al., 2008). While drug-primed reinstatement is not impacted by CRF manipulations, stress-induced reinstatement relies on CRF and norepinephrine signaling and can be blocked by CRF antagonism in the bed nucleus of the stria terminalis (Piazza & Moal, 1998; Stewart, 2000). Stress-induced reinstatement has been demonstrated with numerous drugs of abuse including heroin (Shaham, Erb, & Stewart, 2000), ethanol (Adinoff et al., 2017; Becker, Lopez, & Doremus-Fitzwater, 2011; Lopez, Anderson, & Becker, 2016), nicotine (Bruijnzeel, Prado, & Isaac, 2009; Lee, Calarco, McKee, Mineur, & Picciotto, 2019), and cocaine (Manvich, Stowe, Godfrey, & Weinshenker, 2016; Sinha, Garcia, Paliwal, Kreek, & Rounsaville, 2006). While broadly, mechanisms are the same, there is variability based on the drug and the type of stress used (for review see Mantsch, Baker, Funk, Lê, & Shaham, 2016).

5. Withdrawal-induced depression

While stress can play an important role in precipitating drug intake, drug withdrawal itself can induce negative physical and emotional states that can both drive the transition to addiction, as well as influence the development of depression (Chartoff & Carlezon, 2014). While the physical symptoms of withdrawal vary by drug, withdrawal is often accompanied by a negative emotional state including feelings of irritability, anxiety, dysphoria, and general unease. After repeated cycles of drug use and withdrawal, the allostasis model of addiction indicates that a new baseline point is achieved in which continued drug use must be maintained in order to create negative reinforcement, alleviating the negative state induced by withdrawal, and preserving the modified baseline state (Johnston, Linden, & van den Bree, 2016; Koob & Moal, 2001). This negative emotional state induced by opioids specifically has been termed hyperkatifeia, to contrast with the euphoria of active opioid intoxication (Koob, 2020). The negative state produced after withdrawal is neurobiologically comparable to that described above caused by external stressors, namely the CRF-norepinephrine system activation as well as increased glucocorticoid responding, and glucocorticoid signaling is actually necessary to drive drug intake in some models (Koob, 2009, 2013; Koob & Kreek, 2007; Koob & Moal, 2001). Further, it is thought to result from dysregulation of mesolimbic-accumbens-amygdaloid complex including extended amygdala structures like the BNST, and cortico-striatal-thalamic circuits, both of which regulate depressive-like, anhedonic states as well (Fig. 1) (Koob, 2013; Koob & Moal, 2001). Sensitization of dopaminergic networks in addiction to shifts in baseline activity after repeated cycles of stress and drug use also underlie allostatic regulation of drug intake (Johnston et al., 2016).

The negative state induced by drug withdrawal is so powerful it can induce reductions in reward responding and induce affective symptoms indistinguishable from depression. Clinically, early alcohol withdrawal (first week) is associated with scores on the Beck Depression Inventory of ~20 (15–30 representing moderate depression), and Hamilton Depression Scores of 18, where 20 indicates use of antidepressant medication (Brown & Schuckit, 1988; Koob, 2013; Potokar et al., 1997). While these depression scores decrease with continued abstinence, they remain elevated up to 4 weeks after the conclusion of inpatient treatment (Brown & Schuckit, 1988). Preclinically, withdrawal from alcohol, cocaine, amphetamine, cannabinoids, and nicotine all increase ICSS reward thresholds, an indicator of lowered motivational state as described above (Koob, 2013). This is one of the most consistent findings across drug classes with respect to the impact of drugs of abuse on motivational states, indicating a shared circuitry modified by drug dependence, activated by withdrawal, and controlling motivated behavior, despite the varying mechanisms of action of different drugs. Psychostimulant withdrawal has even been used as a paradigm to preclinically induce depressive-like states (Czéh et al., 2016). Although not previously mentioned here, the medial habenula—interpeduncular nucleus circuit regulates many of the aversive aspects of drug withdrawal across multiple drug classes, as well as aspects of depression and anxiety, and may be an important target for treating comorbid depression and drug use disorders (McLaughlin, Dani, & De Biasi, 2017). The intense distress during withdrawal experienced by chronic users is fundamentally different than offset of an acute dose of a given drug, and is due to neuroplasticity in the circuits mentioned above, which can drive continued drug taking and may explain why individuals with comorbid depression may have a lowered ability to tolerate the negative states associated with withdrawal to achieve abstinence (Brady & Sinha, 2005; Wemm & Sinha, 2019). Depression, substance use, or other forms of stress-inducing trauma all impact the HPA axis, which can cause cross-sensitization of these conditions leading to high rates of comorbidity and poorer treatment outcomes.

6. Therapeutic challenges for comorbid MDD and SUD

Comorbidity of MDD and SUD is consistently linked with poorer treatment outcomes, more treatment resistant depression, poly-drug use, higher symptom levels, increased risk of hospitalization, and overall lower functioning compared to those with only one disorder (Balhara et al., 2017). The higher disease burden is an important contributor to poorer outcomes, however difficulties in appropriately diagnosing co-occurring conditions as well as access to treatment resources are also critical to treatment outcomes. Abuse of illicit drugs is actually linked to shortened time to depression treatment seeking after symptom onset, and a greater perceived need for help (Olfson, Liu, Grant, & Carlos, 2012). Conversely, AUD was specifically linked with longer delays to depression treatment, and reduced perceptions of a need for help (Olfson et al., 2012). Even with these delays, individuals with comorbid AUD and MDD are more likely to seek treatment and utilize more treatment resources than those with AUD alone, possibly due to the pervasive use of alcohol in western society (Chen et al., 2013). Both alcohol and illicit drug users with MDD were more likely to report unmet treatment needs, despite treatment seeking, and cost or other barriers to treatment are often cited as deterrents (Chen et al., 2013; Melchior, Prokofyeva, Younès, Surkan, & Martins, 2014).

This dissatisfaction with treatment is not surprising, as those with AUD and comorbid mood/anxiety disorders entering treatment are more disabled and drank more heavily than those without a comorbid mood disorder, and although treatment reduced drinking overall, those with comorbid mood conditions remained more disabled and were still dinking more 3 months later (Burns, Teesson, & O'Neill, 2005). Treatment outcomes are also poorer for those abusing multiple substances, and in a national longitudinal follow up study, most individuals with one SUD no longer met criteria for SUD 3 years later, while those with multiple SUD still met criteria for SUD on at least one substance at follow up (McCabe & West, 2017). Further, multi-substance users are more likely to have comorbid mood disorders, and while multi-substance users were more likely than those with one SUD to obtain treatment, this still represented only a small fraction of total users seeking treatment (McCabe & West, 2017). These comorbid conditions often afflict high-risk populations, such as the homeless population, those in the criminal justice system, trauma survivors, and those with other health burdens, including a particularly large group also comorbid for HIV/AIDS, although these groups are not always captured in studies of comorbidity or treatment efficacy. While depression is often first treated by a primary care provider, SUD treatment facilities are often siloed off from the rest of the medical community and counselors in these centers are not always certified to provide prescription drugs (National Institute on Drug Abuse, 2018).

With respect to depression treatment, co-occurring SUD makes treatment more difficult, and is highly dependent on if the depression is primary or secondary to the SUD. Individuals seeking treatment for depression with an active or recent SUD were more than twice as likely to have treatment resistant depression (Brenner et al., 2020). Among those with AUD, those with depression secondary to AUD did show depression remission upon alcohol cessation, however depression primary to and preceding the AUD was more common and in this case depression was not helped by alcohol cessation (Swendsen & Merikangas, 2000). Further, antidepressant treatment is less effective for those with secondary SUD, although accurately diagnosing secondary depression is difficult, requiring sobriety to determine if the depression symptoms persist (Pettinati, O'Brien, & Dundon, 2013; Tseng et al., 2017). This is even more complicated by the fact that acute withdrawal is frequently accompanied by mood symptoms very comparable to depression as mentioned above (Quello et al., 2005). The individual variability in both substance use history, including the use of multiple drugs, as well as in depression history, either preceding or resulting from drug use, calls current diagnostic boundaries into question, and indicates the need for more unique treatment approaches depending on the individual (Substance Abuse and Mental Health Services Administration, 2020). Indeed, some theories propose focusing on trait clusters, such as externalizing concurrent with conduct, attention, and self-management issues or internalizing traits concurrent with anxiety, that may underly multiple DSM diagnoses but would require different treatment approaches (Rao, 2010; Uluğ, 2015). Indeed these different traits can even be accompanied by differing HPA axis function even though both groups may present with comorbid MDD and SUD (Rao, 2010).

With respect to treatment, many treatments used for MDD are also effective at curbing SUD, however pharmacotherapies often need to target individual disorders simultaneously. Common behavioral interventions include cognitive behavioral therapy (CBT), dialectical behavioral therapy (DBT), assertive community treatment (ACT), behavioral activation, contingency management, and coping skills (National Institute on Drug Abuse, 2018; Substance Abuse and Mental Health Services Administration, 2020). There is some evidence that CBT may be better for SUD treatment with MDD than the traditional 12-step models, as it provides more individual control over managing mood (Quello et al., 2005), but other studies have found no difference in efficacy, and that independent of treatment program, stressful life events were the biggest predictors of relapse (Tate et al., 2008). For pharmacotherapies, medications for depression and mood management do not have strong effects on SUDs, and combination therapies are more effective (e.g., naltrexone) (Pettinati et al., 2013; Tseng et al., 2017). Pharmacotherapies are especially effective for reducing nicotine use, and antidepressants like bupropion and nortriptyline have therapeutic benefit even if there is no comorbid depression, although this is not broadly applicable to other drugs of abuse (Torrens, Fonseca, Mateu, & Farré, 2005). While naltrexone can be an effective treatment for heroin use, it does not impact comorbid depression symptoms (Ritter, 2002). Interestingly, multiple forms of brain stimulation, such as the invasive deep brain stimulation and non-invasive transcranial magnetic stimulation or transcranial direct current stimulation, are expanding to support cortico-striatal-thalamic and salience network function both as treatment for depression and more recently for SUD as well (Dunlop, Hanlon, & Downar, 2017; Peters et al., 2016). Finally, recent trends are exploring the use of drugs of abuse as novel treatments for treatment resistant depression and addiction.

Transdermal nicotine (patch) is well tolerated and has shown promise for treating cognitive deficits and mood in late-life depression of non-smokers (Gandelman et al., 2018). LSD, psilocybin, and other serotonergic hallucinogens have been used as treatment for both MDD (Baumeister, Barnes, Giaroli, & Tracy, 2014; Reiche et al., 2018) and for addiction (Bogenschutz, 2013; Bogenschutz & Johnson, 2016; Bogenschutz & Pommy, 2012). The most well studied drug of abuse for depression treatment is ketamine, as it has rapid antidepressant effects even in severely treatment resistant patients, although it has not been studied in the context of comorbid disorders (Browne & Lucki, 2013; Duman, Li, Liu, Duric, & Aghajanian, 2012).

7. Conclusion

While much work has been done determining the neurobiological mechanisms driving depression and the transition from casual drug use to abuse, comparatively less work has been done specifically to understand the neurobiology underlying comorbidity. The clearest feature of comorbid patients is their heterogeneity, and how there is evidence in at least some populations for all theories of comorbidity etiology. Some individuals may in fact be using drugs to self-medicate pre-existing depression, while others develop depression secondary to their drug use. Finally, genetic or developmental vulnerabilities, possibly even variance in underlying personality traits, may lead to the development of both disorders. Further research, both preclinically and clinically, is needed to further disentangle these paths to pathology, and eventually to provide more effective, individually tailored treatments for both conditions.

References

Adermark, L., Söderpalm, B., & Burkhardt, J. M. (2014). Brain region specific modulation of ethanol-induced depression of GABAergic neurons in the brain reward system by the nicotine receptor antagonist mecamylamine. *Alcohol*, *48*(5), 455–461. https://doi.org/10.1016/j.alcohol.2014.06.004.

Adinoff, B., Leonard, D., Price, J., Javors, M. A., Walker, R., Brown, E. S., et al. (2017). Adrenocortical sensitivity, moderated by ongoing stress, predicts drinking intensity in alcohol-dependent men. *Psychoneuroendocrinology*, *76*, 67–76. https://doi.org/10.1016/j.psyneuen.2016.10.011.

American Psychiatric Association. (2013). Depressive disorders. In *Diagnostic and statistical manual of mental disorders* (5th ed.). https://doi-org.proxy-hs.researchport.umd.edu/10.1176/appi.books.9780890425596.dsm04. https://doi.org/10.1002/ejoc.201200111.

Anier, K., Malinovskaja, K., Aonurm-Helm, A., Zharkovsky, A., & Kalda, A. (2010). DNA methylation regulates cocaine-induced behavioral sensitization in mice. *Neuropsychopharmacology*, *35*(12), 2450–2461. https://doi.org/10.1038/npp.2010.128.

Audrain-McGovern, J., Leventhal, A. M., & Strong, D. R. (2015). The role of depression in the uptake and maintenance of cigarette smoking. In *Vol. 124. International review of neurobiology* (1st ed.). Elsevier Inc. https://doi.org/10.1016/bs.irn.2015.07.004.

Bagot, R. C., Cates, H. M., Purushothaman, I., Lorsch, Z. S., Walker, D. M., Wang, J., et al. (2016). Circuit-wide transcriptional profiling reveals brain region-specific gene networks regulating depression susceptibility. *Neuron*, *90*(5), 969–983. https://doi.org/10.1016/j.neuron.2016.04.015.

Bagot, R. C., Parise, E. M., Peña, C. J., Zhang, H.-X., Maze, I., Chaudhury, D., et al. (2015). Ventral hippocampal afferents to the nucleus accumbens regulate susceptibility to depression. *Nature Communications*, *6*, 7062. https://doi.org/10.1038/ncomms8062.

Bale, T. L., Abel, T., Akil, H., Carlezon, W. A., Moghaddam, B., Nestler, E. J., et al. (2019). The critical importance of basic animal research for neuropsychiatric disorders. *Neuropsychopharmacology*, *44*, 1349–1353. https://doi.org/10.1038/s41386-019-0405-9.

Balhara, Y. P. S., Kuppili, P. P., & Gupta, R. (2017). Neurobiology of comorbid substance use disorders and psychiatric disorders: Current state of evidence. *Journal of Addictions Nursing*, *28*(1), 11–26. https://doi.org/10.1097/JAN.0000000000000155.

Barker, J. M., Taylor, J. R., De Vries, T. J., & Peters, J. (2015). Brain-derived neurotrophic factor and addiction: Pathological versus therapeutic effects on drug seeking. *Brain Research*, *1628*(Pt A), 68–81. https://doi.org/10.1016/j.brainres.2014.10.058.

Baumeister, D., Barnes, G., Giaroli, G., & Tracy, D. (2014). Classical hallucinogens as antidepressants? A review of pharmacodynamics and putative clinical roles. *Theraputic Advances in Psychopharmacology*, *4*(4), 156–169. https://doi.org/10.1177/2045125314527985.

Becker, H. C., Lopez, M. F., & Doremus-Fitzwater, T. L. (2011). Effects of stress on alcohol drinking: A review of animal studies. *Psychopharmacology*, *218*(1), 131–156. https://doi.org/10.1007/s00213-011-2443-9.

Belzung, C., Turiault, M., & Griebel, G. (2014). Optogenetics to study the circuits of fear- and depression-like behaviors: A critical analysis. *Pharmacology, Biochemistry, and Behavior*, *122C*, 144–157. https://doi.org/10.1016/j.pbb.2014.04.002.

Berton, O., McClung, C. A., Dileone, R. J., Krishnan, V., Renthal, W., Russo, S. J., et al. (2006). Essential role of BDNF in the mesolimbic dopamine pathway in social defeat stress. *Science*, *311*(5762), 864–868. https://doi.org/10.1126/science.1120972.

Blendov, Y. A., Walker, D., Martinez, M., & Harris, R. A. (2006). Reduced alcohol consumption in mice lacking preprodynorphin. *Alochol*, *40*(2), 73–86. https://doi.org/10.1161/CIRCULATIONAHA.110.956839.

Bobadilla, L., Vaske, J., & Asberg, K. (2013). Dopamine receptor (D4) polymorphism is related to comorbidity between marijuana abuse and depression. *Addictive Behaviors*, *38*(10), 2555–2562. https://doi.org/10.1016/j.addbeh.2013.05.014.

Bogenschutz, M. P. (2013). Studying the effects of classic hallucinogens in the treatment of alcoholism: Rationale, methodology, and current research with psilocybin. *Current Drug Abuse Reviews*, *6*, 17–29.

Bogenschutz, M. P., & Johnson, M. W. (2016). Classic hallucinogens in the treatment of addictions. *Progress in Neuropsychopharmacology and Biological Psychiatry*, *64*, 250–258. https://doi.org/10.1016/j.pnpbp.2015.03.002.

Bogenschutz, M. P., & Pommy, J. M. (2012). Therapeutic mechanisms of classic hallucinogens in the treatment of addictions: From indirect evidence to testable hypotheses. *Drug Testing and Analysis*, *4*, 543–555. https://doi.org/10.1002/dta.1376.

Bolton, J. L., Molet, J., Ivy, A., & Baram, T. (2017). New insights into early-life stress and behavioral outcomes. *Current Opinion in Behavioral Sciences*, *14*, 133–139. https://doi.org/10.1016/j.physbeh.2017.03.040.

Boschloo, L., Vogelzangs, N., Smit, J. H., Van Den Brink, W., Veltman, D. J., Beekman, A. T. F., et al. (2011). Comorbidity and risk indicators for alcohol use disorders among persons with anxiety and/or depressive disorders: Findings from the Netherlands study of depression and anxiety (NESDA). *Journal of Affective Disorders*, *131*(1–3), 233–242. https://doi.org/10.1016/j.jad.2010.12.014.

Bouarab, C., Thompson, B., & Polter, A. M. (2019). VTA GABA neurons at the interface of stress and reward. *Frontiers in Neural Circuits*, *13*(December), 1–12. https://doi.org/10.3389/fncir.2019.00078.

Brady, K. T., & Sinha, R. (2005). Co-occurring mental and substance use disorders: The neurobiological effects of chronic stress. *American Journal of Psychiatry*, *162*(8), 1483–1493. https://doi.org/10.1176/appi.ajp.162.8.1483.

Brenner, P., Brandt, L., Li, G., DiBernardo, A., Bodén, R., & Reutfors, J. (2020). Substance use disorders and risk for treatment resistant depression: A population-based, nested case-control study. *Addiction*, *115*(4), 768–777. https://doi.org/10.1111/add.14866.

Briand, L. A., & Blendy, J. A. (2009). Molecular and genetic substrates linking stress and addiction. *Brain Research*, *1314*, 219–234. https://doi.org/10.1016/j.brainres.2009.11.002.

Broom, D. C., Jutkiewicz, E. M., Folk, J. E., Traynor, J. R., Rice, K. C., & Woods, J. H. (2002). Nonpeptidic δ-opioid receptor agonists reduce immobility in the forced swim assay in rats. *Neuropsychopharmacology*, *26*(6), 744–755. https://doi.org/10.1016/S0893-133X(01)00413-4.

Brown, S. A., & Schuckit, M. A. (1988). Changes in depression among abstinent alcoholics. *Journal of Studies on Alcohol*, *49*(5), 412–417. https://doi.org/10.15288/jsa.1988.49.412.

Browne, C. A., & Lucki, I. (2013). Antidepressant effects of ketamine: Mechanisms underlying fast-acting novel antidepressants. *Frontiers in Pharmacology*, *4*, 161. https://doi.org/10.3389/fphar.2013.00161.

Bruijnzeel, A. W., Prado, M., & Isaac, S. (2009). Corticotropin-releasing factor-1 receptor activation mediates nicotine withdrawal-induced deficit in brain reward function and stress-induced relapse. *Biological Psychiatry*, *66*(2), 110–117. https://doi.org/10.1016/j.biopsych.2009.01.010.

Buck, S. A., Torregrossa, M. M., Logan, R. W., & Freyberg, Z. (2020). Roles of dopamine and glutamate co-release in the nucleus accumbens in mediating the actions of drugs of abuse. *The FEBS Journal*. https://doi.org/10.1111/febs.15496.

Burns, L., Teesson, M., & O'Neill, K. (2005). The impact of comorbid anxiety and depression on alcohol treatment outcomes. *Addiction*, *100*(6), 787–796. https://doi.org/10.1111/j.1360-0443.2005.001069.x.

Cadet, J. L. (2016). Epigenetics of stress, addiction, and resilience: Therapeutic implications. *Molecular Neurobiology*, *53*, 545–560. https://doi.org/10.1007/s12035-014-9040-y.

Cai, N., Chang, S., Li, Y., Li, Q., Hu, J., Liang, J., et al. (2015). Molecular signatures of major depression. *Current Biology*, *25*(9), 1146–1156. https://doi.org/10.1016/j.cub.2015.03.008.

Cai, N., Li, Y., Chang, S., Liang, J., Lin, C., Zhang, X., et al. (2015). Genetic control over mtDNA and its relationship to major depressive disorder. *Current Biology*, *25*(24), 3170–3177. https://doi.org/10.1016/j.cub.2015.10.065.

Cao, J. L., Covington, H. E., Friedman, A. K., Wilkinson, M. B., Walsh, J. J., Cooper, D. C., et al. (2010). Mesolimbic dopamine neurons in the brain reward circuit mediate susceptibility to social defeat and antidepressant action. *Journal of Neuroscience*, *30*(49), 16453–16458. https://doi.org/10.1523/JNEUROSCI.3177-10.2010.

Castro, D. C., & Bruchas, M. R. (2019). A motivational and neuropeptidergic hub: Anatomical and functional diversity within the nucleus accumbens shell. *Neuron*, *102*(3), 529–552. https://doi.org/10.1016/j.neuron.2019.03.003.

Cerdá, M., Sagdeo, A., Johnson, J., & Galea, S. (2010). Genetic and environmental influences on psychiatric comorbidity: A systematic review. *Journal of Affective Disorders, 126*(1–2), 14–38. https://doi.org/10.1016/j.jad.2009.11.006.

Challis, C., Beck, S. G., & Berton, O. (2014). Optogenetic modulation of descending prefrontocortical inputs to the dorsal raphe bidirectionally bias socioaffective choices after social defeat. *Frontiers in Behavioral Neuroscience, 8*, 43. https://doi.org/10.3389/fnbeh.2014.00043.

Challis, C., Boulden, J., Veerakumar, A., Espallergues, J., Vassoler, F. M., Christopher Pierce, R., et al. (2013). Raphe GABAergic neurons mediate the acquisition of avoidance after social defeat. *Journal of Neuroscience, 33*(35), 13978–13988. https://doi.org/10.1523/JNEUROSCI.2383-13.2013.

Chartoff, E. H., & Carlezon, W. A. J. (2014). Drug withdrawal conceptualized as a stressor. *Behavioral Pharmacology, 25*, 473–492. https://doi.org/10.1097/FBP.0000000000000080.

Chaudhury, D., Walsh, J. J., Friedman, A. K., Juarez, B., Ku, S. M., Koo, J. W., et al. (2013). Rapid regulation of depression-related behaviours by control of midbrain dopamine neurons. *Nature, 493*(7433), 532–536. https://doi.org/10.1038/nature11713.

Chen, L. Y., Crum, R. M., Martins, S. S., Kaufmann, C. N., Strain, E. C., & Mojtabai, R. (2013). Service use and barriers to mental health care among adults with major depression and comorbid substance dependence. *Psychiatric Services, 64*(9), 863–870. https://doi.org/10.1176/appi.ps.201200289.

Christoffel, D. J., Golden, S. A., Dumitriu, D., Robison, A. J., Janssen, W. G., Ahn, H. F., et al. (2011). IκB kinase regulates social defeat stress-induced synaptic and behavioral plasticity. *Journal of Neuroscience, 31*(1), 314–321. https://doi.org/10.1523/JNEUROSCI.4763-10.2011.

Colston, M. J., Belsher, B. E., Beech, E. H., Curry, J. C., Tyberg, Y. A., Melmed, A. Z., et al. (2016). Posttraumatic stress disorder, depression, and other comorbidities: Clinical and systems approaches to diagnostic uncertainties. *Federal Practitioner, 33*(11), 37–45.

Compton, W. M., Thomas, Y. F., Stinson, F. S., & Grant, B. F. (2007). Prevalence, correlates, disability, and comorbidity of DSM-IV drug abuse and dependence in the United States: Results from the national epidemiological survey on alcohol and related conditions. *Archives of General Psychiatry, 64*, 566–576.

Conway, K. P., Compton, W., Stinson, F. S., & Grant, B. F. (2006). Lifetime comorbidity of DSM-IV mood and anxiety disorders and specific drug use disorders: Results from the National Epidemiologic Survey on Alcohol and Related Conditions. *Journal of Clinical Psychiatry, 67*(2), 247–257. https://doi.org/10.4088/jcp.v67n0211.

Cooper, S. E., Kechner, M., Caraballo-Pérez, D., Kaska, S., Robison, A. J., & Mazei-Robison, M. S. (2017). Comparison of chronic physical and emotional social defeat stress effects on mesocorticolimbic circuit activation and voluntary consumption of morphine. *Scientific Reports*, 7(1), 1–14. https://doi.org/10.1038/s41598-017-09106-3.

Cornelius, J. R., Aizenstein, H. J., & Hariri, A. R. (2010). Amygdala reactivity is inversely related to level of cannabis use in individuals with comorbid cannabis dependence and major depression. *Addictive Behaviors, 35*(6), 644–646. https://doi.org/10.1016/j.addbeh.2010.02.004.

Covington, H. E., Lobo, M. K., Maze, I., Vialou, V., Hyman, J. M., Zaman, S., et al. (2010). Antidepressant effect of optogenetic stimulation of the medial prefrontal cortex. *Journal of Neuroscience, 30*(48), 16082–16090. https://doi.org/10.1523/JNEUROSCI.1731-10.2010.

Covington, H. E., Maze, I., Sun, H., Bomze, H. M., Demaio, K. D., Wu, E. Y., et al. (2011). A role for repressive histone methylation in cocaine-induced vulnerability to stress. *Neuron, 71*(4), 656–670. https://doi.org/10.1016/j.neuron.2011.06.007.

Cranford, J. A., Nolen-Hoeksema, S., & Zucker, R. A. (2011). Alcohol involvement as a function of co-occurring alcohol use disorders and major depressive episode: Evidence from the National Epidemiologic Survey on Alcohol and Related Conditions. *Drug and Alcohol Dependence*, *117*(2–3), 145–151. https://doi.org/10.1016/j.drugalcdep.2011.01.011.

Crombag, H. S., Bossert, J. M., Koya, E., & Shaham, Y. (2008). Context-induced relapse to drug seeking: A review. *Philosophical Transactions of the Royal Society, B: Biological Sciences*, *363*(1507), 3233–3243. https://doi.org/10.1098/rstb.2008.0090.

Czéh, B., Fuchs, E., Wiborg, O., & Simon, M. (2016). Animal models of major depression and their clinical implications. *Progress in Neuro-Psychopharmacology and Biological Psychiatry*, *64*, 293–310. https://doi.org/10.1016/j.pnpbp.2015.04.004.

DeFelipe, M. C., DeCeballos, M. L., Gil, C., & Fuentes, J. A. (1985). Chronic antidepressant treatment increases enkephalin levels in n. accumbens and striatum of the rat. *European Journal of Pharmacology*, *112*, 119–122.

Deisseroth, K., Etkin, A., & Malenka, R. C. (2015). Optogenetics and the circuit dynamics of psychiatric disease. *JAMA: The Journal of the American Medical Association*, *313*(20), 2019–2020. https://doi.org/10.1001/jama.2015.2544.

Der-Avakian, A., & Markou, A. (2012). The neurobiology of anhedonia and other reward-related deficits. *Trends in Neurosciences*, *35*(1), 68–77. https://doi.org/10.1016/j.tins.2011.11.005.

Deroche-Gamonet, V., & Piazza, P. V. (2014). Psychobiology of cocaine addiction: Contribution of a multi-symptomatic animal model of loss of control. *Neuropharmacology*, *76*, 437–449. https://doi.org/10.1016/j.neuropharm.2013.07.014.

Donahue, R. J., Muschamp, J. W., Russo, S. J., Nestler, E. J., & Carlezon, W. A. J. (2014). Effects of striatal ΔFosB overexpression and ketamine on social defeat stress–induced anhedonia in mice. *Biological Psychiatry*, *76*(7), 550–558. https://doi.org/10.1016/j.biopsych.2013.12.014.

Donaldson, Z. R., & Hen, R. (2014). From psychiatric disorders to animal models: A bidirectional and dimensional approach. *Biological Psychiatry*, 77(1), 15–21. https://doi.org/10.1016/j.biopsych.2014.02.004.

Douma, E. H., & de Kloet, E. R. (2020). Stress-induced plasticity and functioning of ventral tegmental dopamine neurons. *Neuroscience and Biobehavioral Reviews*, *108*, 48–77. https://doi.org/10.1016/j.neubiorev.2019.10.015.

Duman, R. S., Li, N., Liu, R., Duric, V., & Aghajanian, G. (2012). Signaling pathways underlying the rapid antidepressant actions of ketamine. *Neuropharmacology*, *62*(1), 35–41. https://doi.org/10.1016/j.neuropharm.2011.08.044.

Duman, R. S., Sanacora, G., & Krystal, J. H. (2019). Altered connectivity in depression: GABA and glutamate neurotransmitter deficits and reversal by novel treatments. *Neuron*, *102*(1), 75–90. https://doi.org/10.1016/j.neuron.2019.03.013.

Dunlop, K., Hanlon, C. A., & Downar, J. (2017). Noninvasive brain stimulation treatments for addiction and major depression. *Annals of the New York Academy of Sciences*, *1394*(1), 31–54. https://doi.org/10.1111/nyas.12985.

Edlund, M. J., Forman-Hoffman, V. L., Winder, C. R., Heller, D. C., Kroutil, L. A., Lipari, R. N., et al. (2015). Opioid abuse and depression in adolescents: Results from the National Survey on Drug Use and Health. *Drug and Alcohol Dependence*, *152*, 131–138. https://doi.org/10.1016/j.drugalcdep.2015.04.010.

Edwards, A. C., Aliev, F., Bierut, L. J., Bucholz, K. K., Edenberg, H., Hesselbrock, V., et al. (2012). Genome-wide association study of comorbid depressive syndrome and alcohol dependence. *Psychiatric Genetics*, *22*(1), 31–41. https://doi.org/10.1097/YPG.0b013e32834acd07.

Esterlis, I., Ph, D., Cosgrove, K. P., Ph, D., Batis, J. C., Ph, D., et al. (2009). GABAA-benzodiazepine receptor availability in smokers and nonsmokers: Relationship to subsyndromal anxiety and depression. *Synapse, 63*(12), 1089–1099. https://doi.org/10.1002/syn.20688.GABA.

Everitt, B. J., & Robbins, T. W. (2016). Drug addiction: Updating actions to habits to compulsions ten years on. *Annual Review of Psychology, 67*, 23–50. https://doi.org/10.1146/annurev-psych-122414-033457.

Ferenczi, E. A., Zalocusky, K. A., Liston, C., Grosenick, L., Warden, M. R., Amatya, D., et al. (2016). Prefrontal cortical regulation of brainwide circuit dynamics and reward-related behavior. *Science, 351*(6268), aac9698. https://doi.org/10.1126/science.aac9698.

Fergusson, D. M., Lynskey, M. T., & Horwood, J. (1996). Comorbidity between depressive disorders and nicotine dependence in a cohort of 16-year-olds. *Archives of General Psychiatry, 53*, 1043–1047.

Fink, D. S., Hu, R., Cerdá, M., Keyes, K. M., Marshall, B. D. L., Galea, S., et al. (2015). Patterns of major depression and nonmedical use of prescription opioids in the United States. *Drug and Alcohol Dependence, 153*, 258–264. https://doi.org/10.1016/j.drugalcdep.2015.05.010.

Flory, J. D., & Yehuda, R. (2015). Comorbidity between post-traumatic stress disorder and major depressive disorder: Alternative explanations and treatment considerations. *Dialogues in Clinical Neuroscience, 17*(2), 141–150.

Fox, M. E., Chandra, R., Menken, M. S., Larkin, E. J., Nam, H., Engeln, M., et al. (2020). Dendritic remodeling of D1 neurons by RhoA/Rho-kinase mediates depression-like behavior. *Molecular Psychiatry, 25*, 1022–1034. https://doi.org/10.1038/s41380-018-0211-5.

Fox, M. E., Figueiredo, A., Menken, M. S., & Lobo, M. K. (2020). Dendritic spine density is increased on nucleus accumbens D2 neurons after chronic social defeat. *Scientific Reports, 10*, 12393. https://doi.org/10.1038/s41598-020-69339-7.

Fox, M. E., & Lobo, M. K. (2019). The molecular and cellular mechanisms of depression: A focus on reward circuitry. *Molecular Psychiatry, 24*, 1798–1815. https://doi.org/10.1038/s41380-019-0415-3.

Francis, T. C., Chandra, R., Friend, D. M., Finkel, E., Dayrit, G., Miranda, J., et al. (2015). Nucleus accumbens medium spiny neuron subtypes mediate depression-related outcomes to social defeat stress. *Biological Psychiatry, 77*(3), 212–222. https://doi.org/10.1016/j.biopsych.2014.07.021.

Francis, T. C., Chandra, R., Gaynor, A., Konkalmatt, P., Metzbower, S. R., Evans, B., et al. (2017). Molecular basis of dendritic atrophy and activity in stress susceptibility. *Molecular Psychiatry, 22*(11), 1512–1519. https://doi.org/10.1038/mp.2017.178.

Francis, T. C., Gaynor, A., Chandra, R., Fox, M. E., & Lobo, M. K. (2019). The selective RhoA inhibitor rhosin promotes stress resiliency through enhancing D1-medium spiny neuron plasticity and reducing hyperexcitability. *Biological Psychiatry, 85*(12), 1001–1010. https://doi.org/10.1016/j.biopsych.2019.02.007.

Francis, T. C., & Lobo, M. K. (2017). Emerging role for nucleus accumbens medium spiny neuron subtypes in depression. *Biological Psychiatry, 81*(8), 645–653. https://doi.org/10.1016/j.biopsych.2016.09.007.

Friedman, A. K., Walsh, J. J., Juarez, B., Ku, S. M., Chaudhury, D., Wang, J., et al. (2014). Enhancing depression mechanisms in midbrain dopamine neurons achieves homeostatic resilience. *Science, 344*(6181), 313–319. https://doi.org/10.1126/science.1249240.

Fuchikami, M., Thomas, A., Liu, R., Wohleb, E. S., Land, B. B., DiLeone, R. J., et al. (2015). Optogenetic stimulation of infralimbic PFC reproduces ketamine's rapid and sustained antidepressant actions. *Proceedings of the National Academy of Sciences of the United States of America, 112*(26), 8106–8111. https://doi.org/10.1073/pnas.1414728112.

Fumagalli, F., Caffino, L., Racagni, G., & Riva, M. A. (2009). Repeated stress prevents cocaine-induced activation of BDNF signaling in rat prefrontal cortex. *European Neuropsychopharmacology*, *19*, 402–408. https://doi.org/10.1016/j.euroneuro.2009.01.003.

Gandelman, J. A., Kang, H., Antal, A., Albert, K., Boyd, B. D., Conley, A. C., et al. (2018). Transdermal nicotine for the treatment of mood and cognitive symptoms in nonsmokers with late-life depression. *Journal of Clinical Psychiatry*, *79*(5), 18m12137. https://doi.org/dx.doi.org/10.4088/JCP.14m09658.

Gilder, D. A., & Ehlers, C. L. (2012). Depression symptoms associated with cannabis dependence in an adolescent American Indian community sample. *The American Journal on Addictions*, *21*(6), 536–543. https://doi.org/10.1038/jid.2014.371.

Gokturk, C., Schultze, S., Nilsson, K. W., Von Knorring, L., Oreland, L., & Hallman, J. (2008). Serotonin transporter (5-HTTLPR) and monoamine oxidase (MAOA) promoter polymorphisms in women with severe alcoholism. *Archives of Women's Mental Health*, *11*, 347–355. https://doi.org/10.1007/s00737-008-0033-6.

Gold, P. W., Machado-Vieira, R., & Pavlatou, M. G. (2015). Clinical and biochemical manifestations of depression: Relation to the neurobiology of stress. *Neural Plasticity*, *2015*, 581976. https://doi.org/10.1155/2015/581976.

Goldner, E. M., Lusted, A., Roerecke, M., Rehm, J., & Fischer, B. (2014). Prevalence of Axis-1 psychiatric (with focus on depression and anxiety) disorder and symptomatology among non-medical prescription opioid users in substance use treatment: Systematic review and meta-analyses. *Addictive Behaviors*, *39*(3), 520–531. https://doi.org/10.1016/j.addbeh.2013.11.022.

Gómez-Coronado, N., Sethi, R., Bortolasci, C. C., Arancini, L., Berk, M., & Dodd, S. (2018). A review of the neurobiological underpinning of comorbid substance use and mood disorders. *Journal of Affective Disorders*, *241*(April), 388–401. https://doi.org/10.1016/j.jad.2018.08.041.

Gong, Q., & He, Y. (2015). Depression, neuroimaging and connectomics: A selective overview. *Biological Psychiatry*, 77(3), 223–235. https://doi.org/10.1016/j.biopsych.2014.08.009.

Granger, A. J., Wallace, M. L., & Sabatini, B. L. (2017). Multi-transmitter neurons in the mammalian central nervous system. *Current Opinion in Neurobiology*, *45*, 85–91. https://doi.org/10.1016/j.conb.2017.04.007.

Grant, B. F. (1995). Comorbidity between DSM-IV drug use disorders and major depression: Results of a national survey of adults. *Journal of Substance Abuse*, 7(4), 481–497. https://doi.org/10.1016/0899-3289(95)90017-9.

Greenberg, P. E., Fournier, A. A., Sisitsky, T., Pike, C. T., & Kessler, R. C. (2015). The economic burden of adults with major depressive disorder in the United States (2005 and 2010). *Journal of Clinical Psychiatry*, *76*(2), 155–162. https://doi.org/10.4088/JCP.14m09298.

Haass-Koffler, C. L., & Bartlett, S. E. (2012). Stress and addiction: Contribution of the corticotropin releasing factor (CRF) system in neuroplasticity. *Frontiers in Molecular Neuroscience*, *5*(Aug 2012), 1–31. https://doi.org/10.3389/fnmol.2012.00091.

Hamilton, P. J., & Nestler, E. J. (2019). Epigenetics and addiction. *Current Opinion in Neurobiology*, *59*, 128–136. https://doi.org/10.1016/j.conb.2019.05.005.

Han, C., Ahn, D., Hahm, W., Nam, J., Park, Y., Lim, S., et al. (2016). Serum levels of growth factors in alcohol-dependent patients according to comorbid depressive symptoms. *Clinical Psychopharmacology and Neuroscience*, *14*(1), 43–48. https://doi.org/10.9758/cpn.2016.14.1.43.

Han, M.-H., & Nestler, E. J. (2017). Neural substrates of depression and resilience. *Neurotherapeutics*, *14*, 677–686. https://doi.org/10.1007/s13311-017-0527-x.

Hansson, A. C., Cippitelli, A., Sommer, W. H., Ciccocioppo, R., & Heilig, M. (2007). Region-specific down-regulation of Crhr1 gene expression in alcohol-preferring msP rats following ad lib access to alcohol. *Addiction Biology*, *12*(1), 30–34. https://doi.org/10.1111/j.1369-1600.2007.00050.x.

Hare, B. D., Shinohara, R., Liu, R. J., Pothula, S., DiLeone, R. J., & Duman, R. S. (2019). Optogenetic stimulation of medial prefrontal cortex Drd1 neurons produces rapid and long-lasting antidepressant effects. *Nature Communications*, *10*(1), 1–12. https://doi.org/10.1038/s41467-018-08168-9.

Hasin, D., & Kilcoyne, B. (2012). Comorbidity of psychiatric and substance use disorders in the United States: Current issues and findings from the NESARC. *Current Opinion in Psychiatry*, *25*(3), 165–171. https://doi.org/10.1097/YCO.0b013e3283523dcc.

Havard, A., Psych, B., Teesson, M., Ph, D., Darke, S., Ph, D., et al. (2006). Depression among heroin users: 12-Month outcomes from the Australian Treatment Outcome Study (ATOS). *Journal of Substance Abuse Treatment*, *30*, 355–362. https://doi.org/10.1016/j.jsat.2006.03.012.

Hayward, M. D., Pintar, J. E., & Low, M. J. (2002). Selective reward deficit in mice lacking β-endorphin and enkephalin. *Journal of Neuroscience*, *22*(18), 8251–8258. https://doi.org/10.1523/jneurosci.22-18-08251.2002.

Heller, E. A., Cates, H. M., Peña, C. J., Sun, H., Shao, N., Feng, J., et al. (2014). Locus-specific epigenetic remodeling controls addiction- and depression-related behaviors. *Nature Neuroscience*, *17*(12), 1720–1727. https://doi.org/10.1038/nn.3871.

Hercher, C., Parent, M., Flores, C., Canetti, L., Turecki, G., & Mechawar, N. (2009). Alcohol dependence-related increase of glial cell density in the anterior cingulate cortex of suicide completers. *Journal of Psychiatry and Neuroscience*, *34*(4), 281–288.

Heshmati, M., Aleyasin, H., Menard, C., Christoffel, D. J., Flanigan, M. E., Pfau, M. L., et al. (2018). Cell-type-specific role for nucleus accumbens neuroligin-2 in depression and stress susceptibility. *Proceedings of the National Academy of Sciences of the United States of America*, *115*(5), 201719014. https://doi.org/10.1073/pnas.1719014115.

Heshmati, M., & Russo, S. J. (2015). Anhedonia and the brain reward circuitry in depression. *Current Behaviorial Neuroscience Reports*, *2*(3), 146–153. https://doi.org/10.1007/s40473-015-0044-3.Anhedonia.

Hodes, G. E., Walker, D. M., Labonté, B., Nestler, E. J., & Russo, S. J. (2017). Understanding the epigenetic basis of sex differences in depression. *Journal of Neuroscience Research*, *95*(1–2), 692–702. https://doi.org/10.1002/jnr.23876.

Hollis, F., & Kabbaj, M. (2014). Social defeat as an animal model for depression. *ILAR Journal*, *55*(2), 221–232. https://doi.org/10.1093/ilar/ilu002.

Hong, M., Ha, T. H., Lee, S., Oh, S., & Myung, W. (2019). Clinical correlates of alcohol use disorder in depressed patients with unipolar and bipolar disorder. *Psychiatry Investigation*, *16*(12), 926–932. https://doi.org/10.30773/pi.2019.0182.

Horger, B. A., Giles, M. K., & Schenk, S. (1992). Preexposure to amphetamine and nicotine predisposes rats to self-administer a low dose of cocaine. *Psychopharmacology*, *107*, 271–276. https://doi.org/10.1007/BF02245147.

Hu, H. (2016). Reward and aversion. *Annual Review of Neuroscience*, *39*, 297–324. https://doi.org/10.1146/annurev-neuro-070815-014106.

Hwa, L. S., Holly, E. N., Debold, J. F., & Miczek, K. A. (2016). Social stress-escalated intermittent alcohol drinking: Modulation by CRF-R1 in the ventral tegmental area and accumbal dopamine in mice. *Psychopharmacology*, *233*(4), 681–690. https://doi.org/10.1007/s00213-015-4144-2.

Ilango, A., & Lobo, M. K. (2015). Shining light on motivation, emotion, and memory processes. *Frontiers in Behavioral Neuroscience*, *9*, 1. https://doi.org/10.3389/fnbeh.2015.00001.

Iñiguez, S. D., Flores-Ramirez, F. J., Riggs, L. M., Alipio, J. B., Garcia-Carachure, I., Hernandez, M. A., et al. (2018). Vicarious social defeat stress induces depression-related outcomes in female mice. *Biological Psychiatry*, *83*(1), 9–17. https://doi.org/10.1016/j.biopsych.2017.07.014.

James, S. L., Abate, D., Abate, K. H., Abay, S. M., Abbafati, C., Abbasi, N., et al. (2018). Global, regional, and national incidence, prevalence, and years lived with disability for 354 Diseases and Injuries for 195 countries and territories, 1990–2017: A systematic analysis for the Global Burden of Disease Study 2017. *The Lancet*, *392*(10159), 1789–1858. https://doi.org/10.1016/S0140-6736(18)32279-7.

Janowsky, D. S., Davis, J. M., El-Yousef, M. K., & Sekerke, H. J. (1972). A cholinergic-adrenergic hypothesis of mania and depression. *The Lancet*, *2*(7778), 632–635.

Johnston, J. H., Linden, D. E. J., & van den Bree, M. B. M. (2016). Combining stress and dopamine based models of addiction: Towards a psycho-neuro-endocrinological theory of addiction. *Current Drug Abuse Reviews*, *9*, 61–74.

Keiflin, R., & Janak, P. H. (2015). Dopamine prediction errors in reward learning and addiction: From theory to neural circuitry. *Neuron*, *88*(2), 247–263. https://doi.org/10.1016/j.neuron.2015.08.037.

Kertes, D. A., Kalsi, G., Prescott, C. A., Kuo, P. H., Patterson, D. G., Walsh, D., et al. (2011). Neurotransmitter and neuromodulator genes associated with a history of depressive symptoms in individuals with alcohol dependence. *Alcoholism: Clinical and Experimental Research*, *35*(3), 496–505. https://doi.org/10.1111/j.1530-0277.2010.01366.x.

Kessler, R. C., Wai, T. C., Demler, O., & Walters, E. E. (2005). Prevalence, severity, and comorbidity of 12-month DSM-IV disorders in the National Comorbidity Survey Replication. *Archives of General Psychiatry*, *62*(6), 617–627. https://doi.org/10.1001/archpsyc.62.6.617.

Khibnik, L. A., Beaumont, M., Doyle, M., Heshmati, M., Slesinger, P. A., Nestler, E. J., et al. (2016). Stress and cocaine trigger divergent and cell type-specific regulation of synaptic transmission at single spines in nucleus accumbens. *Biological Psychiatry*, *79*(11), 898–905. https://doi.org/10.1016/j.biopsych.2015.05.022.

Kim, S., Kwok, S., Mayes, L. C., Potenza, M. N., Rutherford, H. J. V., & Strathearn, L. (2017). Early adverse experience and substance addiction: Dopamine, oxytocin, and glucocorticoid pathways. *Annals of the New York Academy of Sciences*, *1394*(1), 74–91. https://doi.org/10.1111/nyas.13140.

Knowland, D., Lilascharoen, V., Pacia, C. P., Shin, S., Wang, E. H. J., & Lim, B. K. (2017). Distinct ventral pallidal neural populations mediate separate symptoms of depression. *Cell*, *170*(2), 284–297.e18. https://doi.org/10.1016/j.cell.2017.06.015.

Knowland, D., & Lim, B. K. (2018). Circuit-based frameworks of depressive behaviors: The role of reward circuitry and beyond. *Pharmacology Biochemistry and Behavior*, *174*, 42–52. https://doi.org/10.1016/j.pbb.2017.12.010.

Koo, J. W., Labonté, B., Engmann, O., Calipari, E. S., Juarez, B., Lorsch, Z., et al. (2016). Essential role of mesolimbic brain-derived neurotrophic factor in chronic social stress–induced depressive behaviors. *Biological Psychiatry*, *80*(6), 469–478. https://doi.org/10.1016/j.biopsych.2015.12.009.

Koo, J. W., Russo, S. J., Ferguson, D., Nestler, E. J., & Duman, R. S. (2010). Nuclear factor-κB is a critical mediator of stress-impaired neurogenesis and depressive behavior. *Proceedings of the National Academy of Sciences of the United States of America*, *107*(6), 2669–2674. https://doi.org/10.1073/pnas.0910658107.

Koob, G. F. (2009). Dynamics of neuronal circuits in addiction: Reward, antireward, and emotional memory. *Pharmacopsychiatry*, *42*(Suppl. 1), 32–41. https://doi.org/10.1055/s-0029-1216356.

Koob, G. F. (2013). Theoretical frameworks and mechanistic aspects of alcohol addiction: Alcohol addiction as a reward deficit disorder. *Current Topics in Behavioral Neurosciences*, *13*, 3–30. https://doi.org/10.1007/7854.

Koob, G. F. (2020). Neurobiology of opioid addiction: Opponent process, hyperkatifeia, and negative reinforcement. *Biological Psychiatry*, *87*(1), 44–53. https://doi.org/10.1016/j.biopsych.2019.05.023.

Koob, G. F., & Kreek, M. J. (2007). Stress, dysregulation of drug reward pathways, and the transition to drug dependence. *American Journal of Psychiatry*, *164*(8), 1149–1159. https://doi.org/10.1176/appi.ajp.2007.05030503.

Koob, G. F., & Moal, M. L. (2001). Drug addiction, dysregulation of reward, and allostasis. *Neuropsychopharmacology*, *24*(2), 97. https://doi.org/10.1016/S0893-133X(00)00195-0.

Korhonen, T., Loukola, A., Wedenoja, J., Nyman, E., Latvala, A., Broms, U., et al. (2014). Role of nicotine dependence in the association between the dopamine receptor gene DRD3 and major depressive disorder. *PLoS One*, *9*(6), e98199. https://doi.org/10.1371/journal.pone.0098199.

Korosi, A., Shanabrough, M., McClelland, S., Liu, Z.-W., Borok, E., Gao, X.-B., et al. (2010). Early-life experience reduces excitation to stress-responsive hypothalamic neurons and reprograms the expression of corticotropin-releasing hormone. *The Journal of Neuroscience: The Official Journal of the Society for Neuroscience*, *30*(2), 703–713. https://doi.org/10.1523/JNEUROSCI.4214-09.2010.

Kreisel, T., Frank, M. G., Licht, T., Reshef, R., Ben-Menachem-Zidon, O., Baratta, M. V., et al. (2014). Dynamic microglial alterations underlie stress-induced depressive-like behavior and suppressed neurogenesis. *Molecular Psychiatry*, *19*(6), 699–709. https://doi.org/10.1038/mp.2013.155.

Kringelbach, M. L., & Berridge, K. C. (2009). Towards a functional neuroanatomy of pleasure and happiness. *Trends in Cognitive Sciences*, *13*(11), 479–487. https://doi.org/10.1016/j.tics.2009.08.006.

Krishnan, V., Han, M.-H., Graham, D. L., Berton, O., Renthal, W., Russo, S. J., et al. (2007). Molecular adaptations underlying susceptibility and resistance to social defeat in brain reward regions. *Cell*, *131*(2), 391–404. https://doi.org/10.1016/j.cell.2007.09.018.

Krishnan-Sarin, S., Morean, M. E., Camenga, D. R., Cavallo, D. A., & Kong, G. (2015). E-cigarette use among high school and middle school adolescents in connecticut. *Nicotine & Tobacco Research: Official Journal of the Society for Research on Nicotine and Tobacco*, *17*(7), 810–818. https://doi.org/10.1093/ntr/ntu243.

Kumar, A., Choi, K. H., Renthal, W., Tsankova, N. M., Theobald, D. E. H., Truong, H. T., et al. (2005). Chromatin remodeling is a key mechanism underlying cocaine-induced plasticity in striatum. *Neuron*, *48*(2), 303–314. https://doi.org/10.1016/j.neuron.2005.09.023.

Kupchik, Y. M., Brown, R. M., Heinsbroek, J.a., Lobo, M. K., Schwartz, D. J., & Kalivas, P. W. (2015). Coding the direct/indirect pathways by D1 and D2 receptors is not valid for accumbens projections. *Nature Neuroscience*, *18*, 1230–1232. https://doi.org/10.1038/nn.4068.

Kutlu, M. G., Parikh, V., & Gould, T. J. (2015). Nicotine addiction and psychiatric disorders. In M. De Biasi (Ed.), *Vol. 124. Nicotine use in mental illness and neurological disorders* Elsevier Inc., pp. 171–208. https://doi.org/10.1016/bs.irn.2015.08.004.

Lammel, S., Lim, B. K., Ran, C., Huang, K. W., Betley, M. J., Tye, K. M., et al. (2012). Input-specific control of reward and aversion in the ventral tegmental area. *Nature*, *491*(7423), 212–217. https://doi.org/10.1038/nature11527.

LaPlant, Q., Vialou, V., Covington, H. E., Dumitriu, D., Feng, J., Warren, B. L., et al. (2010). Dnmt3a regulates emotional behavior and spine plasticity in the nucleus accumbens. *Nature Neuroscience*, *13*(9), 1137–1143. https://doi.org/10.1038/nn.2619.

Lee, A. M., Calarco, C. A., McKee, S. A., Mineur, Y. S., & Picciotto, M. R. (2019). Variability in nicotine conditioned place preference and stress-induced reinstatement in mice: Effects of sex, initial chamber preference, and guanfacine. *Genes, Brain and Behavior*, *19*(3), e12601. https://doi.org/10.1111/gbb.12601.

LeGates, T. A., Kvarta, M. D., Tooley, J. R., Francis, T. C., Lobo, M. K., Creed, M. C., et al. (2018). Reward behaviour is regulated by the strength of hippocampus–nucleus accumbens synapses. *Nature*, *564*, 258–262. https://doi.org/10.1038/s41586-018-0740-8.

Lenz, J. D., & Lobo, M. K. (2013). Optogenetic insights into striatal function and behavior. *Behavioural Brain Research*, *55*, 44–54. https://doi.org/10.1016/j.bbr.2013.04.018.

Leventhal, A. M., Mooney, M. E., DeLaune, K. A., & Schmitz, J. M. (2006). Using addiction severity profiles to differentiate cocaine-dependent patients with and without comorbid major depression. *American Journal on Addictions*, *15*(5), 362–369. https://doi.org/10.1080/10550490600860148.

Li, X., Venniro, M., & Shaham, Y. (2016). Translational research on incubation of cocaine craving. *JAMA Psychiatry*, *73*(11), 1115–1116. https://doi.org/10.1016/j.physbeh.2017.03.040.

Li, X., & Wolf, M. E. (2015). Multiple faces of BDNF in cocaine addiction. *Behavioral Brain Research*, *279*(February 15), 240–254. https://doi.org/10.1016/j.bbr.2014.11.018.

Lim, B. K., Huang, K. W., Grueter, B.a., Rothwell, P. E., & Malenka, R. C. (2012). Anhedonia requires MC4R-mediated synaptic adaptations in nucleus accumbens. *Nature*, *487*(7406), 183–189. https://doi.org/10.1038/nature11160.

Lin, S. C., Wu, P. L., Ko, H. C., Wu, J. Y. W., Huang, S. Y., Lin, W. W., et al. (2007). Specific personality traits and dopamine, serotonin genes in anxiety-depressive alcoholism among Han Chinese in Taiwan. *Progress in Neuro-Psychopharmacology and Biological Psychiatry*, *31*(7), 1526–1534. https://doi.org/10.1016/j.pnpbp.2007.07.012.

Liu, X., & Weiss, F. (2002). Additive effect of stress and drug cues on reinstatement of ethanol seeking: Exacerbation by history of dependence and role of concurrent activation of corticotropin-releasing factor and opioid mechanisms. *Journal of Neuroscience*, *22*(18), 7856–7861. https://doi.org/10.1523/jneurosci.22-18-07856.2002.

Lobo, M. K., Zaman, S., Damez-Werno, D. M., Koo, J. W., Bagot, R. C., Dinieri, J. A., et al. (2013). ΔFosB induction in striatal medium spiny neuron subtypes in response to chronic pharmacological, emotional, and optogenetic stimuli. *The Journal of Neuroscience: The Official Journal of the Society for Neuroscience*, *33*(47), 18381–18395. https://doi.org/10.1523/JNEUROSCI.1875-13.2013.

Logrip, M. L., Koob, G. F., & Zorrilla, E. P. (2011). Role of corticotropin-releasing factor in drug addiction: Potential for pharmacological intervention. *CNS Drugs*, *25*(4), 271–287. https://doi.org/10.2165/11587790-000000000-00000.

Loonen, A. J. M., & Ivanova, S. A. (2016a). Circuits regulating pleasure and happiness in major depression. *Medical Hypotheses*, *87*, 14–21. https://doi.org/10.1016/j.mehy.2015.12.013.

Loonen, A. J. M., & Ivanova, S. A. (2016b). Circuits regulating pleasure and happiness—Mechanisms of depression. *Frontiers in Human Neuroscience*, *10*(Nov 2016), 571. https://doi.org/10.3389/fnhum.2016.00571.

Lopez, M. F., Anderson, R. I., & Becker, H. C. (2016). Effect of different stressors on voluntary ethanol intake in ethanol-dependent and nondependent C57BL/6J mice. *Alcohol*, *51*, 17–23. https://doi.org/10.1016/j.alcohol.2015.11.010.

Lüscher, C. (2016). The emergence of a circuit model for addiction. *Annual Review of Neuroscience*, *39*, 257–276. https://doi.org/10.1146/annurev-neuro-070815-013920.

Lüthi, A., & Lüscher, C. (2014). Pathological circuit function underlying addiction and anxiety disorders. *Nature Neuroscience, 17*(12), 1635–1643. https://doi.org/10.1038/nn.3849.

Mague, S. D., Pliakas, A. M., Todtenkopf, M. S., Tomasiewicz, H. C., Zhang, Y., Stevens, W. C. J., et al. (2003). Antidepressant-like effects of κ-opioid receptor antagonists in the forced swim test in rats. *Journal of Pharmacology and Experimental Therapeutics, 305*(1), 323–330. https://doi.org/10.1124/jpet.102.046433.

Makino, S., Hashimoto, K., & Gold, P. W. (2002). Multiple feedback mechanisms activating corticotropin-releasing hormone system in the brain during stress. *Pharmacology Biochemistry and Behavior, 73*(1), 147–158. https://doi.org/10.1016/S0091-3057(02)00791-8.

Mantsch, J. R., Baker, D. A., Funk, D., Lê, A. D., & Shaham, Y. (2016). Stress-induced reinstatement of drug seeking: 20 years of progress. *Neuropsychopharmacology, 41*, 335–356. https://doi.org/10.1038/npp.2015.142.

Manvich, D. F., Stowe, T. A., Godfrey, J. R., & Weinshenker, D. (2016). A method for psychosocial stress-induced reinstatement of cocaine seeking in rats. *Biological Psychiatry, 79*(11), 940–946. https://doi.org/10.1016/j.biopsych.2015.07.002.

Martin, L. A., Neighbors, H. W., & Griffith, D. M. (2013). The experience of symptoms of depression in men vs women: Analysis of the national comorbidity survey replication. *JAMA Psychiatry, 70*(10), 1100–1106. https://doi.org/10.1001/jamapsychiatry.2013.1985.

Martins, S. S., Fenton, M. C., Keyes, K. M., Blanco, C., Zhu, H., & Storr, C. L. (2012). Mood and anxiety disorders and their association with non-medical prescription opioid use and prescription opioid-use disorder: Longitudinal evidence from the National Epidemiologic Study on Alcohol and Related Conditions. *Psychological Medicine, 42*(6), 1261–1272. https://doi.org/10.1017/S0033291711002145.

Marton, T. F., & Sohal, V. S. (2016). Of mice, men, and microbial opsins: How optogenetics can help hone mouse models of mental illness. *Biological Psychiatry, 79*(1), 47–52. https://doi.org/10.1016/j.biopsych.2015.04.012.

Mathew, A. R., Hogarth, L., Leventhal, A. M., Cook, J. W., & Hitsman, B. (2017). Cigarette smoking and depression comorbidity: Systematic review and proposed theoretical model. *Addiction, 112*(3), 401–412. https://doi.org/10.1111/add.13604.

McCabe, S. E., & West, B. T. (2017). The 3-year course of multiple substance use disorders in the United States: A national longitudinal study. *Journal of Clinical Psychiatry, 78*(5), e537–e544. https://doi.org/10.4088/JCP.16m10657.

Mccarthy, D. M., Brown, A. N., & Bhide, P. G. (2012). Regulation of BDNF expression by cocaine. *Yale Journal of Biology and Medicine, 85*, 437–446.

McGinty, J. F., Whitfield, T. W., Jr., & Berglind, W. J. (2010). Brain-derived neurotrophic factor and cocaine addiction. *Brain Research, 1314*, 183–193. https://doi.org/10.1016/j.brainres.2009.08.078.

Mcgrath, P. J., & Miller, J. M. (2010). Co-occurring anxiety and depression: Concepts, significance, and treatment implications. In H. B. Simpson, Y. Neria, R. Lewis-Fernandez, & F. Schneier (Eds.), *Anxiety disorders: Theory, research, and clinical perspectives* (pp. 90–102). Cambridge University Press.

McLaughlin, I., Dani, J. A., & De Biasi, M. (2017). The medial habenula and interpeduncular nucleus circuitry is critical in addiction, anxiety, and mood regulation. *Journal of Neurochemistry, 142*, 130–143. https://doi.org/10.1111/jnc.14008.

Melchior, M., Prokofyeva, E., Younès, N., Surkan, P. J., & Martins, S. S. (2014). Treatment for illegal drug use disorders: The role of comorbid mood and anxiety disorders. *BMC Psychiatry, 14*, 89. https://doi.org/10.1186/1471-244X-14-89.

Miczek, K. A., Nikulina, E. M., Shimamoto, A., & Covington, H. E. (2011). Escalated or suppressed cocaine reward, tegmental BDNF, and accumbal dopamine caused by episodic versus continuous social stress in rats. *Journal of Neuroscience*, *31*(27), 9848–9857. https://doi.org/10.1523/JNEUROSCI.0637-11.2011.

Miguel-Hidalgo, J. J., Wei, J., Andrew, M., Overholser, J. C., Jurjus, G., Stockmeier, C. A., et al. (2002). Glia pathology in the prefrontal cortex in alcohol dependence with and without depressive symptoms. *Biological Psychiatry*, *52*, 1121–1133. https://doi.org/10.1016/S0006-3223(02)01439-7.

Mineur, Y. S., Obayemi, A., Wigestrand, M. B., Fote, G. M., Calarco, C. A., Li, A. M., et al. (2013). Cholinergic signaling in the hippocampus regulates social stress resilience and anxiety- and depression-like behavior. *Proceedings of the National Academy of Sciences of the United States of America*, *110*(9), 3573–3578. https://doi.org/10.1073/pnas.1219731110.

Mineur, Y. S., & Picciotto, M. R. (2010). Nicotine receptors and depression: Revisiting and revising the cholinergic hypothesis. *Trends in Pharmacological Sciences*, *31*(12), 580–586. https://doi.org/10.1016/j.tips.2010.09.004.

Molet, J., Maras, P. M., Avishai-Eliner, S., & Baram, T. Z. (2014). Naturalistic rodent models of chronic early-life stress. *Developmental Psychobiology*, *56*(8), 1675–1688. https://doi.org/10.1002/dev.21230.

Monteggia, L. M. (2016). Toward better animal models for molecular psychiatry. *Biological Psychiatry*, *79*(1), 2–3. https://doi.org/10.1016/j.biopsych.2015.10.019.

Moonat, S., & Pandey, S. C. (2012). Stress, epigenetics, and alcoholism. *Alcohol Research*, *34*(4), 495–505.

Morisano, D., Bacher, I., Audrain-McGovern, J., & George, T. P. (2009). Mechanisms underlying the comorbidity of tobacco use in mental health and addictive disorders. *The Canadian Journal of Psychiatry*, *54*(6), 356–367.

Muir, J., Lopez, J., & Bagot, R. C. (2019). Wiring the depressed brain: Optogenetic and chemogenetic circuit interrogation in animal models of depression. *Neuropsychopharmacology*, *44*, 1013–1026. https://doi.org/10.1038/s41386-018-0291-6.

Mukhara, D., Banks, M. L., & Neigh, G. N. (2018). Stress as a risk factor for substance use disorders: A mini-review of molecular mediators. *Frontiers in Behavioral Neuroscience*, *12*, 309. https://doi.org/10.3389/fnbeh.2018.00309.

Nam, H., Chandra, R., Francis, T. C., Dias, C., Cheer, J. F., & Lobo, M. K. (2019). Reduced nucleus accumbens enkephalins underlie vulnerability to social defeat stress. *Neuropsychopharmacology*, *44*(11), 1876–1885. https://doi.org/10.1038/s41386-019-0422-8.

National Institute on Drug Abuse. (2018). *Common comorbidities with substance use disorders* (p. 56). (February), Retrieved from https://www.drugabuse.gov/%0D.

Negus, S. S., & Miller, L. L. (2014). Intracranial self-stimulation to evaluate abuse potential of drugs. *Pharmacological Reviews*, *66*, 869–917.

Nellissery, M., Feinn, R. S., Covault, J., Gelernter, J., Anton, R. F., Pettinati, H., et al. (2003). Alleles of a functional serotonin transporter promoter polymorphism are associated with major depression in alcoholics. *Alcoholism: Clinical and Experimental Research*, *27*(9), 1402–1408. https://doi.org/10.1097/01.ALC.0000085588.11073.BB.

Nestler, E. J. (2012). Transcriptional mechanisms of drug addiction. *Clinical Psychopharmacology and Neuroscience*, *10*(3), 136–143.

Nestler, E. J. (2013). ΔFosB: A molecular switch for reward. *Journal of Drug and Alcohol Research*, *2*, 235651. https://doi.org/10.4303/jdar/235651.

Nestler, E. J. (2015a). Role of the brain's reward circuitry in depression: Transcriptional mechanisms. In *Vol. 124. Nicotine use in mental illness and neurological disorders* (1st ed.). Elsevier Inc. https://doi.org/10.1016/bs.irn.2015.07.003.

Nestler, E. J. (2015b). ΔFosB: A transcriptional regulator of stress and antidepressant response. *European Journal of Pharmacology*, *753*, 66–72. https://doi.org/10.1016/j.ejphar.2014.10.034.

Nestler, E. J., Barrot, M., Dileone, R. J., Eisch, A. J., Gold, S. J., & Monteggia, L. M. (2002). Neurobiology of depression. *Cell*, *34*, 13–25.

Neupane, S. P., Lien, L., Martinez, P., Aukrust, P., Ueland, T., Mollnes, T. E., et al. (2014). High frequency and intensity of drinking may attenuate increased inflammatory cytokine levels of major depression in alcohol-use disorders. *CNS Neuroscience and Therapeutics*, *20*(10), 898–904. https://doi.org/10.1111/cns.12303.

Neupane, S. P., Lien, L., Ueland, T., Mollnes, T. E., Aukrust, P., & Bramness, J. G. (2015). Serum brain-derived neurotrophic factor levels in relation to comorbid depression and cytokine levels in Nepalese men with alcohol-use disorders. *Alcohol*, *49*(5), 471–478. https://doi.org/10.1016/j.alcohol.2015.01.012.

Newman, E. L., Albrechet-Souza, L., Andrew, P. M., Auld, J. G., Burk, K. C., Hwa, L. S., et al. (2018). Persistent escalation of alcohol consumption by mice exposed to brief episodes of social defeat stress: Suppression by CRF-R1 antagonism. *Psychopharmacology*, *235*(6), 1807–1820. https://doi.org/10.1007/s00213-018-4905-9.

Newman, E. L., Leonard, M. Z., Arena, D. T., De Almeida, R. M. M., & Miczek, K. A. (2018). Neurobiology of stress social defeat stress and escalation of cocaine and alcohol consumption: Focus on CRF. *Neurobiology of Stress*, *9*(September), 151–165. https://doi.org/10.1016/j.ynstr.2018.09.007.

Newton, S. S., Thome, J., Wallace, T. L., Shirayama, Y., Schlesinger, L., Sakai, N., et al. (2002). Inhibition of cAMP response element-binding protein or dynorphin in the nucleus accumbens produces an antidepressant-like effect. *Journal of Neuroscience*, *22*(24), 10883–10890. https://doi.org/10.1523/jneurosci.22-24-10883.2002.

Nielsen, D. A., Utrankar, A., Reyes, J. A., Simons, D. D., & Kosten, T. R. (2012). Epigenetics of drug abuse: Predisposition or response. *Pharmacogenetics*, *13*(10), 1149–1160. https://doi.org/10.2217/pgs.12.94.

Nishi, M., Horii-hayashi, N., & Sasagawa, T. (2014). Effects of early life adverse experiences on the brain: Implications from maternal separation models in rodents. *Frontiers in Neuroscience*, *8*, 166. https://doi.org/10.3389/fnins.2014.00166.

O'Connor, E. C., Kremer, Y., Lefort, S., Harada, M., Pascoli, V., Rohner, C., et al. (2015). Accumbal D1R neurons projecting to lateral hypothalamus authorize feeding. *Neuron*, *88*(3), 553–564. https://doi.org/10.1016/j.neuron.2015.09.038.

Olfson, M., Liu, S.-M., Grant, B. F., & Carlos, B. (2012). Influence of comorbid mental disorders on time to treatment seeking for major depressive disorder. *Medical Care*, *50*(3), 227–232. https://doi.org/10.1097/MLR.0b013e318241eb5e.Influence.

Pandya, M., Altinay, M., Malone, D. A., & Anand, A. (2012). Where in the brain is depression? *Current Psychiatry Reports*, *14*(6), 634–642. https://doi.org/10.1007/s11920-012-0322-7.

Papaleo, F., Kieffer, B. L., Tabarin, A., & Contarino, A. (2007). Decreased motivation to eat in μ-opioid receptor-deficient mice. *European Journal of Neuroscience*, *25*(11), 3398–3405. https://doi.org/10.1111/j.1460-9568.2007.05595.x.

Pedrelli, P., Shapero, B., Archibald, A., & Dale, C. (2016). Alcohol use and depression during adolescence and young adulthood: A summary and interpretation of mixed findings. *Current Addiction Reports*, *3*(1), 91–97. https://doi.org/10.1016/j.physbeh.2017.03.040.

Peña, C. J., Kronman, H. G., Walker, D. M., Cates, H. M., Bagot, R. C., Purushothaman, I., et al. (2017). Early life stress confers lifelong stress susceptibility in mice via ventral tegmental area OTX2. *Science*, *356*(6343), 1185–1188. https://doi.org/10.1126/science.aan4491.

Peña, C. J., Smith, M., Ramakrishnan, A., Cates, H. M., Bagot, R. C., Kronman, H. G., et al. (2019). Early life stress alters transcriptomic patterning across reward circuitry in male and female mice. *Nature Communications*, *10*, 5098. https://doi.org/10.1038/s41467-019-13085-6.

Perova, Z., Delevich, K., & Li, B. (2015). Depression of excitatory synapses onto parvalbumin interneurons in the medial prefrontal cortex in susceptibility to stress. *Journal of Neuroscience*, *35*(7), 3201–3206. https://doi.org/10.1523/JNEUROSCI.2670-14.2015.

Peters, S. K., Dunlop, K., & Downar, J. (2016). Cortico-striatal-thalamic loop circuits of the salience network: A central pathway in psychiatric disease and treatment. *Frontiers in Systems Neuroscience*, *10*, 104. https://doi.org/10.3389/fnsys.2016.00104.

Pettinati, H. M., O'Brien, C. P., & Dundon, W. D. (2013). Current status of co-occurring mood and substance use disorders: A new therapeutic target. *American Journal of Psychiatry*, *170*(1), 23–30. https://doi.org/10.1176/appi.ajp.2012.12010112.

Philip, N. S., Carpenter, L. L., Tyrka, A. R., & Price, L. H. (2010). Nicotinic acetylcholine receptors and depression: A review of the preclinical and clinical literature. *Psychopharmacology*, *212*(1), 1–12. https://doi.org/10.1007/s00213-010-1932-6.

Piazza, P. V., & Moal, M. L. (1998). The role of stress in drug self-administration. *Trends in Pharmacological Sciences*, *19*(2), 67–74.

Picciotto, M. R., Lewis, A. S., Van Schalkwyk, G. I., & Mineur, Y. S. (2015). Neuropharmacology mood and anxiety regulation by nicotinic acetylcholine receptors: A potential pathway to modulate aggression and related behavioral states. *Neuropharmacology*, *96*, 235–243. https://doi.org/10.1016/j.neuropharm.2014.12.028.

Pignatelli, M., Tejeda, H. A., Barker, D. J., Bontempi, L., Wu, J., Lopez, A., et al. (2020). Cooperative synaptic and intrinsic plasticity in a disynaptic limbic circuit drive stress-induced anhedonia and passive coping in mice. *Molecular Psychiatry*. https://doi.org/10.1038/s41380-020-0686-8.

Potokar, J., Coupland, N., Glue, P., Groves, S., Malizia, A., Bailey, J., et al. (1997). Flumazenil in alcohol withdrawal: A double-blind placebo-controlled study. *Alcohol and Alcoholism*, *32*(5), 605–611. https://doi.org/10.1093/oxfordjournals.alcalc.a008302.

Quello, S. B., Brady, K. T., & Sonne, S. C. (2005). Mood disorders and substance use disorder: A complex comorbidity. *Science & Practice Perspectives*, *3*(1), 13–21. https://doi.org/10.1151/spp053113.

Rao, U. (2010). Comorbidity between depressive and addictive disorders in adolescents: Role of stress and HPA activity. *US Psychiatry*, *3*, 39–43.

Rao, U., Daley, S. E., & Hammen, C. (1998). Relationship between depression and substance use disorders in adolescent women during the transition to adulthood. *Journal of the American Academy of Child & Adolescent Psychiatry*, *39*(2), 215–222. https://doi.org/10.1097/00004583-200002000-00022.

Rao, U., Hammen, C. L., & Poland, R. E. (2009). Mechanisms underlying the comorbidity between depressive and addictive disorders in adolescents: Interactions between stress and HPA activity. *American Journal of Psychiatry*, *166*, 361–369.

Reiche, S., Hermle, L., Gutwinski, S., Jungaberle, H., Gasser, P., & Majić, T. (2018). Serotonergic hallucinogens in the treatment of anxiety and depression in patients suffering from a life-threatening disease: A systematic review. *Progress in Neuropsychopharmacology and Biological Psychiatry*, *81*, 1–10. https://doi.org/10.1016/j.pnpbp.2017.09.012.

Renthal, W., Maze, I., Krishnan, V., Covington, H. E., Xiao, G., Kumar, A., et al. (2007). Histone deacetylase 5 epigenetically controls behavioral adaptations to chronic emotional stimuli. *Neuron*, *56*(3), 517–529. https://doi.org/10.1016/j.neuron.2007.09.032.

Ritter, A. J. (2002). Naltrexone in the treatment of heroin dependence: Relationship with depression and risk of overdose. *Australian and New Zealand Journal of Psychiatry*, *36*, 224–228.

Robbins, T. W. (2016). Illuminating anhedonia. *Science*, *351*(6268), 24–25. https://doi.org/10.1126/science.aad9698.

Robison, A. J., & Nestler, E. J. (2011). Transcriptional and epigenetic mechanisms of addiction. *Nature Reviews Neuroscience*, *12*(11), 623–637. https://doi.org/10.1038/nrn3111.

Roy, A., Dejong, J., & Ferraro, T. (1991). CSF GABA in depressed patients and normal controls. *Psychological Medicine*, *21*, 613–618. https://doi.org/10.1017/S0033291700022248.
Roy, A., Dejong, J., Lamparski, D., George, T., & Linnoila, M. (1991). Depression among alcoholics: Relationship to clinical and cerebrospinal fluid variables. *Archives of General Psychiatry*, *48*, 428–432.
Russo, S. J., & Nestler, E. J. (2013). The brain reward circuitry in mood disorders. *Nature Reviews Neuroscience*, *14*(9), 609–625. https://doi.org/10.1038/nrn3381.
Savitz, J., & Drevets, W. C. (2009). Bipolar and major depressive disorder: Neuroimaging the developmental-degenerative divide. *Neuroscience and Biobehavioral Reviews*, *33*(5), 699–771. https://doi.org/10.1016/j.neubiorev.2009.01.004.
Scheggi, S., De Montis, M. G., & Gambarana, C. (2018). Making sense of rodent models of anhedonia. *International Journal of Neuropsychopharmacology*, *21*(11), 1049–1065. https://doi.org/10.1093/ijnp/pyy083.
Schepis, T. S., & Hakes, J. K. (2011). Non-medical prescription use increases the risk for the onset and recurrence of psychopathology: Results from the National Epidemiological Survey on Alcohol. *Addiction*, *106*(12), 2146–2155. https://doi.org/10.1111/j.1360-0443.2011.03520.x.
Schmidt, H. D., Mcginty, J. F., West, A. E., & Sadri-Vakili, G. (2014). Epigenetics and psychostimulant addiction. *Psychopharmacology*, *231*, 4031–4039.
Shaham, Y., Erb, S., & Stewart, J. (2000). Stress-induced relapse to heroin and cocaine seeking in rats: A review. *Brain Research Reviews*, *33*, 13–33.
Sheline, Y. I., Wang, P. W., Gado, M. H., Csernansky, J. G., & Vannier, M. W. (1996). Hippocampal atrophy in recurrent major depression. *Proceedings of the National Academy of Sciences of the United States of America*, *93*, 3908–3913.
Shimamoto, A., Holly, E. N., Boyson, C. O., Debold, J. F., & Miczek, K. A. (2015). Individual differences in anhedonic and accumbal dopamine responses to chronic social stress and their link to cocaine self-administration in female rats. *Psychopharmacology*, *232*(4), 825–834. https://doi.org/10.1007/s00213-014-3725-9.
Sinha, R., Garcia, M., Paliwal, P., Kreek, M. J., & Rounsaville, B. J. (2006). Stress-induced cocaine craving and hypothalamic-pituitary-adrenal responses are predictive of cocaine relapse outcomes. *Archives of General Psychiatry*, *63*, 324–331.
Soares-Cunha, C., de Vasconcelos, N. A. P., Coimbra, B., Domingues, A. V., Silva, J. M., Loureiro-Campos, E., et al. (2019). Nucleus accumbens medium spiny neurons subtypes signal both reward and aversion. *Molecular Psychiatry*. https://doi.org/10.1038/s41380-019-0484-3.
Steinberg, E. E., Christoffel, D. J., Deisseroth, K., & Malenka, R. C. (2014). Illuminating circuitry relevant to psychiatric disorders with optogenetics. *Current Opinion in Neurobiology*, *30C*, 9–16. https://doi.org/10.1016/j.conb.2014.08.004.
Steinberg, E. E., Keiflin, R., Boivin, J. R., Witten, I. B., Deisseroth, K., & Janak, P. H. (2013). A causal link between prediction errors, dopamine neurons and learning. *Nature Neuroscience*, *16*(7), 966–973. https://doi.org/10.1038/nn.3413.
Stewart, J. (2000). Pathways to relapse: The neurobiology of drug- and stress-induced relapse to drug taking. *Journal of Psychiatry and Neuroscience*, *25*(1), 125–136.
Su, N., Zhang, L., Fei, F., Hu, H., Wang, K., Hui, H., et al. (2011). The brain-derived neurotrophic factor is associated with alcohol dependence-related depression and antidepressant response. *Brain Research*, *1415*, 119–126. https://doi.org/10.1016/j.brainres.2011.08.005.
Substance Abuse and Mental Health Services Administration. (2018). *Key substance use and mental health indicators in the United States: Results from the 2015 National Survey on Drug Use and Health. Publication No. SMA 16-4984, NSDUH Series H-51* (pp. 1–97). Center for Behavioral Health Statistics and Quality, Substance Abuse and Mental Health Services Administration. https://doi.org/10.1016/j.drugalcdep.2016.10.042.

Substance Abuse and Mental Health Services Administration. (2020). *Substance use disorder treatment for people with co-occurring disorders*. Substance Abuse and Mental Health Services Administration.

Swendsen, J. D., & Merikangas, K. R. (2000). The comorbidity of depression and substance use disorders. *Clinical Psychology Review*, *20*(2), 173–189. https://doi.org/10.1016/S0272-7358(99)00026-4.

Tate, S. R., Wu, J., McQuaid, J. R., Cummins, K., Shriver, C., Krenek, M., et al. (2008). Comorbidity of substance dependence and depression: Role of life stress and self-efficacy in sustaining abstinence. *Psychology of Addictive Behaviors*, *22*(1), 47–57. https://doi.org/10.1037/0893-164X.22.1.47.

Tidey, J. W., & Miczek, K. A. (1996). Social defeat stress selectively alters mesocorticolimbic dopamine release: An in vivo microdialysis study. *Brain Research*, *721*(1–2), 140–149. https://doi.org/10.1016/0006-8993(96)00159-X.

Torrens, M., Fonseca, F., Mateu, G., & Farré, M. (2005). Efficacy of antidepressants in substance use disorders with and without comorbid depression: A systematic review and meta-analysis. *Drug and Alcohol Dependence*, *78*(1), 1–22. https://doi.org/10.1016/j.drugalcdep.2004.09.004.

Torrens, M., Gilchrist, G., & Domingo-Salvany, A. (2011). Psychiatric comorbidity in illicit drug users: Substance-induced versus independent disorders. *Drug and Alcohol Dependence*, *113*(2–3), 147–156. https://doi.org/10.1016/j.drugalcdep.2010.07.013.

Tseng, I., Ganz, A., Mitton, A. G., & Tsuang, J. (2017). Comorbidity of alcohol use disorder and depression: A case report and review of the literature. *Addictive Disorders and Their Treatment*, *16*(3), 121–128. https://doi.org/10.1097/ADT.0000000000000106.

Tye, K. M., & Deisseroth, K. (2012). Optogenetic investigation of neural circuits underlying brain disease in animal models. *Nature Reviews Neuroscience*, *13*(4), 251–266. https://doi.org/10.1038/nrn3171.

Tye, K. M., Mirzabekov, J. J., Warden, M. R., Ferenczi, E. A., Tsai, H., Finkelstein, J., et al. (2013). Dopamine neurons modulate neural encoding and expression of depression-related behaviour. *Nature*, *493*, 537–541. https://doi.org/10.1038/nature11740.

Tynan, R. J., Naicker, S., Hinwood, M., Nalivaiko, E., Buller, K. M., Pow, D. V., et al. (2010). Chronic stress alters the density and morphology of microglia in a subset of stress-responsive brain regions. *Brain, Behavior, and Immunity*, *24*(7), 1058–1068. https://doi.org/10.1016/j.bbi.2010.02.001.

Uluğ, B. D. (2015). The challenge behind the facade: Substance abuse and comorbidities. *Archives of Neuropsychiatry*, *52*(3), 213–215. https://doi.org/10.5152/npa.2015.01082015.

Vargas-Perez, H., Ting-A-Kee, R., Walton, C. H., Hansen, D. M., Razavi, R., Clarke, L., et al. (2009). Ventral tegmental area BDNF induces an opiate-dependent–like reward state in naïve rats. *Science*, *324*, 1732–1734.

Verhagen, M., Meij, V., Der, A., Janzing, J. G. E., Arias-Vásquez, A., Buitelaar, J. K., et al. (2009). Effect of the 5-HTTLPR polymorphism in the serotonin transporter gene on major depressive disorder and related comorbid disorders. *Psychiatric Genetics*, *19*, 39–44. https://doi.org/10.1097/YPG.0b013e3283208061.

Vialou, V., Bagot, R. C., Cahill, M. E., Ferguson, D., Robison, A. J., Dietz, D. M., et al. (2014). Prefrontal cortical circuit for depression- and anxiety-related behaviors mediated by cholecystokinin: Role of ΔFosB. *The Journal of Neuroscience*, *34*(11), 3878–3887. https://doi.org/10.1523/JNEUROSCI.1787-13.2014.

Vialou, V., Robison, A. J., Laplant, Q. C., Covington, H. E., Dietz, D. M., Ohnishi, Y. N., et al. (2010). ΔfosB in brain reward circuits mediates resilience to stress and antidepressant responses. *Nature Neuroscience*, *13*(6), 745–752. https://doi.org/10.1038/nn.2551.

Volkow, N. D. (2004). The reality of comorbidity: Depression and drug abuse. *Biological Psychiatry*, *56*(10), 714–717. https://doi.org/10.1016/j.biopsych.2004.07.007.

Walsh, J. J., Friedman, A. K., Sun, H., Heller, E. A., Ku, S. M., Juarez, B., et al. (2014). Stress and CRF gate neural activation of BDNF in the mesolimbic reward pathway. *Nature Neuroscience*, *17*(1), 27–29. https://doi.org/10.1038/nn.3591.

Warden, M. R., Selimbeyoglu, A., Mirzabekov, J. J., Lo, M., Thompson, K. R., Kim, S. Y., et al. (2012). A prefrontal cortex-brainstem neuronal projection that controls response to behavioural challenge. *Nature*, *492*(7429), 428–432. https://doi.org/10.1038/nature11617.

Warner-Schmidt, J. L., Schmidt, E. F., Marshall, J. J., Rubin, A. J., Arango-lievano, M., Kaplitt, M. G., et al. (2012). Cholinergic interneurons in the nucleus accumbens regulate depression-like behavior. *Proceedings of the National Academy of Sciences of the United States of America*, *109*(28), 11360–11365. https://doi.org/10.1073/pnas.1209293109/-/DCSupplemental.www.pnas.org/cgi/doi/10.1073/pnas.1209293109.

Warren, B. L., Vialou, V. F., Iñiguez, S. D., Alcantara, L. F., Wright, K. N., Feng, J., et al. (2013). Neurobiological sequelae of witnessing stressful events in adult mice. *Biological Psychiatry*, *73*(1), 7–14. https://doi.org/10.1016/j.biopsych.2012.06.006.

Weinberger, A. H., Gbedemah, M., Wall, M. M., Hasin, D. S., Zvolensky, M. J., Chaiton, M., et al. (2017). Depression among non-daily smokers compared to daily smokers and never-smokers in the United States: An emerging problem. *Nicotine & Tobacco Research*, 1062–1072. https://doi.org/10.1093/ntr/ntx009.

Wemm, S. E., & Sinha, R. (2019). Drug-induced stress responses and addiction risk and relapse. *Neurobiology of Stress*, *10*, 100148. https://doi.org/10.1016/j.ynstr.2019.100148.

Wilkinson, M. B., Dias, C., Magida, J., Mazei-Robison, M., Lobo, M., Kennedy, P., et al. (2011). A novel role of the WNT-dishevelled-GSK3β signaling cascade in the mouse nucleus accumbens in a social defeat model of depression. *Journal of Neuroscience*, *31*(25), 9084–9092. https://doi.org/10.1523/JNEUROSCI.0039-11.2011.

Yoo, J. H., Zell, V., Gutierrez-Reed, N., Wu, J., Ressler, R., Shenasa, M. A., et al. (2016). Ventral tegmental area glutamate neurons co-release GABA and promote positive reinforcement. *Nature Communications*, 7(May), 13697. https://doi.org/10.1038/ncomms13697.

Zhang, F. F., Peng, W., Sweeney, J. A., Jia, Z. Y., & Gong, Q. Y. (2018). Brain structure alterations in depression: Psychoradiological evidence. *CNS Neuroscience and Therapeutics*, *24*(11), 994–1003. https://doi.org/10.1111/cns.12835.

Zullig, K. J., & Divin, A. L. (2012). The association between non-medical prescription drug use, depressive symptoms, and suicidality among college students. *Addictive Behaviors*, *37*(8), 890–899. https://doi.org/10.1016/j.addbeh.2012.02.008.

CHAPTER SEVEN

The role of gut-immune-brain signaling in substance use disorders

Kelsey E. Lucerne[a] and Drew D. Kiraly[a,b,*]
[a]Nash Family Department of Neuroscience, Icahn School of Medicine at Mount Sinai, New York, NY, United States
[b]Department of Psychiatry, Icahn School of Medicine at Mount Sinai, New York, NY, United States
*Corresponding author: e-mail address: drew.kiraly@mssm.edu

Contents

Abstract

Substance use disorders (SUDs) are debilitating neuropsychiatric conditions that exact enormous costs in terms of loss of life and individual suffering. While much progress has been made defining the neurocircuitry and intracellular signaling cascades that contribute to SUDs, these studies have yielded limited effective treatment options. This has prompted greater exploration of non-traditional targets in addiction. Emerging data suggest inputs from peripheral systems, such as the immune system and the gut microbiome, impact multiple neuropsychiatric diseases, including SUDs. Until recently the gut microbiome, peripheral immune system, and the CNS have been studied

International Review of Neurobiology, Volume 157
ISSN 0074-7742
https://doi.org/10.1016/bs.irn.2020.09.005

independently; however, current work shows the gut microbiome and immune system critically interact to modulate brain function. Additionally, the gut microbiome and immune system intimately regulate one another via extensive bidirectional communication. Accumulating evidence suggests an important role for gut-immune-brain communication in the pathogenesis of substance use disorders. Thus, a better understanding of gut-immune-brain signaling could yield important insight to addiction pathology and potential treatment options.

Abbreviations

VTA	ventral tegmental area
NAc	nucleus accumbens
PFC	prefrontal cortex
PRR	pattern recognition receptors
DA	dopamine
PAMPs	pathogen-associated molecular patterns
DAMPs	danger-associated molecular patterns
LPS	lipopolysaccharide
PGN	peptidoglycan
TLRs	toll-like receptors
NF-κB	nuclear factor kappa-light-chain-enhancer of activated B cells
MAPK	mitogen-activated protein kinases
IFN	type I interferon
IL-1	interleukin-1
TNF-α	tumor necrosis factor alpha
IFN-γ	interferon gamma, gamma-interferon
IFN-α	interferon alpha
G-CSF	granulocyte colony stimulating factor
GM-CSF	granulocyte-macrophage colony stimulating factor
TGF-β	transforming growth factor-beta
MyD88	myeloid differentiation primary response gene 88
MD2	myeloid differentiation factor 2
CVO	circumventricular organs
NGF	nerve growth factor
BDNF	brain-derived neurotrophic factor
NT3	neurotrophin-3
FGF	fibroblast growth factor
CNTF	ciliary neurotrophic factor
EGF	epidermal growth factor
CCL2 / MCP-1	chemokine ligand 2 / Monocyte chemoattractant protein-1
CX3CL1	C-X3-C Motif Chemokine Ligand 1, Fractalkine
IDO	indoleamine 2,3-dioxygenase
ROS	reactive oxygen species
AMPAR	alpha-amino-3-hydroxy-5-methyl-4-isoxazole propionic acid receptor
GDNF	glial-derived neurotrophic factor
C1q	complement component 1q
c3	complement component 3
CD11B	cluster of differentiation molecule 11B, Integrin alpha M

CD18	integrin beta chain-2
SPARC	secreted protein acidic and rich in cysteine
TSPs	thrombospondins
PET	positron emission tomography
CPP	conditioned place preference
Iba1	ionized calcium binding adaptor molecule 1
CD68	cluster of differentiation 68
METH	methamphetamine
CXCR4	C-X-C chemokine receptor type 4 / fusin / CD184
CCL5	chemokine ligand 5 / RANTES
CCL4 / MIP-1β	chemokine ligand 4/Macrophage inflammatory protein-1β
CCL17	chemokine (C—C motif) ligand 17 / TARC
GFAP	glial fibrillary acidic protein
GALT	gut-associated lymphoid tissues
SFB	segmented filamentous bacterium
SCFA	short-chain fatty acids
GF	germ free
PLC	phospholipase C
PLA2	phospholipase A2

1. Introduction

Substance use disorders (SUDs) are chronic, debilitating disorders that exact an enormous cost in terms of individual suffering and loss of life. These conditions are characterized by early intermittent drug use that transitions to uncontrolled intake, compulsive drug- taking and seeking, withdrawal symptoms, and high rates of relapse despite prolonged periods of abstinence (NIDA, 2020a; Koob & Volkow, 2010). SUDs affect some 31 million people globally (WHO, n.d.) and are one of leading causes of disability in world (NIDA, 2020b). Despite decades of advancements in our understanding of the neurobiology underpinning SUDs, there remain few effective pharmacological treatment options, and those available are ineffective for many or have abuse liability of their own (NIDA, 2020a; Shorter, Domingo, & Kosten, 2015). Thus, more complete understanding of the neurobiology underlying SUDs and development of novel therapeutics is a pressing need (Kuehn, 2012).

Like most neuropsychiatric disorders, the bulk of research in addiction neurobiology has largely focused on mechanisms intrinsic to the brain with an emphasis on neuronal function. However, in recent years, there has been a rapidly growing body of research demonstrating important influences of non-neuronal and peripheral effectors, such as changes in peripheral and central immune system response and effects of the gut microbiome, in driving pathophysiology in models of SUDs (Foster, 2016; Spichak, Dinan, & Cryan, 2019). Immune dysfunction and microbiome dysbiosis can both disrupt CNS homeostasis, affecting neuronal function and consequently behavior (Rankin & Artis, 2018). Mounting evidence demonstrates gut-immune-brain interactions in numerous neuropsychiatric disorders, including depression and anxiety (Peirce & Alviña, 2019), autism spectrum disorders (Osokine & Erlebacher, 2017), and multiple sclerosis (Sauma & Casaccia, 2020). Less well reviewed is the role of gut-immune-brain signaling in SUDs. Multiple classes of abused drugs alter immune function (Levite, 2016; Machelska & Celik, 2020; Zaparte, Schuch, Viola, et al., 2019) in a manner that drives drug-seeking and -taking in rodent models (Felger, Li, Haroon, et al., 2016; Hofford, Russo, & Kiraly, 2018; Karlsson, Schank, Rehman, et al., 2017; Lacagnina, Rivera, & Bilbo, 2017; Northcutt, Hutchinson, Wang, et al., 2015). Further, alterations to the gut microbiome have been demonstrated to drive behavioral effects in multiple models of SUDs (Meckel & Kiraly, 2019; Wang, Liu, Guo, et al., 2018; Wang, Meng, Zhang, et al., 2018). While immune system and gut microbiome changes have significant effects on their own, recent work has begun to examine the intersection between gut microbiome and immune signaling in pathogenesis of substance use disorders, but the literature on this mechanism is sparse (Hofford et al., 2018; Meckel & Kiraly, 2019).

Importantly, while both the microbiome and immune system have important effects brain function independently, the two systems are also highly inter-related and have the potential to affect one another as well as the brain. The gut microbiome has tremendous influence on immune responses throughout the body and vice versa, and shifts in the balance of either of these important system can alter brain function and behavior in important ways (Rankin & Artis, 2018). Only now are we beginning to understand the intricate mechanisms by which the gut microbiome, immune system, and brain communicate to drive health and, when dysfunctional, disease. These brain-body interactions are tremendously complex, and further research into gut-immune-CNS communication is necessary to clarify how it may be leveraged for treatment options.

This review will synthesize the disparate literature on microbiome and immune interactions underlying addictive behaviors to present a model of the gut-immune-brain axis in addiction. We will review the literature of gut-mediated neuroimmune mechanisms of SUDs with a focus on opioids and psychostimulants, as these represent illicit drugs with the highest levels of morbidity and mortality. We will detail both the neuroimmune mechanisms of addiction driven *directly* by drug-induced neuroimmune alterations, and those driven *indirectly* by the gut microbiome-induced immune changes. Together, this perspective will lend insight to important body-brain interactions central to the pathology of SUD. Identifying microbiome-dependent immune factors able to alter behavioral and molecular responses to drugs of abuse is a critical next step in progressing our understanding of addiction neurobiology and a promising avenue of potential treatment.

2. Overview of the immune system

The immune system is comprised of a myriad of cells programmed to detect and respond to both exogenous and endogenous threat signals and promote host homeostasis. Immune cells can be divided into two categories based upon response to infection and trauma: (i) the innate immune system and (ii) the adaptive immune system. The innate immune system refers to a multitude of cellular responses that launch immediate and nonspecific defense mechanisms against injuries, infections, trauma, or xenobiotics (substances not naturally produced by an organism and not expected to be present, such as drugs of abuse). It is composed of cells of the myeloid lineage, including granulocytes, monocytes, macrophages, dendritic cells, and innate lymphocytes (Mayfield, Ferguson, & Harris, 2013; Spits, Artis, Colonna, et al., 2013; Warrington, Watson, Kim, & Antonetti, 2011). These innate immune cells express ***pattern recognition receptors (PRRs)*** which recognize highly conserved molecular patterns from foreign (exogenous) pathogens called pathogen-associated molecular patterns (PAMPs) as well as endogenous signals associated with cell damage, deterioration, or stress called danger-associated molecular patterns (DAMPs) (Tartey & Takeuchi, 2017). PAMPs are often structures that are vital to the survival of a given bacteria, such as ***lipopolysaccharide (LPS),*** the primary component of gram-negative bacteria cell walls, peptidoglycan (PGN), the primary component of gram-positive bacteria cell walls, flagellin, and other microbial products (Franchi, Warner, Viani, & Nuñez, 2009). PRRs are an array of multiple receptors that sense PAMPs and DAMPs to ensure adequate immune response. Common PRRs include ***toll-like receptors (TLRs),*** nucleotide-binding oligomerization domain (NOD)-like receptors, C-type lectin receptors (CLRs), RIG-I-like receptors (RLRs), and AIM2-like receptors (ALRs) (Brubaker, Bonham, Zanoni, & Kagan, 2015). Activation of PRRs induces numerous intracellular-signaling cascades, such as nuclear factor-κB (NF-κB), mitogen-activated protein kinases (MAPKs), and type I interferon (IFN) response, to drive inflammatory response (Franchi et al., 2009). The adaptive immune system confers long-term immunity by recognition of specific antigens. The adaptive immune system consists of antibody-producing B-lymphocytes, and cytokine-producing T-lymphocytes (Warrington et al., 2011).

Both innate and adaptive immune cells utilize cytokines, pleiotropic small proteins that serve as 'immunotransmitters,' to mediate communication and coordinate responses between multiple cells in both the periphery and brain. ***Cytokines*** include ***chemokines***, ***growth factors***, ***interferons (IFNs)*** and ***lymphokines*** (Becher, Spath, & Goverman, 2017), and are central for cellular communication and coordinated cellular responses. They exert their effects by binding to specific receptors expressed by a variety of cell types. Cytokine receptors themselves can also exist in a soluble form, where they can tune the biological activity of circulating cytokines (Miller & Goldsmith, 2017).

Cytokines are typically classified on their ability to initiate or inhibit inflammatory responses. These are referred to as pro-inflammatory or anti-inflammatory cytokines, respectively. Pro-inflammatory cytokines traditionally include ***interleukin***-1 (IL-1), IL-12, IL-18, tumor necrosis factor alpha (TNF-α), gamma-interferon (IFN-γ), and granulocyte-macrophage colony stimulating factor (GM-CSF). Conversely, anti-inflammatory cytokines traditionally refer to IL-4, IL-10, IL-13, IFN-α, and transforming growth factor-β (TGF-β). However, it is worth noting that numerous studies have demonstrated both pro- and anti- inflammatory roles for each of these cytokines depending on the context and cytokine milieu (Balasubbramanian, 2019; Kosloski, Kosmacek, Olson, et al., 2013; Veldhoen, Hocking, Atkins, et al., 2006). Further, in vivo cytokines exist in a heterogenous mixture, and responses to cytokine signaling vary vastly depending on the tissue. Thus, it seems that cytokines are not innately pro-inflammatory or anti-inflammatory, but exert actions in a context, dose, and cell-type -dependent manner (Becher et al., 2017; Cavaillon, 2001). In healthy conditions, cytokines must act in balance to maintain tissue homeostasis. *Dysregulation* of cytokine networks leads to chronic inflammation and disease states, including neuropsychiatric disorders (Hodes, Kana, Menard, et al., 2015; Hofford et al., 2018; Kiraly, Horn, Van Dam, et al., 2017). Cytokine abnormalities have been observed in schizophrenia (Miller & Goldsmith, 2017), depression (Kiraly et al., 2017), autism (Meltzer & Van de Water, 2017) and addiction (Coller & Hutchinson, 2012) and often present as aberrant levels of inflammatory markers. Cytokines also present as an important conduit for immune-brain communication, as multitudes of cytokines are soluble across the blood brain barrier from the periphery, and we will discuss this in more detail in the next section. A thorough characterization could yield insight to immune—brain signaling in psychiatric conditions.

3. Neuroimmune signaling

The unique compartmentalization of the CNS, effectively regulating immune-cell entry, has led to the perception of the brain as an immune-privileged organ (Engelhardt, Vajkoczy, & Weller, 2017). This protection of the brain is achieved through the presence of effective barriers, such as the blood brain barrier (BBB). However, we know that brain immune responses still occur through both systemic and central pathways. Though different from immune responses in other systemic organs, we know that immune responses do occur in the CNS and they are vital for proper brain function. The bidirectional communication that occurs between the brain and immune system is called the ***neuroimmune system***. The neuroimmune system is driven by signals from resident glial cells (namely microglia and astrocytes), signaling from blood-borne cytokines that diffuse from blood to brain (Becher et al., 2017), and migration of peripheral immune cells into the brain via central nervous system lymphatic vessels and regions with BBB permissibility (Louveau, Smirnov, Keyes, et al., 2015). Under healthy conditions, aspects of neuroimmune signaling act to drive tissue homeostasis and cell integrity (Veiga-Fernandes & Artis, 2018). Cells of myeloid lineage, such as dendritic cells, monocytes, and macrophages, patrol the steady-state brain parenchyma and its associated blood vessels, meninges and choroid plexus (Becher et al., 2017). Cytokines are also ever-present in the healthy brain where they can exert functions through the principle immune cells of the brain: microglia and astrocytes. The dysregulation of neuroimmune signaling has been implicated in numerous neuropsychiatric conditions. A dysregulated neuroimmune system alters this homeostasis and often yields neuroinflammation, which appears to be an important mediator of brain pathology in depression (Hodes et al., 2015), schizophrenia (Khandaker, Cousins, Deakin, et al., 2015), multiple sclerosis (Palle, Monaghan, Milne, et al., 2017), and substance use disorders (Hofford et al., 2018). Thus, a more complete understanding of how neuroimmune processes contribute to health and disease in the CNS is vital to appropriate treatment of such conditions.

3.1 Neuroinflammation

Owing to its semi-restricted exposure to blood-borne immune factors, the brain is particularly vulnerable to dysregulated immune signaling. The dysregulation of brain cytokine networks produces increased inflammation called ***neuroinflammation.*** Neuroinflammation is a potential common mediator of numerous neuropsychiatric diseases (Hodes et al., 2015; Miller & Goldsmith, 2017). In the CNS, neuroinflammation results from a disruption in the balance of normal, beneficial immune signaling and a shift toward pathogenic signaling driven by these same pathways. Central to this progression are microglia and astrocytes, immunocompetent cells that surveil the brain environment and respond to injury, infection, or trauma. Microglia are tissue-resident macrophages in the CNS that, under healthy quiescent conditions, regulate neural homeostasis (Prinz & Priller, 2014). Microglia are also the primary responders to PAMPs and DAMPs in the CNS (Salter & Stevens, 2017) poising them as the prime immune-effector cells, however astrocytes also play an important role in CNS immune responses (Geyer, Jacobs, & Hsu, 2019). Astrocytes are stellate cells that bear multiple processes, allowing them to make thousands of contacts with nearby cells to regulate CNS homeostasis and support neuronal function (Allen & Lyons, 2018). Both microglia and astrocytes are able to elicit innate immune responses in the brain in response to PRR activation by PAMPs and DAMPs (Geyer et al., 2019; Salter & Stevens, 2017). This initiates the production and release of cytokines that can go on to exert a multitude of effects.

3.1.1 Toll-like receptors

Of the existing types of PRRs, the best classified in neuroimmune signaling is the TLR family of PRRs. All TLRs are composed of (1) a cytosolic Toll/interleukin-1 receptor (TIR) domain and (2) a leucine-rich ectodomain (Brubaker et al., 2015). These are highly conserved domains that allow TLRs to bind various PAMPs and DAMPs (Piccinini & Midwood, 2010). TLRs are specialized to recognize different signals based upon their residence in either the extracellular (TLR1, TLR2, TLR4, TLR5, and TLR6) or intracellular (TLR3, TLR7, TLR8, and TLR9) compartments (Brubaker et al., 2015). Of the TLR family, TLR4 is the most extensively studied in the CNS for its contributions to neuroimmune signaling and neuroinflammation. In the brain TLR4 is most robustly expressed by microglia (Olson & Miller, 2004), but its expression is also observed on astrocytes (Bowman, Rasley, Tranguch, & Marriott, 2003), oligodendrocytes (Bsibsi, Ravid, Gveric, & van Noort, 2002); and neurons (Tang, Arumugam, Xu, et al., 2007). TLR4 signals through the adaptor protein, myeloid differentiation primary response gene 88 (MyD88), to activate the transcription factor nuclear factor kappa-light-chain-enhancer of activated B cells (NF-κB) (Linker, Cross, & Leslie, 2019). NF-κB is a ubiquitous nuclear transcription factor with important roles in innate immune signaling (Crews, Walter, Coleman, & Vetreno, 2017). NF-κB stimulation promotes expression of both pro- and anti- inflammatory cytokines and expression of their respective receptors (Crews et al., 2017). Once produced, cytokines can go on to signal through their receptors on microglia, astrocytes, or neurons to further affect brain function.

TLR4 has been crucial for studying of the effects of peripheral inflammation on brain function. This has been thoroughly characterized using the TLR4 ligand, LPS. Systemically administered LPS is used to model increased peripheral inflammation. On cell surface, TLR4 forms a complex with Myeloid differentiation factor 2 (MD2) which forms the primary binding site for LPS (Kawai & Akira, 2010). Peripheral LPS can access microglia within the circumventricular organs (CVO), choroid plexus, and other regions with leaky BBB where it binds to TLR4. Systemic LPS activation of microglial TLR4 induces TNF synthesis, and TNF in turn activates proinflammatory signaling cascades in adjacent microglia resulting in progressive inflammatory response throughout the brain parenchyma (Rivest, 2009). This TLR4-based neuroinflammation has been found to contribute to addiction pathology (Crews et al., 2017). Innate immune activation through TLR4 signaling has been implicated in neuroinflammation across various classes of drugs of abuse (Crews et al., 2017; Northcutt et al., 2015). TLR4 knockout male mice are protected from ethanol-induced memory deficits, anxiety-like behaviors, and epigenetic changes to the brains reward circuitry (Pascual, Baliño, Alfonso-Loeches, et al., 2011). Further, blockade of the TLR4 signaling impairs cocaine reinforcement in male rats and mice (Northcutt et al., 2015). Additional work demonstrates opioid activation of TLR4 contributes to drug reward in male mice (Hutchinson, Bland, Johnson, et al., 2007; Wang, Loram, Ramos, et al., 2012). TLRs are known to respond to xenobiotics, and recent evidence suggests drugs of abuse may engage TLR signaling through this manner (Northcutt et al., 2015; Wang, Northcutt, Cochran, et al., 2019).

Moreover, TLR4 activation by drugs of abuse has been demonstrated to trigger glial neuroexcitatory and neuroinflammatory signaling (Crews et al., 2017). Thus, TLRs act as key sensors by which microglia and astrocytes respond to their microenvironment and peripheral signals to propagate neuroinflammation and effect behavioral responses.

3.1.2 Peripheral immune cells and circulating cytokines in the CNS

Blood-borne cytokines can access the brain to initiate neuroinflammation via multiple mechanisms, including (1) active transport across the BBB, (2) passage through the permissive BBB regions in the CVO, and (3) signaling via vagal nerve or other afferents to relevant brain regions such as nucleus of the solitary tract and hypothalamus (Felger & Treadway, 2017). Circulating cytokines IL-1β, IL-1α, IL-6, and TNF-α can cross into the brain through saturable transport (Banks, Kastin, & Broadwell, 1995) where they can act directly on neurons, microglia, and astrocytes (Hodes et al., 2015). Numerous cytokines are known to cross from blood to brain, including G-CSF, GM-CSF, IFN-γ, nerve growth factor (NGF), brain-derived neurotrophic factor (BDNF), neurotrophin-3 (NT3), fibroblast growth factor (FGF), ciliary neurotrophic factor (CNTF), and epidermal growth factor (EGF) to name a few (Banks, Farr, & Morley, 2002). Circulating cytokines have also been demonstrated to alter or disrupt BBB permeability to facilitate passage of other immune cells, such as monocytes, from the blood to brain (Banks et al., 2002). Additionally, there is evidence that both stimulants and opioids can increase permeability of the BBB by affecting tight junction proteins and by other mechanisms (Barr, Brailoiu, & Abood, 2020; Dietrich, 2009; Pimentel, Sivalingam, Doke, & Samikkannu, 2020), and that stimulants in particular can increase migration of leukocytes into the brain parenchyma (Fiala, Gan, & Zhang, 1998). This raises the possibility that some or all of these mechanisms of cytokine migration may be enhanced in the setting of substance use disorders.

Once in the brain, peripheral cytokines can signal through cell surface receptors present on microglia, astrocytes, and neurons. Microglia rapidly activate in response to cytokine signaling, increasing microglial production and release pro-inflammatory cytokines such as TNFα, IL-1, IL-6, CCL2, and CX3CL1, propagating a pro-inflammatory signaling cascade throughout the brain (Laflamme & Rivest, 2001). This occurs through signaling through cytokine receptors as well as TLR signaling previously discussed. Activated microglia also release chemoattractants such as CCL2 (otherwise known as monocyte chemoattractant protein-1 [MCP-1]) and CX3CL1 that promote monocyte trafficking to the site of neuroinflammation (Weber, Godbout, & Sheridan, 2017). During inflammation, tissue-invading monocytes will give rise to tissue-invading macrophages that can propagate neuroinflammation. Both activated microglia and infiltrating macrophages produce an inducible ***nitric oxide synthase***, as well as ***quinolinic acid*** through indoleamine 2,3-dioxygenase (IDO) and kynurenine pathways that promote oxidative stress and the creation of reactive oxygen species (ROS) (Felger & Treadway, 2017). Sustained neuroinflammation leads to BBB breakdown through highly phagocytic neutrophils that produce pro-inflammatory cytokines and granules containing degradative enzymes and toxic molecules (Rumble, Huber, Krishnamoorthy, et al., 2015). This allows infiltration of pro-inflammatory T cells, which can deliver further cytokines to the inflamed CNS such as GM-CSF, IFN-γ, IL-17A, IL-17F, and IL-22 (Becher et al., 2017). In non-pathological states, neuroinflammation is resolved by release of anti-inflammatory cytokines such as TGF-β and IL-10. These findings indicate that the accumulation of peripheral immune cells and cytokines in vascular compartments associated with microglia activation contributes to pathological disease states and high inflammation observed in patients with neuropsychiatric conditions.

Neuroinflammation has important impacts on the brain and behavior, and thus has been implicated in the pathology of numerous neuropsychiatric conditions. Inflammation accompanies conditions such as autism (Meltzer & Van de Water, 2017), schizophrenia (Miller & Goldsmith, 2017), anxiety and depression (Hodes et al., 2015). More recent work has implicated aberrant inflammation in addiction (Hofford et al., 2018). It is known that altered neuroimmune interactions culminate in changes to brain systems that are central to the development of substance use disorders (Hofford et al., 2018). Innate immune activation and subsequent release of pro-inflammatory cytokines and chemokines alter the mesolimbic dopamine system by increasing dopamine clearance, decreasing dopamine receptor expression, and altering dopamine synthesis (Felger & Treadway, 2017). This altered dopamine activity is associated with altered cytokine and chemokine expression in the periphery and brain. Chronic treatment of healthy human patients with IFN-α results in reduced motivation and anhedonia, and this is accompanied by reduced activity in the ventral striatum and decreased dopamine synthesis or release (Capuron, Pagnoni, Drake, et al., 2012). Further work demonstrates that signaling from peripheral cytokines, or from cytokines produced locally by activated microglia or infiltrating macrophages, culminates to decreased dopaminergic signaling in the ventral striatum and increased glutamatergic excitotoxicity (Felger & Treadway, 2017). This demonstrates the importance of balanced cytokine networks to normal brain function and indicates neuroinflammation as a major contributor to brain dysfunction underlying addiction-like states.

3.2 Neuroimmune modulation of synaptic function

Peripheral and local immune signals play essential roles outside of neuroinflammation and are key to appropriately tuning neurotransmitter dynamics and modulating synaptic plasticity. Circulating cytokines that cross from blood to brain activate microglia and astrocytes, and this could contribute to persistent changes in synaptic function. Cytokine receptors are capable of initiating cascades of intracellular signaling upon ligand binding, through which they modulate the release of several neurotransmitters, including glutamate, GABA, and dopamine (Cui, Shurtleff, & Harris, 2014; Felger & Treadway, 2017; Heinisch & Kirby, 2010). Evidence also suggests that pro-inflammatory cytokines play an important role in synaptic plasticity and neuromodulation. Circulating IL-1β, IL-1α, IL-6, and TNF-α can cross into the brain where they drive normal physiological processes and synaptic homeostasis (Gadient & Otten, 1997). More recently TNF-α has been demonstrated to have important roles in neurotransmitter receptor trafficking and homeostatic synaptic plasticity (Pribiag & Stellwagen, 2014). Specifically, TNF-α is central to trafficking of glutamatergic AMPA (alpha-amino-3-hydroxy-5-methyl-4-isoxazole propionic acid) receptors and GABA receptors (Pribiag & Stellwagen, 2013; Stellwagen & Malenka, 2006). TNF-α also appears to be a key regulator of synaptic function in the ventral striatum (including the NAc) (Lewitus, Pribiag, Duseja, et al., 2014). Further, release of the pro-inflammatory cytokine IL-1β contributes to formation of long-term potentiation (Avital, Goshen, Kamsler, et al., 2003; Mori, Nisticò, Mandolesi, et al., 2014). Thus, central and peripheral cytokine signaling plays an important role in normal brain function.

Microglia and astrocytes are also important to neuronal plasticity and circuitry. Microglia and astrocytes express numerous receptors for neurotransmitters, neuropeptides, and neuromodulators, allowing them to sense and respond to neuronal activity (Kettenmann, Kirchhoff, & Verkhratsky, 2013). Neuronal stimulus can activate microglia to generate neurotrophic factors such as nerve growth factor (NGF), neurotrophin (NT)-4/5, TGF-β, glial-derived neurotrophic factor (GDNF), fibroblast growth factor (FGF), and IL-3 (Nakajima, Tohyama, Maeda, et al., 2007). These ***neurotrophic*** factors serve to promote neuronal survival. Microglia also respond to a number of endogenous, secreted and membrane-bound factors that mediate microglial activation states (Hanisch & Kettenmann, 2007) to turn activation "on" and "off". These "on" and "off" signals are reviewed extensively here (Biber, Neumann, Inoue, & Boddeke, 2007). In this review we focus primarily on TLR4 as it has particular relevance to SUD pathology. Any given microglia can also form contacts with multiple synapses in the brain, where they can modify or all-together or eliminate synapses (Kettenmann et al., 2013). Microglia have been demonstrated to eliminate synapses and related neuronal components in an activity-dependent manner, potentially via synapse expression of ***complement proteins*** C1q and C3 to the microglia complement receptor CD11b/CD18 (Lacagnina, Rivera, & Bilbo, 2017; Schafer, Lehrman, Kautzman, et al., 2012). Thus, microglia directly remodel neuronal circuitry, driven by changes in neuronal activity or neuroimmune signaling.

Astrocytes are also central to normal brain function by regulation of synaptic formation, maturation, and remodeling to achieve precise synaptic connectivity of the brain (Paolicelli, Bolasco, Pagani, et al., 2011; Schafer & Stevens, 2013). In the adult human brain, each astrocyte can form as many as 100,000 synaptic contacts where they aid in appropriate neurotransmitter clearance and ion homeostasis (Allen & Eroglu, 2017). These astrocyte-synapse contacts are vital to proper synaptic connectivity, remodeling, and development (Allen & Eroglu, 2017). Purified culture of neurons (i.e., retinal ganglion cells [RGCs]) lacking any glia develop very few functional synapses, but when cultured with a feeder layer of astrocytes these neurons numerous functional synapses (Pfrieger & Barres, 1997; Ullian, Sapperstein, Christopherson, & Barres, 2001). Consequently, a number of astrocyte-secreted synaptogenic molecules have been identified. This includes thrombospondins (TSPs) (Christopherson, Ullian, Stokes, et al., 2005), hevin, and secreted protein acidic and rich in cysteine (SPARC) (Kucukdereli, Allen, Lee, et al., 2011). These molecules are capable of both positive and negative regulation of synapse formation, indicated astrocytes as important in the bidirectional tuning of synaptic activity. Astrocytes and microglia may also interact to drive synaptic modulation. In vitro findings in RGC cultures indicate that astrocyte-produced TGF-β controls neuronal C1q expression to modulate C3-dependent microglial synaptic pruning (Bialas & Stevens, 2013). These findings highlight the critical role of both microglia and astrocytes in synaptic regulation.

The important role of glia in synaptic signaling and plasticity underscores them as key participants in SUD neurobiology. Aberrations in neurotransmission, and consequential dysfunction of glia, may result in the maladaptive restructuring of synapses, thus leading to persistent abnormalities in neuronal activity that contribute to addiction development and maintenance (Lacagnina, Rivera, & Bilbo, 2017). Indeed evidence demonstrates that microglia and astrocytes contribute to the changes in synaptic plasticity in response to drugs of abuse (Lacagnina, Rivera, & Bilbo, 2017). TLRs also play a role outside of immune responses, regulating aspects of neurogenesis, suggesting an important role in neuronal plasticity (Okun, Griffioen, Lathia, et al., 2009) and another mechanism by which altered cytokine signaling at microglia, astrocytes, and neurons contributes to altered synaptic function. Dysregulation of neuroimmune signaling, even in the absence of neuroinflammation, can have widespread effects on neurotransmission, gene expression, and synaptic plasticity (Covelo & Araque, 2018; Eroglu & Barres, 2010; Kettenmann et al., 2013; Pribiag & Stellwagen, 2014; Stellwagen & Malenka, 2006). A complete understanding of how altered neuroimmune signaling by these glial cells contributes to disease pathology is key to developing treatment strategies.

4. Neuroimmune mechanisms of addiction

Substance use disorders are defined by a shift from intermittent drug use to compulsive drug- taking and seeking with a loss of control over intake. This shift from casual drug use to a pathological state is believed to be mediated by aberrant plasticity in brain reward circuits, with persistent changes in gene expression, neurophysiology, and synaptic morphology (Nestler & L€, 2019). The literature on neuroimmune interactions in addiction has focused largely on the mesocorticolimbic dopamine circuitry. This circuit is comprised of dopaminergic projections from the ventral tegmental area (VTA) to the nucleus accumbens (NAc) and prefrontal cortex (PFC) (Lüscher & Malenka, 2011; Russo & Nestler, 2013). Increased dopaminergic activity of VTA to NAc projections is believed to be the shared initial common action across all drugs of abuse (Koob & Volkow, 2010; Nestler & L€, 2019). The mesocorticolimbic dopamine system also is densely connected with glutamatergic regions, including the PFC which sends glutamatergic afferents to the NAc and VTA to further modulate drug taking and seeking behaviors (Russo & Nestler, 2013). While much of the literature we review focuses on these circuits, it is important to remember that the circuit pathology implicated in substance use disorders and addiction-like behaviors is complex and involves multiple cortical regions, the habenula, the dorsal striatum, and the extended amygdala (Reviewed in detail: Koob & Volkow (Koob & Volkow, 2016) and Nestler & Lüscher (Nestler & L€, 2019)).

In addition to these alterations in neural circuits, work from the past two decades has demonstrated the important role of glial cells and immune signaling in addiction neurobiology (Linker et al., 2019; Nestler & L€, 2019). Recent advances that look beyond the direct effects of CNS neurons, including signaling from the peripheral immune system and gut microbiome, have shown the tremendous complexity regulating these biological processes, but also raise new possibilities for translational research in curbing these debilitating disorders (Hofford et al., 2018; Meckel & Kiraly, 2019). Herein, we will provide an overview of neuroimmune mechanisms of psychostimulant and opioid use disorder, focusing on the mesocorticolimbic dopamine system for the sake of brevity.

Inflammatory changes are well characterized to alter dopaminergic signaling in the brain. Dopamine is synthesized by numerous immune effector cells and its receptors are also expressed on these cells (Ferrari, Cosentino, Marino, et al., 2004; Kirillova, Hrutkay, Shurin, et al., 2008) and growing evidence suggests dopamine regulates immune function (Matt & Gaskill, 2019; Sarkar, Basu, Chakroborty, et al., 2010). Increases in dopamine, resultant from drugs of abuse, directly engage dopamine receptors on multiple leukocytes, including T cells, B cells, and monocytes, to increase production of circulating cytokines and other immune factors (Arreola, Alvarez-Herrera, & Pérez-Sánchez, 2016). Findings also show inflammation-associated reductions in dopamine, dopamine metabolites, and striatal release of dopamine (Felger & Treadway, 2017). Given that drugs of abuse exert their rewarding effects through increases to the mesocorticolimbic dopamine system (Coller & Hutchinson, 2012), neuroimmune mechanisms of addiction present as an important neurobiological contributor to addiction. A more complete review of inflammation effect on dopamine can be found elsewhere (Felger & Treadway, 2017).

Inflammation has also been demonstrated to affect other neurotransmitter systems important to the development of addiction. Heightened peripheral inflammation is associated with decreased functional connectivity of the glutamatergic projections from the ventromedial prefrontal cortex (vmPFC) to the ventral striatum (Felger et al., 2016). Signaling from pro-inflammatory cytokines in the brain has also been shown to alter synaptic glutamate homeostasis (Felger & Treadway, 2017) by downregulating glutamate transporters at the synapse (Tilleux & Hermans, 2007). Drugs of abuse, such as cocaine, have been demonstrated to decrease glutamate synaptic transporters as well (Moussawi, Pacchioni, Moran, et al., 2009), and this contributes to relapse-like behaviors (Scofield & Kalivas, 2014). Moreover, the ability of cytokines, microglia, and astrocytes to affect synaptic function in an activity-dependent manner poises the neuroimmune system as a key participant in the neurobiology of SUDs. In the following, we will cover precise neuroimmune mechanisms participating in the development and persistence of SUDs across psychostimulants and opioids.

4.1 Psychostimulants

Psychostimulants, namely cocaine and amphetamine, accounted for about a third (32.9%) of drug overdose deaths in 2017 (Kariisa, Scholl, Wilson, et al., 2019). Despite the pressing need for treatments, there remain no FDA approved pharmacotherapies for the treatment of cocaine- and psychostimulant- use disorders. Thus, much research has been targeted toward elucidating the neurobiology of cocaine and psychostimulant actions. Much of this research has focused on the role of the mesocorticolimbic dopamine system in addiction development and outcomes. The mesocorticolimbic dopamine circuitry is comprised of dopaminergic projections from the ventral tegmental area (VTA) to the nucleus accumbens (NAc) and prefrontal cortex (PFC) (Lüscher & Malenka, 2011; Russo & Nestler, 2013). Increased dopaminergic activity of this circuit, and specifically elevated extracellular dopamine within the NAc, is believed to be the shared initial common action across all drugs of abuse (Koob & Volkow, 2010; Nestler & L€, 2019). Both cocaine and amphetamines enhance dopaminergic transmission in the CNS and periphery by preventing and/or reversing reuptake of dopamine out of the synapse (Kalivas, 2007). Cocaine elevates dopaminergic transmission by blocking the dopamine transporter (DAT)-mediated removal of dopamine out of synapses, and amphetamines block dopamine uptake and reverse dopamine transport by DATs, further elevating synaptic dopamine transmission (Coller & Hutchinson, 2012).

More recent work over the past decade has established an important role for immune signaling in development and pathology of psychostimulant abuse (Coller & Hutchinson, 2012; Fox, D'Sa, Kimmerling, et al., 2012; Hofford et al., 2018). Stimulants have been demonstrated to affect immune signaling, both at the central and peripheral level. Numerous studies demonstrate the pharmacodynamic actions of psychostimulants are also produced, and enhanced, by neuroimmune signaling in key brain reward regions and circuits (Coller & Hutchinson, 2012). Thus, the immunomodulatory effects of psychostimulants are emerging as an important component of SUD pathology, and have strong therapeutic potential. In the following we will review the current clinical and preclinical evidence for immune signaling in psychostimulant use disorder pathology.

4.1.1 Clinical evidence

Early evidence for this emerged from work demonstrating cocaine increased rates of HIV replication in both macrophages and microglia of patients with HIV and worsened peripheral inflammation and disease progression (Tyagi, Weber, Bukrinsky, & Simon, 2016; Webber, Schoenbaum, Gourevitch, et al., 1999). Further work demonstrates cocaine users exhibit altered levels of cytokines, including MCP-1 (aka CCL2), IL-1β, IL-6, IL-17, TNF-α, IL-10, and TGF-α (Araos, Pedraz, Serrano, et al., 2015; Gupta, Sharma, Singh, et al., 2018; Maza-Quiroga, García-Marchena, & Romero-Sanchiz, 2017; Pedraz, Araos, García-Marchena, et al., 2015). Cocaine users were found to have elevated levels of the pro-inflammatory cytokines, such as IL-6 (Levandowski, Hess, Grassi-Oliveira, & de Almeida, 2016; Moreira, Medeiros, Lhullier, et al., 2016), and decreased levels of the anti-inflammatory cytokines, such as IL-10 (Moreira et al., 2016), suggesting pathological cocaine use induces a more pro-inflammatory profile. One key study by Lo lacono et al demonstrates ex vivo and in vivo, as well as in mice and humans, the functional role of peripheral and central immune responses in the neurobehavioral effects of cocaine. Looking at levels of blood leukocyte messenger RNA in humans, this experiment revealed increases in expression of *TLR4*, its inhibitory protein toll interacting protein (*TOLLIP*), and IL-1 receptor accessory protein (*IL1RAP*) in cocaine-addicted subjects (Lo Iacono, Catale, Martini, et al., 2018). Cocaine-dependent subjects also demonstrated increases in pro-inflammatory proteins in the blood such IL-1β, IL-6, TNF-α, IL-12p70, and intercellular adhesion molecule 1 (ICAM-1) regulated by TLR4 (Lo Iacono et al., 2018). Moreover, this study demonstrated that among cocaine-dependent individuals, those that underwent childhood trauma show a higher expression level of mRNAs and pro-inflammatory proteins as compared to cocaine-dependent individuals with no history of childhood trauma (Lo Iacono et al., 2018). This highlights the immune system as a potential mechanism that poises individuals to be more susceptible to developing a SUD. It is important to note that there are a number of studies that conflict with these findings, demonstrating a decrease in pro-inflammatory markers in response to cocaine and a potential immunosuppressive effect of psychostimulants (Halpern, Sholar, Glowacki, et al., 2003; Irwin, Olmos, Wang, et al., 2007). Taken together, this clinical work clearly indicates psychostimulant abuse disrupts the appropriate balance between pro- and anti-inflammatory cytokines of the peripheral milieu, and this perhaps serves to prime an individual toward developing an addictive-like phenotype. Corroborating this is evidence that patients with psychostimulant use disorders demonstrate increases in levels of pro-inflammatory markers in response to presentation of either drug-associated cues (Fox et al., 2012) or unpleasant images (Ersche, Hagan, Smith, et al., 2014). This provides preliminary evidence of peripheral immune dysfunction in SUD and its contribution to addictive-like states.

While this work highlights effects of cocaine on peripheral cytokines, this does not necessarily correlate with central cytokine signaling. However, circulating cytokines are known to infiltrate the brain parenchyma from the periphery (Fig. 1). Furthermore, circulating leukocytes, such as T cells and monocytes, have been demonstrated to extravasate into the brain parenchyma (Hodes et al., 2015; Menard, Pfau, Hodes, et al., 2017). Cocaine use has been shown to affect endothelial cells to disrupt BBB integrity, perhaps allowing for increased cell emigration from the periphery into the brain (Fiala, Gan, & Zhang, 1998). Infiltrating cytokines and leukocytes are known to influence brain function. Other clinical trials report elevated levels of activated microglia and macrophages in the midbrain of post mortem brain tissue of pathological cocaine users (Little, Ramssen, Welchko, et al., 2009). This coincided with a corresponding decrease in dopaminergic neuronal components (Little et al., 2009). Given the importance of microglia in synaptic remodeling, and the key role of synaptic plasticity in SUD pathology, this could be a potential mechanism by which microglia mediate drug dependence. Another study using positron emission tomography (PET) with a radiotracer for activated microglia corroborated these findings, demonstrating increased microglia activation in human methamphetamine abusers that negatively correlates with duration of drug abstinence (Sekine, Ouchi, Sugihara, et al., 2008). Importantly, a similar study looking at chronic cocaine-abusing humans found no differences in microglia activation between controls and drug users (Narendran, Lopresti, Mason, et al., 2014). While informative, it is important to note that these PET studies should be viewed with some caution given discrepancies between their lengths of abstinence and promiscuity

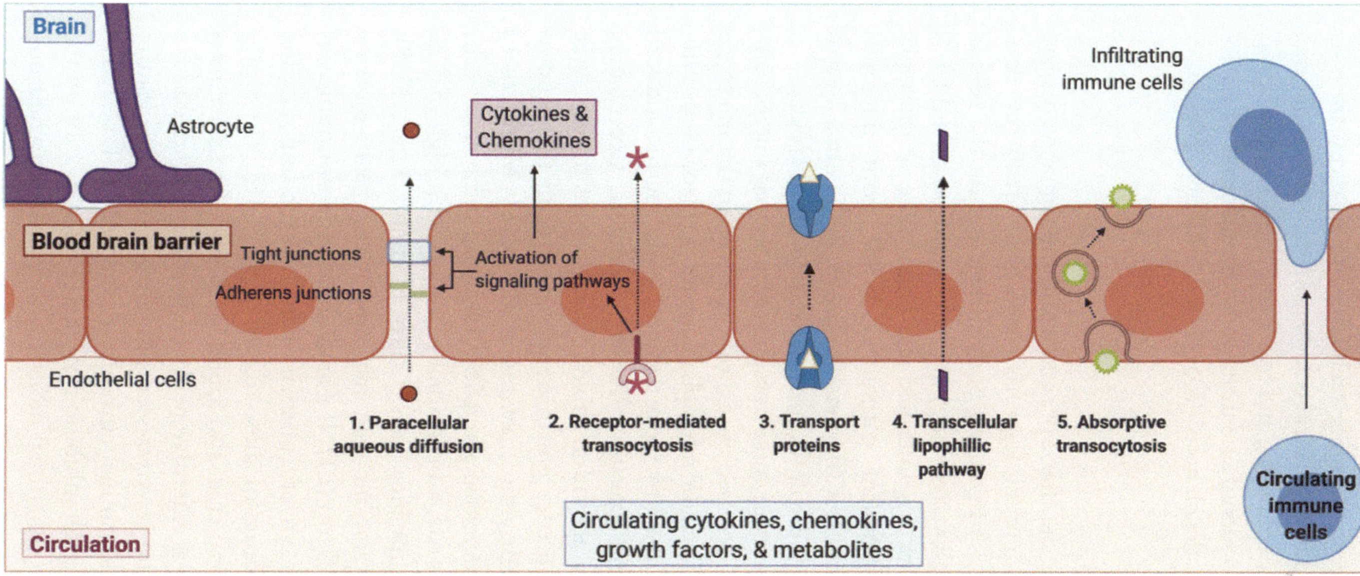

Fig. 1 Mechanisms of BBB transport. There are several mechanisms by which molecules may cross the BBB to enter the brain. (1) The paracellular aqueous diffusion pathway allows for passive movement of water-soluble molecules. This transport is restricted by tight junctions and adherens junctions between endothelial cells. Examples of molecules that utilize the paracellular aqueous diffusion pathway include sucrose, (2) Receptor-mediated transocytosis is utilized by small proteins such as insulin, transferrin, leptin, IgG, and cytokines. (3) Transport proteins help passage molecules such as glucose, SCFAs, creatine and other metabolites, amino acids, and nucleosides. (4) Transcellular lipophilic pathways allow solutes to pass through the endothelial cell wall. This is utilized by lipid-soluble agents, such as O_2 and CO_2, nicotine, barbiturates, and caffeine. (5) Absorptive transocytosis allows passage of larger proteins such as albumin, histone, and other plasma proteins. Circulating immune cells may also extravasate across the BBB into the brain parenchyma, particularly under circumstances with weakened BBB integrity. *Figure was created with BioRender.com with full permission to publish.*

of the PET tracers used (Cosenza-Nashat, Zhao, Suh, et al., 2009). Regardless, the use of different imaging techniques to quantify the role neuroimmune signaling in addicts, and across addiction-states, has promising potential to improve our understanding and treatment of SUDs. The applications of human imaging to identify neuroimmune mechanisms of addiction is reviewed in further detail here (Woodcock, Hillmer, Mason, & Cosgrove, 2019).

4.1.2 Preclinical evidence

Much of the work examining the effects of central immune signaling on psychostimulant behavior and neurobiology has come from preclinical data. Recent years have brought a burgeoning body of research on the role of neuroimmune signaling in addictive-like phenotypes and neurobiological substrates. Considerable work has indicated the importance of microglia and astrocytes to psychostimulant pathology. While both microglia (Sekine et al., 2008) and astrocytes (Bowers & Kalivas, 2003) are activated by psychostimulants, herein we will focus our attention on microglia.

Cocaine administration in male mice upregulated activated microglia in the striatum as indicated by increased co-expression between the microglia label Iba1 and CD68—a marker of microglia activation (Liao, Guo, Niu, et al., 2016). This is corroborated by work demonstrating cocaine-induced increases in striatal microglia activation (Lewitus, Konefal, Greenhalgh, et al., 2016). This microglia activation drives production of the cytokine TNF-α, which in turn suppresses glutamatergic synapses in the NAc to attenuate cocaine-related behaviors (Lewitus et al., 2016). More recent demonstrates increases in microglial activation following intravenous self-administration of cocaine in rats, a behavioral paradigm with high translational relevance to human substance users (Cotto, Li, Tuma, et al., 2018). This study also found cocaine-induced increases in microglia-secreted brain-derived neurotrophic factor (BDNF), an important neurotrophic signaling molecule involved in synaptic plasticity that has been implicated in addiction pathology across drugs of abuse (Cotto et al., 2018). The aforementioned study by Lo lacono et al. also found increases microglia activation within the VTA as well as extensive peripheral cytokine production in response to cocaine exposure in stress-susceptible mice, and this lead to subsequent reduction in dopamine currents in the VTA (Lo Iacono et al., 2018). A pharmacological inhibition of microglial activation in the VTA reversed the observed functional deficits in dopaminergic neurons, highlighting microglia as contributing to cocaine-induced pathology

(Lo Iacono et al., 2018). Methamphetamine (METH) has also been well-established to stimulate microglia activation throughout the brain in a dose-dependent manner (Thanos, Kim, & Delis, 2016). METH also specifically degrades dopaminergic terminals within the striatum, and this process is associated with METH-induced microglial activation (Thomas, Walker, Benjamins, et al., 2004). Moreover, studies indicate that inhibition of microglia, either by the antibiotic ***minocycline*** or by the phosphodiesterase inhibitor/TLR4 antagonist ***ibudilast***, reduces the rewarding properties of psychostimulants (Attarzadeh-Yazdi, Arezoomandan, & Haghparast, 2014; Chen, Uz, & Manev, 2009; Snider, Hendrick, & Beardsley, 2013). More work is needed to identify the precise mechanisms by which psychostimulants act to induce microglia activation. A more in-depth review on the role of glia in psychostimulant abuse can be found here (Lacagnina, Rivera, & Bilbo, 2017). Altogether these studies suggest an important role for microglia and microglia-derived factors in regulating psychostimulant pathology and behavior.

One potential mechanism by which neuroimmune signaling modulates psychostimulant activity is via TLR4 signaling, a PRR robustly expressed by microglia in the CNS (Zhang & Wang, 2014). Northcutt et al. (Northcutt et al., 2015) demonstrated cocaine directly binds the TLR4 receptor. Further work has demonstrated that the psychostimulant methamphetamine (METH) engages TLR4 signaling, as well (Northcutt et al., 2015; Wang et al., 2019). This TLR4 signaling importantly contributes to addictive-like states. Cocaine-TLR4 signaling is critical for cocaine-induced elevations in dopamine in the NAc, as well as formation of cocaine conditioned place preference and cocaine self-administration (Northcutt et al., 2015). Subsequent work has shown that TLR4 is important for NMDA-dependent synaptic plasticity in the NAc (Kashima & Grueter, 2017). Moreover, the pharmacological inhibition of TLR4 in the VTA prevents drug-primed cocaine seeking in a model of cocaine self-administration (Brown, Levis, O'Neill, et al., 2018). Furthermore, METH induces activation of microglial TLR4 in the VTA, resulting in increased transcription of IL-6, elevated dopamine in the NAc, and formation of cocaine conditioned place preference (Wang et al., 2019). However, a study by Tanda et al. notes that the TLR4 antagonists (+)-naloxone and (+)-naltrexone do not specifically effect drug-associated neurochemical changes or behavioral outcomes (Tanda, Mereu, Hiranita, et al., 2016). Their group observed TLR4 antagonist did not decrease dopamine increases in response to acute drug, nor did it have behavioral effects specifically confined to drugs of abuse

(Tanda et al., 2016). While more work is needed to clarify the role of TLR4 in psychostimulant neuropathological outcomes, TLR4 represents a novel target that has strong promise as a treatment of psychostimulant disorders. At the very least, research into TLR4 provides new insight to the neurobiological mechanisms underlying drug abuse.

Psychostimulants also profoundly affect release of inflammatory mediators, such as cytokines. Recent efforts have been made toward defining the interactions between cytokine and cocaine effects, and how these may drive drug reward, motivation, and decision making to mediate the development and expression SUDs. Cocaine and METH have long been known to drive production of various cytokines (Wang, Huang, & Watson, 1994), such as IL-1β (Brown et al., 2018), IL-6 (Gonçalves, Martins, Ferreira, et al., 2008; Sil, Niu, Tom, et al., 2019), TNF-α (Gonçalves et al., 2008; Lewitus et al., 2016), and many more. Work from our group demonstrated that the cytokine granulocyte-colony stimulating factor (G-CSF) is upregulated in the periphery and in the brain following multiple cocaine treatment paradigms, and this increase in G-CSF is correlated with addictive-like behaviors (Calipari, Godino, & Peck, 2018). G-CSF is increased in the NAc following cocaine, as well as by selective activation of mPFC to NAc projections (Calipari, Godino, & Peck, 2018). G-CSF was also found to alter protein expression in the VTA related to synaptic plasticity and neuronal morphology (Mervosh, Wilson, & Rauniyar, 2018). Further work reveals G-CSF acts as a modulator of the mesolimbic dopamine circuit, where sustained increases in circulating G-CSF increased evoked dopamine release in the NAc (Kutlu, Brady, Peck, et al., 2018). Moreover, sex-specific mechanisms of G-CSF on dopamine and the rewarding properties of cocaine have been identified, where the effects of G-CSF in females was found to be modulated by ovarian hormone cycles (Brady, Hofford, Tat, et al., 2019). Finally, our group examined the effect of G-CSF administration to rats during intravenous cocaine self-administration. We demonstrated that G-CSF during drug extinction accelerated the rate of cocaine extinction, whereas G-CSF during drug abstinence attenuated cue-induced cocaine-seeking, suggesting G-CSF may function to enhance learning and memory processes related to cocaine use (Hofford, Euston, Wilson, et al., 2020). Altogether, this work represents some of the first endeavoring to extensively define the relationship between peripheral and central cytokine signaling on psychostimulant addictive-like phenomena.

Another cytokine that has received recent attention for its role in psychostimulant addictive-like phenomena is TNF-α. Microglia-derived

TNF-α has been shown to initiate the internalization of AMPA glutamate receptors in NAc neurons (Lewitus et al., 2014; Lewitus et al., 2016). NAc neurons are primarily excited by AMPA-type glutamate receptors, highlighting them as essential in the development and maintenance of maladaptive drug-associated behaviors. Psychostimulants have been thoroughly classified as affecting AMPAR dynamics in the NAc, thus affecting drug-associated behaviors (Conrad, Tseng, Uejima, et al., 2008). Microglial TNF-α thus represents a mechanism by which neuroimmune signaling may affect psychostimulant pathology. Lewitus et al demonstrated that repeated cocaine exposure induces microglial TNF-α production, which subsequently decreases AMPA receptor expression in the NAc core (Lewitus et al., 2016). This decrease in AMPA receptor expression in the NAc suppresses glutamatergic synaptic strength, effectively limiting the development of cocaine-induced locomotor sensitization (Lewitus et al., 2016). More recent work demonstrated that male rats experiencing an early life stressor exhibited elevations in striatal TNF-α, and this increase coincided with a reduction in levels of AMPA-type receptors in the striatum (Ganguly, Honeycutt, Rowe, et al., 2019). This males also demonstrated an increased preference for drug in a model of cocaine conditioned place preference (CPP), and treatment with the TLR4 antagonist ibudilast or a blocker of soluble TNF rescued both the AMPA receptor as well as the behavioral effect (Ganguly et al., 2019). This suggests a mechanistic link between TNF signaling and drug-cue and context conditioning. Moreover, TNF-α is increased following METH in an IL-6-dependent manner, and this increase appears to be protective against METH-induced microglial cell death (Coelho-Santos, Gonçalves, Fontes-Ribeiro, & Silva, 2012). This suggests these systems are important to minimizing the neurotoxic effects of METH, highlighting them as potential therapeutic targets. Cumulatively, this work represents a mechanism by which TNF-α may protect against development of maladaptive drug neuropathology. Together this work reframes the neurobiological underpinnings of psychostimulant addiction to include the role of central immune signaling.

4.2 Opioids

Opioids are a class of drugs including prescription painkillers, heroin and other morphine-derived drugs. These drugs alter reward-processing neurocircuitry, primarily the mesocorticolimbic dopamine system. There are multiple classes of opioid receptor, including μ, δ, κ, and nociceptin receptors. Opioids exert their anti-nociceptive effects by signaling at spinal and supraspinal sites expressing μ-, δ-, and κ- opioid receptors (Ossipov, Lai, King, et al., 2004). The periaqueductal gray (PAG), rostral ventromedial medulla (RVM), and locus coeruleus are opioid receptor-enriched regions that exert descending control of pain (Ossipov et al., 2004). Opioids also mediate analgesic affects via signaling at nociceptive C- and A δ fibers in the dorsal horn of the spinal cord (Millan, 2002; Ossipov et al., 2004). In contrast, opioid's reinforcing effects are largely attributed to signaling at the μ opioid receptor in brain reward regions (Matthes, Maldonado, Simonin, et al., 1996). Signaling through μ receptors acts to suppress local tonic GABAergic inhibition of dopaminergic neurons in the VTA, the consequence of which is increased dopamine release in the NAc through disinhibition (Coller & Hutchinson, 2012). Opioids also exert reinforcing effects in the NAc through signaling at local μ- and κ-opioid receptors located at dopamine terminals, further regulating dopamine release (Browne, Godino, Salery, & Nestler, 2020; Przewlocki, 2004). Opioid receptors are also expressed on non-neuronal cell types in the NAc (Machelska & Celik, 2020) that to contribute to changes in opioid reward (Schwarz, Smith, & Bilbo, 2013; Taylor, Castonguay, Taylor, et al., 2015). Through these diverse actions, opioids induce persistent adaptations to cellular function that result in the pathological drug- taking and seeking characteristic of addiction.

Understanding neuroimmune mechanisms contributing to opioid abuse pathogenesis is especially vital as opioids are often prescribed to individuals with chronic illnesses and weakened immune systems. Moreover, opioids are well known to affect immune cell function, playing an important role in infection and immunosuppression and their receptors have been detected on both astrocytes and microglia (Machelska & Celik, 2020). Herein we will review neuroimmune processes involved in opioid use disorder neuropathology and addictive-like behaviors.

4.2.1 Clinical evidence

Despite being broadly prescribed to patients for pain management, evidence suggests opioids worsen overall health outcomes. This has been demonstrated in particular in patients with HIV, where opioid users are more susceptible to HIV infection and disease complications (Donahoe & Vlahov, 1998; Risdahl, Khanna, Peterson, & Molitor, 1998). This has been partially attributed to the immunosuppressive effects of opioids (Plein & Rittner, 2018). A recent study found that HIV-positive heroin users exhibited decreased retrovirus restriction factors tripartite motif-containing (TRIM)

22, TRIM5α, apolipoprotein B mRNA editing enzyme catalytic polypeptide-like 3G (APOBEC3G), and IFN-α (Zhu, Liu, Mu, et al., 2017). This finding demonstrated how opioid immunosuppression contributes to a decreased ability to combat retroviral infection. Moreover, opioid users also display impaired function of monocytes/macrophages (Carrera, Catala, Monedero, et al., 1992), natural killer cells (Castilla-Cortázar, Castilla, & Gurpegui, 1998), granulocytes (Makman, Bilfinger, & Stefano, 1995) and T-cells (Govitrapong, Suttitum, Kotchabhakdi, & Uneklabh, 1998). Additionally, numerous immune cells produce endogenous opioids as well as express opioid receptors, highlighting the immune system in the actions of opioids.

The role of cytokine signaling in response to opioids has returned mixed results. Heroin users have been found to have elevated levels of circulating TNF-α, TGF-α and IL-8, suggestive of increased systemic inflammation (Piepenbrink, Samuel, & Zheng, 2016). Supporting this is another study that demonstrated increased neutrophil-lymphocyte ratios (NLR) and platelet-lymphocyte ratios (PLR) in male heroin users, a marker of increased systemic inflammation (Cıcek, Demırel, Cıcek, et al., 2018). This work indicates that increased inflammatory processes play a potential role in the pathophysiology of opioid use and dependence. Further work demonstrates μ opioid receptor agonism increases production of multiple cytokines, including TGF-β1, CCL5 and CXCR4 (Happel, Steele, Finley, et al., 2008). Moreover, morphine has been shown to stimulate production of the cytokine CCL2 in human neurons via a μ opioid receptor-related mechanism (Rock, Hu, Sheng, & Peterson, 2006). Despite this evidence for increased immune activation in response to opioids, there is also a large body of evidence demonstrating suppressed immune responses as a result of opioids. One study determined that healthy controls administered morphine had decreased cytotoxic natural killer (NK) cell responses compared to controls (Yeager, Colacchio, Yu, et al., 1995). In support of this, data suggest suppression of human NK cells across different classes of opioids, and that this effect is mediated by μ- and κ-receptor agonism (Maher, Walia, & Heller, 2019). Additionally, following an immune challenge via in vitro stimulation with LPS, cells from heroin abusers exhibit decreased levels of pro-inflammatory cytokines such as IL-1β, IL-6, TNF-α, and IFN-γ (Meijerink, Indrati, & Utami, 2015). More recently, data show that opioid abusers with HIV have reduced production of IL-10, IL-8, IL-6, IL-1α and TNF-α in response to an immune challenge by LPS when compared with HIV patients without an opioid use disorder (Underwood, Nguyen,

Uebelhoer, et al., 2020). Levels of TNF-α and IL-6 are negatively correlated with cognitive effects of opioid use disorder, indicating their importance in opioid user cognitive processes (Wang, Lee, Chang, et al., 2018), and elevated levels of IL-6 was correlated with poor methadone maintenance therapy outcomes (Lu, Wang, Lee, et al., 2019). In summary, opioid use induces disparate effects on immune signaling depending on drug class and regimen, but overall immune function plays a key role in opioid pathology. To this end, numerous trials have begun investigating the efficacy of glial inhibitors, such as ibudilast, in decreasing opioid addiction and physical dependence (Bachtell, Jones, Heinzerling, et al., 2017).

4.2.2 Preclinical evidence

While the neuronal actions of opioids remain a large focus of addiction literature, emerging evidence implicates an important role for glial cells and their mediated immune responses in opioid addiction pathology. Numerous studies have established morphine-induced increases in both glial fibrillary acidic protein (GFAP) and ionized calcium binding adaptor molecule 1 (Iba1) in rodents, markers of astrocyte or microglia activation, respectively (Beitner-Johnson, Guitart, & Nestler, 1993; García-Pérez, Luisa Laorden, Núñez, & Victoria Milanés, 2014; Goins & Bajic, 2018; Hutchinson, Lewis, Coats, et al., 2009; Marie-Claire, Courtin, Roques, & Noble, 2004; Zhang, Cui, Cui, et al., 2012). Morphine also increased brain expression of [^{18}F]DPA-714, a PET biomarker of glial activation in non-human primates (Liu, Guo, Zhang, et al., 2018). In addition to cell-surface markers of glial activation, opioid administration also increases a number of cytokines associated with glia activation, including IL-1β, TNF-α, CCL2 (MCP-1) [reviewed in (Lacagnina, Rivera, & Bilbo, 2017)], IFN-γ, IL-6, IL-10, CCL4, CCL17 [reviewed in (Zhang, Largent-Milnes, & Vanderah, 2020)], as well as numerous intracellular signaling pathways related to immune gene systems such as (extracellular signal-related kinase (ERK)/mitogen-activated protein kinase (MAPK) pathways, NF-κB pathways, activator protein-1 (AP-1)) [reviewed in (Coller & Hutchinson, 2012). Indeed, opioids have been extensively demonstrated to significantly affect gene expression profiles of glial cells to promote inflammation/immune pathways (García-Pérez et al., 2014; Hutchinson et al., 2009; Schwarz, Hutchinson, & Bilbo, 2011; Zhang, Liang, Levran, et al., 2017). Moreover, application of glial activation inhibitors attenuated both acquisition and maintenance of morphine CPP, as well as dopamine release in NAc (Hutchinson, Northcutt, Chao, et al., 2008; Narita,

Miyatake, Narita, et al., 2006; Williams, Probert, Stepankova, et al., 2006; Zhang et al., 2012). Indeed, the inhibition of microglia or astrocytes has been shown to decrease a variety of opioid-induced behaviors and side-effects, such as development of tolerance (Eidson & Murphy, 2013; Fukagawa, Koyama, Kakuyama, & Fukuda, 2013; Hutchinson, Coats, Lewis, et al., 2008; Raghavendra, Tanga, & DeLeo, 2004; Song & Zhao, 2001), morphine reward (Hutchinson, Northcutt, et al., 2008; Zhang et al., 2012), reinstatement of morphine CPP (Schwarz et al., 2011), and physical dependence (Bland, Hutchinson, Maier, et al., 2009; Hutchinson et al., 2009). It is important to note that the contributions of glia activation to opioid-induced behaviors appears brain-region dependent, and a protective role of microglial activation to opioid pathology has been described in the hippocampus (Rivera, Hanamsagar, Kan, et al., 2019). The opioid-induced activation of glial cells could be mediated by binding at various opioid receptors expressed by glial cells. Previous work has shown that all four members of opioid receptors (μ, δ, κ and nociceptin receptors) are expressed by glial cells (Chao, Gekker, Hu, et al., 1996; Dobrenis, Makman, & Stefano, 1995; Eriksson, Hansson, & Rönnbäck, 1992; Fu, Zhu, Wang, & Wu, 2007; Gurwell, Duncan, Maderspach, et al., 1996; Maduna, Audouard, Dembélé, et al., 2019; Meyer, Paisley, Mohamed, et al., 2017; Mika, Popiolek-Barczyk, Rojewska, et al., 2014; Nam, Han, Lee, et al., 2018; Shrivastava, Cabrera, Chastain, et al., 2017; Woo, Bae, Nam, et al., 2018). However, whether or not these receptors mediate opioid-induced glial activation remains unclear. There are some discrepancies regarding whether glia actually express opioid receptors, as some reports show that glia do not express opioid receptors (Corder, Tawfik, Wang, et al., 2017; Kao, Zhao, Lee, et al., 2012). Moreover, traditionally opioid receptor activation has been associated with immunosuppression, which is not in line with the observed glial activation and related immune responses.

Recently, growing evidence has suggested TLR4 signaling as a potential mechanism by which opioids trigger neuroimmune signaling. Initial evidence for this arose from in silico docking simulations that predicted opioid binding at the TLR4-MD2 complex, the same location where LPS binds TLR4 (Hutchinson, Lewis, Coats, et al., 2010; Hutchinson, Zhang, Shridhar, et al., 2010). This finding was later confirmed in vitro (Wang et al., 2012). This is in agreement with other studies demonstrating opioid-induced activation of signaling cascades downstream of TLR4 such as mitogen-activated protein kinases (MAPKs), including p38, JNK, and ERK (Wang et al., 2012; Zhang et al., 2012). Moreover, genetic deletion of TLR4 blocks this opioid-induced MAPK (Wang et al., 2012; Zhang et al., 2012). Further work demonstrates opioid TLR4 activation drives glial activation, as well, and this effect can be suppressed by TLR4 antagonists (Eidson & Murphy, 2013; Hutchinson, Zhang, et al., 2010; Wang et al., 2012; Zhang et al., 2012). Evidence supports that opioid-TLR4 signaling influences behavioral responses to opioids, as well. Administering the TLR4 antagonist (+)-naloxone prior to morphine exposure prevents development of morphine CPP (Hutchinson, Northcutt, Hiranita, et al., 2012). Likewise, transgenic mice with a knockdown of either TLR4 or the TLR downstream adaptor protein, MyD88, failed to establish CPP for oxycodone (Hutchinson et al., 2012). Additionally, chronic antagonism of TLR4 via prolonged minipump administration of (+)-naloxone during heroin withdrawal significantly reduced cue-induced reinstatement (Theberge, Li, Kambhampati, et al., 2013). Notably, other data demonstrate brain-region specific protective effects of TLR signaling. In the hippocampus, microglial deletion of *MyD88* enhanced opioid addiction-like behaviors such as indicated through prolonged extinction and enhanced reinstatement (Rivera et al., 2019). This suggests a potential protective role for hippocampal microglia *MyD88* signaling in reward learning and maintenance (Valentinova, Tchenio, Trusel, et al., 2019). Moreover, other work suggests that TLR4 signaling is not involved in the development of morphine tolerance, suggesting other mechanisms independent of TLR4 signaling could contribute to opioid use pathophysiology (Fukagawa et al., 2013). Evidence supporting this shows TLR4 is not required for opioid-induced tolerance or physical dependence (Mattioli, Leduc-Pessah, & Skelhorne-Gross, 2014) and the TLR4 antagonists (+)-naloxone and (+)-naltrexone do not affect opioid abuse-related behaviors at clinically relevant doses (Tanda et al., 2016). Altogether, this data highlights the need for further investigation of TLR signaling to elucidate how it may contribute to neuroimmune signaling in opioid pathology.

Glial activation, and the resultant neuroimmune signaling, contribute to pathogenesis of opioid abuse largely through central cytokine signaling. Some of the first evidence for functional consequences of opioid-induced neuroimmune signaling came from studies showing morphine CPP, as well as maintenance and reinstatement of morphine CPP, was augmented by intra-NAc infusions of astrocyte-conditioned medium (ACM), which consists of various cytokines (Arezoomandan, Khodagholi, & Haghparast, 2016; Narita et al., 2006). Chronic morphine has been shown to elevate levels of circulating IL-1β (Wang et al., 2012), IL-6 (Meng, Banerjee, & Li, 2015),

IL-17A (Meng, Banerjee, & Li, 2015), and acute morphine elevated levels of IL-10 (Schwarz et al., 2011), TNF-α, IL-2, and IFN-γ (Pacifici, Di Carlo, Bacosi, et al., 2000) in rodents. Blockade of the pro-inflammatory cytokine IL-1β or increasing of the anti-inflammatory cytokine IL-10 in the NAc of opioid-treated rats effectively reduced morphine CPP as well (Lim, Kim, McCabe, et al., 2014; Schwarz et al., 2011). Moreover, intra-NAc injections of plasmid DNA encoding IL-10 in male rats reduce self-administration of remifentanil self-administration, an ultra-short-acting synthetic opioid, in a dose-dependent manner (Lacagnina, Kopec, Cox, et al., 2017). TNF-α has also been shown to contribute to the negative symptoms associated during morphine withdrawal (Valentinova et al., 2019), and a dominant negative version of TNF reduces opioid tolerance (Eidson, Inoue, Young, et al., 2017). Additionally, the inhibition of IL-6 or fractalkine, otherwise known as CX3CL1, reduce morphine tolerance (Hutchinson, Coats, et al., 2008). Thus, it appears that pro-inflammatory cytokine signaling promotes opioid neuropathology. Further evidence demonstrating pro-inflammatory potentiation of opioid-induced reward supports this idea. Rats given IFN-α displayed augmented relapse-like behaviors during reinstatement of morphine-CPP (Lang, Wang, Zhai, et al., 2009). Altogether these data indicate the important contribution of neuroimmune signaling to opioid abuse and emphasize the necessity of continued exploration of these largely novel role neuroimmune signaling plays in addiction development and maintenance.

5. Mechanisms of gut microbiome-brain signaling

Recent work has established the importance of peripheral signals from the gut microbiome in proper brain development, function, and behavior (Dinan & Cryan, 2017; Lu, Synowiec, & Lu, 2018). The gut microbiome communicates bidirectionally with the central nervous system (CNS), and this so called 'gut-brain-axis' has been implicated in a variety of neuropsychiatric disorders including autism, major depressive disorder, anxiety, multiple sclerosis, and more recently, addiction (Hofford et al., 2018; Lee, Menezes, Umesaki, & Mazmanian, 2011; Meckel & Kiraly, 2019; Scriven, Dinan, Cryan, & Wall, 2018). However, the exact mechanisms mediating this gut-brain communication are yet to be fully elucidated. Potential mechanisms include neural, endocrine, metabolic, and immune mechanisms (Dinan & Cryan, 2017; El Aidy, Dinan, & Cryan, 2015). Given the intimate relationship between immune function and the microbiome, the gut-immune-brain axis emerges as a promising mechanism mediating the gut microbiome's effects on the brain and behavior. Additionally, critical periods of brain development closely parallel the dynamic developmental periods of the immune system and gut microbiome, poising them all to be closely related across the lifespan.

5.1 Gut microbiome regulation of the immune system

The gastrointestinal tract is home to hundreds of trillions of microorganisms. These microbial communities inhabiting the gut are termed the ***microbiota***, and the genes they transcribe are termed the ***microbiome*** *(*Clemente, Ursell, Parfrey, & Knight, 2012*)*. The gut microbiome is capable of effecting numerous physiological processes, including metabolism, and immune cell function (Chu, Murdock, Jing, et al., 2019). The gut microbiome shares a particularly close relationship with the host immune system. The microbiome has co-evolved alongside the mammalian immune system over a millennium to support development and function. In addition to harboring the highest density of commensal microorganisms of the body, the gut also hosts and the largest amount lymphoid tissue in the body (Huh & Veiga-Fernandes, 2020; Sorini, Cardoso, Gagliani, & Villablanca, 2018; Vighi, Marcucci, Sensi, et al., 2008). Perhaps unsurprisingly, then, the gut microbiome plays a fundamental role in the development, training, and function of the host immune system (Belkaid & Hand, 2014).

The gut microbiome affects local and systemic immune responding through two primary mechanisms. First, we will discuss gut-immune interactions that are caused by commensal bacteria signaling to ***Gut-Associated Lymphoid Tissues (GALTs)***. GALTs are comprised of numerous immune cells, including innate lymphoid cells (Hepworth, Fung, Masur, et al., 2015; Lehmann, von Burg, Ivanek, et al., 2020; Mortha, Chudnovskiy, Hashimoto, et al., 2014), T cells (Sorini, Cardoso, Gagliani, & Villablanca, 2018), B cells (Mora, Iwata, Eksteen, et al., 2006), macrophages (Farache, Zigmond, Shakhar, & Jung, 2013; Zigmond & Jung, 2013), and dendritic cells (Dillon, Lee, Kotter, et al., 2016; Jiao, Wu, Huntington, & Zhang, 2020). These cells sense and respond to the commensal microbes of the gut with great specificity and are crucial to maintaining immune tolerance and supporting immune homeostasis. Evidence demonstrates direct gut-immune interactions that are regulated by commensal microbes in the gut. For example, Ivanov et al demonstrated that the bacteria segmented filamentous bacterium (SFB) is sufficient to induce production of T helper 17 cells (Th17 cells) that produce the cytokines IL-17 and IL-22 (Atarashi, Tanoue,

Ando, et al., 2015; Ivanov, Atarashi, Manel, et al., 2009). Atarashi et al demonstrated that bacteria of the genus Clostridium promotes T regulatory cell (Treg) accumulation and production of the anti-inflammatory cytokines transforming growth factor-β (TGF-β) and interleukin-10 (IL-10) (Atarashi, Tanoue, Oshima, et al., 2013; Atarashi, Tanoue, Shima, et al., 2011). Importantly, these microbe-induced immune responses are not limited to the local environment, but affect systemic immune responses through cytokine release into circulation (Sorini, Cardoso, Gagliani, & Villablanca, 2018). Circulating cytokines are important mediators of the gut-immune–brain communication, as they can reach be transported from blood to brain to induce neuroimmune signaling that is associated with changes brain function and behavior. Evidence suggests that TLRs can be activated by intestinal microbiota signaling, making TLRs important for microbiome-mediated immune response (Rogier, Koenders, & Abdollahi-Roodsaz, 2015). TLR signaling is also important for maintenance of gut barrier functions, which prevents translocation of bacteria into the blood and subsequent inflammation (Rogier et al., 2015). Thus TLRs represent a channel between microbiome signaling and immune system homeostasis (Kelly, Kennedy, Cryan, et al., 2015).

Another key mechanism by which the gut microbiome affects immune function is via the production of bioactive metabolic byproducts. Evidence demonstrates that microbial metabolites produced from the gut called ***short-chain fatty acids (SCFAs)*** regulate immune responses of GALTs via epigenetic mechanisms (Jiao et al., 2020). SCFAs are produced by microbial anaerobic fermentation of non-digestible fibers in the gut (Louis & Flint, 2017) and have been demonstrated to have extensive health benefits, largely from their anti-inflammatory effects. The SCFA butyrate, a by-product of starch fermentation, facilitates production of Treg cells in the local gut environment as well as the periphery (Arpaia, Campbell, Fan, et al., 2013; Furusawa, Obata, Fukuda, et al., 2013) promoting anti-inflammatory responses. The SCFA propionate has also been found to prevent the systemic and central inflammation associated with multiple sclerosis pathogenesis (Duscha, Gisevius, Hirschberg, et al., 2020; Haghikia, Jörg, Duscha, et al., 2015). SCFAs are also released into circulation, where they can have far-reaching effects on immune functions in different tissues, particularly the brain. In addition to SCFAs, gut commensal bacteria can convert ***bile acids*** into various biologically active substances, such as hormones, important in shaping host innate immune response. The bacterial transformation of bile acids seems crucial for gut microbe control of Treg cells homeostasis in the gut and periphery, another key mechanism by which microbial signals drive systemic host immune responses (Campbell, McKenney, & Konstantinovsky, 2020; Song, Sun, Oh, et al., 2020). Altogether, this demonstrates the central role of the gut microbiome in regulating immune responses.

5.2 Gut-immune-brain connection

Primary evidence for behavioral modulation by gut-immune signaling came from ***germ-free (GF) mice***. GF mice have no commensal bacteria and are raised in a sterile/gnotobiotic environment (Foster, 2016). GF mice display abnormal systemic immune systems with a reduction in numbers B cells and T cells (Foster, 2016). When inoculated with microbes at 3 weeks of age, the systemic immune system of GF mice remains permanently altered, with changed cytokine production and immune system reactivity (Hansen, Nielsen, & Kverka, 2012). GF mice also display robust behavioral differences. They exhibit enhanced stress reactivity and altered anxiety-like behaviors (Neufeld, Kang, Bienenstock, & Foster, 2011a) and this coincides with altered gene expression in the amygdala and hippocampus (Neufeld, Kang, Bienenstock, & Foster, 2011b). Germ free mice also exhibit compromised BBB integrity (Braniste, Al-Asmakh, Kowal, et al., 2014) and impaired development and function of the innate immune cells of the brain (Erny, De Angelis, Jaitin, et al., 2015). These results support the idea that gut microbiome signaling through immune cells critically effects brain function.

More insight to the effects of gut microbiome dysbiosis on immune and brain functions has come from rodent models with antibiotic-induced depletion of their gut microbiome. While these antibiotic treatments differ in recipes and regimens, they still yield important insight to how precise temporal manipulations of the gut can affect brain and behavior. One such study revealed that early-life exposure to penicillin had long-term effects, including diminished sociability and anxiety-like behaviors, as well as heightened aggression that coincided with increased cytokine expression in the frontal cortex and impaired BBB integrity, allowing immune cell extravasation to the brain (Leclercq, Mian, Stanisz, et al., 2017). Indeed, antibiotic-induced perturbations to the gut microbiome leads to altered pro-inflammatory immune responses (Erny et al., 2015). Antibiotic depletion of gut microbes has also been shown to increase markers of neuroinflammation as well as alter microglial morphology (Jang, Lee, Jang, et al., 2018; Minter, Hinterleitner, Meisel, et al., 2017; Minter, Zhang,

Leone, et al., 2016). Both GF mice and mice with antibiotic-induced microbiome depletion, exhibit altered microglial inflammatory gene expression profiles, altered morphology, and attenuated threat responses (Erny et al., 2015).

Further links between gut-immune-brain signaling come from patients with chronic inflammatory disorders of the gastrointestinal such as Inflammatory bowel disease (IBD). Patients with IBD exhibit altered gut microbiomes alongside dysregulated T cell networks, levels of cytokines, cytokine receptors, and regulators of cytokine signaling (Chen & Sundrud, 2016; Glassner, Abraham, & Quigley, 2020). Importantly, patients with IBD display reduced cognitive and psychomotor function (Tadin Hadjina, Zivkovic, Matetic, et al., 2019), highlighting the importance of gut-immune-brain interactions. The communication between gut-immune-brain signaling thus has been suggested to be mediated by two potential mechanisms: (i) microbiome-induced altering to levels of circulating cytokines that are transported from blood to brain or (ii) gut-derived metabolites, such as SCFAs, that diffuse to the brain where they effect gene transcription and function of resident immune cells, such as microglia. SCFAs from the gut are known to cross the BBB where they can directly affect glial and neuronal function. Signaling from SCFAs has been shown to closely regulate microglia homeostasis, maturation and function (Erny et al., 2015). Further work demonstrated stress-induced increases in intestinal permeability, stress responsivity, and long-term anhedonia (van de Wouw, Boehme, Lyte, et al., 2018). Importantly, all of these changes were ameliorated by SCFA supplementation (van de Wouw et al., 2018). Notably, these two mechanisms are not mutually exclusive, as SCFA signaling also regulates local immune responses in the gut and cytokine production. These studies lend novel insight as to how gut-immune-brain signaling may occur, and the conceivable importance of this communication in neuropsychiatric conditions (Fig. 2). The role of SCFA as a mechanism for gut-brain communication is discussed further in Box 1.

6. Gut microbiome and substance use disorders

Evidence culminating over the past several years suggests the gut microbiome plays a key role in the neurobiology of addiction. Despite the compelling and growing clinical and preclinical evidence for gut-brain communication in addiction, mechanistic studies remain limited. Thus, the exact mechanisms mediating gut-brain communication in the context of

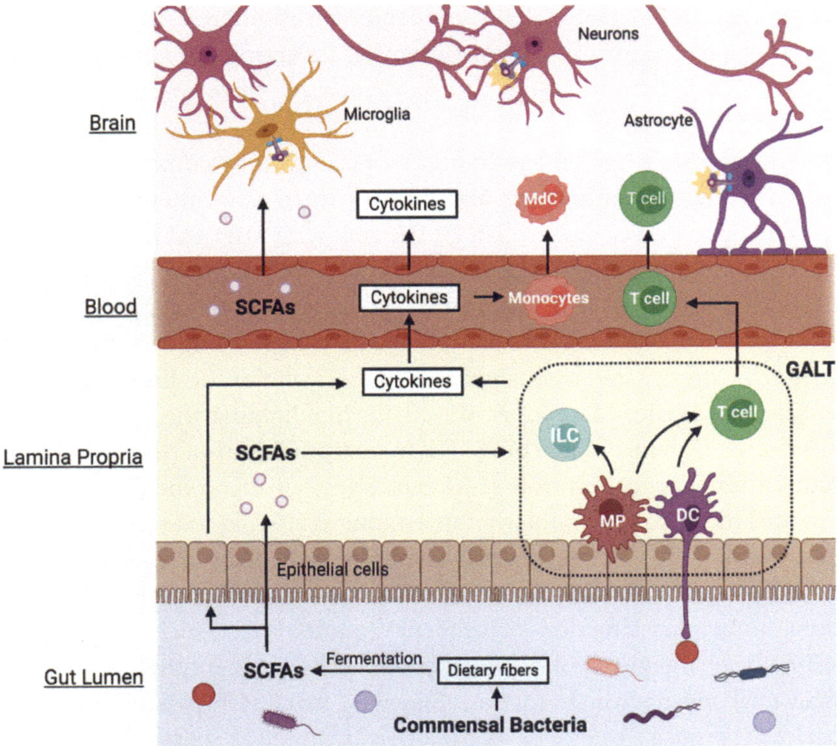

Fig. 2 A potential mechanism for gut-immune-brain communication. Resident microbes of the gastrointestinal tract are able to affect local and systemic immune responses. These immune responses are primarily initiated by the gut-associated lymphoid tissues (GALT), the largest mass of lymphoid tissue in the body. It consists of T cells, B cells, macrophages, dendritic cells, and innate lymphoid cells. Importantly, the GALT plays a central role in mediating immune homeostasis, balancing tolerance and immunity to antigens. Cells of the GALT produce immune responses in response to signals from commensal gut microbes, namely through modulating levels of systemic cytokines and circulating immune cells, which can cross the BBB to affect brain function. Cytokine receptors are expressed by microglia, astrocytes, and neurons. Immune responses stimulated by the gut may also aide immune cell extravasation into the brain parenchyma. Commensal gut microbes also drive production of dietary metabolites called SCFAs. SCFAs can modulate local immune responses in the gut. SCFAs also cross the BBB, and have been evidenced to affect microglia and neurons, potentially through epigenetic mechanisms. *Figure was created with BioRender.com with full permission to publish.*

BOX 1 Short chain fatty acids

Short chain fatty acids (SCFAs) are metabolic byproducts derived from colon microbial carbohydrate fermentation of dietary fibers that escape digestion in the upper gastrointestinal tract (Koh, De Vadder, Kovatcheva-Datchary, & Bäckhed, 2016). The most abundant SCFAs include butyrate, propionate, and acetate. These SCFAs have numerous roles in health and disease, and can act as energy substrates, inhibit histone deacetylases, and activate G protein-coupled receptors (GPCRs) (Koh et al., 2016). In the gut, SCFAs act locally to promote immune homeostasis and gut barrier integrity (Morrison & Preston, 2016; Smith, Howitt, Panikov, et al., 2013). In circulation, SCFAs regulate systemic immune function by stimulating T regulatory cells (Treg) differentiation and by mediating the production of cytokines (Silva, Bernardi, & Frozza, 2020). Circulating SCFAs, mostly butyrate but propionate and acetate as well, are also known to readily cross the BBB via the transport proteins located on endothelial cells, where they go on to exert effects directly on microglia and neurons (Joseph, Depp, Shih, et al., 2017). SCFAs have been shown to have widespread effects on brain function, highlighting these metabolites as key candidate mediators of gut-brain communication (reviewed extensively here: Dalile et al. (Dalile, Van Oudenhove, Vervliet, & Verbeke, 2019)). Microbiome-derived SCFAs play key roles in maintaining BBB integrity (Braniste et al., 2014) and mediating appropriate maturation and function of microglia (Erny et al., 2015). Recent work from our group demonstrated a potential role for SCFAs in a model of SUD. Our group found an antibiotic-induced depletion of the gut microbiome rendered male mice more susceptible to developing maladaptive cocaine behaviors, and repletion with SCFAs normalized drug behaviors (Kiraly et al., 2016). Other studies indicate that systemic SCFA administration produces antidepressant-like effects in rats, as well as effecting central neurotransmission and BDNF expression (Sun, Wang, Hong, et al., 2016). While the precise mechanisms by which SCFAs alter brain function are not fully understood, one promising mechanism is via their role as histone deacetylase inhibitors (HDACs), thus promoting histone acetylation and gene expression (Stilling, van de Wouw, Clarke, et al., 2016). SCFAs further promote acetylation activity by increasing activity of histone acetyltransferases (HATs) (Dalile et al., 2019). Altogether, this results in an increase of transcriptionally active chromatin. SCFAs may also affect brain function by binding to GPCRs expressed throughout the body, including in the central nervous system such as the free fatty acid receptor 2 and 3 (FFAR2 and FFAR3), GPR109a/HCAR2 (hydrocarboxylic acid receptor), and GPR164 (Silva et al., 2020). Once activated, these GPCRs initiate signaling cascades that include PLC, MAPKs, PLA2, and NF-κB pathways (Dalile et al., 2019). Moreover, SCFAs can directly activate vagal afferents via FFAR signaling, thereby communicating to the brain. SCFAs can also stimulate vagal pathways by promoting production of gut hormones, such as glucagon-like peptide 1 (GLP1) and peptide YY (PYY), from the enteroendocrine L cells (Dalile et al., 2019). Another potential mechanism by which SCFAs alter in brain function and behavior via effects on other gut–brain signaling pathways, such as the immune and endocrine systems (Clarke, Stilling, Kennedy, et al., 2014; Dalile et al., 2019). Thus, further studies are needed to elucidate the precise mechanisms by which microbiome derived SCFAs exert their effects on brain and behavior.

substance use disorders is still an area of active research. Potential mechanisms for this gut-brain communication include the immune system, neuroactive metabolites, or direct signaling via neural afferents. Notably, all of these systems exhibit vast sexual dimorphisms, and these may contribute to the observed sex differences in SUDs (Becker & Chartoff, 2019). Sex differences in the gut microbiome are beginning to be classified (Jašarević, Morrison, & Bale, 2016), and how these contribute to divergent neurobiology in addiction is still an area of active research (reviewed in (Jašarević, Morrison, & Bale, 2016)). While much of the work to date examining for gut-brain communication in addiction has been performed in males, one recent study from Peterson et al demonstrates that gut microbiome is associated with aspects of addiction-phenotype, such as impulsivity, in a sex-dependent manner in male and female rats (Peterson, Richards, Meyer, et al., 2020). Further work is needed to identify the contribution of sex differences in the microbiome to addiction-like behaviors. The existent addiction work investigating effects of microbiome-brain signaling in females, and any *sec* differences, will be indicated. Here in we will provide an overview of the data that demonstrate gut-brain communication may be contributing to the pathogenesis of substance use disorders, focusing in psychostimulants and opioids. A more extensive review on the role of the gut microbiome in substance use disorders can be found elsewhere (reviewed in (Meckel & Kiraly, 2019)).

6.1 Psychostimulants

While the clinical evidence for the role of microbiome in psychostimulant use disorder remains sparse, existent research suggests a critical role for gut-brain signaling in psychostimulant use disorder. One study by Volpe et al. (Volpe, Ward, Mwamburi, et al., 2014) investigated the composition of the microbiome in patients with and without HIV who qualified for cocaine use disorder compared to healthy controls. This study found cocaine use disorder was an independent predictor of gut microbial dysbiosis, with cocaine users exhibiting a decreased relative abundance of Firmicutes and increased relative abundance of Bacteroidetes (Volpe et al., 2014). Moreover, cocaine users had a strong trend toward increased levels of bacterial DNA in their serum, suggestive of increased bacterial translocation from the gut microbiome which may result in a pro-inflammatory state (Volpe et al., 2014). While this clinical data is limited, there is growing evidence from translational preclinical literature demonstrating a pivotal role for the gut microbiome in behavioral responses to drugs of abuse. One such study by Scorza et al. (Scorza, Piccini, Martínez Busi, et al., 2019) examined the effects of volatilized cocaine as well as phenacetin and caffeine, two common adulterants in cocaine, on the on the structure, diversity, and functionality of the gut microbiome in male rats. This study is especially translationally relevant, as volatilized cocaine with adulterants is most similar to cocaine is frequently smoked by individuals with substance use disorder. Cocaine, as well as phenacetin, induced significant shifts in fecal microbial communities (Scorza et al., 2019). Moreover, fecal samples from cocaine treated rats revealed increased levels of microbial genes related to the synthesis of the SCFA butyrate (Scorza et al., 2019). While further research is necessary to determine if these shifts in microbial populations affects behavioral responses to cocaine, this represents some of the first evidence of psychostimulant-induced microbiome dysbiosis.

Recent work from our group aimed to define the behavioral and neurobiological consequences of an antibiotic-induced gut microbiome knockdown in a mouse model of cocaine use disorder. We placed male mice on a cocktail of broad spectrum antibiotics to significantly reduce total bacteria as well as bacterial species diversity in the gut microbiome, without altering cocaine metabolism or serum corticosterone levels (Kiraly, Walker, Calipari, et al., 2016). This gut microbiome depletion increased sensitivity to cocaine in both CPP and locomotor sensitization (Kiraly et al., 2016). Notably, supplementation of SCFAs to antibiotic treated mice was sufficient to lower drug preference in CPP back to control levels, indicating the importance of SCFAs in the behavioral effects of gut microbiome (Kiraly et al., 2016). Moreover, depletion of the gut microbiome also altered gene expression in the NAc, notably increasing levels of *Bdnf*, *Drd1*, *Drd2*, and *Gria2* AMPA receptors (Kiraly et al., 2016). This is in line with other literature showing the gut microbiome affects expression of BDNF and dopamine receptors in the brain (Bercik, Denou, Collins, et al., 2011; Desbonnet, Clarke, Traplin, et al., 2015; Heijtz, Wang, Anuar, et al., 2011; Savignac, Corona, Mills, et al., 2013). Importantly, both BDNF and dopamine receptors contribute to neuropathology of various drugs of abuse (Graham, Edwards, Bachtell, et al., 2007). Thus, further investigation into the causal links between the gut microbiome and BDNF and dopamine signaling in the brain is needed. It is also important to note that another study by Lee et al. (Lee, Vuong, Nusbaum, et al., 2018) found cocaine CPP was decreased when male mice were pretreated with antibiotics prior to CPP testing. These two studies utilized differing antibiotic regimens and treatment intervals, highlighting the importance of duration and extent of microbiome depletion in drug-related behaviors and the complexity of this gut-brain connection. Another study by Ning et al. (Ning, Gong, Xie, & Ma, 2017) found that METH administration lead to modest increases in relative abundance of several bacteria populations. Significant decreases in caecal levels of the SCFA propionate were also observed (Ning et al., 2017). While more work is needed, mounting evidence indicates the importance of gut microbiome on behavioral response to psychostimulants.

6.2 Opioids

While a number of clinical studies exist connecting gut microbial dysbiosis and opioid use, these studies are often limited in their interpretation by confounding variables such as underlying patient comorbidities, concomitant non-opioid drug use, and varied opioid doses and duration of use (Acharya, Betrapally, Gillevet, et al., 2017; Barengolts, Green, & Eisenberg, 2018; Vincent, Miller, Edens, et al., 2016; Xu, Xie, Wang, et al., 2017; Zhernakova, Kurilshikov, Bonder, et al., 2016). Work by Acharya et al. (Acharya et al., 2017) found that male and female patients with hepatic encephalopathy (HE) it was opioid use, and not liver disease, that was predictive of readmission. Further, chronic opioid use was associated with significant dysbiosis of the gut microbiome (Acharya et al., 2017). Other studies have found opioid use to be significantly associated with an increase in in alpha diversity, potentially caused by opioid-induced delays in colon transit time that leads to increased growth of gut bacteria (Meckel & Kiraly, 2019; Vincent et al., 2016; Zhernakova et al., 2016). A study by Xu et al. (Xu et al., 2017) discovered significant alterations in the intestinal microbiome of male patients with various SUDs, but no alterations specific to heroin, methamphetamine, or ephedrine were present. This could be due to multiple confounding factors, including differences in lifestyles and diets of patients with SUDs compared to healthy controls (Xu et al., 2017). Putative metagenomes analysis of the microbiome from patients with SUDs also exhibited an increase in pathways predicted to be involved in translation, cell growth and death, and DNA replication and repair, whereas pathways related to cellular processes and signaling, and metabolism were decreased (Xu et al., 2017). Thus, SUDs not only alter gut microbial communities, but also their functional properties.

Recent preclinical studies in rodent models have confirmed much of these clinical observations and has begun to elucidate the mechanisms by which this affects drug behavior. One such study found that that morphine treatment rapidly induced dramatic alterations in the gut microbiome, in as little as 1 day of morphine pellet implantation (Wang, Meng, et al., 2018). Experimenter administered morphine increased the systemic spread of bacteria through a compromised intestinal epithelial barrier (Banerjee, Sindberg, Wang, et al., 2016; Kang, Mischel, Bhave, et al., 2017; Meng, Banerjee, & Li, 2015). Further, this morphine-induced bacterial spread and intestinal epithelial barrier compromise was found to be dependent on the pro-inflammatory IL-17A, where treatment with IL-17A reduced these effects as well as decreased IL-6 production (Meng, Banerjee, & Li, 2015). Morphine treatment has been shown to reduce levels of the gram-negative Bacteroidetes, and conversely, expand levels of the gram-positive Firmicutes (Banerjee et al., 2016; Kang et al., 2017; Meng, Banerjee, & Li, 2015; Meng, Yu, & Ma, 2013; Wang, Meng, et al., 2018). Notably, reductions in the ratio of Bacteroidetes to Firmicutes has been implicated with increased systemic inflammation in obese patients (Verdam, Fuentes, & De Jonge, 2013). Other work found the administration of the gram-positive bacteria *Enterococcus faecalus* was sufficient to accelerate the development of morphine tolerance (Wang, Meng, et al., 2018). Agreeing with this work, studies found that administration of oral antibiotics to which lower systemic bacterial spread attenuated morphine tolerance (Kang et al., 2017). Studies in non-human primates have demonstrated that morphine-induced alterations to microbial communities and metabolic profiles is maintained with long-term morphine treatment (Sindberg, Callen, Banerjee, et al., 2019). Thus, opioid-induced changes to the gut microbiome offer a novel mechanism by which opioid abuse and dependence form.

7. Conclusion

In this review we have provided an overview of the neuroimmune mechanisms contributing to both psychostimulant and opioid disorders and highlighted the involvement of the gut microbiome. While the interactions between the brain and the immune system are tremendously complex, they warrant further investigation to gain mechanistic understanding of neuroimmune mechanisms in the development and persistence of substance use disorders. Based upon our overview, it is clear that various substances engage immune signaling pathways to alter the brain and behavior. Further research is needed to gain better insight to the nuanced effects of these drugs on neurobiological mechanisms and behavioral outcomes.

From this analysis of the literature, it is clear the distinction between immune responses as being either 'pro-inflammatory' or 'anti-inflammatory' is not sufficient and does not accurately capture the nuanced properties of these molecules (i.e. the pro-inflammatory cytokine IL-1β induces an enhanced behavioral response to both cocaine and opioids (Hutchinson et al., 2012; Northcutt et al., 2015) whereas the pro-inflammatory cytokine TNF-α dampens cocaine responses (Lewitus et al., 2016). Indeed, many studies have demonstrated both pro- and anti- inflammatory roles for various cytokines, depending on the context and cytokine milieu (Balasubbramanian, 2019; Kosloski et al., 2013; Veldhoen et al., 2006). Therefore, a better classification of neuroimmune responses is needed.

While our understanding of the role neuroimmune signaling, and the gut microbiome, play in substance use disorders is rapidly expanding, it is still far from complete. Neuroimmune pathways involved in psychostimulant pathology remain preliminary, and future studies must be geared toward specific intracellular signaling pathways. Additionally, while much of the work now published has focused on the mesocorticolimbic system, further research into other brain regions affected by drugs of abuse represents a key target for clarifying the roles of these influences on substance use disorders. Further, studies must quantify neuroimmune participation in response to drugs across different phases of the addiction cycle, including during relapse/reinstatement, to gain a full picture of what is happening. Additionally, our understanding of how the gut microbiome plays a role in addiction is still in its early stages. It's known that SUD can have profound effects on patients diets, and diet composition and nutritional status critically modify gut microbiome composition (Sandhu, Sherwin, Schellekens, et al., 2017). Thus, it is possible that dietary differences could contribute to the distinct intestinal microbiome profile in SUDs (Xu et al., 2017); however, preclinical research on rodents with controlled diets suggests drug abuse itself is capable of modifying microbiome composition (Scorza et al., 2019; Wang, Meng, et al., 2018). Regardless, this work highlights the important impact of the microbiome SUD pathology, indicating the possibility of nutritional interventions as tractable treatments for SUDs.

From the work summarized here, the gut microbiome crucially influences behavioral response to drugs of abuse. While the precise mechanisms mediating gut-brain communication in addiction is unclear, given the extensive gut-immune interactions outlined, the immune system represents a promising candidate. Going forward a more nuanced understanding of which bacterial populations, metabolites, cytokines and immune response affect the behavioral response to drugs of abuse will be essential. It is crucial the field integrate a more complete perspective of addiction biology, extending it to microbiome and immune system, to develop a full understanding of addiction neurobiology and potential treatment avenues.

Acknowledgments

Preparation of this manuscript was supported by NIH grants DA044308, DA051551, and DA049568 to D.D.K. as well as a NARSAD Young Investigator Award also to D.D.K. We would like to thank Dr. Rebecca Hofford and Ms. Katherine Meckel for thoughtful input on manuscript content and layout.

Glossary

Lipopolysaccharide (LPS) A component of Gram-negative bacteria cell wall that serves as a protective barrier for bacteria. LPS is a strong mediator of inflammation and septic shock. In research, administration of purified LPS can induce a robust immune response without making the animal sick (as there is no bacteria present, just the LPS).

Pattern recognition receptors (PRRs) A receptor of the innate immune system specialized to recognize molecular patterns that are conserved amongst foreign pathogens and self-derived molecules from damaged cells, called Pathogen-Associated Molecular Patterns (PAMPs) and damage-associated molecules patterns (DAMPs), respectively.

Toll-like receptors (TLRs) A family of pattern recognition receptor comprised of 10 members (TLR1–TLR10) in human and 12 (TLR1–TLR9, TLR11–TLR13) in mouse. They are membrane-spanning receptors that are localized to either the cell surface or to endosomal vesicles.

Cytokines Refers to a broad class of small proteins, peptides, or glycoproteins (~5–20 kDa) that are secreted, typically by immune cells, and signal to and affect the function of other cells by binding to cell surface receptors. Cytokines include interleukins, lymphokines and cell signal molecules. Cytokines function largely to modulate inflammatory and immune responses.

Chemokines A class of cytokines that function to direct chemotaxis of responsive cells, facilitating cell migration.

Growth factors A class of cytokines that stimulate the growth, proliferation, differentiation, and survival of cells.

Interferons (IFNs) A class of cytokines with three main classes (alpha, beta, and gamma) that is produced in response to a virus to inhibit virus replication.

Lymphokines A subset of cytokines produced by immune cells known as lymphocytes, which include T cells, B cells, and Natural Killer cells.

Interleukin (IL) Refers to a large group of cytokines that was first observed to be produced by white blood cells, including lymphocytes, granulocytes, monocytes, and macrophages. Interleukins are now known to be produced by many other body cells.

Neuroimmune system The interface of CNS and immune system that functions to protect neurons from pathogens, but also affect neuronal processes underlying memory, emotion and cognition. The neuroimmune system seems to be critically involved in numerous neuropsychiatric conditions, and plays a central role in brain development, homeostasis, and response to injury.

Neuroinflammation An inflammatory response within the nervous tissues mediated by brain's innate immune system in response to an inflammatory challenge/ insult/injury.

Nitric oxide synthase

A family of enzymes that catalyze the production of nitric oxide (NO), a chemical that can participate in immune responses generated by phagocytes (monocytes, macrophages, and neutrophils).

Quinolinic acid (QUIN or QA) Also called pyridine-2,3-dicarboxylic acid, is a metabolite of tryptophan and the biosynthetic precursor to nicotine. QUIN is also neurotoxic in high concentrations and may play a role in neurodegenerative disorders.

Neurotrophic A class of neuropeptides, such as nerve growth factor (NGF), that mediates the growth, differentiation, and survival of developing and mature neurons.

Complement proteins An intricate system of more than 30 proteins that function as a part of the innate immune system to promote clearance of infectious microorganisms.

Intravenous self-administration An operant conditioning procedure in which animals are trained to lever press/respond to obtain infusion of drug (nicotine, cocaine, heroin, etc). This model closely replicates the human condition of drug taking, seeking, and dependence, making it a useful tool for preclinical investigation.

Minocycline A broad-spectrum tetracycline antibiotic that functions in the brain to inhibit microglial activation.

Ibudilast A phosphodiesterase inhibitor that also inhibits glial cell activation.

Gut-associated lymphoid tissues (GALTs) Refers to the organization of lymphoid tissues found in the gastrointestinal tract that serves to regulate the mucosal immune system of the gut, preventing invasion of infectious microbes.

Microbiota The resident microorganisms of a particular environment (e.g. the gastrointestinal tract) including bacteria, viruses, and fungi.

Microbiome The aggregate genetic material of the resident microorganisms comprising a given microbiota. Often used interchangeably with *microbiota*.

Bile acids Hydroxylated steroids, synthesized in the liver that function predominantly to help digest fats.

Short chain fatty acids Metabolic byproducts derived from colon microbial carbohydrate fermentation of dietary fibers that escape digestion in the upper gastrointestinal tract, including butyrate, propionate, and acetate.

Germ-free (GF) animals Differ from conventional animals in that they have no microorganisms living in or on them. They also lack all systemic responses to commensal microorganisms. This makes GF animals highly valuable to the study of how microbiota affect vast physiological functions.

References

Acharya, C., Betrapally, N. S., Gillevet, P. M., et al. (2017). Chronic opioid use is associated with altered gut microbiota and predicts readmissions in patients with cirrhosis. *Alimentary Pharmacology & Therapeutics*, *45*, 319–331. https://doi.org/10.1111/apt.13858.

Allen, N. J., & Eroglu, C. (2017). Cell biology of astrocyte-synapse interactions. *Neuron*, *96*, 697–708.

Allen, N. J., & Lyons, D. A. (2018). Glia as architects of central nervous system formation and function. *Science*, *362*, 181–185.

Araos, P., Pedraz, M., Serrano, A., et al. (2015). Plasma profile of pro-inflammatory cytokines and chemokines in cocaine users under outpatient treatment: Influence of cocaine symptom severity and psychiatric co-morbidity. *Addiction Biology, 20*, 756–772. https://doi.org/10.1111/adb.12156.

Arezoomandan, R., Khodagholi, F., & Haghparast, A. (2016). Administration of the glial condition medium in the nucleus accumbens prolong maintenance and intensify reinstatement of morphine-seeking behavior. *Neurochemical Research, 41*, 855–868. https://doi.org/10.1007/s11064-015-1762-3.

Arpaia, N., Campbell, C., Fan, X., et al. (2013). Metabolites produced by commensal bacteria promote peripheral regulatory T-cell generation. *Nature, 504*, 451–455. https://doi.org/10.1038/nature12726.

Arreola, R., Alvarez-Herrera, S., Pérez-Sánchez, G., et al. (2016). Immunomodulatory effects mediated by dopamine. *Journal of Immunology Research, 2016*, 3160486. https://doi.org/10.1155/2016/3160486. Epub 2016 Oct 4.

Atarashi, K., Tanoue, T., Ando, M., et al. (2015). Th17 cell induction by adhesion of microbes to intestinal epithelial cells. *Cell, 163*, 367–380. https://doi.org/10.1016/J.CELL.2015.08.058.

Atarashi, K., Tanoue, T., Oshima, K., et al. (2013). Treg induction by a rationally selected mixture of clostridia strains from the human microbiota. *Nature, 500*, 232–236. https://doi.org/10.1038/nature12331.

Atarashi, K., Tanoue, T., Shima, T., et al. (2011). Induction of colonic regulatory T cells by indigenous clostridium species. *Science, 331*, 337–341. https://doi.org/10.1126/science.1198469.

Attarzadeh-Yazdi, G., Arezoomandan, R., & Haghparast, A. (2014). Minocycline, an antibiotic with inhibitory effect on microglial activation, attenuates the maintenance and reinstatement of methamphetamine-seeking behavior in rat. *Progress in Neuro-Psychopharmacology & Biological Psychiatry, 53*, 142–148. https://doi.org/10.1016/j.pnpbp.2014.04.008.

Avital, A., Goshen, I., Kamsler, A., et al. (2003). Impaired interleukin-1 signaling is associated with deficits in hippocampal memory processes and neural plasticity. *Hippocampus, 13*, 826–834. https://doi.org/10.1002/hipo.10135.

Bachtell, R. K., Jones, J. D., Heinzerling, K. G., et al. (2017). Glial and neuroinflammatory targets for treating substance use disorders. *Drug and Alcohol Dependence, 180*, 156–170.

Balasubbramanian, D. (2019). Is IL-12 pro-inflammatory or anti-inflammatory? Depends on the blood pressure. *Cardiovascular Research, 115*, 998–999. https://doi.org/10.1093/cvr/cvz028.

Banerjee, S., Sindberg, G., Wang, F., et al. (2016). Opioid-induced gut microbial disruption and bile dysregulation leads to gut barrier compromise and sustained systemic inflammation. *Mucosal Immunology, 9*, 1418–1428. https://doi.org/10.1038/mi.2016.9.

Banks, W. A., Farr, S. A., & Morley, J. E. (2002). Entry of blood-borne cytokines into the central nervous system: Effects on cognitive processes. *Neuroimmunomodulation, 10*, 319–327. https://doi.org/10.1159/000071472.

Banks, W. A., Kastin, A. J., & Broadwell, R. D. (1995). Passage of cytokines across the blood-brain barrier. *Neuroimmunomodulation, 2*, 241–248.

Barengolts, E., Green, S. J., Eisenberg, Y., et al. (2018). Gut microbiota varies by opioid use, circulating leptin and oxytocin in African American men with diabetes and high burden of chronic disease. *PLoS One, 13*(3), e0194171. https://doi.org/10.1371/journal.pone.0194171.

Barr, J. L., Brailoiu, G. C., Abood, M. E., et al. (2020). Acute cocaine administration alters permeability of blood-brain barrier in freely-moving rats—Evidence using miniaturized fluorescence microscopy. *Drug and Alcohol Dependence, 206*, 107637. https://doi.org/10.1016/j.drugalcdep.2019.107637. Epub 2019 Oct 18.

Becher, B., Spath, S., & Goverman, J. (2017). Cytokine networks in neuroinflammation. *Nature Reviews. Immunology*, *17*, 49–59.

Becker, J. B., & Chartoff, E. (2019). Sex differences in neural mechanisms mediating reward and addiction. *Neuropsychopharmacology*, *44*, 166–183.

Beitner-Johnson, D., Guitart, X., & Nestler, E. J. (1993). Glial fibrillary acidic protein and the mesolimbic dopamine system: Regulation by chronic morphine and Lewis-Fischer strain differences in the rat ventral tegmental area. *Journal of Neurochemistry*, *61*, 1766–1773. https://doi.org/10.1111/j.1471-4159.1993.tb09814.x.

Belkaid, Y., & Hand, T. W. (2014). Role of the microbiota in immunity and inflammation. *Cell*, *157*, 121–141. https://doi.org/10.1016/J.CELL.2014.03.011.

Bercik, P., Denou, E., Collins, J., et al. (2011). The intestinal microbiota affect central levels of brain-derived neurotropic factor and behavior in mice. *Gastroenterology*, *141*, 599–609. https://doi.org/10.1053/j.gastro.2011.04.052.

Bialas, A. R., & Stevens, B. (2013). TGF-β signaling regulates neuronal C1q expression and developmental synaptic refinement. *Nature Neuroscience*, *16*, 1773–1782. https://doi.org/10.1038/nn.3560.

Biber, K., Neumann, H., Inoue, K., & Boddeke, H. W. G. M. (2007). Neuronal "On" and "Off" signals control microglia. *Trends in Neurosciences*, *30*, 596–602.

Bland, S. T., Hutchinson, M. R., Maier, S. F., et al. (2009). The glial activation inhibitor AV411 reduces morphine-induced nucleus accumbens dopamine release. *Brain, Behavior, and Immunity*, *23*, 492–497. https://doi.org/10.1016/j.bbi.2009.01.014.

Bowers, M. S., & Kalivas, P. W. (2003). Forebrain astroglial plasticity is induced following withdrawal from repeated cocaine administration. *The European Journal of Neuroscience*, *17*, 1273–1278. https://doi.org/10.1046/j.1460-9568.2003.02537.x.

Bowman, C. C., Rasley, A., Tranguch, S. L., & Marriott, I. (2003). Cultured astrocytes express toll-like receptors for bacterial products. *Glia*, *43*, 281–291. https://doi.org/10.1002/glia.10256.

Brady, L. J., Hofford, R. S., Tat, J., et al. (2019). Granulocyte-colony stimulating factor alters the pharmacodynamic properties of cocaine in female mice. *ACS Chemical Neuroscience*, *10*, 4213–4220. https://doi.org/10.1021/acschemneuro.9b00309.

Braniste, V., Al-Asmakh, M., Kowal, C., et al. (2014). The gut microbiota influences blood-brain barrier permeability in mice. *Science Translational Medicine*, *6*, 263ra158. https://doi.org/10.1126/scitranslmed.3009759.

Brown, K. T., Levis, S. C., O'Neill, C. E., et al. (2018). Innate immune signaling in the ventral tegmental area contributes to drug-primed reinstatement of cocaine seeking. *Brain, Behavior, and Immunity*, *67*, 130–138. https://doi.org/10.1016/j.bbi.2017.08.012.

Browne, C. J., Godino, A., Salery, M., & Nestler, E. J. (2020). Epigenetic mechanisms of opioid addiction. *Biological Psychiatry*, *87*, 22–33. https://doi.org/10.1016/j.biopsych.2019.06.027.

Brubaker, S. W., Bonham, K. S., Zanoni, I., & Kagan, J. C. (2015). Innate immune pattern recognition: A cell biological perspective. *Annual Review of Immunology*, *33*, 257–290. https://doi.org/10.1146/annurev-immunol-032414-112240.

Bsibsi, M., Ravid, R., Gveric, D., & van Noort, J. M. (2002). Broad expression of toll-like receptors in the human central nervous system. *Journal of Neuropathology and Experimental Neurology*, *61*, 1013–1021. https://doi.org/10.1093/jnen/61.11.1013.

Calipari, E. S., Godino, A., Peck, E. G., et al. (2018). Granulocyte-colony stimulating factor controls neural and behavioral plasticity in response to cocaine. *Nature Communications*, *9*(1), 9. https://doi.org/10.1038/s41467-017-01881-x.

Campbell, C., McKenney, P. T., Konstantinovsky, D., et al. (2020). Bacterial metabolism of bile acids promotes generation of peripheral regulatory T cells. *Nature*, *581*(7809), 475–479. https://doi.org/10.1038/s41586-020-2193-0.

Capuron, L., Pagnoni, G., Drake, D. F., et al. (2012). Dopaminergic mechanisms of reduced basal ganglia responses to hedonic reward during interferon alfa administration. *Archives of General Psychiatry*, *69*, 1044–1053. https://doi.org/10.1001/archgenpsychiatry.2011.2094.

Carrera, J., Catala, J. C., Monedero, P., et al. (1992). Depression of the mononuclear phagocyte system caused by high doses of narcotics. *Revista de Medicina de la Universidad de Navarra*, 119–125.

Castilla-Cortázar, I., Castilla, A., & Gurpegui, M. (1998). Opioid peptides and immunodysfunction in patients with major depression and anxiety disorders. *Journal of Physiology and Biochemistry*, *54*, 203–216.

Cavaillon, J. M. (2001). Pro- versus anti-inflammatory cytokines: Myth or reality. *Cellular and Molecular Biology (Noisy-le-Grand, France)*, *47*, 695–702.

Chao, C. C., Gekker, G., Hu, S., et al. (1996). κ opioid receptors in human microglia downregulate human immunodeficiency virus 1 expression. *Proceedings of the National Academy of Sciences of the United States of America*, *93*, 8051–8056. https://doi.org/10.1073/pnas.93.15.8051.

Chen, M. L., & Sundrud, M. S. (2016). Cytokine networks and T-cell subsets in inflammatory bowel diseases. *Inflammatory Bowel Diseases*, *22*, 1157–1167.

Chen, H., Uz, T., & Manev, H. (2009). Minocycline affects cocaine sensitization in mice. *Neuroscience Letters*, *452*, 258–261. https://doi.org/10.1016/j.neulet.2009.01.078.

Christopherson, K. S., Ullian, E. M., Stokes, C. C. A., et al. (2005). Thrombospondins are astrocyte-secreted proteins that promote CNS synaptogenesis. *Cell*, *120*, 421–433. https://doi.org/10.1016/j.cell.2004.12.020.

Chu, C., Murdock, M. H., Jing, D., et al. (2019). The microbiota regulate neuronal function and fear extinction learning. *Nature*, *574*, 543–548. https://doi.org/10.1038/s41586-019-1644-y.

Cıcek, E., Demırel, B., Cıcek, I. E., et al. (2018). Increased neutrophil-lymphocyte and platelet-lymphocyte ratios in male heroin addicts: A prospective controlled study. *Clinical Psychopharmacology and Neuroscience*, *16*, 190–196. https://doi.org/10.9758/cpn.2018.16.2.190.

Clarke, G., Stilling, R. M., Kennedy, P. J., et al. (2014). Minireview: Gut microbiota: The neglected endocrine organ. *Molecular Endocrinology*, *28*, 1221–1238.

Clemente, J. C., Ursell, L. K., Parfrey, L. W., & Knight, R. (2012). The impact of the gut microbiota on human health: An integrative view. *Cell*, *148*, 1258–1270. https://doi.org/10.1016/J.CELL.2012.01.035.

Coelho-Santos, V., Gonçalves, J., Fontes-Ribeiro, C., & Silva, A. P. (2012). Prevention of methamphetamine-induced microglial cell death by TNF-α and IL-6 through activation of the JAK-STAT pathway. *Journal of Neuroinflammation*, *9*, 565. https://doi.org/10.1186/1742-2094-9-103.

Coller, J. K., & Hutchinson, M. R. (2012). Implications of central immune signaling caused by drugs of abuse: Mechanisms, mediators and new therapeutic approaches for prediction and treatment of drug dependence. *Pharmacology & Therapeutics*, *134*, 219–245.

Conrad, K. L., Tseng, K. Y., Uejima, J. L., et al. (2008). Formation of accumbens GluR2-lacking AMPA receptors mediates incubation of cocaine craving. *Nature*, *454*, 118–121. https://doi.org/10.1038/nature06995.

Corder, G., Tawfik, V. L., Wang, D., et al. (2017). Loss of μ opioid receptor signaling in nociceptors, but not microglia, abrogates morphine tolerance without disrupting analgesia. *Nature Medicine*, *23*, 164–173. https://doi.org/10.1038/nm.4262.

Cosenza-Nashat, M., Zhao, M. L., Suh, H. S., et al. (2009). Expression of the translocator protein of 18 kDa by microglia, macrophages and astrocytes based on immunohistochemical localization in abnormal human brain. *Neuropathology and Applied Neurobiology*, *35*, 306–328. https://doi.org/10.1111/j.1365-2990.2008.01006.x.

Cotto, B., Li, H., Tuma, R. F., et al. (2018). Cocaine-mediated activation of microglia and microglial MeCP2 and BDNF production. *Neurobiology of Disease*, *117*, 28–41. https://doi.org/10.1016/j.nbd.2018.05.017.

Covelo, A., & Araque, A. (2018). Neuronal activity determines distinct gliotransmitter release from a single astrocyte. *eLife*, 7, e32237. https://doi.org/10.7554/eLife.32237.

Crews, F. T., Walter, T. J., Coleman, L. G., & Vetreno, R. P. (2017). Toll-like receptor signaling and stages of addiction. *Psychopharmacology*, *234*, 1483–1498.

Cui, C., Shurtleff, D., & Harris, R. A. (2014). Neuroimmune mechanisms of alcohol and drug addiction. *International Review of Neurobiology*, *118*, 1–12. https://doi.org/10.1016/B978-0-12-801284-0.00001-4.

Dalile, B., Van Oudenhove, L., Vervliet, B., & Verbeke, K. (2019). The role of short-chain fatty acids in microbiota–gut–brain communication. *Nature Reviews. Gastroenterology & Hepatology*, *16*, 461–478.

Desbonnet, L., Clarke, G., Traplin, A., et al. (2015). Gut microbiota depletion from early adolescence in mice: Implications for brain and behaviour. *Brain, Behavior, and Immunity*, *48*, 165–173. https://doi.org/10.1016/j.bbi.2015.04.004.

Dietrich, J. B. (2009). Alteration of blood-brain barrier function by methamphetamine and cocaine. *Cell and Tissue Research*, *336*, 385–392.

Dillon, S. M., Lee, E. J., Kotter, C. V., et al. (2016). Gut dendritic cell activation links an altered colonic microbiome to mucosal and systemic T-cell activation in untreated HIV-1 infection. *Mucosal Immunology*, *9*, 24–37. https://doi.org/10.1038/mi.2015.33.

Dinan, T. G., & Cryan, J. F. (2017). The microbiome-gut-brain axis in health and disease. *Gastroenterology Clinics of North America*, *46*, 77–89.

Dobrenis, K., Makman, M. H., & Stefano, G. B. (1995). Occurrence of the opiate alkaloid-selective μ3 receptor in mammalian microglia, astrocytes and Kupffer cells. *Brain Research*, *686*, 239–248. https://doi.org/10.1016/0006-8993(95)00452-V.

Donahoe, R. M., & Vlahov, D. (1998). Opiates as potential cofactors in progression of HIV-1 infections to AIDS. *Journal of Neuroimmunology*, *83*, 77–87. https://doi.org/10.1016/S0165-5728(97)00224-5.

Duscha, A., Gisevius, B., Hirschberg, S., et al. (2020). Propionic acid shapes the multiple sclerosis disease course by an immunomodulatory mechanism. *Cell*, *180*, 1067–1080.e16. https://doi.org/10.1016/j.cell.2020.02.035.

Eidson, L. N., Inoue, K., Young, L. J., et al. (2017). Toll-like receptor 4 mediates morphine-induced neuroinflammation and tolerance via soluble tumor necrosis factor signaling. *Neuropsychopharmacology*, *42*, 661–670. https://doi.org/10.1038/npp.2016.131.

Eidson, L. N., & Murphy, A. Z. (2013). Blockade of toll-like receptor 4 attenuates morphine tolerance and facilitates the pain relieving properties of morphine. *The Journal of Neuroscience*, *33*, 15952–15963. https://doi.org/10.1523/JNEUROSCI.1609-13.2013.

El Aidy, S., Dinan, T. G., & Cryan, J. F. (2015). Gut microbiota: The conductor in the Orchestra of immune-neuroendocrine communication. *Clinical Therapeutics*, *37*, 954–967.

Engelhardt, B., Vajkoczy, P., & Weller, R. O. (2017). The movers and shapers in immune privilege of the CNS. *Nature Immunology*, *18*, 123–131.

Eriksson, P. S., Hansson, E., & Rönnbäck, L. (1992). δ and k opiate receptors in primary astroglial cultures part II: Receptor sets in cultures from various brain regions and interactions with ß-receptor activated cyclic AMP. *Neurochemical Research*, *17*, 545–551. https://doi.org/10.1007/BF00968781.

Erny, D., De Angelis, A. L. H., Jaitin, D., et al. (2015). Host microbiota constantly control maturation and function of microglia in the CNS. *Nature Neuroscience*, *18*, 965–977. https://doi.org/10.1038/nn.4030.

Eroglu, C., & Barres, B. A. (2010). Regulation of synaptic connectivity by glia. *Nature*, *468*, 223–231.

Ersche, K. D., Hagan, C. C., Smith, D. G., et al. (2014). Aberrant disgust responses and immune reactivity in cocaine-dependent men. *Biological Psychiatry, 75*, 140–147. https://doi.org/10.1016/j.biopsych.2013.08.004.

Farache, J., Zigmond, E., Shakhar, G., & Jung, S. (2013). Contributions of dendritic cells and macrophages to intestinal homeostasis and immune defense. *Immunology and Cell Biology, 91*, 232–239.

Felger, J. C., Li, Z., Haroon, E., et al. (2016). Inflammation is associated with decreased functional connectivity within corticostriatal reward circuitry in depression. *Molecular Psychiatry, 21*, 1358–1365. https://doi.org/10.1038/mp.2015.168.

Felger, J. C., & Treadway, M. T. (2017). Inflammation effects on motivation and motor activity: Role of dopamine. *Neuropsychopharmacology, 42*, 216–241.

Ferrari, M., Cosentino, M., Marino, F., et al. (2004). Dopaminergic D1-like receptor-dependent inhibition of tyrosine hydroxylase mRNA expression and catecholamine production in human lymphocytes. *Biochemical Pharmacology, 67*, 865–873. https://doi.org/10.1016/j.bcp.2003.10.004.

Fiala, M., Gan, X. H., Zhang, L., et al. (1998). Cocaine enhances monocyte migration across the blood-brain barrier. Cocaine's connection to aids dementia and vasculitis? *Advances in Experimental Medicine and Biology, 437*, 199–205. https://doi.org/10.1007/978-1-4615-5347-2_22.

Foster, J. A. (2016). Gut microbiome and behavior: Focus on neuroimmune interactions. In J. F. Cryan, & G Clarke (Eds.), *International review of neurobiology* (pp. 49–65). Academic Press Inc.

Fox, H. C., D'Sa, C., Kimmerling, A., et al. (2012). Immune system inflammation in cocaine dependent individuals: Implications for medications development. *Human Psychopharmacology, 27*, 156–166. https://doi.org/10.1002/hup.1251.

Franchi, L., Warner, N., Viani, K., & Nuñez, G. (2009). Function of nod-like receptors in microbial recognition and host defense. *Immunological Reviews, 227*, 106–128.

Fu, X., Zhu, Z. H., Wang, Y. Q., & Wu, G. C. (2007). Regulation of proinflammatory cytokines gene expression by nociceptin/orphanin FQ in the spinal cord and the cultured astrocytes. *Neuroscience, 144*, 275–285. https://doi.org/10.1016/j.neuroscience.2006.09.016.

Fukagawa, H., Koyama, T., Kakuyama, M., & Fukuda, K. (2013). Microglial activation involved in morphine tolerance is not mediated by toll-like receptor 4. *Journal of Anesthesia, 27*, 93–97. https://doi.org/10.1007/s00540-012-1469-4.

Furusawa, Y., Obata, Y., Fukuda, S., et al. (2013). Commensal microbe-derived butyrate induces the differentiation of colonic regulatory T cells. *Nature, 504*, 446–450. https://doi.org/10.1038/nature12721.

Gadient, R. A., & Otten, U. H. (1997). Interleukin-6 (IL-6)—A molecule with both beneficial and destructive potentials. *Progress in Neurobiology, 52*, 379–390.

Ganguly, P., Honeycutt, J. A., Rowe, J. R., et al. (2019). Effects of early life stress on cocaine conditioning and AMPA receptor composition are sex-specific and driven by TNF. *Brain, Behavior, and Immunity, 78*, 41–51. https://doi.org/10.1016/j.bbi.2019.01.006.

García-Pérez, D., Luisa Laorden, M., Núñez, C., & Victoria Milanés, M. (2014). Glial activation and midkine and pleiotrophin transcription in the ventral tegmental area are modulated by morphine administration. *Journal of Neuroimmunology, 274*, 244–248. https://doi.org/10.1016/j.jneuroim.2014.07.017.

Geyer, S., Jacobs, M., & Hsu, N.-J. (2019). Immunity against bacterial infection of the central nervous system: An astrocyte perspective. *Frontiers in Molecular Neuroscience, 12*, 57. https://doi.org/10.3389/fnmol.2019.00057.

Glassner, K. L., Abraham, B. P., & Quigley, E. M. M. (2020). The microbiome and inflammatory bowel disease. *The Journal of Allergy and Clinical Immunology, 145*, 16–27. https://doi.org/10.1016/j.jaci.2019.11.003.

Goins, E. C., & Bajic, D. (2018). Astrocytic hypertrophy in the rat ventral tegmental area following chronic morphine differs with age. *Journal of Neurology and Neurorehabilitation Research*, *3*, 14.

Gonçalves, J., Martins, A., Ferreira, R., et al. (2008). Methamphetamine-induced early increase of IL-6 and TNF-α mRNA expression in the mouse brain. *Annals of the New York Academy of Sciences*, *1139*, 103–111. https://doi.org/10.1196/annals.1432.043.

Govitrapong, P., Suttitum, T., Kotchabhakdi, N., & Uneklabh, T. (1998). Alterations of immune functions in heroin addicts and heroin withdrawal subjects. *The Journal of Pharmacology and Experimental Therapeutics*, *286*, 883–889.

Graham, D. L., Edwards, S., Bachtell, R. K., et al. (2007). Dynamic BDNF activity in nucleus accumbens with cocaine use increases self-administration and relapse. *Nature Neuroscience*, *10*, 1029–1037. https://doi.org/10.1038/nn1929.

Gupta, K., Sharma, R., Singh, V., et al. (2018). Intravenous cocaine results in an acute decrease in levels of biomarkers of vascular inflammation in humans. *Cardiovascular Toxicology*, *18*, 295–303. https://doi.org/10.1007/s12012-017-9440-0.

Gurwell, J. A., Duncan, M. J., Maderspach, K., et al. (1996). κ-Opioid receptor expression defines a phenotypically distinct subpopulation of astroglia: Relationship to Ca2+ mobilization, development, and the antiproliferative effect of opioids. *Brain Research*, *737*, 175–187. https://doi.org/10.1016/0006-8993(96)00728-7.

Haghikia, A., Jörg, S., Duscha, A., et al. (2015). Dietary fatty acids directly impact central nervous system autoimmunity via the small intestine. *Immunity*, *43*, 817–829. https://doi.org/10.1016/j.immuni.2015.09.007.

Halpern, J. H., Sholar, M. B., Glowacki, J., et al. (2003). Diminished interleukin-6 response to proinflammatory challenge in men and women after intravenous cocaine administration. *The Journal of Clinical Endocrinology and Metabolism*, *88*, 1188–1193. https://doi.org/10.1210/jc.2002-020804.

Hanisch, U. K., & Kettenmann, H. (2007). Microglia: Active sensor and versatile effector cells in the normal and pathologic brain. *Nature Neuroscience*, *10*, 1387–1394.

Hansen, C. H. F., Nielsen, D. S., Kverka, M., et al. (2012). Patterns of early gut colonization shape future immune responses of the host. *PLoS One*, 7(3), e34043. https://doi.org/10.1371/journal.pone.0034043. Epub 2012 Mar 27.

Happel, C., Steele, A. D., Finley, M. J., et al. (2008). DAMGO-induced expression of chemokines and chemokine receptors: The role of TGF- 1. *Journal of Leukocyte Biology*, *83*, 956–963. https://doi.org/10.1189/jlb.1007685.

Heijtz, R. D., Wang, S., Anuar, F., et al. (2011). Normal gut microbiota modulates brain development and behavior. *Proceedings of the National Academy of Sciences of the United States of America*, *108*, 3047–3052. https://doi.org/10.1073/pnas.1010529108.

Heinisch, S., & Kirby, L. G. (2010). SDF-1α/CXCL12 enhances GABA and glutamate synaptic activity at serotonin neurons in the rat dorsal raphe nucleus. *Neuropharmacology*, *58*, 501–514. https://doi.org/10.1016/j.neuropharm.2009.08.022.

Hepworth, M. R., Fung, T. C., Masur, S. H., et al. (2015). Group 3 innate lymphoid cells mediate intestinal selection of commensal bacteria-specific CD4+ T cells. *Science*, *348*, 1031–1035. https://doi.org/10.1126/science.aaa4812.

Hodes, G. E., Kana, V., Menard, C., et al. (2015). Neuroimmune mechanisms of depression. *Nature Neuroscience*, *18*, 1386–1393. https://doi.org/10.1038/nn.4113.

Hofford, R. S., Euston, T. J., Wilson, R. S., et al. (2020). *Granulocyte-Colony stimulating factor reduces cocaine-seeking and downregulates glutamatergic synaptic proteins in medial prefrontal cortex*. Cold Spring Harbor Laboratory.

Hofford, R. S., Russo, S. J., & Kiraly, D. D. (2018). Neuroimmune mechanisms of psychostimulant and opioid use disorders. *The European Journal of Neuroscience*, *50*, 2562–2573. https://doi.org/10.1111/ejn.14143.

Huh, J. R., & Veiga-Fernandes, H. (2020). Neuroimmune circuits in inter-organ communication. *Nature Reviews. Immunology*, *20*, 217–228.

Hutchinson, M. R., Bland, S. T., Johnson, K. W., et al. (2007). Opioid-induced glial activation: Mechanisms of activation and implications for opioid analgesia, dependence, and reward. *The Scientific World Journal*, 7, 98–111. https://doi.org/10.1100/tsw.2007.230.

Hutchinson, M. R., Coats, B. D., Lewis, S. S., et al. (2008). Proinflammatory cytokines oppose opioid-induced acute and chronic analgesia. *Brain, Behavior, and Immunity*, *22*, 1178–1189. https://doi.org/10.1016/j.bbi.2008.05.004.

Hutchinson, M. R., Lewis, S. S., Coats, B. D., et al. (2009). Reduction of opioid withdrawal and potentiation of acute opioid analgesia by systemic AV411 (ibudilast). *Brain, Behavior, and Immunity*, *23*, 240–250. https://doi.org/10.1016/j.bbi.2008.09.012.

Hutchinson, M. R., Lewis, S. S., Coats, B. D., et al. (2010). Possible involvement of toll-like receptor 4/myeloid differentiation factor-2 activity of opioid inactive isomers causes spinal proinflammation and related behavioral consequences. *Neuroscience*, *167*, 880–893. https://doi.org/10.1016/j.neuroscience.2010.02.011.

Hutchinson, M. R., Northcutt, A. L., Chao, L. W., et al. (2008). Minocycline suppresses morphine-induced respiratory depression, suppresses morphine-induced reward, and enhances systemic morphine-induced analgesia. *Brain, Behavior, and Immunity*, *22*, 1248–1256. https://doi.org/10.1016/j.bbi.2008.07.008.

Hutchinson, M. R., Northcutt, A. L., Hiranita, T., et al. (2012). Opioid activation of toll-like receptor 4 contributes to drug reinforcement. *The Journal of Neuroscience*, *32*, 11187–11200. https://doi.org/10.1523/JNEUROSCI.0684-12.2012.

Hutchinson, M. R., Zhang, Y., Shridhar, M., et al. (2010). Evidence that opioids may have toll-like receptor 4 and MD-2 effects. *Brain, Behavior, and Immunity*, *24*, 83–95. https://doi.org/10.1016/j.bbi.2009.08.004.

Irwin, M. R., Olmos, L., Wang, M., et al. (2007). Cocaine dependence and acute cocaine induce decreases of monocyte proinflammatory cytokine expression across the diurnal period: Autonomic mechanisms. *The Journal of Pharmacology and Experimental Therapeutics*, *320*, 507–515. https://doi.org/10.1124/jpet.106.112797.

Ivanov, I. I., Atarashi, K., Manel, N., et al. (2009). Induction of intestinal Th17 cells by segmented filamentous bacteria. *Cell*, *139*, 485–498. https://doi.org/10.1016/J.CELL.2009.09.033.

Jang, H. M., Lee, H. J., Jang, S. E., et al. (2018). Evidence for interplay among antibacterial-induced gut microbiota disturbance, neuro-inflammation, and anxiety in mice. *Mucosal Immunology*, *11*, 1386–1397. https://doi.org/10.1038/s41385-018-0042-3.

Jašarević, E., Morrison, K. E., & Bale, T. L. (2016). Sex differences in the gut microbiome—Brain axis across the lifespan. *Philosophical Transactions of the Royal Society of London. Series B, Biological Sciences*, *371*(1688), 20150122. https://doi.org/10.1098/rstb.2015.0122. Epub 2016 Feb 1. PMID: 26833840; PMCID: PMC4785905.

Jiao, Y., Wu, L., Huntington, N. D., & Zhang, X. (2020). Crosstalk between gut microbiota and innate immunity and its implication in autoimmune diseases. *Frontiers in Immunology*, *11*, 282.

Joseph, J., Depp, C., Shih, P. A. B., et al. (2017). Modified mediterranean diet for enrichment of short chain fatty acids: Potential adjunctive therapeutic to target immune and metabolic dysfunction in schizophrenia? *Frontiers in Neuroscience*, *11*, 155. https://doi.org/10.3389/fnins.2017.00155.

Kalivas, P. W. (2007). Cocaine and amphetamine-like psychostimulants: Neuro circuitry and glutamate neuroplasticity. *Dialogues in Clinical Neuroscience*, *9*, 389–397.

Kang, M., Mischel, R. A., Bhave, S., et al. (2017). The effect of gut microbiome on tolerance to morphine mediated antinociception in mice. *Scientific Reports*, 7, 42658. https://doi.org/10.1038/srep42658.

Kao, S. C., Zhao, X., Lee, C. Y., et al. (2012). Absence of μ opioid receptor mRNA expression in astrocytes and microglia of rat spinal cord. *Neuroreport*, *23*, 378–384. https://doi.org/10.1097/WNR.0b013e3283522e1b.

Kariisa, M., Scholl, L., Wilson, N., et al. (2019). Drug overdose deaths involving cocaine and psychostimulants with abuse potential—United States, 2003–2017. *MMWR. Morbidity and Mortality Weekly Report*, *68*, 388–395. https://doi.org/10.15585/mmwr.mm6817a3.

Karlsson, C., Schank, J. R., Rehman, F., et al. (2017). Proinflammatory signaling regulates voluntary alcohol intake and stress-induced consumption after exposure to social defeat stress in mice. *Addiction Biology*, *22*, 1279–1288. https://doi.org/10.1111/adb.12416.

Kashima, D. T., & Grueter, B. A. (2017). Toll-like receptor 4 deficiency alters nucleus accumbens synaptic physiology and drug reward behavior. *Proceedings of the National Academy of Sciences of the United States of America*, *114*, 8865–8870. https://doi.org/10.1073/pnas.1705974114.

Kawai, T., & Akira, S. (2010). The role of pattern-recognition receptors in innate immunity: Update on toll-like receptors. *Nature Immunology*, *11*(5), 373–384.

Kelly, J. R., Kennedy, P. J., Cryan, J. F., et al. (2015). Breaking down the barriers: The gut microbiome, intestinal permeability and stress-related psychiatric disorders. *Frontiers in Cellular Neuroscience*, *9*, 392.

Kettenmann, H., Kirchhoff, F., & Verkhratsky, A. (2013). Microglia: New roles for the synaptic stripper. *Neuron*, *77*, 10–18.

Khandaker, G. M., Cousins, L., Deakin, J., et al. (2015). Inflammation and immunity in schizophrenia: Implications for pathophysiology and treatment. *The Lancet Psychiatry*, *2*, 258–270.

Kiraly, D. D., Horn, S. R., Van Dam, N. T., et al. (2017). Altered peripheral immune profiles in treatment-resistant depression: Response to ketamine and prediction of treatment outcome. *Translational Psychiatry*, *7*, e1065. https://doi.org/10.1038/tp.2017.31.

Kiraly, D. D., Walker, D. M., Calipari, E. S., et al. (2016). Alterations of the host microbiome affect behavioral responses to cocaine. *Scientific Reports*, *6*, 35455. https://doi.org/10.1038/srep35455.

Kirillova, G. P., Hrutkay, R. J., Shurin, M. R., et al. (2008). Dopamine receptors in human lymphocytes: Radioligand binding and quantitative RT-PCR assays. *Journal of Neuroscience Methods*, *174*, 272–280. https://doi.org/10.1016/j.jneumeth.2008.07.018.

Koh, A., De Vadder, F., Kovatcheva-Datchary, P., & Bäckhed, F. (2016). From dietary fiber to host physiology: Short-chain fatty acids as key bacterial metabolites. *Cell*, *165*, 1332–1345.

Koob, G. F., & Volkow, N. D. (2010). Neurocircuitry of addiction. *Neuropsychopharmacology*, *35*, 217–238.

Koob, G. F., & Volkow, N. D. (2016). Neurobiology of addiction: A neurocircuitry analysis. *The Lancet Psychiatry*, *3*, 760–773.

Kosloski, L. M., Kosmacek, E. A., Olson, K. E., et al. (2013). GM-CSF induces neuroprotective and anti-inflammatory responses in 1-methyl-4-phenyl-1,2,3,6-tetrahydropyridine intoxicated mice. *Journal of Neuroimmunology*, *265*, 1–10. https://doi.org/10.1016/J.JNEUROIM.2013.10.009.

Kucukdereli, H., Allen, N. J., Lee, A. T., et al. (2011). Control of excitatory CNS synaptogenesis by astrocyte-secreted proteins hevin and SPARC. *Proceedings of the National Academy of Sciences of the United States of America*, *108*, E440–E449. https://doi.org/10.1073/pnas.1104977108.

Kuehn, B. M. (2012). WHO documents worldwide need for better drug abuse treatment—And access to it. *JAMA*, *308*, 442–443.

Kutlu, M. G., Brady, L. J., Peck, E. G., et al. (2018). Granulocyte colony stimulating factor enhances reward learning through potentiation of mesolimbic dopamine system function. *The Journal of Neuroscience*, *38*, 8845–8859. https://doi.org/10.1523/JNEUROSCI.1116-18.2018.

Lacagnina, M. J., Kopec, A. M., Cox, S. S., et al. (2017). Opioid self-administration is attenuated by early-life experience and gene therapy for anti-inflammatory IL-10 in the nucleus accumbens of male rats. *Neuropsychopharmacology*, *42*, 2128–2140. https://doi.org/10.1038/npp.2017.82.

Lacagnina, M. J., Rivera, P. D., & Bilbo, S. D. (2017). Glial and neuroimmune mechanisms as critical modulators of drug use and abuse. *Neuropsychopharmacology*, *42*, 156–177.

Laflamme, N., & Rivest, S. (2001). Toll-like receptor 4: The missing link of the cerebral innate immune response triggered by circulating gram-negative bacterial cell wall components. *The FASEB Journal*, *15*, 155–163. https://doi.org/10.1096/fj.00-0339com.

Lang, J. Y., Wang, J. S., Zhai, H. F., et al. (2009). Interferon-alpha reinstates morphine-conditioned place preference through opioid receptors in rats. *Behavioural Pharmacology*, *20*, 166–173. https://doi.org/10.1097/FBP.0b013e32832a805e.

Leclercq, S., Mian, F. M., Stanisz, A. M., et al. (2017). Low-dose penicillin in early life induces long-term changes in murine gut microbiota, brain cytokines and behavior. *Nature Communications*, *8*, 1–12. https://doi.org/10.1038/ncomms15062.

Lee, Y. K., Menezes, J. S., Umesaki, Y., & Mazmanian, S. K. (2011). Proinflammatory T-cell responses to gut microbiota promote experimental autoimmune encephalomyelitis. *Proceedings of the National Academy of Sciences of the United States of America*, *108*(Suppl. 1), 4615–4622. https://doi.org/10.1073/pnas.1000082107.

Lee, K., Vuong, H. E., Nusbaum, D. J., et al. (2018). The gut microbiota mediates reward and sensory responses associated with regimen-selective morphine dependence. *Neuropsychopharmacology*, *43*, 2606–2614. https://doi.org/10.1038/s41386-018-0211-9.

Lehmann, F. M., von Burg, N., Ivanek, R., et al. (2020). Microbiota-induced tissue signals regulate ILC3-mediated antigen presentation. *Nature Communications*, *11*, 1794. https://doi.org/10.1038/s41467-020-15612-2.

Levandowski, M. L., Hess, A. R. B., Grassi-Oliveira, R., & de Almeida, R. M. M. (2016). Plasma interleukin-6 and executive function in crack cocaine-dependent women. *Neuroscience Letters*, *628*, 85–90. https://doi.org/10.1016/j.neulet.2016.06.023.

Levite, M. (2016). Dopamine and T cells: Dopamine receptors and potent effects on T cells, dopamine production in T cells, and abnormalities in the dopaminergic system in T cells in autoimmune, neurological and psychiatric diseases. *Acta Physiologica*, *216*, 42–89.

Lewitus, G. M., Konefal, S. C., Greenhalgh, A. D., et al. (2016). Microglial TNF-α suppresses cocaine-induced plasticity and behavioral sensitization. *Neuron*, *90*, 483–491. https://doi.org/10.1016/j.neuron.2016.03.030.

Lewitus, G. M., Pribiag, H., Duseja, R., et al. (2014). An adaptive role of TNFα in the regulation of striatal synapses. *The Journal of Neuroscience*, *34*, 6146–6155. https://doi.org/10.1523/JNEUROSCI.3481-13.2014.

Liao, K., Guo, M., Niu, F., et al. (2016). Cocaine-mediated induction of microglial activation involves the ER stress-TLR2 axis. *Journal of Neuroinflammation*, *13*, 33. https://doi.org/10.1186/s12974-016-0501-2. PMID: 26860188; PMCID: PMC4748483.

Lim, G., Kim, H., McCabe, M. F., et al. (2014). A leptin-mediated central mechanism in analgesia-enhanced opioid reward in rats. *The Journal of Neuroscience*, *34*, 9779–9788. https://doi.org/10.1523/JNEUROSCI.0386-14.2014.

Linker, K. E., Cross, S. J., & Leslie, F. M. (2019). Glial mechanisms underlying substance use disorders. *The European Journal of Neuroscience*, *50*, 2574–2589. https://doi.org/10.1111/ejn.14163.

Little, K. Y., Ramssen, E., Welchko, R., et al. (2009). Decreased brain dopamine cell numbers in human cocaine users. *Psychiatry Research*, *168*, 173–180. https://doi.org/10.1016/j.psychres.2008.10.034.

Liu, Y., Guo, W., Zhang, Y., et al. (2018). Decreased resting-state interhemispheric functional connectivity correlated with neurocognitive deficits in drug-naive first-episode adolescent-onset schizophrenia. *The International Journal of Neuropsychopharmacology*, *21*, 33–41. https://doi.org/10.1093/IJNP.

Lo Iacono, L., Catale, C., Martini, A., et al. (2018). From traumatic childhood to cocaine abuse: The critical function of the immune system. *Biological Psychiatry*, *84*, 905–916. https://doi.org/10.1016/j.biopsych.2018.05.022.

Louis, P., & Flint, H. J. (2017). Formation of propionate and butyrate by the human colonic microbiota. *Environmental Microbiology*, *19*, 29–41. https://doi.org/10.1111/1462-2920.13589.

Louveau, A., Smirnov, I., Keyes, T. J., et al. (2015). Structural and functional features of central nervous system lymphatic vessels. *Nature*, *523*, 337–341. https://doi.org/10.1038/nature14432.

Lu, J., Synowiec, S., Lu, L., et al. (2018). Microbiota influence the development of the brain and behaviors in C57BL/6J mice. *PLoS One*, *13*(8), e0201829. https://doi.org/10.1371/journal.pone.0201829.

Lu, R. B., Wang, T. Y., Lee, S. Y., et al. (2019). Correlation between interleukin-6 levels and methadone maintenance therapy outcomes. *Drug and Alcohol Dependence*, *204*, 107516. https://doi.org/10.1016/j.drugalcdep.2019.06.018.

Lüscher, C., & Malenka, R. C. (2011). Drug-evoked synaptic plasticity in addiction: From molecular changes to circuit remodeling. *Neuron*, *69*, 650–663.

Machelska, H., & Celik, M.Ö. (2020). Opioid receptors in immune and glial cells—Implications for pain control. *Frontiers in Immunology*, *11*, 1–13. https://doi.org/10.3389/fimmu.2020.00300. 300.

Maduna, T., Audouard, E., Dembélé, D., et al. (2019). Microglia express mu opioid receptor: Insights from transcriptomics and fluorescent reporter mice. *Frontiers in Psychiatry*, *10*, 726. https://doi.org/10.3389/fpsyt.2018.00726.

Maher, D. P., Walia, D., & Heller, N. M. (2019). Suppression of human natural killer cells by different classes of opioids. *Anesthesia and Analgesia*, *128*, 1013–1021. https://doi.org/10.1213/ANE.0000000000004058.

Makman, M. H., Bilfinger, T. V., & Stefano, G. B. (1995). Human granulocytes contain an opiate alkaloid-selective receptor mediating inhibition of cytokine-induced activation and chemotaxis. *Journal of Immunology*, *154*, 1323–1330. https://doi.org/10.1155/2020/4139093.

Marie-Claire, C., Courtin, C., Roques, B. P., & Noble, F. (2004). Cytoskeletal genes regulation by chronic morphine treatment in rat striatum. *Neuropsychopharmacology*, *29*, 2208–2215. https://doi.org/10.1038/sj.npp.1300513.

Matt, S. M., & Gaskill, P. J. (2019). Where is dopamine and how do immune cells see it?: Dopamine-mediated immune cell function in health and disease. *Journal of Neuroimmune Pharmacology*, *15*, 114–164. https://doi.org/10.1007/s11481-019-09851-4.

Matthes, H. W. D., Maldonado, R., Simonin, F., et al. (1996). Loss of morphine-induced analgesia, reward effect and withdrawal symptoms in mice lacking the μ-opioid-receptor gene. *Nature*, *383*, 822–823. https://doi.org/10.1038/383819a0.

Mattioli, T. A., Leduc-Pessah, H., Skelhorne-Gross, G., et al. (2014). Toll-like receptor 4 mutant and null mice retain morphine-induced tolerance, hyperalgesia, and physical dependence. *PLoS One*, *9*(5), e97361. https://doi.org/10.1371/journal.pone.0097361.

Mayfield, J., Ferguson, L., & Harris, R. A. (2013). Neuroimmune signaling: A key component of alcohol abuse. *Current Opinion in Neurobiology*, *23*, 513–520.

Maza-Quiroga, R., García-Marchena, N., Romero-Sanchiz, P., et al. (2017). Evaluation of plasma cytokines in patients with cocaine use disorders in abstinence identifies transforming growth factor alpha (tgfα) as a potential biomarker of consumption and dual diagnosis. *PeerJ*, *5*, e3926. https://doi.org/10.7717/peerj.3926.

Meckel, K. R., & Kiraly, D. D. (2019). A potential role for the gut microbiome in substance use disorders. *Psychopharmacology*, *236*, 1513–1530.

Meijerink, H., Indrati, A., Utami, F., et al. (2015). Heroin use is associated with suppressed pro-inflammatory cytokine response after LPS exposure in HIV-infected individuals. *PLoS One*, *10*(4), e0122822. https://doi.org/10.1371/journal.pone.0122822. PMID: 25830312; PMCID: PMC4382331.

Meltzer, A., & Van de Water, J. (2017). The role of the immune system in autism spectrum disorder. *Neuropsychopharmacology*, *42*, 284–298. https://doi.org/10.1038/npp.2016.158.

Menard, C., Pfau, M. L., Hodes, G. E., et al. (2017). Social stress induces neurovascular pathology promoting depression. *Nature Neuroscience*, *20*, 1752–1760. https://doi.org/10.1038/s41593-017-0010-3.

Meng, J., Banerjee, S., Li, D., et al. (2015). Opioid exacerbation of gram-positive sepsis, induced by gut microbial modulation, is rescued by IL-17A neutralization. *Scientific Reports*, *5*, 10918. https://doi.org/10.1038/srep10918.

Meng, J., Yu, H., Ma, J., et al. (2013). Morphine induces bacterial translocation in mice by compromising intestinal barrier function in a TLR-dependent manner. *PLoS One*, *8*(1), e54040. https://doi.org/10.1371/journal.pone.0054040. Epub 2013 Jan 18. PMID: 23349783; PMCID: PMC3548814.

Mervosh, N. L., Wilson, R., Rauniyar, N., et al. (2018). Granulocyte-colony-stimulating factor alters the proteomic landscape of the ventral tegmental area. *Proteomes*, *6*(4), 35. https://doi.org/10.3390/proteomes6040035. PMID: 30249060; PMCID: PMC6313867.

Meyer, L. C., Paisley, C. E., Mohamed, E., et al. (2017). Novel role of the nociceptin system as a regulator of glutamate transporter expression in developing astrocytes. *Glia*, *65*, 2003–2023. https://doi.org/10.1002/glia.23210.

Mika, J., Popiolek-Barczyk, K., Rojewska, E., et al. (2014). Delta-opioid receptor analgesia is independent of microglial activation in a rat model of neuropathic pain. *PLoS One*, *9*, e104420. https://doi.org/10.1371/journal.pone.0104420.

Millan, M. J. (2002). Descending control of pain. *Progress in Neurobiology*, *66*, 355–474.

Miller, B. J., & Goldsmith, D. R. (2017). Towards an immunophenotype of schizophrenia: Progress, potential mechanisms, and future directions. *Neuropsychopharmacology*, *42*, 299–317.

Minter, M. R., Hinterleitner, R., Meisel, M., et al. (2017). Antibiotic-induced perturbations in microbial diversity during post-natal development alters amyloid pathology in an aged APPSWE/PS1ΔE9 murine model of Alzheimer's disease. *Scientific Reports*, 7, 1–18. https://doi.org/10.1038/s41598-017-11047-w.

Minter, M. R., Zhang, C., Leone, V., et al. (2016). Antibiotic-induced perturbations in gut microbial diversity influences neuro-inflammation and amyloidosis in a murine model of Alzheimer's disease. *Scientific Reports*, *6*, 1–12. https://doi.org/10.1038/srep30028.

Mora, J. R., Iwata, M., Eksteen, B., et al. (2006). Generation of gut-homing IgA-secreting B cells by intestinal dendritic cells. *Science*, *314*, 1157–1160. https://doi.org/10.1126/science.1132742.

Moreira, F. P., Medeiros, J. R. C., Lhullier, A. C., et al. (2016). Cocaine abuse and effects in the serum levels of cytokines IL-6 and IL-10. *Drug and Alcohol Dependence*, *158*, 181–185. https://doi.org/10.1016/j.drugalcdep.2015.11.024.

Mori, F., Nisticò, R., Mandolesi, G., et al. (2014). Interleukin-1β promotes long-term potentiation in patients with multiple sclerosis. *Neuromolecular Medicine*, *16*, 38–51. https://doi.org/10.1007/s12017-013-8249-7.

Morrison, D. J., & Preston, T. (2016). Formation of short chain fatty acids by the gut microbiota and their impact on human metabolism. *Gut Microbes*, 7, 189–200.

Mortha, A., Chudnovskiy, A., Hashimoto, D., et al. (2014). Microbiota-dependent crosstalk between macrophages and ILC3 promotes intestinal homeostasis. *Science*, *343*, 1249288. https://doi.org/10.1126/science.1249288.

Moussawi, K., Pacchioni, A., Moran, M., et al. (2009). N-Acetylcysteine reverses cocaine-induced metaplasticity. *Nature Neuroscience*, *12*, 182–189. https://doi.org/10.1038/nn.2250.
Nakajima, K., Tohyama, Y., Maeda, S., et al. (2007). Neuronal regulation by which microglia enhance the production of neurotrophic factors for GABAergic, catecholaminergic, and cholinergic neurons. *Neurochemistry International*, *50*, 807–820. https://doi.org/10.1016/j.neuint.2007.02.006.
Nam, M. H., Han, K. S., Lee, J., et al. (2018). Expression of μ-opioid receptor in CA1 hippocampal astrocytes. *Experimental Neurobiology*, *27*, 120–128. https://doi.org/10.5607/en.2018.27.2.120.
Narendran, R., Lopresti, B. J., Mason, N. S., et al. (2014). Cocaine abuse in humans is not associated with increased microglial activation: An 18-kDa translocator protein positron emission tomography imaging study with [11C]PBR28. *The Journal of Neuroscience*, *34*, 9945–9950. https://doi.org/10.1523/JNEUROSCI.0928-14.2014.
Narita, M., Miyatake, M., Narita, M., et al. (2006). Direct evidence of astrocytic modulation in the development of rewarding effects induced by drugs of abuse. *Neuropsychopharmacology*, *31*, 2476–2488. https://doi.org/10.1038/sj.npp.1301007.
Nestler, E. J., & L€, U. C. (2019). The molecular basis of drug addiction: Linking epigenetic to synaptic and circuit mechanisms. *Neuron*, *102*, 48–59. https://doi.org/10.1016/j.neuron.2019.01.016.
Neufeld, K.-A. M., Kang, N., Bienenstock, J., & Foster, J. A. (2011a). Effects of intestinal microbiota on anxiety-like behavior. *Communicative & Integrative Biology*, *4*, 492. https://doi.org/10.4161/CIB.4.4.15702.
Neufeld, K. M., Kang, N., Bienenstock, J., & Foster, J. A. (2011b). Reduced anxiety-like behavior and central neurochemical change in germ-free mice. *Neurogastroenterology and Motility*, *23*, 255–e119. https://doi.org/10.1111/j.1365-2982.2010.01620.x.
NIDA. (2020a). *The science of drug use and addiction: The basics*. June 25, Retrieved from https://www.drugabuse.gov/publications/media-guide/science-drug-use-addiction-basic.
NIDA. (2020b). *Trends & statistics*. June 25, Retrieved from https://www.drugabuse.gov/drug-topics/trends-statistics#supplemental-references-for-economic-costs.
Ning, T., Gong, X., Xie, L., & Ma, B. (2017). Gut microbiota analysis in rats with methamphetamine-induced conditioned place preference. *Frontiers in Microbiology*, *8*, 1620. https://doi.org/10.3389/fmicb.2017.01620. PMCID: PMC5575146.
Northcutt, A. L., Hutchinson, M. R., Wang, X., et al. (2015). DAT isn't all that: Cocaine reward and reinforcement require toll-like receptor 4 signaling. *Molecular Psychiatry*, *20*, 1525–1537. https://doi.org/10.1038/mp.2014.177.
Okun, E., Griffioen, K. J., Lathia, J. D., et al. (2009). Toll-like receptors in neurodegeneration. *Brain Research Reviews*, *59*, 278–292.
Olson, J. K., & Miller, S. D. (2004). Microglia initiate central nervous system innate and adaptive immune responses through multiple TLRs. *Journal of Immunology*, *173*, 3916–3924. https://doi.org/10.4049/jimmunol.173.6.3916.
Osokine, I., & Erlebacher, A. (2017). Inflammation and autism: From maternal gut to fetal brain. *Trends in Molecular Medicine*, *23*, 1070–1071.
Ossipov, M. H., Lai, J., King, T., et al. (2004). Antinociceptive and nociceptive actions of opioids. *Journal of Neurobiology*, *61*, 126–148. https://doi.org/10.1002/neu.20091.
Pacifici, R., Di Carlo, S., Bacosi, A., et al. (2000). Pharmacokinetics and cytokine production in heroin and morphine-treated mice. *International Journal of Immunopharmacology*, *22*, 603–614. https://doi.org/10.1016/S0192-0561(00)00023-0.
Palle, P., Monaghan, K. L., Milne, S. M., et al. (2017). Cytokine signaling in multiple sclerosis and its therapeutic applications. *Medical Science*, *5*, 23. https://doi.org/10.3390/medsci5040023.

Paolicelli, R. C., Bolasco, G., Pagani, F., et al. (2011). Synaptic pruning by microglia is necessary for normal brain development. *Science*, *333*, 1456–1458. https://doi.org/10.1126/science.1202529.

Pascual, M., Baliño, P., Alfonso-Loeches, S., et al. (2011). Impact of TLR4 on behavioral and cognitive dysfunctions associated with alcohol-induced neuroinflammatory damage. *Brain, Behavior, and Immunity*, *25*, S80–S91. https://doi.org/10.1016/j.bbi.2011.02.012.

Pedraz, M., Araos, P., García-Marchena, N., et al. (2015). Sex differences in psychiatric comorbidity and plasma biomarkers for cocaine addiction in abstinent cocaine-addicted subjects in outpatient settings. *Frontiers in Psychiatry*, *6*, 17. https://doi.org/10.3389/fpsyt.2015.00017. PMID: 25762940; PMCID: PMC4329735.

Peirce, J. M., & Alviña, K. (2019). The role of inflammation and the gut microbiome in depression and anxiety. *Journal of Neuroscience Research*, *97*, 1223–1241.

Peterson, V. L., Richards, J. B., Meyer, P. J., et al. (2020). Sex-dependent associations between addiction-related behaviors and the microbiome in outbred rats. *eBioMedicine*, *55*, 102769. https://doi.org/10.1016/j.ebiom.2020.102769.

Pfrieger, F. W., & Barres, B. A. (1997). Synaptic efficacy enhanced by glial cells in vitro. *Science*, *277*, 1684–1687. https://doi.org/10.1126/science.277.5332.1684.

Piccinini, A., & Midwood, K. (2010). DAMPening inflammation by modulating TLR signalling. *Mediators of Inflammation*, *2010*, 672395. https://doi.org/10.1155/2010/672395.

Piepenbrink, M. S., Samuel, M., Zheng, B., et al. (2016). Humoral dysregulation associated with increased systemic inflammation among injection heroin users. *PLoS One*, *11*(7), e0158641. https://doi.org/10.1371/journal.pone.0158641. PMID: 27379802; PMCID: PMC4933366.

Pimentel, E., Sivalingam, K., Doke, M., & Samikkannu, T. (2020). Effects of drugs of abuse on the blood-brain barrier: A brief overview. *Frontiers in Neuroscience*, *14*, 513. https://doi.org/10.3389/fnins.2020.00513.

Plein, L. M., & Rittner, H. L. (2018). Opioids and the immune system—Friend or foe. *British Journal of Pharmacology*, *175*, 2717–2725.

Pribiag, H., & Stellwagen, D. (2013). Tnf-α downregulates inhibitory neurotransmission through protein phosphatase 1-dependent trafficking of GABAA receptors. *The Journal of Neuroscience*, *33*, 15879–15893. https://doi.org/10.1523/JNEUROSCI.0530-13.2013.

Pribiag, H., & Stellwagen, D. (2014). Neuroimmune regulation of homeostatic synaptic plasticity. *Neuropharmacology*, *78*, 13–22. https://doi.org/10.1016/j.neuropharm.2013.06.008.

Prinz, M., & Priller, J. (2014). Microglia and brain macrophages in the molecular age: From origin to neuropsychiatric disease. *Nature Reviews. Neuroscience*, *15*, 300–312.

Przewlocki, R. (2004). Opioid abuse and brain gene expression. *European Journal of Pharmacology*, *500*, 331–349.

Raghavendra, V., Tanga, F. Y., & DeLeo, J. A. (2004). Attenuation of morphine tolerance, withdrawal-induced hyperalgesia, and associated spinal inflammatory immune responses by propentofylline in rats. *Neuropsychopharmacology*, *29*, 327–334. https://doi.org/10.1038/sj.npp.1300315.

Rankin, L. C., & Artis, D. (2018). Beyond host defense: Emerging functions of the immune system in regulating complex tissue physiology. *Cell*, *173*, 554–567.

Risdahl, J. M., Khanna, K. V., Peterson, P. K., & Molitor, T. W. (1998). Opiates and infection. *Journal of Neuroimmunology*, *83*, 4–18. https://doi.org/10.1016/S0165-5728(97)00216-6.

Rivera, P. D., Hanamsagar, R., Kan, M. J., et al. (2019). Removal of microglial-specific MyD88 signaling alters dentate gyrus doublecortin and enhances opioid addiction-like behaviors. *Brain, Behavior, and Immunity*, *76*, 104–115. https://doi.org/10.1016/j.bbi.2018.11.010.

Rivest, S. (2009). Regulation of innate immune responses in the brain. *Nature Reviews. Immunology*, *9*, 429–439.

Rock, R. B., Hu, S., Sheng, W. S., & Peterson, P. K. (2006). Morphine stimulates CCL2 production by human neurons. *Journal of Neuroinflammation*, *3*, 32. https://doi.org/10.1186/1742-2094-3-32.

Rogier, R., Koenders, M. I., & Abdollahi-Roodsaz, S. (2015). Toll-like receptor mediated modulation of T cell response by commensal intestinal microbiota as a trigger for autoimmune arthritis. *Journal of Immunology Research*, *2015*, 527696. https://doi.org/10.1155/2015. Epub 2015 Feb 23. PMID: 25802876; PMCID: PMC4352938.

Rumble, J. M., Huber, A. K., Krishnamoorthy, G., et al. (2015). Neutrophil-related factors as biomarkers in EAE and MS. *The Journal of Experimental Medicine*, *212*, 23–35. https://doi.org/10.1084/jem.20141015.

Russo, S. J., & Nestler, E. J. (2013). The brain reward circuitry in mood disorders. *Nature Reviews. Neuroscience*, *14*, 609–625.

Salter, M. W., & Stevens, B. (2017). Microglia emerge as central players in brain disease. *Nature Medicine*, *23*, 1018–1027.

Sandhu, K. V., Sherwin, E., Schellekens, H., et al. (2017). Feeding the microbiota-gut-brain axis: Diet, microbiome, and neuropsychiatry. *Translational Research*, *179*, 223–244.

Sarkar, C., Basu, B., Chakroborty, D., et al. (2010). The immunoregulatory role of dopamine: An update. *Brain, Behavior, and Immunity*, *24*, 525–528.

Sauma, S., & Casaccia, P. (2020). Gut-brain communication in demyelinating disorders. *Current Opinion in Neurobiology*, *62*, 92–101.

Savignac, H. M., Corona, G., Mills, H., et al. (2013). Prebiotic feeding elevates central brain derived neurotrophic factor, N-methyl-d-aspartate receptor subunits and d-serine. *Neurochemistry International*, *63*, 756–764. https://doi.org/10.1016/j.neuint.2013.10.006.

Schafer, D. P., Lehrman, E. K., Kautzman, A. G., et al. (2012). Microglia sculpt postnatal neural circuits in an activity and complement-dependent manner. *Neuron*, *74*, 691–705. https://doi.org/10.1016/j.neuron.2012.03.026.

Schafer, D. P., & Stevens, B. (2013). Phagocytic glial cells: Sculpting synaptic circuits in the developing nervous system. *Current Opinion in Neurobiology*, *23*, 1034–1040.

Schwarz, J. M., Hutchinson, M. R., & Bilbo, S. D. (2011). Early-life experience decreases drug-induced reinstatement of morphine CPP in adulthood via microglial-specific epigenetic programming of anti-inflammatory IL-10 expression. *The Journal of Neuroscience*, *31*, 17835–17847. https://doi.org/10.1523/JNEUROSCI.3297-11.2011.

Schwarz, J. M., Smith, S. H., & Bilbo, S. D. (2013). FACS analysis of neuronal-glial interactions in the nucleus accumbens following morphine administration. *Psychopharmacology*, *230*, 525–535. https://doi.org/10.1007/s00213-013-3180-z.

Scofield, M. D., & Kalivas, P. W. (2014). Astrocytic dysfunction and addiction: Consequences of impaired glutamate homeostasis. *The Neuroscientist*, *20*, 610–622.

Scorza, C., Piccini, C., Martínez Busi, M., et al. (2019). Alterations in the gut microbiota of rats chronically exposed to volatilized cocaine and its active adulterants caffeine and phenacetin. *Neurotoxicity Research*, *35*, 111–121. https://doi.org/10.1007/s12640-018-9936-9.

Scriven, M., Dinan, T. G., Cryan, J. F., & Wall, M. (2018). Neuropsychiatric disorders: Influence of gut microbe to brain signalling. *Diseases (Basel, Switzerland)*, *6*(3), 78. https://doi.org/10.3390/diseases6030078.

Sekine, Y., Ouchi, Y., Sugihara, G., et al. (2008). Methamphetamine causes microglial activation in the brains of human abusers. *The Journal of Neuroscience*, *28*, 5756–5761. https://doi.org/10.1523/JNEUROSCI.1179-08.2008.

Shorter, D., Domingo, C. B., & Kosten, T. R. (2015). Emerging drugs for the treatment of cocaine use disorder: A review of neurobiological targets and pharmacotherapy. *Expert Opinion on Emerging Drugs*, *20*, 15–29. https://doi.org/10.1517/14728214.2015.985203.

Shrivastava, P., Cabrera, M. A., Chastain, L. G., et al. (2017). Mu-opioid receptor and delta-opioid receptor differentially regulate microglial inflammatory response to control proopiomelanocortin neuronal apoptosis in the hypothalamus: Effects of neonatal alcohol. *Journal of Neuroinflammation*, *14*, 83. https://doi.org/10.1186/s12974-017-0844-3.

Sil, S., Niu, F., Tom, E., et al. (2019). Cocaine mediated neuroinflammation: Role of dysregulated autophagy in pericytes. *Molecular Neurobiology*, *56*, 3576–3590. https://doi.org/10.1007/s12035-018-1325-0.

Silva, Y. P., Bernardi, A., & Frozza, R. L. (2020). The role of short-chain fatty acids from gut microbiota in gut-brain communication. *Frontiers in Endocrinology*, *11*, 25.

Sindberg, G. M., Callen, S. E., Banerjee, S., et al. (2019). Morphine potentiates dysbiotic microbial and metabolic shifts in acute SIV infection. *Journal of Neuroimmune Pharmacology*, *14*, 200–214. https://doi.org/10.1007/s11481-018-9805-6.

Smith, P. M., Howitt, M. R., Panikov, N., et al. (2013). The microbial metabolites, short-chain fatty acids, regulate colonic T reg cell homeostasis. *Science*, *341*, 569–573. https://doi.org/10.1126/science.1241165.

Snider, S. E., Hendrick, E. S., & Beardsley, P. M. (2013). Glial cell modulators attenuate methamphetamine self-administration in therat. *European Journal of Pharmacology*, *701*, 124–130. https://doi.org/10.1016/j.ejphar.2013.01.016.

Song, X., Sun, X., Oh, S. F., et al. (2020). Microbial bile acid metabolites modulate gut RORγ+ regulatory T cell homeostasis. *Nature*, *577*, 410–415. https://doi.org/10.1038/s41586-019-1865-0.

Song, P., & Zhao, Z. Q. (2001). The involvement of glial cells in the development of morphine tolerance. *Neuroscience Research*, *39*, 281–286. https://doi.org/10.1016/S0168-0102(00)00226-1.

Sorini, C., Cardoso, R. F., Gagliani, N., & Villablanca, E. J. (2018). Commensal bacteria-specific CD4 + T cell responses in health and disease. *Frontiers in Immunology*, *9*, 2667. https://doi.org/10.3389/fimmu.2018.02667.

Spichak, S., Dinan, T. G., & Cryan, J. F. (2019). Gut-neuroimmune interactions: The unexpected role of the immune system in brain development. *Biochemistry (London)*, *41*(1), 36–41. https://doi.org/10.1042/BIO04101036.

Spits, H., Artis, D., Colonna, M., et al. (2013). Innate lymphoid cells-a proposal for uniform nomenclature. *Nature Reviews. Immunology*, *13*, 145–149. https://doi.org/10.1038/nri3365.

Stellwagen, D., & Malenka, R. C. (2006). Synaptic scaling mediated by glial TNF-α. *Nature*, *440*, 1054–1059. https://doi.org/10.1038/nature04671.

Stilling, R. M., van de Wouw, M., Clarke, G., et al. (2016). The neuropharmacology of butyrate: The bread and butter of the microbiota-gut-brain axis? *Neurochemistry International*, *99*, 110–132.

Sun, J., Wang, F., Hong, G., et al. (2016). Antidepressant-like effects of sodium butyrate and its possible mechanisms of action in mice exposed to chronic unpredictable mild stress. *Neuroscience Letters*, *618*, 159–166. https://doi.org/10.1016/j.neulet.2016.03.003.

Tadin Hadjina, I., Zivkovic, P. M., Matetic, A., et al. (2019). Impaired neurocognitive and psychomotor performance in patients with inflammatory bowel disease. *Scientific Reports*, *9*, 1–13. https://doi.org/10.1038/s41598-019-50192-2.

Tanda, G., Mereu, M., Hiranita, T., et al. (2016). Lack of specific involvement of (+)-Naloxone and (+)-Naltrexone on the reinforcing and neurochemical effects of cocaine and opioids. *Neuropsychopharmacology*, *41*, 2772–2781. https://doi.org/10.1038/npp.2016.91.

Tang, S. C., Arumugam, T. V., Xu, X., et al. (2007). Pivotal role for neuronal toll-like receptors in ischemic brain injury and functional deficits. *Proceedings of the National Academy of Sciences of the United States of America*, *104*, 13798–13803. https://doi.org/10.1073/pnas.0702553104.

Tartey, S., & Takeuchi, O. (2017). Pathogen recognition and toll-like receptor targeted therapeutics in innate immune cells. *International Reviews of Immunology, 36*, 57–73. https://doi.org/10.1080/08830185.2016.1261318.

Taylor, A. M. W., Castonguay, A., Taylor, A. J., et al. (2015). Microglia disrupt mesolimbic reward circuitry in chronic pain. *The Journal of Neuroscience, 35*, 8442–8450. https://doi.org/10.1523/JNEUROSCI.4036-14.2015.

Thanos, P. K., Kim, R., Delis, F., et al. (2016). Chronic methamphetamine effects on brain structure and function in rats. *PLoS One, 11*(6), e0155457. https://doi.org/10.1371/journal.pone.0155457.

Theberge, F. R., Li, X., Kambhampati, S., et al. (2013). Effect of chronic delivery of the toll-like receptor 4 antagonist (+)-naltrexone on incubation of heroin craving. *Biological Psychiatry, 73*, 729–737. https://doi.org/10.1016/j.biopsych.2012.12.019.

Thomas, D. M., Walker, P. D., Benjamins, J. A., et al. (2004). Methamphetamine neurotoxicity in dopamine nerve endings of the striatum is associated with microglial activation. *The Journal of Pharmacology and Experimental Therapeutics, 311*, 1–7. https://doi.org/10.1124/jpet.104.070961.

Tilleux, S., & Hermans, E. (2007). Neuroinflammation and regulation of glial glutamate uptake in neurological disorders. *Journal of Neuroscience Research, 85*, 2059–2070. https://doi.org/10.1002/jnr.21325.

Tyagi, M., Weber, J., Bukrinsky, M., & Simon, G. L. (2016). The effects of cocaine on HIV transcription. *Journal of Neurovirology, 22*, 261–274.

Ullian, E. M., Sapperstein, S. K., Christopherson, K. S., & Barres, B. A. (2001). Control of synapse number by glia. *Science, 291*, 657–661. https://doi.org/10.1126/science.291.5504.657.

Underwood, M. L., Nguyen, T., Uebelhoer, L. S., et al. (2020). Altered monocyte phenotype and dysregulated innate cytokine responses among people living with HIV and opioid-use disorder. *AIDS, 34*, 177–188. https://doi.org/10.1097/QAD.0000000000002416.

Valentinova, K., Tchenio, A., Trusel, M., et al. (2019). Morphine withdrawal recruits lateral habenula cytokine signaling to reduce synaptic excitation and sociability. *Nature Neuroscience, 22*, 1053–1056. https://doi.org/10.1038/s41593-019-0421-4.

van de Wouw, M., Boehme, M., Lyte, J. M., et al. (2018). Short-chain fatty acids: Microbial metabolites that alleviate stress-induced brain–gut axis alterations. *The Journal of Physiology, 596*, 4923–4944. https://doi.org/10.1113/JP276431.

Veiga-Fernandes, H., & Artis, D. (2018). Neuronal-immune system cross-talk in homeostasis. *Science, 359*, 1465–1466. https://doi.org/10.1126/science.aap9598.

Veldhoen, M., Hocking, R. J., Atkins, C. J., et al. (2006). TGFβ in the context of an inflammatory cytokine milieu supports de novo differentiation of IL-17-producing T cells. *Immunity, 24*, 179–189. https://doi.org/10.1016/j.immuni.2006.01.001.

Verdam, F. J., Fuentes, S., De Jonge, C., et al. (2013). Human intestinal microbiota composition is associated with local and systemic inflammation in obesity. *Obesity (Silver Spring), 21*(12), E607–E615. https://doi.org/10.1002/oby.20466. Epub 2013 Jun 22. PMID: 23526699

Vighi, G., Marcucci, F., Sensi, L., et al. (2008). Allergy and the gastrointestinal system. *Clinical and Experimental Immunology, 153*, 3–6.

Vincent, C., Miller, M. A., Edens, T. J., et al. (2016). Bloom and bust: Intestinal microbiota dynamics in response to hospital exposures and Clostridium difficile colonization or infection. *Microbiome, 4*, 12. https://doi.org/10.1186/s40168-016-0156-3. PMID: 26975510; PMCID: PMC4791782.

Volpe, G. E., Ward, H., Mwamburi, M., et al. (2014). Associations of cocaine use and HIV infection with the intestinal microbiota, microbial translocation, and inflammation. *Journal of Studies on Alcohol and Drugs, 75*, 347–357. https://doi.org/10.15288/jsad.2014.75.347.

Wang, Y., Huang, D. S., & Watson, R. R. (1994). In vivo and in vitro cocaine modulation on production of cytokines in C57BL/6 mice. *Life Sciences*, *54*, 401–411. https://doi.org/10.1016/0024-3205(94)00698-9.

Wang, T. Y., Lee, S. Y., Chang, Y. H., et al. (2018). Correlation of cytokines, BDNF levels, and memory function in patients with opioid use disorder undergoing methadone maintenance treatment. *Drug and Alcohol Dependence*, *191*, 6–13. https://doi.org/10.1016/j.drugalcdep.2018.06.024.

Wang, G., Liu, Q., Guo, L., et al. (2018). Gut microbiota and relevant metabolites analysis in alcohol dependent mice. *Frontiers in Microbiology*, *9*, 1874. https://doi.org/10.3389/fmicb.2018.01874.

Wang, X., Loram, L. C., Ramos, K., et al. (2012). Morphine activates neuroinflammation in a manner parallel to endotoxin. *Proceedings of the National Academy of Sciences of the United States of America*, *109*, 6325–6330. https://doi.org/10.1073/pnas.1200130109.

Wang, F., Meng, J., Zhang, L., et al. (2018). Morphine induces changes in the gut microbiome and metabolome in a morphine dependence model. *Scientific Reports*, *8*, 3596. https://doi.org/10.1038/s41598-018-21915-8.

Wang, X., Northcutt, A. L., Cochran, T. A., et al. (2019). Methamphetamine activates toll-like receptor 4 to induce central immune signaling within the ventral tegmental area and contributes to extracellular dopamine increase in the nucleus accumbens shell. *ACS Chemical Neuroscience*, *10*, 3622–3634. https://doi.org/10.1021/acschemneuro.9b00225.

Warrington, R., Watson, W., Kim, H. L., & Antonetti, F. R. (2011). An introduction to immunology and immunopathology. *Allergy, Asthma & Clinical Immunology*, 7, S1. https://doi.org/10.1186/1710-1492-7-s1-s1.

Webber, M. P., Schoenbaum, E. E., Gourevitch, M. N., et al. (1999). A prospective study of HIV disease progression in female and male drug users. *AIDS*, *13*, 257–262. https://doi.org/10.1097/00002030-199902040-00014.

Weber, M. D., Godbout, J. P., & Sheridan, J. F. (2017). Repeated social defeat, neuroinflammation, and behavior: Monocytes carry the signal. *Neuropsychopharmacology*, *42*, 46–61.

WHO|Facts and figures. (n.d.). Retrieved October 2, 2020b, from https://www.who.int/substance_abuse/facts/en/.

Williams, A. M., Probert, C. S. J., Stepankova, R., et al. (2006). Effects of microflora on the neonatal development of gut mucosal T cells and myeloid cells in the mouse. *Immunology*, *119*, 470. https://doi.org/10.1111/J.1365-2567.2006.02458.X.

Woo, D. H., Bae, J. Y., Nam, M.-H., et al. (2018). Activation of astrocytic μ-opioid receptor elicits fast glutamate release through TREK-1-containing K2P channel in hippocampal astrocytes. *Frontiers in Cellular Neuroscience*, *12*, 319. https://doi.org/10.3389/fncel.2018.00319.

Woodcock, E. A., Hillmer, A. T., Mason, G. F., & Cosgrove, K. P. (2019). Imaging biomarkers of the neuroimmune system among substance use disorders: A systematic review. *Molecular Neuropsychiatry*, *5*, 125–146. https://doi.org/10.1159/000499621.

Xu, Y., Xie, Z., Wang, H., et al. (2017). Bacterial diversity of intestinal microbiota in patients with substance use disorders revealed by 16S rRNA gene deep sequencing. *Scientific Reports*, 7, 3628. https://doi.org/10.1038/s41598-017-03706-9.

Yeager, M. P., Colacchio, T. A., Yu, C. T., et al. (1995). Morphine inhibits spontaneous and cytokine-enhanced natural killer cell cytotoxicity in volunteers. *Anesthesiology*, *83*, 500–508. https://doi.org/10.1097/00000542-199509000-00008.

Zaparte, A., Schuch, J. B., Viola, T. W., et al. (2019). Cocaine use disorder is associated with changes in Th1/Th2/Th17 cytokines and lymphocytes subsets. *Frontiers in Immunology*, *10*, 2435. https://doi.org/10.3389/fimmu.2019.02435.

Zhang, X. Q., Cui, Y., Cui, Y., et al. (2012). Activation of p38 signaling in the microglia in the nucleus accumbens contributes to the acquisition and maintenance of morphine-

induced conditioned place preference. *Brain, Behavior, and Immunity*, *26*, 318–325. https://doi.org/10.1016/j.bbi.2011.09.017.
Zhang, H., Largent-Milnes, T. M., & Vanderah, T. W. (2020). Glial neuroimmune signaling in opioid reward. *Brain Research Bulletin*, *155*, 102–111.
Zhang, Y., Liang, Y., Levran, O., et al. (2017). Alterations of expression of inflammation/immune-related genes in the dorsal and ventral striatum of adult C57BL/6J mice following chronic oxycodone self-administration: A RNA sequencing study. *Psychopharmacology*, *234*, 2259–2275. https://doi.org/10.1007/s00213-017-4657-y.
Zhang, L., & Wang, C. C. (2014). Inflammatory response of macrophages in infection. *Hepatobiliary & Pancreatic Diseases International*, *13*, 138–152.
Zhernakova, A., Kurilshikov, A., Bonder, M. J., et al. (2016). Population-based metagenomics analysis reveals markers for gut microbiome composition and diversity. *Science*, *352*, 565–569. https://doi.org/10.1126/science.aad3369.
Zhu, J. W., Liu, F. L., Mu, D., et al. (2017). Heroin use is associated with lower levels of restriction factors and type I interferon expression and facilitates HIV-1 replication. *Microbes and Infection*, *19*, 288–294. https://doi.org/10.1016/j.micinf.2017.01.002.
Zigmond, E., & Jung, S. (2013). Intestinal macrophages: Well educated exceptions from the rule. *Trends in Immunology*, *34*, 162–168.

CHAPTER EIGHT

Bidirectional causality between addiction and cognitive deficits

Patrick R. Melugin, Suzanne O. Nolan, and Cody A. Siciliano*
Department of Pharmacology, Vanderbilt Brain Institute, Vanderbilt Center for Addiction Research, Vanderbilt University, Nashville, TN, United States
*Corresponding author: e-mail address: cody.siciliano@vanderbilt.edu

Contents

Abstract

Cognitive deficits are highly comorbid with substance use disorders. Deficits span multiple cognitive domains, are associated with disease severity across substance classes, and persist long after cessation of substance use. Furthermore, recovery of cognitive function during protracted abstinence is highly predictive of treatment adherence, relapse, and overall substance use disorder prognosis, suggesting that addiction may be best characterized as a disease of executive dysfunction. While the association between cognitive deficits and substance use disorders is clear, determining causalities is made difficult by the complex interplay between these variables. Cognitive dysfunction present prior to first drug use can act as a risk factor for substance use initiation, likelihood of pathology, and disease trajectory. At the same time, substance use can directly cause cognitive impairments even in individuals without preexisting deficits. Thus, parsing preexisting risk factors from substance-induced adaptations, and how they may interact, poses significant challenges. Here, focusing on psychostimulants and alcohol, we review evidence from clinical literature implicating cognitive deficits as a risk factor for addiction, a consequence of substance use, and the role the prefrontal cortex plays in these phenomena. We then review corresponding preclinical literature, highlighting the high degree of congruency between animal and human studies, and emphasize the unique opportunity that animal models provide to test causality

International Review of Neurobiology, Volume 157
ISSN 0074-7742
https://doi.org/10.1016/bs.irn.2020.11.001

between cognitive phenotypes and substance use, and to investigate the underlying neurobiology at a cellular and molecular level. Together, we provide an accessible resource for assessing the validity and utility of forward- and reverse-translation between these clinical and preclinical literatures.

1. Introduction

Substance use disorders (SUDs) represent a significant global public health crisis. In the United States alone, the societal cost of SUDs has grown to be in excess of 740 billion dollars annually (NIDA, 2020) and worldwide SUDs exceed all other mental health disorders in regard to premature mortality due to illness (Whiteford et al., 2013). Many treatments for SUDs focus on the management of withdrawal symptoms and the reduction of craving (Beck, Wright, Newman, & Liese, 1993; Haass-Koffler, Leggio, & Kenna, 2014; Myrick, Brady, & Malcolm, 2001; Perkins, Conklin, & Levine, 2008; Phillips, Epstein, & Preston, 2014; Veilleux, Colvin, Anderson, York, & Heinz, 2010), sequelae that are also heavily emphasized in the preclinical addiction literature (Becker, 2000; Lichtman & Martin, 2002; Markou et al., 1993; Markou & Koob, 1991; Venniro, Caprioli, & Shaham, 2016). Although treating withdrawal symptoms in abstinent patients is a critical step toward recovery, relapse often occurs long after these symptoms have subsided and therefore there is increasing interest in other mechanisms that outlast these processes. An ever-growing body of clinical research indicates that the dysregulation of executive function, a diverse set of cognitive processes responsible for purposeful, goal-directed behavior (Diamond, 2013), plays a fundamental role in the development and maintenance of SUDs. The prevalence of cognitive comorbidities across substance classes, their role in the initiation and maintenance of harmful patterns of substance use, and the long-lasting nature of these deficits compared to other consequences of drug use suggests that SUD may be best characterized as a disease of executive dysfunction.

Although it is indisputable that cognitive deficits are comorbid with SUDs, the relationship between the two is complex. It has long been held that prolonged exposure to substances of abuse, such as alcohol and psychostimulants, promote neuroadaptations that produce cognitive abnormalities. However, substantial evidence suggests that executive dysfunction is also a risk factor for SUDs, therefore making it difficult to determine the direction and nature of causality in between these variables in clinical populations. Given this complexity, animal models, where risk factors and exposure can be fully measured and controlled, offer unique advantages for probing the relationship between cognitive function and substance use.

Here we review the clinical evidence for the integral role of cognitive deficits in both the development and maintenance of stimulant use disorder (StUD) and alcohol use disorders (AUD), as well as survey insights from the clinical and preclinical literatures regarding the underlying neurobiological mechanisms mediating these deficits. We begin with an overview of clinical research examining the interplay between SUDs and cognitive deficits (for more exhaustive reviews see Bates, Bowden, & Barry, 2002; Domínguez-Salas, Díaz-Batanero, Lozano-Rojas, & Verdejo-García, 2016). Guided by this research, we then examine evidence from animal studies and discuss potential neural circuit mechanisms underlying cognitive deficits in SUDs. We focus on the prefrontal cortex (PFC) given its role in mediating cognitive processes and the extensive evidence linking substance use to dysregulation of PFC function and structure (for review of subcortical structures role in addiction see Everitt & Robbins, 2013; Koob & Volkow, 2010; Yager, Garcia, Wunsch, & Ferguson, 2015), and limit our discussion to literature examining alcohol and psychostimulant use (for review of other compounds see Crean, Crane, & Mason, 2011; Gruber, Silveri, & Yurgelun-Todd, 2007; Jasinska, Zorick, Brody, & Stein, 2014; Klugman & Gruzelier, 2003). While cognitive deficits are commonly observed across SUDs (Rolland et al., 2019), we focus on psychostimulants and alcohol because they are well characterized in both the clinical and preclinical literatures, and therefore provide a suitable platform for comparing human and animal studies.

We highlight similarities in findings between preclinical and clinical investigations, which support the validity of animal models, and discuss areas that are understudied regarding the neuronal basis of these phenomena. Our goal is to provide a broad, accessible introduction to evidence from animal and human studies as to the relationships between cognitive function and SUDs, and to provide a resource on the translatability of specific animal models by comparing these literatures.

2. Cognitive deficits and SUDs: Clinical findings

SUDs are often associated with cognitive impairments spanning multiple domains including attention, memory, and executive function (Rolland et al., 2019; Stavro, Pelletier, & Potvin, 2013; Verdejo-Garcia & Rubenis, 2020). Estimates indicate that between 30 and 80% of individuals with AUD (Bruijnen et al., 2019; Fein, Bachman, Fisher, & Davenport, 1990;

Løberg & Miller, 1986; Martin, Adinoff, Weingartner, Mukherjee, & Eckardt, 1986; Meek, Clark, & Solana, 1989; Morgenstern & Bates, 1999) and 30–50% of individuals with StUD (Bruijnen et al., 2019; O'Malley, Adamse, Heaton, & Gawin, 1992; Rippeth et al., 2004; Vonmoos et al., 2013) exhibit some degree of cognitive impairment. Moreover, deficits are observed in laboratory-based decision-making tasks utilizing non-drug reinforcers demonstrating that these cognitive deficits represent a fundamental dysregulation of function that is not restricted to drug-associated contexts or decisions involving drug (Bechara et al., 2001; Brière et al., 2019; Stout, Busemeyer, Lin, Grant, & Bonson, 2004; Verdejo-Garcia et al., 2007). Interest in substance-induced cognitive impairments has been further driven by findings that deficits, particularly in cognitive processes associated with executive function, are linked to poor clinical outcomes (Bates, Pawlak, Tonigan, & Buckman, 2006; Czapla et al., 2016; Domínguez-Salas et al., 2016; Goncalves et al., 2017). For example, deficits in basic (e.g., response inhibition) and higher-order executive functions (e.g., problem solving, decision making) are associated with early relapse and poor treatment adherence in AUD and StUD populations (Rolland et al., 2019; Rubenis, Fitzpatrick, Lubman, & Verdejo-Garcia, 2019; Stevens et al., 2014).

The PFC, a neocortical structure composed of several functionally and structurally diverse subregions, is critically involved in executive function (Diamond, 2013; Robbins, 1998; Stuss, 2011). In healthy adults, greater PFC volume and thickness is associated with better executive performance across a variety of neuropsychological tests (Burzynska et al., 2012; Yuan & Raz, 2014). In addition, the PFC is disproportionately impacted by prolonged use of alcohol and psychostimulants as compared to other brain regions (Chanraud et al., 2007; Ersche, Williams, Robbins, & Bullmore, 2013; Goldstein & Volkow, 2011; Mackey & Paulus, 2013; Oscar-Berman, Kirkley, Gansler, & Couture, 2004; Pfefferbaum, Sullivan, Rosenbloom, Mathalon, & Lim, 1998; Volkow, Mullani, Gould, Adler, & Krajewski, 1988). Significant physiological and morphological differences are commonly observed in the PFC of individuals with AUD or StUD, often coinciding with impaired executive function (Chanraud et al., 2007; Dao-Castellana et al., 1998; Fein, Di Sclafani, & Meyerhoff, 2002; Goldstein & Volkow, 2002; Hanlon, Dufault, Wesley, & Porrino, 2011; Kim et al., 2006; Le Berre et al., 2014).

While it is clear that individuals with AUD and StUD display deficits in cognitive function concomitant with structural and functional dysregulation of regions involved with these processes, the extent to which deficits are caused by substance use can be difficult to parse. In fact, extant clinical research suggests that deficits in executive function present prior to first drug use (i.e., preexisting) serve as a risk factor that facilitates substance use initiation and confers vulnerability to subsequent development of SUDs. At the same time, these preexisting deficits may be further exacerbated by substance-induced neuroadaptations (Fig. 1). Thus, to understand the complex interplay between cognitive deficits and SUDs, both directions of causality must be considered. Below we discuss the clinical evidence (1) implicating cognitive deficits as a risk factor for SUDs, and (2) as a consequence of prolonged substance use. We do not intend to be exhaustive in our review of the clinical literature; rather, we aim to provide an accessible overview for researchers outside of the immediate field, and to motivate our further discussion of questions that can be uniquely addressed by animal models (see Section 2).

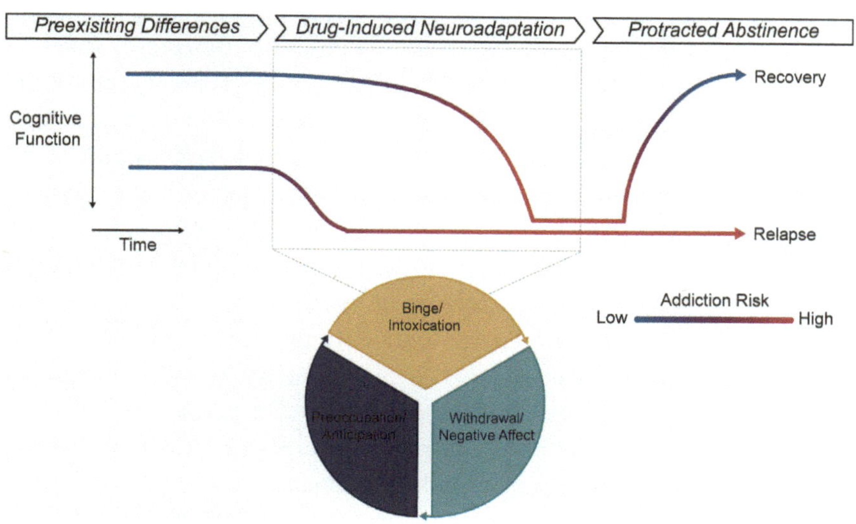

Fig. 1 Interplay between cognitive function, drug use, and addiction outcomes. Variation in cognitive function prior to initial substance use is a critical factor influencing the likelihood of initiation and maintenance of harmful patterns of substance use. Substance-induced neuroadaptations may exacerbate premorbid cognitive deficits or promote the development of cognitive deficits in cognitively normal individuals. Deficits far outlast acute phases of withdrawal and associated negative affect, serving as an enduring risk factor in addiction recovery, and recovery of cognitive function during abstinence is associated with lower rates of relapse; therfore, treatments targeting cognitive dysfunction may significantly improve clinical outcomes.

3. Clinical evidence linking executive dysfunction and addiction vulnerability

Many individuals are exposed to or actively use illicit drugs and alcohol, however only a relatively small subset of these individuals will develop a SUD in their lifetime (SAMHSA, 2019). Thus, to understand, treat, and prevent SUDs, it is imperative to identify behavioral and neurobiological features that may promote the initiation of substance use as well as those that confer resilience or susceptibility to the subsequent development of SUDs. Premorbid deficits in response inhibition (i.e., inhibitory control), one of the core executive functions alongside working memory and cognitive flexibility, have been consistently implicated in SUD vulnerability. Prospective studies have found that relatively poor performance on measures of response inhibition in individuals with minimal substance use history is predictive of the frequency and intensity of later substance use (Nigg et al., 2006; Squeglia, Jacobus, Nguyen-Louie, & Tapert, 2014). Moreover, neuroimaging studies have consistently documented PFC hypofunction in individuals with SUDs during tasks involving response inhibition (Bolla et al., 2004; Fu et al., 2008; Morein-Zamir, Simon Jones, Bullmore, Robbins, & Ersche, 2013; Salo, Ursu, Buonocore, Leamon, & Carter, 2009) and it is thought that decreased activation of the frontal lobe during these tasks may also serve as an important functional marker for susceptibility to problematic substance use. For example, during a go/no-go task in which respondents were required to respond to certain stimuli (go trials) and inhibit responding to others (no-go trials), decreased activity in multiple subregions of the PFC among adolescents was associated with transition to heavy alcohol use ~3 years later (Norman et al., 2011). Similarly, lower activation in the PFC during this task also has been shown to predict heightened substance use and SUD symptoms at an 18-month follow-up (Mahmood et al., 2013). Importantly, differences in task performance and PFC activation in these studies have typically been most pronounced during "no-go" trials, where the subject is required to inhibit a response, suggesting that vulnerability to SUDs is related to deficits in response inhibition rather than poor task performance in general.

For many individuals, substance use initiation begins during adolescence (Johnston et al., 2019), a critical transition period characterized by significant physical, social, emotional, and cognitive development as well as continued brain maturation including the refinement of connectivity in the PFC

(Spear, 2013). Differences in cognitive performance across development provide a unique opportunity to examine the impact of cognitive function on vulnerability to SUDs. For a variety of substances including alcohol and psychostimulants, age of first substance use is correlated with SUD prevalence whereby younger age of onset is associated with a greater likelihood of developing a SUD (Chen, Storr, & Anthony, 2009; Dawson, Goldstein, Chou, Ruan, & Grant, 2008; Flory, Lynam, Milich, Leukefeld, & Clayton, 2004; King & Chassin, 2007; Li, Duncan, & Hops, 2001; Lopez-Quintero et al., 2011). Although many PFC-mediated cognitive processes continue to mature throughout adolescence, the protracted development of cognitive control, juxtaposed with a fully developed and hyperactive reward system, has been proposed to be an underlying factor in adolescent SUD vulnerability (Hammond, Mayes, & Potenza, 2014). Indeed, adolescents display high attraction to rewarding activities associated with novelty and sensation seeking and are less sensitive to negative outcomes of risky choices (Somerville & Casey, 2010). Moreover, prospective studies have demonstrated that high reward-seeking in conjunction with poorer working memory performance is predictive of early substance use progression and the ultimate development of SUDs (Khurana et al., 2015; Khurana, Romer, Betancourt, & Hurt, 2017). While these studies implicate PFC-mediated cognitive processes as important factors in vulnerability to problematic drug use, they also further highlight the difficulty of precisely parsing causal relationships between these variables in human subjects.

Many of the cognitive phenotypes implicated in addiction vulnerability are heritable traits and linked to family history of SUDs (Chassin, Pitts, & Prost, 2002; Clark, Cornelius, Kirisci, & Tarter, 2005; Friedman et al., 2008; Hill, Shen, Lowers, & Locke, 2000). Family history of SUDs is associated with deficits across multiple domains of executive function including cognitive flexibility, response inhibition, and working memory (Corral, Holguín, & Cadaveira, 2003; Ersche, Jones, et al., 2012; Ersche, Turton, et al., 2012; Habeych, Folan, Luna, & Tarter, 2006; Nigg et al., 2006). For example, individuals from families with a history of alcoholism display less improvement in cognitive flexibility during development relative to those without a family history of alcoholism (Corral et al., 2003) as well as aberrant patterns of frontal lobe activation during response inhibition tasks (Schweinsburg et al., 2004; Silveri, Rogowska, McCaffrey, & Yurgelun-Todd, 2011). Along with cognitive/behavioral traits, marked morphological abnormalities have been observed in individuals with a family history of alcoholism including reductions in brain volume and cortical thickness in frontal regions such as the orbitofrontal cortex (OFC; Henderson et al., 2018; Hill et al., 2009). A recent, landmark longitudinal study revealed that reduced gray matter brain volume in the dorsolateral PFC and insular cortex, brain regions involved in cognitive processes including decision making, reasoning, response inhibition, and working memory (Krawczyk, 2002), was predictive of the initiation of alcohol drinking during adolescence and increased alcohol use during early adulthood (Baranger et al., 2020). These associations were found to be due to shared genetic factors and that genomic risk for greater alcohol use was enriched in genes that were preferentially expressed in the dorsolateral PFC (Baranger et al., 2020). Similarly, lower gray matter volume in the medial PFC of occasional stimulant users predicted future escalation of stimulant use (Becker et al., 2015), suggesting that deficits in cortical function confer vulnerability to SUD across substance classes.

Together, there is extensive evidence demonstrating that executive dysfunction as well as structural and functional abnormalities in the PFC often serve as premorbid factors that facilitate problematic substance use. However, there is a clear need for further prospective investigations of prexisting individual differences in executive function, their role in SUD vulnerability, and how these phenotypes interact with substance-induced alterations in PFC function.

4. Clinical evidence for substance-induced impairments in executive function

Although it is accepted that prolonged substance use contributes to functional and structural brain changes and subsequent cognitive impairments observed in SUDs, many clinical studies have relied heavily on cross-sectional designs that cannot fully dissociate the impact of drug-induced adaptations from preexisting differences. A prominent and often replicated finding is that greater substance use is associated with decline in brain volume in cortical regions as well as with extent of cognitive dysfunction (Albein-Urios, Martinez-González, Lozano, Clark, & Verdejo-García, 2012; Bechara et al., 2001; Bolla, Funderburk, & Cadet, 2000; Fein, Klein, & Finn, 2004; Verdejo-Garcia et al., 2007). For example, heavy alcohol drinking in humans has been linked to widespread reductions in brain volume (Bjork, Grant, & Hommer, 2003; Paul et al., 2008; Pfefferbaum et al., 1998), decreased gray matter volume in the frontal lobe (Cardenas, Studholme, Meyerhoff, Song, & Weiner, 2005; Kubota et al., 2001; Pfefferbaum et al., 1998), and decreased neuronal density in the OFC (Miguel-Hidalgo, Overholser, Meltzer, Stockmeier, & Rajkowska, 2006). Similarly, psychostimulant use is also associated with reductions in gray matter volume in multiple PFC subregions (Ersche et al., 2013). While there is no question that there is a robust relationship between excessive drug use, dysregulated cortical function, and cognitive deficits, as discussed in the previous section, it is important to consider that some of these effects may be driven by preexisting differences or may be a result of complex interactions with latent traits which ultimately determine the extent of cognitive dysregulation and propensity to develop a SUD.

One notable study examined cognitive performance in both recreational and dependent cocaine users who differed markedly in their patterns and amount of cocaine use (Vonmoos et al., 2013). Dependent cocaine users exhibited deficits in cognitive performance, including performance on tasks of executive function, relative to recreational users and cocaine-naive individuals. Interestingly, cognitive performance in recreational cocaine users was intermediate to that of cocaine-dependent and cocaine-naive individuals (Vonmoos et al., 2013). In order to parse the contribution of preexisting and substance-induced effects, a handful of longitudinal studies have sought to determine whether the degree of substance use is associated with morphological changes and cognitive deficits over time. A 5-year longitudinal study that tracked cortical gray matter and ventricular changes in men with AUD found that the amount of alcohol consumed throughout this period was predictive of sulci expansion and cortical gray matter loss that were most prominent in prefrontal and frontal regions (Pfefferbaum et al., 1998). Furthermore, high levels of daily alcohol consumption in older adults has been linked to a significant decline in global cognition, executive function, and memory over a 10-year period relative to light-to-moderate drinkers and alcohol abstainers (Sabia et al., 2014). Finally, in cocaine users, increased cocaine use throughout a 1-year period was associated with a reduction in cognitive performance with the greatest impairments observed in working memory (Vonmoos et al., 2014).

Perhaps the strongest approach to isolating drug-induced volumetric and cognitive deficits in SUDs from preexisting differences comes from studies demonstrating that these deficits attenuate during protracted abstinence. In AUD, moderate impairments across multiple cognitive domains including executive functions, though persistent during early and intermediate sobriety, become less severe after a year (Stavro et al., 2013). This is in line with findings that alcohol-dependent individuals with an average of 6 years of sobriety display normal cognitive functioning (Fein, Torres, Price, & Di Sclafani, 2006). Improvements in verbal memory, executive function, and social cognition have been observed in methamphetamine-dependent individuals after 6 months of continued abstinence (Zhong et al., 2016). In cocaine users, decreased cocaine use has been associated with minor improvements in domains such as attention, declarative memory, and executive function, with the greatest improvements observed in individuals remaining abstinent at a one-year follow-up (Vonmoos et al., 2014).

In contrast to the delayed recovery of cognitive impairments in substance users, regional brain volume increases are evident during early abstinence and continue to improve over time. For example, volume increases occur during alcohol abstinence in multiple brain regions including the dorsolateral PFC, OFC, anterior cingulate cortex, and insula as early as 1 month (Zou, Durazzo, & Meyerhoff, 2018), though improvements in cognitive function do not typically appear until later (discussed above). Importantly, while brain volume continues to increase in these brain regions during long-term abstinence from alcohol, rates of increase are at least 2.5 times higher during short-term vs. long-term abstinence indicating rapid abstinence-induced recovery of frontal and insular brain volumes among individuals with AUD (Zou et al., 2018). Furthermore, increases in gray matter in multiple cortical regions including the anterior and posterior cingulate, insula, and PFC are apparent during early cocaine abstinence and volumetric deficits are fully recovered in as little as 35 weeks after cessation of use (Connolly, Bell, Foxe, & Garavan, 2013). These findings suggest that cognitive improvements during abstinence may be contingent on the recovery of brain regions that subserve affected cognitive domains, but that behavioral recovery may lag significantly behind these morphological changes. Additionally, while evidence indicates that abstinence promotes the recovery of these domains which may, in turn, reduce the likelihood of relapse (Rolland et al., 2019; Rubenis et al., 2019; Stevens et al., 2014), questions remain as to whether comorbid cognitive impairments in SUDs are fully reversible and to what extent residual impairments are substance-induced. It is conceivable that preexisting deficits in executive function are still present throughout prolonged abstinence and therefore not only represent a risk factor for the development of SUDs but also a persistent obstacle in the path to recovery.

5. Insights from preclinical models

As summarized above, there is strong evidence supporting causality for cognitive deficits as a risk factor for SUDs and a consequence of substance use that maintains harmful behavioral patterns. Both directions of causality are accepted to be true. However, parsing the precise contributions of preexisting and substance-induced deficits to SUDs, and investigating the neurobiology underlying the two phenomena, remains a challenge. The use of human subjects poses many roadblocks in the attempt to understand the bidirectional relationship between cognitive dysfunction and SUDs due to a multitude of factors, including the cross-sectional nature of most studies, reliance on self-reports of prior substance use, and the sheer number of variables that must be controlled for. Preclinical studies in animal models, though not without their own limitations, can provide further insight into this relationship whereby a substance naive population can be studied longitudinally, while allowing complete experimenter control over the nature of substance exposure and measurements of cognitive dysregulation. Here, we review the wealth of preclinical literature that has contributed to our understanding of this complex relationship and highlight the congruency between animal and clinical studies.

6. Phenotypic predictors of SUD vulnerability in animal models

While the majority of preclinical literature has examined neuroadaptations underlying drug-induced cognitive impairments, there has also been substantial work probing the role of preexisting variation in a wide array of behavioral traits as predictors of later drug and alcohol self-administration behaviors. These investigations have implicated various aspects of cognitive function such as response inhibition and cognitive flexibility, corroborating and extending clinical findings linking preexisting cognitive deficits with vulnerability to SUDs (for an overview of the tasks used, see Fig. 2). Importantly, in preclinical studies history of substance use is assured, rather than self-reported, and propensity for substance use can be assessed independent from availability due to environmental factors.

As discussed previously, cognitive flexibility, a critical component of executive function that is broadly defined as the ability to make behavioral

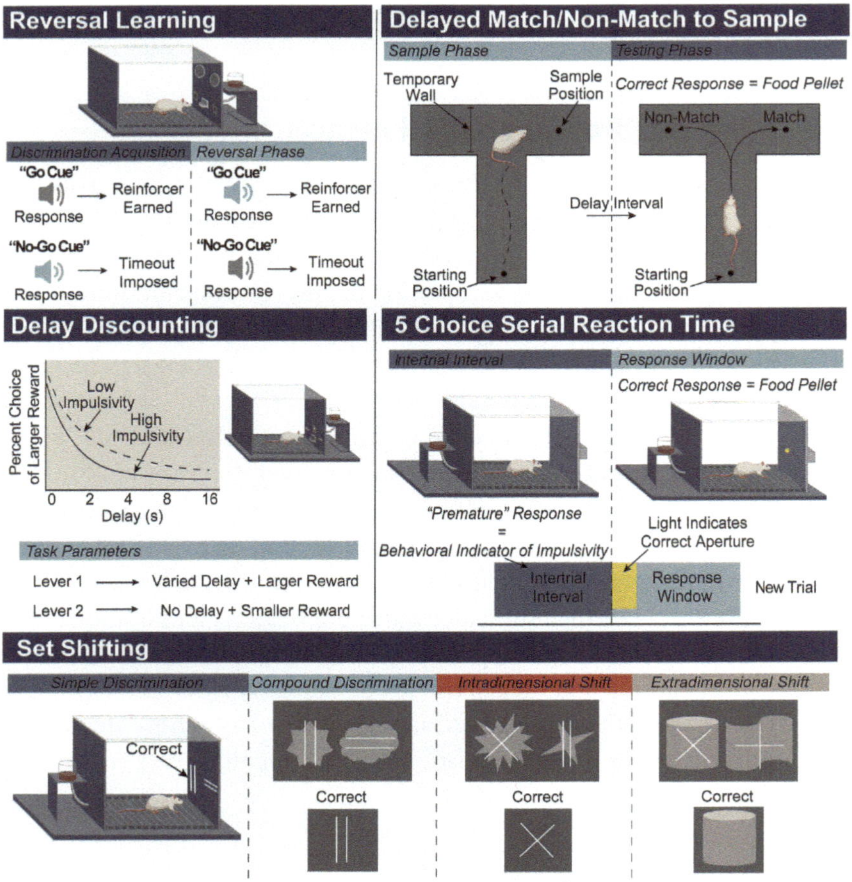

Fig. 2 Behavioral tasks commonly used to assess cognitive function in rodents. Reversal learning tests how quickly rodents can learn a contingency switch as a measure of cognitive flexibility. In the delayed match/non-match to sample task, which is used to measure working memory, animals are shown a sample stimulus and are required to remember the sample throughout a delay period. To receive a reward, they must then return to the sample stimulus (delayed match to sample task) or avoid the sample (delayed non-match to sample task). In delay discounting, animals choose between responding for a small reward delivered immediately or a larger reward delivered after a delay period. Here, impulsivity is measured by how strongly reward magnitude is discounted as a function of wait time. The five-choice serial reaction time task can also be used to assess impulsivity. In this task, animals must withhold responding until the correct response is indicated by brief illumination of a port. Animals must quickly respond after the presentation and the frequency of premature responses is used as a measure of impulsivity. Set shifting is a complex task that requires the subject to learn a set of rules presented within complex stimuli (e.g., the presence of parallel lines indicates the target stimulus) while ignoring irrelevant features (e.g., the shape of the visual stimulus). Intradimensional shifts where the initial rule is altered (e.g., perpendicular lines become the target) or extradimensional shifts where the initial rule is replaced by an entirely new rule (e.g., stimulus shape becomes the relevant feature) are introduced accross sessions and the animals' ability to update behavior is assessed as a measure of cognitive flexibility. Note that many variations of these tasks exist, and exact methodologies can differ greatly.

adaptations in response to changing environmental contingencies (Ragozzino, Detrick, & Kesner, 1999), is impaired in humans with AUD (Fein et al., 2006; Goudriaan, Oosterlaan, Beurs, & Brink, 2006) and StUD (Ersche, Roiser, Robbins, & Sahakian, 2008; Kim et al., 2006). In both humans and animals, commonly used tests of cognitive flexibility include reversal learning and set shifting (explained below). Importantly, evidence from lesion studies of rodents, humans, and non-human primates indicate that performance on these tasks is dependent on the OFC and dorsolateral PFC (medial PFC in rodents) (Bissonette et al., 2008; Dias, Robbins, & Roberts, 1996a, 1996b; McAlonan & Brown, 2003), indicating that these cognitive constructs have validity for cross-species comparisons.

In animal models, reversal learning is typically assessed in tasks wherein the animal learns that one type of response yields a reinforcer (e.g., food) while the other yields a negative consequence (e.g., a timeout period where food cannot be obtained or delivery of an electrical shock). After meeting a learning criterion, typically a high percentage of correct responses, the contingencies are then switched and the ability to quickly update behavioral strategies to match the new rules is used as a measure of cognitive flexibility (Fig. 2). Low cognitive flexibility indicated by poor performance in reversal learning tasks predicts various aspects of drug-taking behaviors in multiple model species [reviewed in (Izquierdo & Jentsch, 2012)]. For example, a higher latency to reach criterion after a contingency switch (i.e., poor performance) is associated with faster acquisition of cocaine self-administration and higher administration rates across sessions in mice (Cervantes, Laughlin, & Jentsch, 2013). Similarly, strain variation across recombinant inbred mice shows phenotypic overlap between poor performance in reversal learning and high levels of alcohol self-administration and cue-induced reinstatement of alcohol seeking (Laughlin, Grant, Williams, & Jentsch, 2011; Loos, Staal, Smit, De Vries, & Spijker, 2013). Together, these findings provide further support for the idea that an impaired ability to update responding following a contingency switch is a predictor of heightened self-administration of both cocaine and alcohol.

Similarly, performance on set-shifting tasks is also predictive of future aspects of substance use in preclinical models. In this task, animals learn by trial and error to attend to a relevant cue in the environment (e.g., shape) while ignoring irrelevant cues (e.g., color) in order to receive a food reward. During an intradimensional shift, animals must maintain the attentional set by learning to ignore a previously reward-paired stimuli (e.g., triangle) to attend to another stimulus within the same dimension (e.g., circle). This attentional set is then challenged during the extradimensional shift wherein a previously irrelevant dimension (e.g., color) is now relevant. Importantly, this task can be easily applied across multiple species, including in humans as a variation of the Wisconsin Card Sorting Test, therefore presenting a highly translatable approach (Brown & Tait, 2016). In nonhuman primates, lower performance in set-shifting tasks is associated with higher overall preference for alcohol, reduced latency to reach maximum alcohol consumption, and higher overall intake (Shnitko, Gonzales, & Grant, 2019). While these results provide strong evidence that relationships between cognitive performance and SUD vulnerability can be faithfully recapitulated in a range of models, relatively little preclinical work has focused on neural circuit mechanisms mediating these effects prior to first drug use.

Response inhibition, the ability to suppress prepotent responses in order to select more appropriate goal-directed behaviors (Diamond, 2013), has been linked in both clinical and preclinical studies to measures of impulsivity, suggesting they may have similar underlying neural mechanisms (Dalley, Everitt, & Robbins, 2011; Franken, van Strien, Nijs, & Muris, 2008; Horn, Dolan, Elliott, Deakin, & Woodruff, 2003; Izquierdo & Jentsch, 2012). Furthermore, like reversal learning performance, behavioral measures of impulsivity also have well-studied relationships with aspects of SUD vulnerability in animal models, both for psychostimulants and alcohol [reviewed in (Belin, Belin-Rauscent, Everitt, & Dalley, 2016; Winstanley, Olausson, Taylor, & Jentsch, 2010)]. Performance in delay discounting tasks represents one commonly utilized measurement of impulsivity in both clinical and preclinical studies, which is shown to be predictive of aspects of substance use (Vanderveldt, Oliveira, & Green, 2016). In this paradigm, impulsivity is defined as the extent to which time reduces the subjective value (or reward magnitude) of a reinforcer. This is typically assessed by allowing the choice between a smaller reward that will be delivered immediately after the choice is made, or a larger reward that is delivered after a delay period. By varying the length of the delay, discounting of reward value as a function of time is assessed as a measure of impulsivity (Fig. 2) (Bizot, Le Bihan, Puech, Hamon, & Thiébot, 1999; Thiébot, Le Bihan, Soubrié, & Simon, 1985; Vanderveldt et al., 2016).

Early studies in the field found that impulsive performance in a delay discounting paradigm predicts higher alcohol consumption in rats (Poulos, Le, & Parker, 1995; Poulos, Parker, & Lê, 1998). Furthermore, when animals are selectively bred for alcohol preference, high-alcohol-preferring mouse strains exhibit steeper delay discounting curves (i.e., small, immediate rewards are preferred over large, delayed rewards), even when tested prior to first alcohol exposure (Oberlin & Grahame, 2009; Wilhelm & Mitchell, 2008). Impulsivity can also be measured using the five-choice serial reaction time task wherein an animal is required to select the correct nose-poke port following a brief presentation of a visual cue indicating which option is correct on that trial (typically a light in the response aperture), and withhold responding during the intertrial interval to maximize the number of reinforcers earned. However, it should be noted that impulsivity in this task, as measured by premature, anticipatory responses during the intertrial interval, does not appear to be related to alcohol preference (Peña-Oliver et al., 2015). Altogether, preexisting differences in impulsivity prior to alcohol exposure are predictive of subsequent AUD-like phenotypes in preclinical models, mirroring results in humans, but this relationship may be specific to subdomains within impulsivity or to sensitive to the methodology used to assess impulsivity.

Individual differences in impulsivity also map onto aspects of rodent psychostimulant use. In this body of work, multiple subdomains of impulsivity have been considered, such as attentional impulsivity/impulsive choice and motor impulsivity/inhibitory failure [(Robinson et al., 2009); reviewed in (Dalley et al., 2011)]. In rodents, attentional impulsivity is often defined as steeper delay discounting curves, while motor impulsivity is typically defined in reaction time tasks as high levels of anticipatory/premature motor response behaviors prior to the presentation of stimuli indicating that a reinforcer is available. For cocaine, high attentional impulsivity in delay discounting predicts faster escalation of self-administration (Anker, Perry, Gliddon, & Carroll, 2009; Perry, Larson, German, Madden, & Carroll, 2005; Perry, Nelson, & Carroll, 2008). Moreover, high motor impulsivity in the five-choice serial reaction time task predicts risk to develop compulsive cocaine-taking in outbred rats (Belin, Mar, Dalley, Robbins, & Everitt, 2008; Dalley et al., 2007), mirroring the clinical findings (Nigg et al., 2006). Impulsive performance in this task also correlates strongly with several facets of pathology-like psychostimulant-taking behaviors, including extinction resistance (Broos, Diergaarde, Schoffelmeer, Pattij, & De Vries, 2012), insensitivity to punishment (Belin et al., 2008), as well as higher reinstatement of cocaine-taking after punishment-induced abstinence (Economidou, Pelloux, Robbins, Dalley, & Everitt, 2009).

In sum, individual differences in motor and attentional components of impulsivity appear to predict multiple aspects of drug-taking behavior across substance classes. Specifically, variation in attentional impulsivity has a strong relationship with the preliminary aspects of drug-taking, such as preference and escalation, while motor impulsivity predicts compulsive aspects of drug-taking, such as persistent drug use despite negative consequences. Collectively, these findings suggest that the two proposed components of impulsive behavior are indeed dissociable and should likely be examined separately. In further support of this notion, increased motor impulsivity in the five-choice serial reaction time task fails to predict alcohol preference (Peña-Oliver et al., 2015). Based on the above findings from the cocaine literature, motor impulsivity may instead be associated with likelihood of persistent alcohol seeking, though this remains to be tested. Overall, these results provide strong support for the impact of underlying impulsivity and cognitive inflexibility on vulnerability to addiction and highlight the need for further research into the neurobiological mechanisms underlying these relationships.

7. Substance-induced impairments in cognitive function in animal models

Animal models have provided valuable insight into the nature of substance-induced deficits in PFC function and associated cognitive deficits. Importantly, these studies can definitively parse preexisting traits from substance-induced deficits using longitudinal approaches where genetic and environmental backgrounds can be fully observed and controlled. Findings from this research have revealed that several facets of executive function shown to be impaired in individuals with SUD including cognitive flexibility (Chung et al., 2007; Errico, King, Lovallo, & Parsons, 2002; Ersche et al., 2008), response inhibition (Goudriaan et al., 2006; Lawrence, Luty, Bogdan, Sahakian, & Clark, 2009; Monterosso, Aron, Cordova, Xu, & London, 2005), and working memory (Albein-Urios et al., 2012; Ambrose, Bowden, & Whelan, 2001; Gonzalez, Bechara, & Martin, 2007), are similarly disrupted in rodents and non-human primates following exposure to substances of abuse.

Impaired cognitive flexibility has been observed in alcohol-exposed rodents across ages of exposure, length of exposure, and withdrawal time points. Chronic intermittent ethanol (CIE) exposure is a well-validated model of alcohol dependence where subjects are exposed to multiple cycles of high concentrations of alcohol vapor followed by abstinence/withdrawal periods (Avegno & Gilpin, 2019; Becker & Lopez, 2004; Gilpin, Richardson, Cole, & Koob, 2008; Lopez & Becker, 2005; Vendruscolo & Roberts, 2013). In rats, CIE exposure during adolescence produces set-shifting impairments in adulthood, enhanced alcohol-seeking behavior, and volumetric reductions in brain regions including the neocortex, thalamus, and hippocampus (Gass et al., 2014). Similar studies conducted in both mice and rats have also revealed that CIE exposure causes impairments in set-shifting performance assessed during brief abstinence (Kroener et al., 2012; Trantham-Davidson et al., 2014) while studies involving voluntary alcohol drinking have revealed reversal learning deficits in rats assessed following more prolonged abstinence periods (Charlton et al., 2019). Notably, these reversal learning deficits coincided with significant cortical cell loss in the OFC, medial PFC, and motor cortex (Charlton et al., 2019).

Similar to alcohol, exposure to psychostimulants produces persistent impairments in cognitive flexibility in rodents and non-human primates. In monkeys, both short-term (2 weeks) or long-term (~5 years) exposure to experimenter-administered or self-administered cocaine, respectively, induced deficits in reversal learning (Gould, Gage, & Nader, 2012; Jentsch, Olausson, De La Garza, & Taylor, 2002). Interestingly, deficits in reversal learning following short-term exposure to cocaine were still evident after a month of abstinence (Jentsch et al., 2002). These findings are in agreement with studies in rats in which experimenter-administered or self-administered cocaine led to reversal learning deficits which persisted 1–3 months after drug cessation (Calu et al., 2007; Schoenbaum, Saddoris, Ramus, Shaham, & Setlow, 2004). Notably, the degree of reversal learning impairment induced by cocaine was indistinguishable from that induced by bilateral OFC lesions (Calu et al., 2007; Schoenbaum et al., 2004; Schoenbaum, Setlow, Nugent, Saddoris, & Gallagher, 2003), suggesting that psychostimulants result in OFC-mediated deficits in cognitive flexibility in both rodents and non-human primates. In addition, performance on tasks of cognitive flexibility thought to be primarily mediated by the medial PFC (i.e., set shifting) are similarly disrupted in psychostimulant-exposed rodents. For example, in rats, methamphetamine self-administration induced set-shifting deficits that were comparable to those observed in animals with lesions to the dorsomedial PFC (Parsegian, Glen, Lavin, & See, 2011). Although it is tempting to conclude from any set of studies that drug-induced deficits in cognitive function can be localized to a single brain region, it is much more likely that deficits in the function of many regions converge to produce rigid behavioral strategies.

In addition to deficits in cognitive flexibility, animal models have also faithfully reproduced substance-induced deficits in working memory commonly observed in individuals with AUD and StUD during protracted abstinence (Chanraud et al., 2007; Fein et al., 2006; Gonzalez et al., 2007; Goudriaan et al., 2006; Rendell, Mazur, & Henry, 2009). Working memory refers to a cognitive system that allows for the temporary storage and manipulation of limited amounts of information required to perform complex cognitive tasks including planning, reasoning, and problem-solving (Cowan, 2014). Lesion studies in rodents and non-human primates have revealed that working memory is dependent on the dorsolateral PFC (medial PFC in rodents) (Tsutsui, Oyama, Nakamura, & Iijima, 2016). In both humans and animals, delayed match to sample and delayed non-match to sample tasks are commonly used to assess working memory performance. These tasks can vary significantly in exact methodology (Fitz, Gibbs, & Johnson, 2008; George et al., 2012; George, Mandyam, Wee, & Koob, 2008; Herndon, Moss, Rosene, & Killiany, 1997; Porter et al., 2011), but typically involve presenting the subject with a stimulus (in rodents this is often one arm of a T-maze) referred to as the sample. In a delayed match to sample task, the subject must then remember the sample stimulus through a given delay period before returning to the sample stimulus to receive a reward. In a delayed non-match to sample task, to receive a reward the subject must avoid the sample stimulus (for example by choosing the opposite arm of the T-maze) (Fig. 2). A similar task, termed spontaneous alternation, is also commonly used in rodents and relies on novelty rather than a food reward to reinforce behavior (Hughes, 2004).

In mice, voluntary alcohol drinking produced persistent deficits in working memory (Dominguez et al., 2017). Notably, these deficits were only evident in animals that underwent a prolonged period of abstinence and not those with continued alcohol access at the time of testing (Dominguez et al., 2017). Impairments in working memory performance have also been observed in rats on a delayed non-match to sample task during acute abstinence after intermittent, but not continuous, access to alcohol; however, these deficits were no longer apparent when tested during more prolonged abstinence periods (George et al., 2012). These findings suggest that alcohol-induced deficits in working memory in rodents may be most pronounced during acute withdrawal, as compared to during maintenance of drinking, but dissipate over protracted abstinence.

Regarding psychostimulants, long-term cocaine self-administration in non-human primates induced deficits on a delayed match to sample task when tested during acute abstinence (Porter et al., 2011). Furthermore, escalation of cocaine intake during extended access cocaine self-administration in rats was associated with working memory deficits on a delayed non-match to sample task, while limited access self-administration was not (George et al., 2008). Working memory deficits during protracted abstinence from cocaine correlates with a lower density of neurons and oligodendrocytes in the dorsomedial PFC and oligodendrocytes in the OFC (George et al., 2008). As mentioned earlier, deficits in executive functions including working memory gradually improve with extended abstinence in human cocaine users (Vonmoos et al., 2014). Several studies in rodents and non-human primates have also revealed improvements in cocaine-induced working memory deficits following prolonged abstinence. For example, persistent impairments in working memory on a delayed alternation task, a measure of spatial working memory, were observed following short-term cocaine self-administration in rats that dissipated after 6 weeks of abstinence (Fijał, Nowak, Leśkiewicz, Budziszewska, & Filip, 2015). Additionally, non-human primates with long-term exposure to self-administered cocaine displayed improvements in working memory on a delayed match to sample task by 30 days of abstinence (Gould et al., 2012).

Taken together, these findings indicate that both alcohol and psychostimulant exposure across multiple animal models can induce significant impairments in PFC-mediated executive functions. This is consistent with clinical studies in humans that have found that individuals with SUDs commonly display signs of executive dysfunction and confirms that prolonged substance abuse induces cognitive deficits in addition to those that may be preexisting. In several animal studies, substance-induced impairments were comparable to those observed in animals that received selective damage to regions of the PFC that subserve these functions. Furthermore, multiple animal studies reported substance-induced impairments coincident with loss of cortical volume and reductions in cellular density in the PFC (Charlton et al., 2019; Gass et al., 2014; George et al., 2008), consistent with clinical observations (Fein et al., 2002; Kim et al., 2006; Le Berre et al., 2014). This is in agreement with studies revealing that individuals with SUDs display reduced markers of neuronal integrity and viability in the frontal lobe and PFC (Chang, Ernst, Strickland, & Mehringer, 1999; Ernst, Chang, Leonido-Yee, & Speck, 2000; Xia et al., 2012) as well as reductions in brain volume and density in these regions (Hanlon et al., 2011; Matochik, London, Eldreth, Cadet, & Bolla, 2003; Nakama et al., 2011; Pfefferbaum et al., 1998; Pfefferbaum, Sullivan, Mathalon, & Lim, 1997). This suggests that substance-induced alterations to cognitive performance in animal models is associated with morphological and functional alterations to PFC with a high degree of similarity to alterations observed in individuals with SUDs, further supporting the utility of animal models for mechanistic investigations of these phenomena.

Notably, the aforementioned animal studies utilized a variety of methods to expose animals to alcohol and psychostimulants, and there is significant debate within the field as to which methods are most relevant for human addiction. While there is extensive nuances in the models used, and even how the same model/procedure is performed between laboratories, one major distinguishing feature is whether the drug is used voluntarily (i.e., the animal chooses to consume the drug, by pressing a lever to activate a syringe pump or licking a spout) or non-voluntarily (i.e., the experimenter chooses if the animal is exposed, via injection or filling the homecage with vaporized drug). It is clear that the complexities of SUDs cannot be captured in a single animal model, and that diversity of models to address specific aspects of SUDs is a strength of the preclinical literature. That said, the differential impact of various exposure methods on behavioral and neuronal function is not trivial. Indeed, contingent vs. non-contingent drug exposure can produce distinct neuroadaptations and cognitive deficits (Schweppe et al., 2020; Wiskerke, Schoffelmeer, & De Vries, 2016). Taken together, the ability to specifically isolate and test specific aspects of SUDs is a unique strength of animal models, but the validity and caveats of these procedures should continue to be a matter of discussion and results that are consistent across models should be emphasized.

8. Circuit mechanisms of substance-induced cognitive deficits

As summarized in the sections above, the structure and function of PFC is heavily impacted by psychostimulant and alcohol exposure. Animal models provide an avenue to investigate the specific neural circuits and transmitter systems involved in these adaptations at the cellular and molecular level (Siciliano & Tye, 2019; Spanagel, 2017). The degree of similarity in the behavioral impact of chronic drug use in humans and animals, discussed above, gives confidence that these models are appropriate for investigating the neurobiology underlying these phenomena. Here we provide a brief overview of circuit-specific drug-induced adaptations that have been linked to cognitive deficits in animal models and highlight further convergence with findings in human subjects.

The mesolimbic dopamine system has been extensively implicated in the reinforcing and rewarding aspects of drugs and alcohol as well as craving and seeking (Di Chiara & Imperato, 1988; Nestler, 2005; Siciliano, Calipari, Ferris, & Jones, 2015). In contrast, mesocortical dopamine projections that originate in the midbrain and terminate in the PFC are believed to serve a crucial role in many cognitive processes and in deficits induced by chronic psychostimulant or alcohol use. The PFC, which is composed mostly of glutamatergic pyramidal neurons and GABAergic interneurons, relies on a balance of excitatory and inhibitory neurotransmission for normal physiological and cognitive function. Activity in both of these cell types is modulated via release of dopamine from presynaptic terminals arising from cell bodies in the midbrain. Once released, dopamine acts on D1 and D2-type dopamine receptors located both postsynaptically on interneurons and pyramidal neurons and presynaptically on cortical inputs from upstream regions and interlaminar lateral connectivity (Benes & Berretta, 2001; Gao, Krimer, & Goldman-Rakic, 2001; Paspalas & Goldman-Rakic, 2005; Seamans & Yang, 2004; Vander Weele, Siciliano, & Tye, 2019). Human neuroimaging studies have consistently demonstrated that individuals with SUDs exhibit abnormal patterns of PFC activity under baseline conditions, while performing cognitively demanding tasks, and during exposure to drugs and drug-associated cues that often differ between subregions (Bolla et al., 2003; Childress et al., 1999; Ersche et al., 2005; Goldstein et al., 2007; Goldstein & Volkow, 2002; Grant et al., 1996; Kilts et al., 2001; London, Ernst, Grant, Bonson, & Weinstein, 2000; Paulus et al., 2002; Salo et al., 2009; Volkow et al., 1992, 2005). For example, stimulant-dependent individuals commonly display greater activation in the OFC and reduced activation in the medial and dorsolateral PFC during decision making tasks (Bolla et al., 2003; Ersche et al., 2005). Such observations may be due, in part, to substance-induced alterations in dopamine signaling (Volkow et al., 2001, 1993, 1996) and further evidence suggests that preexisting variation in dopamine signaling may confer susceptibility or resiliency to SUDs in humans (Volkow et al., 2006).

In animal models, dysregulated dopamine release and receptor function in the medial PFC is associated with impairments in multiple cognitive domains including working memory, cognitive flexibility, and decision making (Floresco, 2013). Notably, mesolimbic and mesocortical dopamine neurons undergo contrasting plasticity after exposure to substances of abuse (Lammel, Ion, Roeper, & Malenka, 2011). Growing evidence suggests that alterations in dopamine release in the PFC in conjunction with a reduced ability of dopamine receptors to modulate neuronal activity in this region are a major factor mediating substance-induced deficits in executive function. For example, chronic alcohol exposure is associated with reduced firing of midbrain dopamine neurons and decreased dopamine transmission in the dorsolateral PFC, medial PFC, and OFC (Narendran et al., 2014; Trantham-Davidson & Chandler, 2015) whereas repeated exposure to methamphetamine is associated with progressive cell death and reductions in dopamine terminal density in the medial PFC (Kadota & Kadota, 2004). Amphetamine and cocaine have also been demonstrated to reduce the influence of midbrain dopamine neuron activation on postsynaptic excitability in medial PFC neurons (Nogueira, Kalivas, & Lavin, 2006; Tse, Cantor, & Floresco, 2011). Together these data suggest that alcohol and psychostimulants significantly disrupt dopaminergic regulation of the PFC. Moreover, reduced expression and impaired function of dopamine receptor subtypes in the medial PFC have been linked to deficits in executive function in cocaine and alcohol-exposed rodents during abstinence (Briand et al., 2008; Trantham-Davidson et al., 2014). These substance-induced deficits are likely mediated, in part, by impaired dopamine receptor modulation of excitatory inputs to the PFC. For example, repeated amphetamine exposure has been shown to diminish the ability of D1 and D2 receptors to modulate basolateral amygdala inputs to medial PFC during abstinence in rodents that exhibited impaired decision making (Tse et al., 2011). Similar alterations may also disrupt hippocampal- PFC circuits involved in working memory (Seamans, Floresco, & Phillips, 1998).

Changes in dopaminergic activity as well as many other systems converge to ultimately produce dysregulated and aberrant activity of cortical neurons. Indeed, methamphetamine- induced alterations in neuronal firing patterns in the dorsomedial PFC have been associated with impaired set-shifting performance and enhanced drug seeking (Parsegian et al., 2011) while rescue of cocaine-induced reductions in medial PFC excitability decreased compulsive drug-taking behavior (Chen et al., 2013). Similarly, rescue of cocaine-induced reductions in OFC excitability ameliorates deficits in Pavlovian summation in rats, thought to be a measure of insight or reasoning (Lucantonio et al., 2014). Interestingly, activity and synaptic connectivity in distinct PFC subregions are uniquely affected by psychostimulants such as cocaine and amphetamine. For example, repeated exposure to amphetamine produced enhanced inhibitory and excitatory responses in the medial PFC and OFC, respectively, that were associated with progressive impairments in instrumental responding (Homayoun & Moghaddam, 2006). Furthermore, amphetamine and or cocaine exposure increase the number of dendritic branches and the density of dendritic spines on pyramidal neurons in the medial PFC and medium spiny neurons in the nucleus accumbens, effects that are still evident after a month of abstinence (Crombag, Gorny, Li, Kolb, & Robinson, 2005; Robinson, Gorny, Mitton, & Kolb, 2001; Robinson & Kolb, 1997, 1999). This is in contrast to the OFC where amphetamine reduces spine density (Crombag et al., 2005). Mirroring findings in clinical studies examining risk factors for SUDs (Baranger et al., 2020), the activity in medial PFC during initiation of alcohol use is a marker of compulsive drinking vulnerability in mice (Siciliano et al., 2019).

One plausible take away from these findings is that substance-induced alterations to PFC structure and function commonly found in SUDs are, in part, a consequence of impaired dopamine neuromodulation of PFC neuronal networks and region-specific changes in synaptic connectivity. The combined influence of these changes could drastically impact excitatory drive in the PFC leading to behavioral inflexibility and impaired decision making that may ultimately undermine rehabilitation efforts and increase risk of relapse. While preclinical studies have consistently demonstrated that executive functions are compromised in a variety of animal models following both short- and long-term exposure to alcohol and psychostimulants, a more robust understanding of the neural circuitry involved and how substance-induced neuroadaptive changes evolve over time is needed.

9. Concluding remarks

While animal studies have greatly expanded our knowledge of the circuit mechanisms mediating drug and alcohol-induced cognitive deficits, this realm is dramatically understudied in the preclinical literature in comparison with drug-taking and seeking processes. Both clinical and preclinical studies are needed to develop a comprehensive understanding of how addiction impacts the brain to elucidate what areas of cognition are both predictive of subsequent use and most affected by continued use. Extensive clinical research so far has cogently demonstrated that PFC-mediated executive dysfunction and impulsivity are most heavily impacted by continued substance use, and cross-sectional work further indicates dysfunction in these faculties may confer individual vulnerability to SUD development, particularly during adolescence when substance use is most often initiated. Meanwhile, preclinical studies can effectively recapitulate these phenotypes in animal models, allowing us to address questions of causality, illuminate underlying mechanisms, as well as characterize the longitudinal nature of these deficits.

However, there is still much we do not understand. While years of research have illuminated how drugs of abuse hijack natural reinforcement circuitry and drive maladaptive behaviors, there is still a lack of understanding why only a subset of individuals that experiment with these drugs will go on to develop an addicted phenotype. There is a strong need for more longitudinal work from the clinical research side, which contends with issues of subject recruitment and continued retention, to understand the complex and bidirectional factors that lead an individual to develop these problematic patterns of use and identify how substances can confer vulnerability to relapse through exacerbated executive dysfunction. Understanding the longitudinal nature of these cognitive deficits, both prior to and after exposure, can then inform experimental design in the preclinical realm, and lead to greater understanding and enhanced treatment development. Additionally, in the preclinical field, much work is needed to delineate the distinct nature of the relationships among the various classes of drugs and across the lifespan. Due to technical limitations, including issues with ecological validity in behavioral task design and lack of complete understanding for the mechanisms of these functions, While there is much more work that needs to be done, communication between preclinical and clinical researchers will be critical in the search for potential treatments for this set of disorders.

Funding and disclosure

The authors have no conflicts to report. C.A.S. is supported by a NIH grant R00 DA04510 (NIDA), and a NARSAD Young Investigator Award (Brain and Behavior Research Foundation). P.R.M. is supported by an institutional T32 (T32MH064913–17). S.O·N is supported by an institutional T32 (T32MH065215). This work was also funded by an Alkermes Pathways Research Award grant, an independent competitive grants program supported by Alkermes (C.A.S.).

References

Albein-Urios, N., Martinez-González, J. M., Lozano, O., Clark, L., & Verdejo-García, A. (2012). Comparison of impulsivity and working memory in cocaine addiction and pathological gambling: Implications for cocaine-induced neurotoxicity. *Drug and Alcohol Dependence, 126*, 1–6.

Ambrose, M. L., Bowden, S. C., & Whelan, G. (2001). Working memory impairments in alcohol-dependent participants without clinical amnesia. *Alcoholism, Clinical and Experimental Research, 25*, 185–191.

Anker, J. J., Perry, J. L., Gliddon, L. A., & Carroll, M. E. (2009). Impulsivity predicts the escalation of cocaine self-administration in rats. *Pharmacology, Biochemistry, and Behavior, 93*, 343–348.

Avegno, E. M., & Gilpin, N. W. (2019). Inducing alcohol dependence in rats using chronic intermittent exposure to alcohol vapor. *Bio-Protocol, 9*, e3222.

Baranger, D. A. A., Demers, C. H., Elsayed, N. M., Knodt, A. R., Radtke, S. R., Desmarais, A., et al. (2020). Convergent evidence for predispositional effects of brain gray matter volume on alcohol consumption. *Biological Psychiatry, 87*, 645–655.

Bates, M. E., Bowden, S. C., & Barry, D. (2002). Neurocognitive impairment associated with alcohol use disorders: Implications for treatment. *Experimental and Clinical Psychopharmacology, 10*, 193–212.

Bates, M. E., Pawlak, A. P., Tonigan, J. S., & Buckman, J. F. (2006). Cognitive impairment influences drinking outcome by altering therapeutic mechanisms of change. *Psychology of Addictive Behaviors: Journal of the Society of Psychologists in Addictive Behaviors, 20*, 241–253.

Bechara, A., Dolan, S., Denburg, N., Hindes, A., Anderson, S. W., & Nathan, P. E. (2001). Decision-making deficits, linked to a dysfunctional ventromedial prefrontal cortex, revealed in alcohol and stimulant abusers. *Neuropsychologia, 39*, 376–389.

Beck, A. T., Wright, F. D., Newman, C. F., & Liese, B. S. (1993). *Cognitive therapy of substance abuse*. New York: Guilford Press.

Becker, B., Wagner, D., Koester, P., Tittgemeyer, M., Mercer-Chalmers-Bender, K., Hurlemann, R., et al. (2015). Smaller amygdala and medial prefrontal cortex predict escalating stimulant use. *Brain: A Journal of Neurology, 138*, 2074–2086.

Becker, H. C. (2000). Animal models of alcohol withdrawal. *Alcohol Research & Health: The Journal of the National Institute on Alcohol Abuse and Alcoholism, 24*, 105–113.

Becker, H. C., & Lopez, M. F. (2004). Increased ethanol drinking after repeated chronic ethanol exposure and withdrawal experience in C57BL/6 mice. *Alcoholism, Clinical and Experimental Research, 28*, 1829–1838.

Belin, D., Belin-Rauscent, A., Everitt, B. J., & Dalley, J. W. (2016). In search of predictive endophenotypes in addiction: Insights from preclinical research. *Genes, Brain, and Behavior, 15*, 74–88.

Belin, D., Mar, A. C., Dalley, J. W., Robbins, T. W., & Everitt, B. J. (2008). High impulsivity predicts the switch to compulsive cocaine-taking. *Science, 320*, 1352–1355.

Benes, F. M., & Berretta, S. (2001). GABAergic interneurons: Implications for understanding schizophrenia and bipolar disorder. *Neuropsychopharmacology: Official Publication of the American College of Neuropsychopharmacology, 25*, 1–27.

Bissonette, G. B., Martins, G. J., Franz, T. M., Harper, E. S., Schoenbaum, G., & Powell, E. M. (2008). Double dissociation of the effects of medial and orbital prefrontal cortical lesions on attentional and affective shifts in mice. *The Journal of Neuroscience: The Official Journal of the Society for Neuroscience, 28*, 11124–11130.

Bizot, J. C., Le Bihan, C., Puech, A. J., Hamon, M., & Thiébot, M. H. (1999). Serotonin and tolerance to delay of reward in rats. *Psychopharmacology, 146*, 400–412.

Bjork, J. M., Grant, S. J., & Hommer, D. W. (2003). Cross-sectional volumetric analysis of brain atrophy in alcohol dependence: Effects of drinking history and comorbid substance use disorder. *The American Journal of Psychiatry, 160*, 2038–2045.

Bolla, K., Ernst, M., Kiehl, K., Mouratidis, M., Eldreth, D., Contoreggi, C., et al. (2004). Prefrontal cortical dysfunction in abstinent cocaine abusers. *The Journal of Neuropsychiatry and Clinical Neurosciences, 16*, 456–464.

Bolla, K. I., Eldreth, D. A., London, E. D., Kiehl, K. A., Mouratidis, M., Contoreggi, C., et al. (2003). Orbitofrontal cortex dysfunction in abstinent cocaine abusers performing a decision-making task. *NeuroImage, 19*, 1085–1094.

Bolla, K. I., Funderburk, F. R., & Cadet, J. L. (2000). Differential effects of cocaine and cocaine alcohol on neurocognitive performance. *Neurology, 54*, 2285–2292.

Briand, L. A., Flagel, S. B., Garcia-Fuster, M. J., Watson, S. J., Akil, H., Sarter, M., et al. (2008). Persistent alterations in cognitive function and prefrontal dopamine D2 receptors following extended, but not limited, access to self-administered cocaine. *Neuropsychopharmacology: Official Publication of the American College of Neuropsychopharmacology*, *33*, 2969–2980.
Brière, M., Tocanier, L., Allain, P., Le Gal, D., Allet, G., Gorwood, P., et al. (2019). Decision-making measured by the Iowa Gambling Task in patients with alcohol use disorders choosing harm reduction versus relapse prevention program. *European Addiction Research*, *25*, 182–190.
Broos, N., Diergaarde, L., Schoffelmeer, A. N., Pattij, T., & De Vries, T. J. (2012). Trait impulsive choice predicts resistance to extinction and propensity to relapse to cocaine seeking: A bidirectional investigation. *Neuropsychopharmacology*, *37*, 1377–1386.
Brown, V. J., & Tait, D. S. (2016). Attentional set-shifting across species. *Current Topics in Behavioral Neurosciences*, *28*, 363–395.
Bruijnen, C. J. W. H., Dijkstra, B. A. G., Walvoort, S. J. W., Markus, W., VanDerNagel, J. E. L., Kessels, R. P. C., et al. (2019). Prevalence of cognitive impairment in patients with substance use disorder. *Drug and Alcohol Review*, *38*, 435–442.
Burzynska, A. Z., Nagel, I. E., Preuschhof, C., Gluth, S., Bäckman, L., Li, S.-C., et al. (2012). Cortical thickness is linked to executive functioning in adulthood and aging. *Human Brain Mapping*, *33*, 1607–1620.
Calu, D. J., Stalnaker, T. A., Franz, T. M., Singh, T., Shaham, Y., & Schoenbaum, G. (2007). Withdrawal from cocaine self-administration produces long-lasting deficits in orbitofrontal-dependent reversal learning in rats. *Learning & Memory (Cold Spring Harbor, N.Y.)*, *14*, 325–328.
Cardenas, V. A., Studholme, C., Meyerhoff, D. J., Song, E., & Weiner, M. W. (2005). Chronic active heavy drinking and family history of problem drinking modulate regional brain tissue volumes. *Psychiatry Research*, *138*, 115–130.
Cervantes, M. C., Laughlin, R. E., & Jentsch, J. D. (2013). Cocaine self-administration behavior in inbred mouse lines segregating different capacities for inhibitory control. *Psychopharmacology*, *229*, 515–525.
Chang, L., Ernst, T., Strickland, T., & Mehringer, C. M. (1999). Gender effects on persistent cerebral metabolite changes in the frontal lobes of abstinent cocaine users. *The American Journal of Psychiatry*, *156*, 716–722.
Chanraud, S., Martelli, C., Delain, F., Kostogianni, N., Douaud, G., Aubin, H.-J., et al. (2007). Brain morphometry and cognitive performance in detoxified alcohol-dependents with preserved psychosocial functioning. *Neuropsychopharmacology: Official Publication of the American College of Neuropsychopharmacology*, *32*, 429–438.
Charlton, A. J., May, C., Luikinga, S. J., Burrows, E. L., Kim, J. H., Lawrence, A. J., et al. (2019). Chronic voluntary alcohol consumption causes persistent cognitive deficits and cortical cell loss in a rodent model. *Scientific Reports*, *9*, 1–19.
Chassin, L., Pitts, S. C., & Prost, J. (2002). Binge drinking trajectories from adolescence to emerging adulthood in a high-risk sample: Predictors and substance abuse outcomes. *Journal of Consulting and Clinical Psychology*, *70*, 67–78.
Chen, B. T., Yau, H.-J., Hatch, C., Kusumoto-Yoshida, I., Cho, S. L., Hopf, F. W., et al. (2013). Rescuing cocaine-induced prefrontal cortex hypoactivity prevents compulsive cocaine seeking. *Nature*, *496*, 359–362.
Chen, C.-Y., Storr, C. L., & Anthony, J. C. (2009). Early-onset drug use and risk for drug dependence problems. *Addictive Behaviors*, *34*, 319–322.
Childress, A. R., Mozley, P. D., McElgin, W., Fitzgerald, J., Reivich, M., & O'Brien, C. P. (1999). Limbic activation during cue-induced cocaine craving. *The American Journal of Psychiatry*, *156*, 11–18.

Chung, A., Lyoo, I. K., Kim, S. J., Hwang, J., Bae, S. C., Sung, Y. H., et al. (2007). Decreased frontal white-matter integrity in abstinent methamphetamine abusers. *The International Journal of Neuropsychopharmacology, 10*, 765–775.

Clark, D. B., Cornelius, J. R., Kirisci, L., & Tarter, R. E. (2005). Childhood risk categories for adolescent substance involvement: A general liability typology. *Drug and Alcohol Dependence, 77*, 13–21.

Connolly, C. G., Bell, R. P., Foxe, J. J., & Garavan, H. (2013). Dissociated grey matter changes with prolonged addiction and extended abstinence in cocaine users. *PLoS One, 8*, e59645.

Corral, M., Holguín, S. R., & Cadaveira, F. (2003). Neuropsychological characteristics of young children from high-density alcoholism families: A three-year follow-up. *Journal of Studies on Alcohol, 64*, 195–199.

Cowan, N. (2014). Working memory underpins cognitive development, learning, and education. *Educational Psychology Review, 26*, 197–223.

Crean, R. D., Crane, N. A., & Mason, B. J. (2011). An evidence based review of acute and long-term effects of cannabis use on executive cognitive functions. *Journal of Addiction Medicine, 5*, 1–8.

Crombag, H. S., Gorny, G., Li, Y., Kolb, B., & Robinson, T. E. (2005). Opposite effects of amphetamine self-administration experience on dendritic spines in the medial and orbital prefrontal cortex. *Cerebral Cortex, 15*, 341–348.

Czapla, M., Simon, J. J., Richter, B., Kluge, M., Friederich, H.-C., Herpertz, S., et al. (2016). The impact of cognitive impairment and impulsivity on relapse of alcohol-dependent patients: Implications for psychotherapeutic treatment. *Addiction Biology, 21*, 873–884.

Dalley, J. W., Everitt, B. J., & Robbins, T. W. (2011). Impulsivity, compulsivity, and top-down cognitive control. *Neuron, 69*, 680–694.

Dalley, J. W., Fryer, T. D., Brichard, L., Robinson, E. S. J., Theobald, D. E. H., Lääne, K., et al. (2007). Nucleus accumbens D2/3 receptors predict trait impulsivity and cocaine reinforcement. *Science, 315*, 1267–1270.

Dao-Castellana, M. H., Samson, Y., Legault, F., Martinot, J. L., Aubin, H. J., Crouzel, C., et al. (1998). Frontal dysfunction in neurologically normal chronic alcoholic subjects: Metabolic and neuropsychological findings. *Psychological Medicine, 28*, 1039–1048.

Dawson, D. A., Goldstein, R. B., Chou, S. P., Ruan, W. J., & Grant, B. F. (2008). Age at first drink and the first incidence of adult-onset DSM-IV alcohol use disorders. *Alcoholism, Clinical and Experimental Research, 32*, 2149–2160.

Di Chiara, G., & Imperato, A. (1988). Drugs abused by humans preferentially increase synaptic dopamine concentrations in the mesolimbic system of freely moving rats. *Proceedings of the National Academy of Sciences of the United States of America, 85*, 5274–5278.

Diamond, A. (2013). Executive functions. *Annual Review of Psychology, 64*, 135–168.

Dias, R., Robbins, T. W., & Roberts, A. C. (1996a). Dissociation in prefrontal cortex of affective and attentional shifts. *Nature, 380*, 69–72.

Dias, R., Robbins, T. W., & Roberts, A. C. (1996b). Primate analogue of the Wisconsin card sorting test: Effects of excitotoxic lesions of the prefrontal cortex in the marmoset. *Behavioral Neuroscience, 110*, 872–886.

Dominguez, G., Belzung, C., Pierard, C., David, V., Henkous, N., Decorte, L., et al. (2017). Alcohol withdrawal induces long-lasting spatial working memory impairments: Relationship with changes in corticosterone response in the prefrontal cortex. *Addiction Biology, 22*, 898–910.

Domínguez-Salas, S., Díaz-Batanero, C., Lozano-Rojas, O. M., & Verdejo-García, A. (2016). Impact of general cognition and executive function deficits on addiction treatment outcomes: Systematic review and discussion of neurocognitive pathways. *Neuroscience and Biobehavioral Reviews, 71*, 772–801.

Economidou, D., Pelloux, Y., Robbins, T. W., Dalley, J. W., & Everitt, B. J. (2009). High impulsivity predicts relapse to cocaine-seeking after punishment-induced abstinence. *Biological Psychiatry*, *65*, 851–856.

Ernst, T., Chang, L., Leonido-Yee, M., & Speck, O. (2000). Evidence for long-term neurotoxicity associated with methamphetamine abuse: A 1H MRS study. *Neurology*, *54*, 1344–1349.

Errico, A. L., King, A. C., Lovallo, W. R., & Parsons, O. A. (2002). Cortisol dysregulation and cognitive impairment in abstinent male alcoholics. *Alcoholism, Clinical and Experimental Research*, *26*, 1198–1204.

Ersche, K. D., Fletcher, P. C., Lewis, S. J. G., Clark, L., Stocks-Gee, G., London, M., et al. (2005). Abnormal frontal activations related to decision-making in current and former amphetamine and opiate dependent individuals. *Psychopharmacology*, *180*, 612–623.

Ersche, K. D., Jones, P. S., Williams, G. B., Turton, A. J., Robbins, T. W., & Bullmore, E. T. (2012). Abnormal brain structure implicated in stimulant drug addiction. *Science*, *335*, 601–604.

Ersche, K. D., Roiser, J. P., Robbins, T. W., & Sahakian, B. J. (2008). Chronic cocaine but not chronic amphetamine use is associated with perseverative responding in humans. *Psychopharmacology*, *197*, 421–431.

Ersche, K. D., Turton, A. J., Chamberlain, S. R., Müller, U., Bullmore, E. T., & Robbins, T. W. (2012). Cognitive dysfunction and anxious-impulsive personality traits are endophenotypes for drug dependence. *The American Journal of Psychiatry*, *169*, 926–936.

Ersche, K. D., Williams, G. B., Robbins, T. W., & Bullmore, E. T. (2013). Meta-analysis of structural brain abnormalities associated with stimulant drug dependence and neuroimaging of addiction vulnerability and resilience. *Current Opinion in Neurobiology*, *23*, 615–624.

Everitt, B. J., & Robbins, T. W. (2013). From the ventral to the dorsal striatum: Devolving views of their roles in drug addiction. *Neuroscience & Biobehavioral Reviews*, *37*, 1946–1954.

Fein, G., Bachman, L., Fisher, S., & Davenport, L. (1990). Cognitive impairments in abstinent alcoholics. *Western Journal of Medicine*, *152*, 531–537.

Fein, G., Di Sclafani, V., & Meyerhoff, D. J. (2002). Prefrontal cortical volume reduction associated with frontal cortex function deficit in 6-week abstinent crack-cocaine dependent men. *Drug and Alcohol Dependence*, *68*, 87–93.

Fein, G., Klein, L., & Finn, P. (2004). Impairment on a simulated gambling task in long-term abstinent alcoholics. *Alcoholism, Clinical and Experimental Research*, *28*, 1487–1491.

Fein, G., Torres, J., Price, L. J., & Di Sclafani, V. (2006). Cognitive performance in long-term abstinent alcoholic individuals. *Alcoholism, Clinical and Experimental Research*, *30*, 1538–1544.

Fijał, K., Nowak, E., Leśkiewicz, M., Budziszewska, B., & Filip, M. (2015). Working memory deficits and alterations of ERK and CREB phosphorylation following withdrawal from cocaine self-administration. *Pharmacological Reports*, *67*, 881–889.

Fitz, N. F., Gibbs, R. B., & Johnson, D. A. (2008). Selective lesion of septal cholinergic neurons in rats impairs acquisition of a delayed matching to position T-maze task by delaying the shift from a response to a place strategy. *Brain Research Bulletin*, *77*, 356–360.

Floresco, S. B. (2013). Prefrontal dopamine and behavioral flexibility: Shifting from an "inverted-U" toward a family of functions. *Frontiers in Neuroscience*, *7*, 62.

Flory, K., Lynam, D., Milich, R., Leukefeld, C., & Clayton, R. (2004). Early adolescent through young adult alcohol and marijuana use trajectories: Early predictors, young adult outcomes, and predictive utility. *Development and Psychopathology*, *16*, 193–213.

Franken, I. H. A., van Strien, J. W., Nijs, I., & Muris, P. (2008). Impulsivity is associated with behavioral decision-making deficits. *Psychiatry Research*, *158*, 155–163.

Friedman, N. P., Miyake, A., Young, S. E., DeFries, J. C., Corley, R. P., & Hewitt, J. K. (2008). Individual differences in executive functions are almost entirely genetic in origin. *Journal of Experimental Psychology. General*, *137*, 201–225.

Fu, L., Bi, G., Zou, Z., Wang, Y., Ye, E., Ma, L., et al. (2008). Impaired response inhibition function in abstinent heroin dependents: An fMRI study. *Neuroscience Letters*, *438*, 322–326.

Gao, W. J., Krimer, L. S., & Goldman-Rakic, P. S. (2001). Presynaptic regulation of recurrent excitation by D1 receptors in prefrontal circuits. *Proceedings of the National Academy of Sciences of the United States of America*, *98*, 295–300.

Gass, J. T., Glen, W. B., McGonigal, J. T., Trantham-Davidson, H., Lopez, M. F., Randall, P. K., et al. (2014). Adolescent alcohol exposure reduces behavioral flexibility, promotes disinhibition, and increases resistance to extinction of ethanol self-administration in adulthood. *Neuropsychopharmacology*, *39*, 2570–2583.

George, O., Mandyam, C. D., Wee, S., & Koob, G. F. (2008). Extended access to cocaine self- administration produces long-lasting prefrontal cortex-dependent working memory impairments. *Neuropsychopharmacology: Official Publication of the American College of Neuropsychopharmacology*, *33*, 2474–2482.

George, O., Sanders, C., Freiling, J., Grigoryan, E., Vu, S., Allen, C. D., et al. (2012). Recruitment of medial prefrontal cortex neurons during alcohol withdrawal predicts cognitive impairment and excessive alcohol drinking. *Proceedings of the National Academy of Sciences of the United States of America*, *109*, 18156–18161.

Gilpin, N. W., Richardson, H. N., Cole, M., & Koob, G. F. (2008). Vapor inhalation of alcohol in rats. *Current Protocols in Neuroscience*, *44*, 9.29.1–9.29.19.

Goldstein, R. Z., Alia-Klein, N., Tomasi, D., Zhang, L., Cottone, L. A., Maloney, T., et al. (2007). Is decreased prefrontal cortical sensitivity to monetary reward associated with impaired motivation and self-control in cocaine addiction? *The American Journal of Psychiatry*, *164*, 43–51.

Goldstein, R. Z., & Volkow, N. D. (2002). Drug addiction and its underlying neurobiological basis: Neuroimaging evidence for the involvement of the frontal cortex. *The American Journal of Psychiatry*, *159*, 1642–1652.

Goldstein, R. Z., & Volkow, N. D. (2011). Dysfunction of the prefrontal cortex in addiction: Neuroimaging findings and clinical implications. *Nature Reviews. Neuroscience*, *12*, 652–669.

Goncalves, P. D., Ometto, M., Malbergier, A., Martins, P., Beraldo, L., Santos, B., et al. (2017). Executive functioning and outpatient treatment adherence after intensive inpatient care in cocaine dependence: A six-month follow-up study. *Drug and Alcohol Dependence*, *171*, e75.

Gonzalez, R., Bechara, A., & Martin, E. M. (2007). Executive functions among individuals with methamphetamine or alcohol as drugs of choice: Preliminary observations. *Journal of Clinical and Experimental Neuropsychology*, *29*, 155–159.

Goudriaan, A. E., Oosterlaan, J., Beurs, E. D., & Brink, W. V. D. (2006). Neurocognitive functions in pathological gambling: A comparison with alcohol dependence, Tourette syndrome and normal controls. *Addiction*, *101*, 534–547.

Gould, R. W., Gage, H. D., & Nader, M. A. (2012). Effects of chronic cocaine self-administration on cognition and cerebral glucose utilization in rhesus monkeys. *Biological Psychiatry*, *72*, 856–863.

Grant, S., London, E. D., Newlin, D. B., Villemagne, V. L., Liu, X., Contoreggi, C., et al. (1996). Activation of memory circuits during cue-elicited cocaine craving. *Proceedings of the National Academy of Sciences of the United States of America*, *93*, 12040–12045.

Gruber, S. A., Silveri, M. M., & Yurgelun-Todd, D. A. (2007). Neuropsychological consequences of opiate use. *Neuropsychology Review*, *17*, 299–315.

Haass-Koffler, C. L., Leggio, L., & Kenna, G. A. (2014). Pharmacological approaches to reducing craving in patients with alcohol use disorders. *CNS Drugs*, *28*, 343–360.

Habeych, M. E., Folan, M. M., Luna, B., & Tarter, R. E. (2006). Impaired oculomotor response inhibition in children of alcoholics: The role of attention deficit hyperactivity disorder. *Drug and Alcohol Dependence*, *82*, 11–17.

Hammond, C. J., Mayes, L. C., & Potenza, M. N. (2014). Neurobiology of adolescent substance use and addictive behaviors: Treatment implications. *Adolescent Medicine: State of the Art Reviews*, *25*, 15–32.

Hanlon, C. A., Dufault, D. L., Wesley, M. J., & Porrino, L. J. (2011). Elevated gray and white matter densities in cocaine abstainers compared to current users. *Psychopharmacology*, *218*, 681–692.

Henderson, K. E., Vaidya, J. G., Kramer, J. R., Kuperman, S., Langbehn, D. R., & O'Leary, D. S. (2018). Cortical thickness in adolescents with a family history of alcohol use disorder. *Alcoholism, Clinical and Experimental Research*, *42*, 89–99.

Herndon, J. G., Moss, M. B., Rosene, D. L., & Killiany, R. J. (1997). Patterns of cognitive decline in aged rhesus monkeys. *Behavioural Brain Research*, *87*, 25–34.

Hill, S. Y., Shen, S., Lowers, L., & Locke, J. (2000). Factors predicting the onset of adolescent drinking in families at high risk for developing alcoholism. *Biological Psychiatry*, *48*, 265–275.

Hill, S. Y., Wang, S., Kostelnik, B., Carter, H., Holmes, B., McDermott, M., et al. (2009). Disruption of orbitofrontal cortex laterality in offspring from multiplex alcohol dependence families. *Biological Psychiatry*, *65*, 129–136.

Homayoun, H., & Moghaddam, B. (2006). Progression of cellular adaptations in medial prefrontal and orbitofrontal cortex in response to repeated amphetamine. *The Journal of Neuroscience: The Official Journal of the Society for Neuroscience*, *26*, 8025–8039.

Horn, N. R., Dolan, M., Elliott, R., Deakin, J. F. W., & Woodruff, P. W. R. (2003). Response inhibition and impulsivity: An fMRI study. *Neuropsychologia*, *41*, 1959–1966.

Hughes, R. N. (2004). The value of spontaneous alternation behavior (SAB) as a test of retention in pharmacological investigations of memory. *Neuroscience and Biobehavioral Reviews*, *28*, 497–505.

Izquierdo, A., & Jentsch, J. D. (2012). Reversal learning as a measure of impulsive and compulsive behavior in addictions. *Psychopharmacology*, *219*, 607–620.

Jasinska, A. J., Zorick, T., Brody, A. L., & Stein, E. A. (2014). Dual role of nicotine in addiction and cognition: A review of neuroimaging studies in humans. *Neuropharmacology*, *84*, 111–122.

Jentsch, J. D., Olausson, P., De La Garza, R., & Taylor, J. R. (2002). Impairments of reversal learning and response perseveration after repeated, intermittent cocaine administrations to monkeys. *Neuropsychopharmacology: Official Publication of the American College of Neuropsychopharmacology*, *26*, 183–190.

Johnston, L. D., Miech, R. A., O'Malley, P. M., Bachman, J. G., Schulenberg, J. E., & Patrick, M. E. (2019). *Monitoring the future national survey results on drug use, 1975–2018: Overview, key findings on adolescent drug use.* Institute for Social Research.

Kadota, T., & Kadota, K. (2004). Neurotoxic morphological changes induced in the medial prefrontal cortex of rats behaviorally sensitized to methamphetamine. *Archives of Histology and Cytology*, *67*, 241–251.

Khurana, A., Romer, D., Betancourt, L. M., Brodsky, N. L., Giannetta, J. M., & Hurt, H. (2015). Experimentation versus progression in adolescent drug use: A test of an emerging neurobehavioral imbalance model. *Development and Psychopathology*, *27*, 901–913.

Khurana, A., Romer, D., Betancourt, L. M., & Hurt, H. (2017). Working memory ability and early drug use progression as predictors of adolescent substance use disorders. *Addiction*, *112*, 1220–1228.

Kilts, C. D., Schweitzer, J. B., Quinn, C. K., Gross, R. E., Faber, T. L., Muhammad, F., et al. (2001). Neural activity related to drug craving in cocaine addiction. *Archives of General Psychiatry*, *58*, 334–341.

Kim, S. J., Lyoo, I. K., Hwang, J., Chung, A., Hoon Sung, Y., Kim, J., et al. (2006). Prefrontal grey-matter changes in short-term and long-term abstinent methamphetamine abusers. *International Journal of Neuropsychopharmacology, 9*, 221–228.

King, K. M., & Chassin, L. (2007). A prospective study of the effects of age of initiation of alcohol and drug use on young adult substance dependence. *Journal of Studies on Alcohol and Drugs, 68*, 256–265.

Klugman, A., & Gruzelier, J. (2003). Chronic cognitive impairment in users of "ecstasy" and cannabis. *World Psychiatry, 2*, 184–190.

Koob, G. F., & Volkow, N. D. (2010). Neurocircuitry of addiction. *Neuropsychopharmacology, 35*, 217–238.

Krawczyk, D. C. (2002). Contributions of the prefrontal cortex to the neural basis of human decision making. *Neuroscience and Biobehavioral Reviews, 26*, 631–664.

Kroener, S., Mulholland, P. J., New, N. N., Gass, J. T., Becker, H. C., & Chandler, L. J. (2012). Chronic alcohol exposure alters behavioral and synaptic plasticity of the rodent prefrontal cortex. *PLoS One*, 7, e37541.

Kubota, M., Nakazaki, S., Hirai, S., Saeki, N., Yamaura, A., & Kusaka, T. (2001). Alcohol consumption and frontal lobe shrinkage: Study of 1432 non-alcoholic subjects. *Journal of Neurology, Neurosurgery, and Psychiatry, 71*, 104–106.

Lammel, S., Ion, D. I., Roeper, J., & Malenka, R. C. (2011). Projection-specific modulation of dopamine neuron synapses by aversive and rewarding stimuli. *Neuron, 70*, 855–862.

Laughlin, R. E., Grant, T. L., Williams, R. W., & Jentsch, J. D. (2011). Genetic dissection of behavioral flexibility: Reversal learning in mice. *Biological Psychiatry, 69*, 1109–1116.

Lawrence, A. J., Luty, J., Bogdan, N. A., Sahakian, B. J., & Clark, L. (2009). Impulsivity and response inhibition in alcohol dependence and problem gambling. *Psychopharmacology, 207*, 163–172.

Le Berre, A. P., Rauchs, G., La Joie, R., Mézenge, F., Boudehent, C., Vabret, F., et al. (2014). Impaired decision-making and brain shrinkage in alcoholism. *European Psychiatry: The Journal of the Association of European Psychiatrists, 29*, 125–133.

Li, F., Duncan, T. E., & Hops, H. (2001). Examining developmental trajectories in adolescent alcohol use using piecewise growth mixture modeling analysis. *Journal of Studies on Alcohol, 62*, 199–210.

Lichtman, A. H., & Martin, B. R. (2002). Marijuana withdrawal syndrome in the animal model. *The Journal of Clinical Pharmacology, 42*, 20S–27S.

Løberg, T., & Miller, W. R. (1986). Personality, cognitive, and neuropsychological correlates of harmful alcohol consumption: A cross-national comparison of clinical samples. *Annals of the New York Academy of Sciences, 472*, 75–97.

London, E. D., Ernst, M., Grant, S., Bonson, K., & Weinstein, A. (2000). Orbitofrontal cortex and human drug abuse: Functional imaging. *Cerebral Cortex, 10*, 334–342.

Loos, M., Staal, J., Smit, A. B., De Vries, T. J., & Spijker, S. (2013). Enhanced alcohol self-administration and reinstatement in a highly impulsive, inattentive recombinant inbred mouse strain. *Frontiers in Behavioral Neuroscience*, 7, 151.

Lopez, M. F., & Becker, H. C. (2005). Effect of pattern and number of chronic ethanol exposures on subsequent voluntary ethanol intake in C57BL/6J mice. *Psychopharmacology, 181*, 688–696.

Lopez-Quintero, C., Pérez de los Cobos, J., Hasin, D. S., Okuda, M., Wang, S., Grant, B. F., et al. (2011). Probability and predictors of transition from first use to dependence on nicotine, alcohol, cannabis, and cocaine: Results of the National Epidemiologic Survey on Alcohol and Related Conditions (NESARC). *Drug and Alcohol Dependence, 115*, 120–130.

Lucantonio, F., Takahashi, Y., Hoffman, A. F., Chang, C. Y., Chaudhary, S., Shaham, Y., et al. (2014). Orbitofrontal activation restores insight lost after cocaine use. *Nature Neuroscience, 17*, 1092–1099.

Mackey, S., & Paulus, M. (2013). Are there volumetric brain differences associated with the use of cocaine and amphetamine-type stimulants? *Neuroscience and Biobehavioral Reviews*, *37*, 300–316.

Mahmood, O. M., Goldenberg, D., Thayer, R., Migliorini, R., Simmons, A. N., & Tapert, S. F. (2013). Adolescents' fMRI activation to a response inhibition task predicts future substance use. *Addictive Behaviors*, *38*, 1435–1441.

Markou, A., & Koob, G. F. (1991). Postcocaine anhedonia. An animal model of cocaine withdrawal. *Neuropsychopharmacology: Official Publication of the American College of Neuropsychopharmacology*, *4*, 17–26.

Markou, A., Weiss, F., Gold, L. H., Caine, S. B., Schulteis, G., & Koob, G. F. (1993). Animal models of drug craving. *Psychopharmacology*, *112*, 163–182.

Martin, P. R., Adinoff, B., Weingartner, H., Mukherjee, A. B., & Eckardt, M. J. (1986). Alcoholic organic brain disease: Nosology and pathophysiologic mechanisms. *Progress in Neuro-Psychopharmacology & Biological Psychiatry*, *10*, 147–164.

Matochik, J. A., London, E. D., Eldreth, D. A., Cadet, J.-L., & Bolla, K. I. (2003). Frontal cortical tissue composition in abstinent cocaine abusers: A magnetic resonance imaging study. *NeuroImage*, *19*, 1095–1102.

McAlonan, K., & Brown, V. J. (2003). Orbital prefrontal cortex mediates reversal learning and not attentional set shifting in the rat. *Behavioural Brain Research*, *146*, 97–103.

Meek, P. S., Clark, H. W., & Solana, V. L. (1989). Neurocognitive impairment: The unrecognized component of dual diagnosis in substance abuse treatment. *Journal of Psychoactive Drugs*, *21*, 153–160.

Miguel-Hidalgo, J. J., Overholser, J. C., Meltzer, H. Y., Stockmeier, C. A., & Rajkowska, G. (2006). Reduced glial and neuronal packing density in the orbitofrontal cortex in alcohol dependence and its relationship with suicide and duration of alcohol dependence. *Alcoholism, Clinical and Experimental Research*, *30*, 1845–1855.

Monterosso, J. R., Aron, A. R., Cordova, X., Xu, J., & London, E. D. (2005). Deficits in response inhibition associated with chronic methamphetamine abuse. *Drug and Alcohol Dependence*, *79*, 273–277.

Morein-Zamir, S., Simon Jones, P., Bullmore, E. T., Robbins, T. W., & Ersche, K. D. (2013). Prefrontal hypoactivity associated with impaired inhibition in stimulant-dependent individuals but evidence for hyperactivation in their unaffected siblings. *Neuropsychopharmacology*, *38*, 1945–1953.

Morgenstern, J., & Bates, M. E. (1999). Effects of executive function impairment on change processes and substance use outcomes in 12-step treatment. *Journal of Studies on Alcohol*, *60*, 846–855.

Myrick, H., Brady, K. T., & Malcolm, R. (2001). New developments in the pharmacotherapy of alcohol dependence. *The American Journal on Addictions*, *10*, s3–s15.

Nakama, H., Chang, L., Fein, G., Shimotsu, R., Jiang, C. S., & Ernst, T. (2011). Methamphetamine users show greater than normal age-related cortical gray matter loss. *Addiction*, *106*, 1474–1483.

Narendran, R., Mason, N. S., Paris, J., Himes, M. L., Douaihy, A. B., & Frankle, W. G. (2014). Decreased prefrontal cortical dopamine transmission in alcoholism. *American Journal of Psychiatry*, *171*, 881–888.

Nestler, E. J. (2005). Is there a common molecular pathway for addiction? *Nature Neuroscience*, *8*, 1445–1449.

NIDA. (2020). *Trends & statistics*. Retrieved from https://www.drugabuse.gov/drug-topics/trends-statistics.

Nigg, J. T., Wong, M. M., Martel, M. M., Jester, J. M., Puttler, L. I., Glass, J. M., et al. (2006). Poor response inhibition as a predictor of problem drinking and illicit drug use in adolescents at risk for alcoholism and other substance use disorders. *Journal of the American Academy of Child and Adolescent Psychiatry*, *45*, 468–475.

Nogueira, L., Kalivas, P. W., & Lavin, A. (2006). Long-term neuroadaptations produced by withdrawal from repeated cocaine treatment: Role of dopaminergic receptors in modulating cortical excitability. *Journal of Neuroscience, 26*, 12308–12313.

Norman, A. L., Pulido, C., Squeglia, L. M., Spadoni, A. D., Paulus, M. P., & Tapert, S. F. (2011). Neural activation during inhibition predicts initiation of substance use in adolescence. *Drug and Alcohol Dependence, 119*, 216–223.

Oberlin, B. G., & Grahame, N. J. (2009). High-alcohol preferring mice are more impulsive than low-alcohol preferring mice as measured in the delay discounting task. *Alcoholism, Clinical and Experimental Research, 33*, 1294–1303.

O'Malley, S., Adamse, M., Heaton, R. K., & Gawin, F. H. (1992). Neuropsychological impairment in chronic cocaine abusers. *The American Journal of Drug and Alcohol Abuse, 18*, 131–144.

Oscar-Berman, M., Kirkley, S. M., Gansler, D. A., & Couture, A. (2004). Comparisons of Korsakoff and non-Korsakoff alcoholics on neuropsychological tests of prefrontal brain functioning. *Alcoholism, Clinical and Experimental Research, 28*, 667–675.

Parsegian, A., Glen, W. B., Lavin, A., & See, R. E. (2011). Methamphetamine self-administration produces attentional set-shifting deficits and alters prefrontal cortical neurophysiology in rats. *Biological Psychiatry, 69*, 253–259.

Paspalas, C. D., & Goldman-Rakic, P. S. (2005). Presynaptic D1 dopamine receptors in primate prefrontal cortex: Target-specific expression in the glutamatergic synapse. *The Journal of Neuroscience, 25*, 1260–1267.

Paul, C. A., Au, R., Fredman, L., Massaro, J. M., Seshadri, S., Decarli, C., et al. (2008). Association of alcohol consumption with brain volume in the Framingham study. *Archives of Neurology, 65*, 1363–1367.

Paulus, M. P., Hozack, N. E., Zauscher, B. E., Frank, L., Brown, G. G., Braff, D. L., et al. (2002). Behavioral and functional neuroimaging evidence for prefrontal dysfunction in methamphetamine-dependent subjects. *Neuropsychopharmacology: Official Publication of the American College of Neuropsychopharmacology, 26*, 53–63.

Peña-Oliver, Y., Giuliano, C., Economidou, D., Goodlett, C. R., Robbins, T. W., Dalley, J. W., et al. (2015). Alcohol-preferring rats show goal oriented behaviour to food incentives but are neither sign-trackers nor impulsive. *PLoS One, 10*, e0131016.

Perkins, K. A., Conklin, C. A., & Levine, M. D. (2008). *Cognitive-behavioral therapy for smoking cessation: A practical guidebook to the most effective treatments* (p. xxii, 258). New York, NY, US: Routledge/Taylor & Francis Group.

Perry, J. L., Larson, E. B., German, J. P., Madden, G. J., & Carroll, M. E. (2005). Impulsivity (delay discounting) as a predictor of acquisition of IV cocaine self-administration in female rats. *Psychopharmacology, 178*, 193–201.

Perry, J. L., Nelson, S. E., & Carroll, M. E. (2008). Impulsive choice as a predictor of acquisition of IV cocaine self-administration and reinstatement of cocaine-seeking behavior in male and female rats. *Experimental and Clinical Psychopharmacology, 16*, 165–177.

Pfefferbaum, A., Sullivan, E. V., Mathalon, D. H., & Lim, K. O. (1997). Frontal lobe volume loss observed with magnetic resonance imaging in older chronic alcoholics. *Alcoholism, Clinical and Experimental Research, 21*, 521–529.

Pfefferbaum, A., Sullivan, E. V., Rosenbloom, M. J., Mathalon, D. H., & Lim, K. O. (1998). A controlled study of cortical gray matter and ventricular changes in alcoholic men over a 5-year interval. *Archives of General Psychiatry, 55*, 905–912.

Phillips, K. A., Epstein, D. H., & Preston, K. L. (2014). Psychostimulant addiction treatment. *Neuropharmacology, 87*, 150–160.

Porter, J. N., Olsen, A. S., Gurnsey, K., Dugan, B. P., Jedema, H. P., & Bradberry, C. W. (2011). Chronic cocaine self-administration in rhesus monkeys: Impact on associative learning, cognitive control, and working memory. *The Journal of Neuroscience, 31*, 4926–4934.

Poulos, C. X., Le, A. D., & Parker, J. L. (1995). Impulsivity predicts individual susceptibility to high levels of alcohol self-administration. *Behavioural Pharmacology*, *6*, 810–814.
Poulos, C. X., Parker, J. L., & Lê, D. A. (1998). Increased impulsivity after injected alcohol predicts later alcohol consumption in rats: Evidence for "loss-of-control drinking" and marked individual differences. *Behavioral Neuroscience*, *112*, 1247–1257.
Ragozzino, M. E., Detrick, S., & Kesner, R. P. (1999). Involvement of the prelimbic–infralimbic areas of the rodent prefrontal cortex in behavioral flexibility for place and response learning. *The Journal of Neuroscience*, *19*, 4585–4594.
Rendell, P. G., Mazur, M., & Henry, J. D. (2009). Prospective memory impairment in former users of methamphetamine. *Psychopharmacology*, *203*, 609–616.
Rippeth, J. D., Heaton, R. K., Carey, C. L., Marcotte, T. D., Moore, D. J., Gonzalez, R., et al. (2004). Methamphetamine dependence increases risk of neuropsychological impairment in HIV infected persons. *Journal of the International Neuropsychological Society*, *10*, 1–14.
Robbins, T. W. (1998). Dissociating executive functions of the prefrontal cortex. In *The prefrontal cortex: Executive and cognitive functions* (pp. 117–130). New York, NY, US: Oxford University Press.
Robinson, E. S. J., Eagle, D. M., Economidou, D., Theobald, D. E. H., Mar, A. C., Murphy, E. R., et al. (2009). Behavioural characterisation of high impulsivity on the 5-choice serial reaction time task: Specific deficits in "waiting" versus "stopping". *Behavioural Brain Research*, *196*, 310–316.
Robinson, T. E., Gorny, G., Mitton, E., & Kolb, B. (2001). Cocaine self-administration alters the morphology of dendrites and dendritic spines in the nucleus accumbens and neocortex. *Synapse*, *39*, 257–266.
Robinson, T. E., & Kolb, B. (1997). Persistent structural modifications in nucleus accumbens and prefrontal cortex neurons produced by previous experience with amphetamine. *The Journal of Neuroscience: The Official Journal of the Society for Neuroscience*, *17*, 8491–8497.
Robinson, T. E., & Kolb, B. (1999). Alterations in the morphology of dendrites and dendritic spines in the nucleus accumbens and prefrontal cortex following repeated treatment with amphetamine or cocaine. *The European Journal of Neuroscience*, *11*, 1598–1604.
Rolland, B., D'Hondt, F., Montègue, S., Brion, M., Peyron, E., D'Aviau de Ternay, J., et al. (2019). A patient-tailored evidence-based approach for developing early neuropsychological training programs in addiction settings. *Neuropsychology Review*, *29*, 103–115.
Rubenis, A. J., Fitzpatrick, R. E., Lubman, D. I., & Verdejo-Garcia, A. (2019). Working memory predicts methamphetamine hair concentration over the course of treatment: Moderating effect of impulsivity and implications for dual-systems model. *Addiction Biology*, *24*, 145–153.
Sabia, S., Elbaz, A., Britton, A., Bell, S., Dugravot, A., Shipley, M., et al. (2014). Alcohol consumption and cognitive decline in early old age. *Neurology*, *82*, 332–339.
Salo, R., Ursu, S., Buonocore, M. H., Leamon, M. H., & Carter, C. (2009). Impaired prefrontal cortical function and disrupted adaptive cognitive control in methamphetamine abusers: A functional magnetic resonance imaging study. *Biological Psychiatry*, *65*, 706–709.
SAMHSA. (2019). *Key substance use and mental health indicators in the United States: Results from the 2018 National Survey on Drug Use and Health (HHS Publication No. PEP19-5068, NSDUH Series H-54)*. Rockville, MD: Center for Behavioral Health Statistics and Quality, Substance Abuse and Mental Health Services Administration. Retrieved from https://www.samhsa.gov/data/.
Schoenbaum, G., Saddoris, M. P., Ramus, S. J., Shaham, Y., & Setlow, B. (2004). Cocaine-experienced rats exhibit learning deficits in a task sensitive to orbitofrontal cortex lesions. *The European Journal of Neuroscience*, *19*, 1997–2002.
Schoenbaum, G., Setlow, B., Nugent, S. L., Saddoris, M. P., & Gallagher, M. (2003). Lesions of orbitofrontal cortex and basolateral amygdala complex disrupt acquisition of odor-guided discriminations and reversals. *Learning & Memory*, *10*, 129–140.

Schweinsburg, A. D., Paulus, M. P., Barlett, V. C., Killeen, L. A., Caldwell, L. C., Pulido, C., et al. (2004). An FMRI study of response inhibition in youths with a family history of alcoholism. *Annals of the New York Academy of Sciences*, *1021*, 391–394.

Schweppe, C. A., Burzynski, C., Jayanthi, S., Ladenheim, B., Cadet, J. L., Gardner, E. L., et al. (2020). Neurochemical and behavioral comparisons of contingent and non-contingent methamphetamine exposure following binge or yoked long-access self-administration paradigms. *Psychopharmacology*, *237*, 1989–2005.

Seamans, J. K., Floresco, S. B., & Phillips, A. G. (1998). D1 receptor modulation of hippocampal–prefrontal cortical circuits integrating spatial memory with executive functions in the rat. *Journal of Neuroscience*, *18*, 1613–1621.

Seamans, J. K., & Yang, C. R. (2004). The principal features and mechanisms of dopamine modulation in the prefrontal cortex. *Progress in Neurobiology*, *74*, 1–58.

Shnitko, T. A., Gonzales, S. W., & Grant, K. A. (2019). Low cognitive flexibility as a risk for heavy alcohol drinking in non-human primates. *Alcohol*, *74*, 95–104.

Siciliano, C. A., Calipari, E. S., Ferris, M. J., & Jones, S. R. (2015). Adaptations of presynaptic dopamine terminals induced by psychostimulant self-administration. *ACS Chemical Neuroscience*, *6*, 27–36.

Siciliano, C. A., Noamany, H., Chang, C.-J., Brown, A. R., Chen, X., Leible, D., et al. (2019). A cortical-brainstem circuit predicts and governs compulsive alcohol drinking. *Science*, *366*, 1008–1012.

Siciliano, C. A., & Tye, K. M. (2019). Leveraging calcium imaging to illuminate circuit dysfunction in addiction. *Alcohol*, *74*, 47–63.

Silveri, M. M., Rogowska, J., McCaffrey, A., & Yurgelun-Todd, D. A. (2011). Adolescents at risk for alcohol abuse demonstrate altered frontal lobe activation during stroop performance. *Alcoholism, Clinical and Experimental Research*, *35*, 218–228.

Somerville, L. H., & Casey, B. (2010). Developmental neurobiology of cognitive control and motivational systems. *Current Opinion in Neurobiology*, *20*, 236–241.

Spanagel, R. (2017). Animal models of addiction. *Dialogues in Clinical Neuroscience*, *19*, 247–258.

Spear, L. P. (2013). Adolescent neurodevelopment. *The Journal of Adolescent Health : Official Publication of the Society for Adolescent Medicine*, *52*, S7–13.

Squeglia, L. M., Jacobus, J., Nguyen-Louie, T. T., & Tapert, S. F. (2014). Inhibition during early adolescence predicts alcohol and marijuana use by late adolescence. *Neuropsychology*, *28*, 782–790.

Stavro, K., Pelletier, J., & Potvin, S. (2013). Widespread and sustained cognitive deficits in alcoholism: A meta-analysis. *Addiction Biology*, *18*, 203–213.

Stevens, L., Verdejo-García, A., Goudriaan, A. E., Roeyers, H., Dom, G., & Vanderplasschen, W. (2014). Impulsivity as a vulnerability factor for poor addiction treatment outcomes: A review of neurocognitive findings among individuals with substance use disorders. *Journal of Substance Abuse Treatment*, *47*, 58–72.

Stout, J. C., Busemeyer, J. R., Lin, A., Grant, S. J., & Bonson, K. R. (2004). Cognitive modeling analysis of decision-making processes in cocaine abusers. *Psychonomic Bulletin & Review*, *11*, 742–747.

Stuss, D. T. (2011). Functions of the frontal lobes: Relation to executive functions. *Journal of the International Neuropsychological Society*, *17*, 759–765.

Thiébot, M. H., Le Bihan, C., Soubrié, P., & Simon, P. (1985). Benzodiazepines reduce the tolerance to reward delay in rats. *Psychopharmacology*, *86*, 147–152.

Trantham-Davidson, H., Burnett, E. J., Gass, J. T., Lopez, M. F., Mulholland, P. J., Centanni, S. W., et al. (2014). Chronic alcohol disrupts dopamine receptor activity and the cognitive function of the medial prefrontal cortex. *The Journal of Neuroscience: The Official Journal of the Society for Neuroscience*, *34*, 3706–3718.

Trantham-Davidson, H., & Chandler, L. J. (2015). Alcohol-induced alterations in dopamine modulation of prefrontal activity. *Alcohol*, *49*, 773–779.

Tse, M. T. L., Cantor, A., & Floresco, S. B. (2011). Repeated amphetamine exposure disrupts dopaminergic modulation of amygdala–prefrontal circuitry and cognitive/emotional functioning. *Journal of Neuroscience*, *31*, 11282–11294.

Tsutsui, K.-I., Oyama, K., Nakamura, S., & Iijima, T. (2016). Comparative overview of visuospatial working memory in monkeys and rats. *Frontiers in Systems Neuroscience*, *10*, 99.

Vander Weele, C. M., Siciliano, C. A., & Tye, K. M. (2019). Dopamine tunes prefrontal outputs to orchestrate aversive processing. *Brain Research*, *1713*, 16–31.

Vanderveldt, A., Oliveira, L., & Green, L. (2016). Delay discounting: Pigeon, rat, human—Does it matter? *Journal of Experimental Psychology: Animal Learning and Cognition*, *42*, 141–162.

Veilleux, J. C., Colvin, P. J., Anderson, J., York, C., & Heinz, A. J. (2010). A review of opioid dependence treatment: Pharmacological and psychosocial interventions to treat opioid addiction. *Clinical Psychology Review*, *30*, 155–166.

Venniro, M., Caprioli, D., & Shaham, Y. (2016). Animal models of drug relapse and craving: From drug priming-induced reinstatement to incubation of craving after voluntary abstinence. *Progress in Brain Research*, *224*, 25–52.

Verdejo-Garcia, A., Benbrook, A., Funderburk, F., David, P., Cadet, J.-L., & Bolla, K. I. (2007). The differential relationship between cocaine use and marijuana use on decision-making performance over repeat testing with the Iowa Gambling Task. *Drug and Alcohol Dependence*, *90*, 2–11.

Verdejo-Garcia, A., & Rubenis, A. J. (2020). Chapter 11—Cognitive deficits in people with stimulant use disorders. In A. Verdejo-Garcia (Ed.), *Cognition and addiction* (pp. 155–163). Academic Press.

Vendruscolo, L., & Roberts, A. (2013). Operant alcohol self-administration in dependent rats: Focus on the vapor model. *Alcohol*, *48*, 277–286.

Volkow, N. D., Chang, L., Wang, G. J., Fowler, J. S., Ding, Y. S., Sedler, M., et al. (2001). Low level of brain dopamine D2 receptors in methamphetamine abusers: Association with metabolism in the orbitofrontal cortex. *The American Journal of Psychiatry*, *158*, 2015–2021.

Volkow, N. D., Fowler, J. S., Wang, G. J., Hitzemann, R., Logan, J., Schlyer, D. J., et al. (1993). Decreased dopamine D2 receptor availability is associated with reduced frontal metabolism in cocaine abusers. *Synapse (New York, N.Y.)*, *14*, 169–177.

Volkow, N. D., Hitzemann, R., Wang, G. J., Fowler, J. S., Wolf, A. P., Dewey, S. L., et al. (1992). Long-term frontal brain metabolic changes in cocaine abusers. *Synapse*, *11*, 184–190.

Volkow, N. D., Mullani, N., Gould, K. L., Adler, S., & Krajewski, K. (1988). Cerebral blood flow in chronic cocaine users: A study with positron emission tomography. *The British Journal of Psychiatry*, *152*, 641–648.

Volkow, N. D., Wang, G.-J., Begleiter, H., Porjesz, B., Fowler, J. S., Telang, F., et al. (2006). High levels of dopamine D2 receptors in unaffected members of alcoholic families: Possible protective factors. *Archives of General Psychiatry*, *63*, 999–1008.

Volkow, N. D., Wang, G.-J., Fowler, J. S., Logan, J., Hitzemann, R., Ding, Y.-S., et al. (1996). Decreases in dopamine receptors but not in dopamine transporters in alcoholics. *Alcoholism: Clinical and Experimental Research*, *20*, 1594–1598.

Volkow, N. D., Wang, G.-J., Ma, Y., Fowler, J. S., Wong, C., Ding, Y.-S., et al. (2005). Activation of orbital and medial prefrontal cortex by methylphenidate in cocaine-addicted subjects but not in controls: Relevance to addiction. *The Journal of Neuroscience*, *25*, 3932–3939.

Vonmoos, M., Hulka, L. M., Preller, K. H., Jenni, D., Baumgartner, M. R., Stohler, R., et al. (2013). Cognitive dysfunctions in recreational and dependent cocaine users: Role of attention-deficit hyperactivity disorder, craving and early age at onset. *The British Journal of Psychiatry: the Journal of Mental Science*, *203*, 35–43.

Vonmoos, M., Hulka, L. M., Preller, K. H., Minder, F., Baumgartner, M. R., & Quednow, B. B. (2014). Cognitive impairment in cocaine users is drug-induced but partially reversible: Evidence from a longitudinal study. *Neuropsychopharmacology: Official Publication of the American College of Neuropsychopharmacology*, *39*, 2200–2210.

Whiteford, H. A., Degenhardt, L., Rehm, J., Baxter, A. J., Ferrari, A. J., Erskine, H. E., et al. (2013). Global burden of disease attributable to mental and substance use disorders: Findings from the Global Burden of Disease Study 2010. *The Lancet*, *382*, 1575–1586.

Wilhelm, C. J., & Mitchell, S. H. (2008). Rats bred for high alcohol drinking are more sensitive to delayed and probabilistic outcomes. *Genes, Brain, and Behavior*, *7*, 705–713.

Winstanley, C. A., Olausson, P., Taylor, J. R., & Jentsch, J. D. (2010). Insight into the relationship between impulsivity and substance abuse from studies using animal models. *Alcoholism, Clinical and Experimental Research*, *34*, 1306–1318.

Wiskerke, J., Schoffelmeer, A. N. M., & De Vries, T. J. (2016). Response contingency directs long-term cocaine-induced neuroplasticity in prefrontal and striatal dopamine terminals. *European Neuropsychopharmacology*, *26*, 1667–1672.

Xia, Y., Ma, D., Hu, J., Tang, C., Wu, Z., Liu, L., et al. (2012). Effect of metabotropic glutamate receptor 3 genotype on N-acetylaspartate levels and neurocognition in non- smoking, active alcoholics. *Behavioral and Brain Functions*, *8*, 42.

Yager, L. M., Garcia, A. F., Wunsch, A. M., & Ferguson, S. M. (2015). The ins and outs of the striatum: Role in drug addiction. *Neuroscience*, *301*, 529–541.

Yuan, P., & Raz, N. (2014). Prefrontal cortex and executive functions in healthy adults: A meta-analysis of structural neuroimaging studies. *Neuroscience and Biobehavioral Reviews*, *42*, 180–192.

Zhong, N., Jiang, H., Du, J., Zhao, Y., Sun, H., Xu, D., et al. (2016). The cognitive impairments and psychological wellbeing of methamphetamine dependent patients compared with health controls. *Progress in Neuro-Psychopharmacology & Biological Psychiatry*, *69*, 31–37.

Zou, X., Durazzo, T. C., & Meyerhoff, D. J. (2018). Regional brain volume changes in alcohol- dependent individuals during short-term and long-term abstinence. *Alcoholism, Clinical and Experimental Research*, *42*, 1062–1072.

CHAPTER NINE

Drug addiction co-morbidity with alcohol: Neurobiological insights

M. Adrienne McGinn*,†, Caroline B. Pantazis*,†, Brendan J. Tunstall, Renata C.N. Marchette, Erika R. Carlson, Nadia Said, George F. Koob, and Leandro F. Vendruscolo

Intramural Research Program, National Institute on Drug Abuse, National Institutes of Health, Baltimore, MD, United States

*Corresponding authors: e-mail address: adrienne.mcginn@nih.gov; caroline.pantazis@nih.gov

Contents

Abstract

Addiction is a chronic disorder that consists of a three-stage cycle of binge/intoxication, withdrawal/negative affect, and preoccupation/anticipation. These stages involve, respectively, neuroadaptations in brain circuits involved in incentive salience and habit formation, stress surfeit and reward deficit, and executive function. Much research on

† These authors contributed equally to this work.

International Review of Neurobiology, Volume 157
ISSN 0074-7742
https://doi.org/10.1016/bs.irn.2020.11.002

addiction focuses on the neurobiology underlying single drug use. However, alcohol use disorder (AUD) can be co-morbid with substance use disorder (SUD), called dual dependence. The limited epidemiological data on dual dependence indicates that there is a large population of individuals suffering from addiction who are dependent on more than one drug and/or alcohol, yet dual dependence remains understudied in addiction research. Here, we review neurobiological data on neurotransmitter and neuropeptide systems that are known to contribute to addiction pathology and how the involvement of these systems is consistent or divergent across drug classes. In particular, we highlight the dopamine, opioid, corticotropin-releasing factor, norepinephrine, hypocretin/orexin, glucocorticoid, neuroimmune signaling, endocannabinoid, glutamate, and GABA systems. We also discuss the limited research on these systems in dual dependence. Collectively, these studies demonstrate that the use of multiple drugs can produce neuroadaptations that are distinct from single drug use. Further investigation into the neurobiology of dual dependence is necessary to develop effective treatments for addiction to multiple drugs.

1. Introduction

Alcohol and other addictive drugs are initially used because of their pleasurable effects (i.e., positive reinforcement). Chronic use leads to compulsive drug seeking and taking and an inability to limit intake. A negative emotional state also emerges during abstinence that includes anxiety, irritability, and dysphoria. The presence of this state is hypothesized to motivate further consumption of alcohol and drugs because these drugs, at least transiently, can alleviate unpleasant emotional states (i.e., negative reinforcement).

Addiction has been conceptualized as a three-stage cycle (Koob, Buck, et al., 2014; Koob & Le Moal, 1997) of *binge/intoxication* (heavy drug/alcohol consumption), *withdrawal/negative affect* (anxiety, pain, irritability, dysphoria), and *preoccupation/anticipation* (craving/relapse) that repeats with increasing intensity as addiction progresses (Fig. 1). These stages involve alterations in multiple neurotransmitter systems in brain circuits that are involved in incentive salience/habit formation (basal ganglia), reward deficit/stress surfeit (extended amygdala), and executive function (prefrontal cortex).

2. Epidemiology

Alcohol use disorder (AUD) can be co-morbid with other substance use disorders, referred to as dual dependence or homotypic co-morbidity. Data from the National Survey on Drug Use and Health in 2018 revealed that 14.8 million individuals aged 12 and older in the US suffer from AUD,

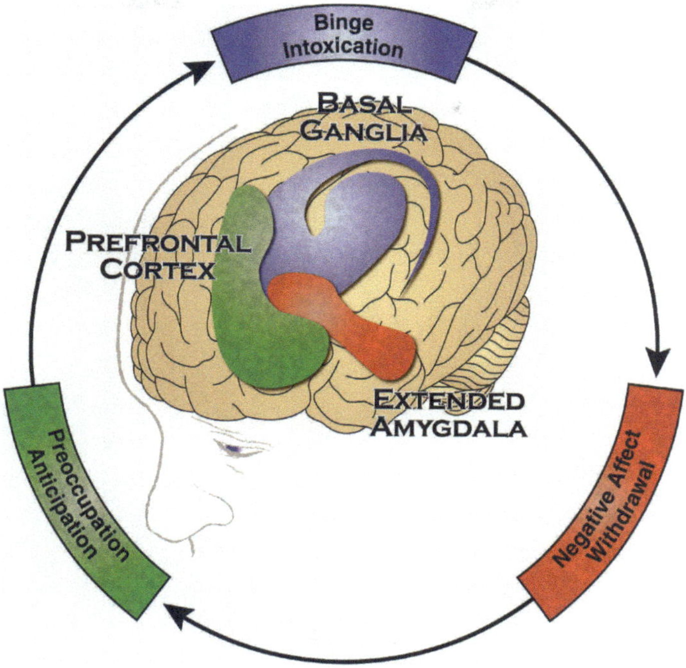

Fig. 1 The three stages of the addiction cycle and associated brain regions. *Modified with permission from Koob, G. F., Arends, M. A., & Le Moal, M. (2014). Drugs, addiction, and the brain. Academic Press.*

representing 5.4% of the US population (Substance Abuse and Mental Health Services Administration (SAMHSA), 2018). Data from the 2012–13 National Epidemiologic Survey on Alcohol and Related Conditions-III showed that AUD was associated with other substance use disorders, including nicotine use disorder, and this association was stronger in individuals with severe AUD (Grant et al., 2015). The presence of dual dependence in respondents was associated with a family history of co-morbid AUD and at least one additional substance use disorder (Jung, Goldstein, & Grant, 2016). The 2012–13 NESARC indicated that 46.5% of the individuals who met DSM-V criteria for AUD also had one or more concurrent substance use disorders, specifically, 31.5% with co-morbid nicotine use disorder, 3.2% with co-morbid cannabis use disorder, and 6.1% with a co-morbid substance use disorder involving other drugs such as opioids or stimulants (Saha et al., 2018) (Fig. 2). Data from the 2012–13 NESARC also revealed that 59.4% of men and 59.5% of women who

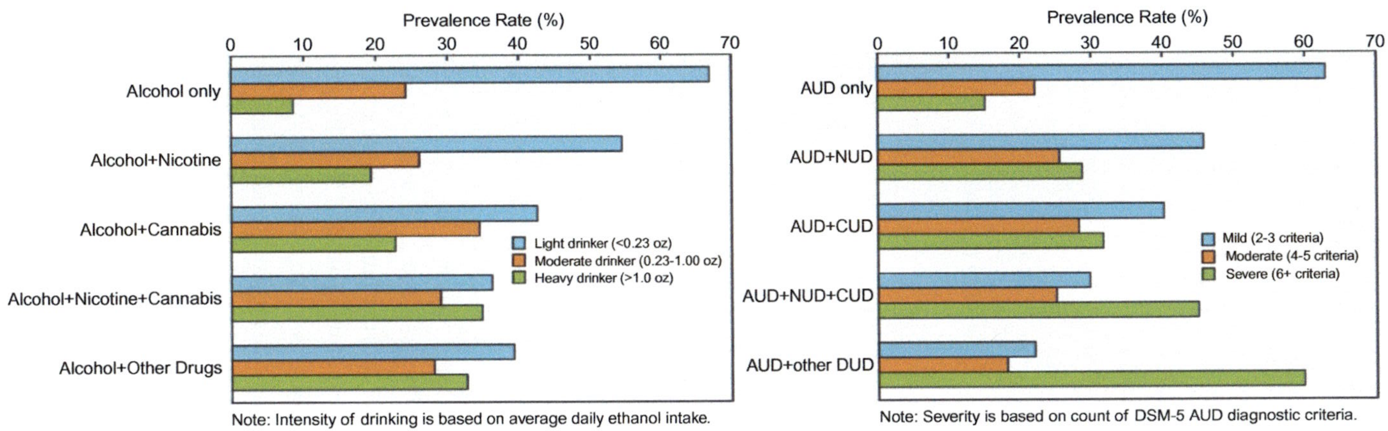

Fig. 2 Prevalence rates of alcohol and drug use or alcohol use disorder and other substance use disorders from the National epidemiologic Survey on Alcohol and Related Conditions-III. Left graph: the prevalence of light, moderate, or heavy drinking among respondents who reported past-year alcohol or alcohol and drug consumption. Right graph: the prevalence of mild, moderate, or severe DSM-5 diagnosis among respondents who reported an alcohol use disorder only or alcohol use disorder and another concurrent substance use disorder diagnosis. Alcohol use disorder (AUD), nicotine use disorder (NUD), cannabis use disorder (CUD), substance use disorder (DUD). *Modified from Saha, T. D., Grant, B. F., Chou, S. P., Kerridge, B. T., Pickering, R. P., & Ruan, W. J. (2018). Concurrent use of alcohol with other drugs and DSM-5 alcohol use disorder comorbid with other drug use disorders: Sociodemographic characteristics, severity, and psychopathology.* Drug and Alcohol Dependence, 187, *261–269, with permission.*

met DSM-5 diagnostic criteria for cannabis use disorder also had AUD, indicating a high prevalence of this co-morbidity, regardless of sex (Kerridge, Pickering, Chou, Saha, & Hasin, 2018).

Collectively, these findings indicate that a large population of individuals suffering from another substance use disorder may have dual dependence with AUD. Across these studies there emerges a typical ordering of drugs in which the appearance of the use of certain drugs is exceedingly rare in the absence of the co-use of other drugs. For example, cases of "pure" cannabis use disorder or gambling disorder are especially rare (Cohn, Johnson, Rath, & Villanti, 2016; Dash et al., 2019).

Among adolescent drug users, co-morbidity of alcohol, tobacco/nicotine, and cannabis use is common (Cohn et al., 2016; Daw, Nowotny, & Boardman, 2013) across countries. For example, in North American indigenous adolescents (Hautala, Sittner, & Walls, 2019), a national sample of US young adults (Cohn et al., 2016, 2018), a national sample of New Zealand youth (White, Walton, & Walker, 2015), and Canadian youth (Leatherdale, Hammond, & Ahmed, 2008), the co-morbidity of tobacco and/or alcohol and/or cannabis use ranged from 10% to 60%. These findings among adolescent populations indicate that polydrug use trends begin early in drug taking history.

2.1 Pattern and etiology of co-morbidity

The co-use of alcohol and other drugs may occur for the purpose of enhancing the intoxicating effects or the "high" of one or both drugs. Co-use of alcohol and cannabis to deliberately produce additive or synergistic effects is common, and their co-use leads to greater subjective feelings of general intoxication (Sokolovsky, Gunn, Micalizzi, White, & Jackson, 2020). Heavy drinking tends to occur on cannabis use days, particularly in those with co-morbid AUD and cannabis use disorder (Metrik, Gunn, Jackson, Sokolovsky, & Borsari, 2018). More alcohol may be consumed on days when cannabis is used by recreational users, but less alcohol might be consumed among those using medical marijuana, specifically those that indicate they use cannabis to substitute for alcohol consumption (Gunn, Jackson, Borsari, & Metrik, 2019). Alcohol and cannabis use disorders appear to be largely dissociable disorders, although the withdrawal syndrome of each compound was noted as a potential bridge between these conditions in a network analysis (Baggio et al., 2018). Note, however, at the population level, in Washington state in the 2 years following introduction of legal stores for recreational cannabis (2014–16), no significant increase in co-use of alcohol and cannabis or in total alcohol consumption was observed despite a significant increase in cannabis use (Subbaraman & Kerr, 2020). As cannabis legislation continues to change and access to both medical and recreational cannabis increases in the United States, it will be important to monitor patterns of drug use and how this may or may not change patterns of use for commonly co-used drugs.

Enhanced intoxication due to additive intoxicating effects of certain drugs may greatly increase the risk of negative effects, and these risks may be more difficult for the user to anticipate. This has practical implications not only for risk to the individual but also for the risk of societal harm. In a study of polydrug users within a large sample of individuals involved in road accidents in Virginia Beach (5,375 in the "crash" group, 7,415 in the "noncrash" control condition, collected from the same roads at comparable times of day and traffic conditions), it was clear that opioid-benzodiazepine co-users were at the highest risk of involvement in road accidents and had the highest likelihood of the presence of alcohol in their breath. They were also the only group with significantly elevated risk of breath alcohol concentrations above the legal limit. The enhanced risk of a road accident is likely due to the additive sedative effects of these drugs and additional effects that impair driving, including fatigue, dizziness, and distorted vision. We did not find information on whether the combination of opioids and psychostimulants (speedball) synergistically increases motor impairment, however see Leri, Bruneau, & Stewart, 2003 for review.

Drugs may also be used to mitigate undesired effects of one or both drugs. For example, alcohol and nicotine are known to potentiate each other's effects (i.e., cross-sensitization), but nicotine can additionally counteract the sedative effects of alcohol intoxication (i.e., cross-tolerance; Adams, 2017). Subjective reports reflect the cross-sensitization of alcohol and nicotine in that these two drugs have synergistic effects on feelings of being "buzzed" and "dizzy." For example, participants reported increased pleasure of cigarettes with the consumption of alcohol, and higher levels of craving for both alcohol and tobacco were found with co-use (Piasecki et al., 2011). It is likely that both of these processes (reward and craving) interact in the ethology of concurrent drinking of alcohol and smoking of cigarettes.

Chronic pain may be a potential risk factor for polydrug misuse and co-morbid substance use disorders. Between 2015 and 2017, approximately one-quarter (26.4%) of individuals with opioid use disorder (OUD) experienced co-morbid AUD (Jones & McCance-Katz, 2019). Among individuals seeking treatment for OUD, approximately two-thirds reported chronic pain (64.4%; 61.8% of which had chronic pain prior to OUD diagnosis). Approximately one-third (33.4%) of those with concurrent OUD and chronic pain also reported a co-morbid AUD diagnosis (Hser et al., 2017). Notably, pain-related anxiety was positively associated with alcohol and opioid co-use, as well as concurrent misuse of alcohol and opioids (LaRowe et al., 2020). Although acutely the combination of alcohol and opioids may be additive in suppressing pain and pain-related anxiety, this trend is particularly alarming, as they are contraindicated because of negative health effects, including potentiation of CNS-depressant effects and increased risk of respiratory depression/coma/death with severe intoxication (LaRowe et al., 2020). There is increasing interest in understanding the neurobiological interactions between chronic pain states and alcohol or opioid use (Edwards, Vendruscolo, Gilpin, Wojnar, & Witkiewitz, 2020; McGinn, Edwards, & Edwards, 2020; Pahng, Paulsen, McGinn, Edwards, & Edwards, 2017). Although little is known about the unique neurobiological adaptations that occur with co-abuse of alcohol and opioids, it is hypothesized that co-abuse of alcohol and opioids leads to more extensive neurobiological adaptations (Hood, Leyrer-Jackson, & Olive, 2020; Witkiewitz & Vowles, 2018). Perhaps consequently, the treatment of co-morbid AUD and OUD is more difficult, such that it is associated with worse treatment outcomes relative to either substance alone (Hood et al., 2020; Witkiewitz & Vowles, 2018).

Given the complex nature of drug interactions in polydrug use, methods to study these interactions are needed. Behavioral economics is one approach that is useful for determining the nature of such interactions. For example, using a hypothetical purchase task, it was found that among individuals that misuse alcohol, cocaine is a complement to alcohol (i.e., as alcohol prices increase, alcohol consumption decreases and cocaine consumption also decreases), but that alcohol also functions as a substitute for cocaine (i.e., as cocaine prices increase, cocaine consumption decreases, and alcohol consumption increases; Petry, 2001). This kind of analysis involves determining the "elasticity" of demand by measuring consumption of a reinforcer across a range of prices (i.e., creating a "demand curve"). By normalizing this consumption to baseline consumption when the reinforcer is freely available (i.e., at a theoretical price of zero), the elasticity of demand can be determined independent of scalar properties of the reinforcer (e.g., the number of food pellets delivered per reinforcement or the unit dose of drug, both of which influence the number of reinforcers that will be consumed by the organism at baseline). This kind of analysis is powerful because it allows the comparison of demand for reinforcement across different types of reinforcers (including drug and non-drug reinforcers), different groups of animals with different consumption baselines (e.g., drug-dependent and non-dependent groups), and across species (including rodent, non-human primates, and humans; Hursh, Raslear, Shurtleff, Bauman, & Simmons, 1988; Hursh & Silberberg, 2008; Hursh & Winger, 1995; Lea, 1978; (McConnell et al., 2021). Understanding these interactions is critical in designing behavioral (e.g., voluntary abstinence) or pharmacological treatments (e.g., substitution of the drug effect via agonism, or blockade of the drug effect via antagonism) that anticipate the consequences of changing the consumption of one drug by a polydrug user. For example, based on the data of Petry (2001), treating cocaine use disorder (CUD) co-morbid with AUD might lead to a redistribution of behavior toward alcohol. In contrast, treating AUD may in and of itself be effective in reducing cocaine use associated with CUD.

The study of polydrug use, misuse, and comorbid substance use disorders poses unique challenges. In general, it appears that the treatment of two or more concurrent substance use disorders is more difficult than treatment of a single substance use disorder. For example, co-morbid cigarette smoking with AUD is estimated to be as high as 80% in treatment seekers (Kalman, Kim, DiGirolamo, Smelson, & Ziedonis, 2010). There is a strong association between nicotine and alcohol use through early life and the continued use of both (McPherson et al., 2018; Witkiewitz & Vowles, 2018). A similar relationship was true for alcohol and cannabis use in early life through young adulthood (Duncan et al., 2015), and studies have demonstrated heavier use of both alcohol and cannabis in dual-diagnoses compared to AUD or CUD alone (Hayaki, Anderson, & Stein, 2016). Nicotine use disorder is also more severe among those with AUD (Romberger & Grant, 2004), and worse withdrawal is observed during withdrawal from nicotine and alcohol relative to withdrawal from nicotine alone (Choi et al., 2014). Continued cannabis smoking was reported to worsen tobacco cessation success (alcohol use did not matter; McClure et al., 2020). Increased cocaine and heroin craving and actual use among polydrug users with OUD were associated with the consumption of alcohol, even though the sample specifically excluded heavy drinkers (Preston, Jobes, Phillips, & Epstein, 2016). Naltrexone was effective in reducing alcohol consumption in people with AUD but was not effective in people with co-morbid AUD and CUD (Hersh, Van Kirk, & Kranzler, 1998). Finally, varenicline is the first treatment to have shown efficacy in treating a poly-substance use disorder: varenicline reduced both alcohol and tobacco use in heavy smoking individuals with AUD (Mitchell, Teague, Kayser, Bartlett, & Fields, 2012; Litten et al., 2013). Varenicline was equally effective in reducing alcohol consumption in individuals that do or do not smoke tobacco (Litten et al., 2013). Varenicline was also an effective treatment for tobacco/nicotine use disorder and AUD (Fagerstrom & Hughes, 2008; Litten et al., 2013).

The potential interactions of multiple co-used drugs and their cessation make this a difficult area to conduct research. An additional challenge is that the combination of drugs can create molecules that are also pharmacologically active (e.g., cocaethylene resulting from the combination of alcohol and cocaine). At present, the best approach to achieving drug abstinence in people with poly-substance use disorders or engaging in polydrug misuse is unclear. Even in the common case of concurrent AUD and tobacco smoking , there has been much debate on the appropriate approach to treatment, with some studies supporting the beneficial effect of concurrent treatment of AUD and tobacco smoking, whereas studies support the treatment of AUD first (i.e., patients are allowed to smoke while in treatment for AUD), followed by smoking treatment. Cooney et al. (2015) directly compared drinking outcomes in an intensive outpatient AUD treatment program where the participants were randomized to a concurrent smoking cessation intervention or to delayed smoking cessation. Their results showed that daily smoking abstinence improved several alcohol-related outcomes, such as lower alcohol consumption and urge to drink, lower negative affect, and lower positive alcohol outcome expectancies, thus supporting the concurrent treatment of AUD and smoking (Cooney et al., 2015).

2.2 Gateway hypothesis

The original gateway hypothesis proposed that drug use follows a sequential pattern, whereby licit drugs like alcohol and cigarettes are "gateway drugs" to marijuana, increasing the likelihood of its use, and marijuana use leads to an increased likelihood of the use of illicit drugs like cocaine and heroin (Kandel, Yamaguchi, & Chen, 1992). The gateway hypothesis developed from correlational studies of reported drug initiation and use, in which few adolescents experimented with marijuana without first experimenting with alcoholic beverages or cigarettes (Kandel, 1975; Kandel & Faust, 1975). The age of initiation and the extent of drug use were also predictors of progression, such that individuals that started using licit drugs at a young age and/or with increased frequency of use were more likely to escalate to the use of illicit drugs later in life (Kandel et al., 1992).

Data from both humans with substance use disorders and rodent models of alcohol and drug use indicate that a form of the gateway hypothesis is supported, but in a broader developmental context. It appears that exposure to addictive drugs in general at an early age can facilitate the development of drug misuse and addiction later in life (Chassin, Presson, Sherman, & Edwards, 1990; Grant & Dawson, 1998; Hingson, Heeren, Zakocs, Winter, & Wechsler, 2003), but the use of legal drugs like alcohol and tobacco/nicotine does not necessarily precede a transition to the use of illicit drugs. A study of adults that heavily used cocaine or heroin reported that this majority of this group began using heroin or cocaine without prior use of alcohol or marijuana (Mackesy-Amiti, Fendrich, & Goldstein, 1997). Other studies have similarly reported the existence of populations with substance use disorders without prior use or misuse of marijuana (Tarter, Vanyukov, Kirisci, Reynolds, & Clark, 2006; Wu, Pilowsky, & Schlenger, 2005). Epidemiological studies have importantly highlighted associations between the use of different drugs, but these types of studies are unable to identify whether and to what extent drug use may alter brain circuitry in a manner that motivates the progression to the use of illicit, "harder" drugs or other addictive drugs in general (Kandel & Kandel, 2014).

Rodent models of drug use could elucidate how drug history impacts responding to other drugs of abuse later in life. However, the results of studies conducted thus far have been inconclusive. Some studies have shown that adolescent exposure to legal drugs can sensitize behavioral responses to illicit drugs using conditioned place-preference (Hutchison & Riley, 2012; Levine et al., 2011; Molet, Herve, Thiebot, Hamon, & Lanfumey, 2013). Similarly, mice that were exposed to alcohol prenatally and postnatally as pups self-administered more cocaine and had a greater preference for a cocaine-paired chamber in a place-conditioning test compared to non-alcohol-exposed control mice (Cantacorps, Montagud-Romero, Lujan, & Valverde, 2020). However, these studies measured drug seeking behavior after passive administration of illicit drugs, which makes it challenging to determine if contingent administration of one drug primes reward circuitry and alters reinforcement or tolerance to a second drug.

Other studies using operant measures of drug consumption have shown that self-administration of one drug may sensitize responding to a second drug, possibly by altering brain reward circuitry. Adolescent alcohol exposure transiently increased cocaine self-administration in female mice (Mateos-Garcia et al., 2015). As an example that supports a broader gateway hypothesis, systemic nicotine pretreatment increased alcohol drinking in alcohol-naïve rats (Doyon et al., 2013; Smith, Horan, Gaskin, & Amit, 1999) and alcohol-experienced non-dependent rats (Leao et al., 2015; Montanari et al., 2020). Systemic nicotine pretreatment accelerated the escalation of alcohol intake in rats undergoing chronic intermittent alcohol vapor exposure (Leao et al., 2015). Chronic daily nicotine treatment also increased progressive-ratio responding as well as alcohol intake despite adverse consequences in alcohol-dependent rats (Leao et al., 2015). Moreover, rats with a history of nicotine exposure during adolescence increased alcohol self-administration in adulthood. Adolescent nicotine exposure in these animals also caused GABAergic inhibition of dopamine (DA) neurons in the lateral VTA in response to alcohol in adulthood (Thomas et al., 2018). However, a rodent study of sequential drug use found that a history of alcohol drinking in late adolescence did not influence cocaine self-administration behaviors, including acquisition, self-administration across a range of doses, extinction, or reinstatement (Fredriksson et al., 2017). Neurobiological insights from rodent studies may shed light on whether the use of one drug can "prime" reward and stress circuitry and promote the use of another drug through direct cross-reinforcement or cross-reinforcement due to drug history (Tunstall, Vendruscolo, & Worthington, 2020).

A more modern and broad version of the gateway hypothesis is that there are genetic, environmental (e.g., drug availability and price), and neurobiological factors that predispose certain individuals to addiction (Vanyukov et al., 2012). The earlier the adolescent exposure to any drug, the more it increases the likelihood of a substance use disorder later in life (Grant & Dawson, 1998). Current ongoing longitudinal studies in humans, such as the Adolescent Brain Cognitive Development (ABCD) study (2020), will give valuable information in this regard.

Clearly, chronic drug and alcohol use produces neuroadaptations that render vulnerable individuals susceptible to developing a substance use disorder (Koob & Volkow, 2016). Chronic use of both licit and illicit drugs drives allostatic processes that lead to dependence and compulsive drug taking/seeking (Koob & Schulkin, 2019). These processes impact numerous neuromodulatory and neurotransmitter systems that are largely consistent across drug classes (reviewed here). These changes lead to decreased reward function and pathological motivation, as well as dysregulated emotional processing and emotional and physical withdrawal states (Koob & Le Moal, 2001). The degree of perturbation is dependent on drug type and severity of misuse and impacted by co-occurring substance use or mental health disorders. In this way, repeated stressors and challenges create an allostatic load that contributes to the development of an addiction disease state.

2.3 Self-medication hypothesis

Individuals susceptible to addiction may take drugs of abuse to self-medicate. The self-medication hypothesis of substance use disorders suggests that drug taking is driven by the desire to control or reduce negative affective states like pain, anxiety, aggression, or dysphoria. These states may be the result of a pre-existing mental health disorder like depression or post-traumatic stress disorder (Khantzian, 1997). However, repeated drug taking to cope with these issues leads to a regression in the ability to self-regulate emotions and develop natural coping strategies for aversive emotions (Khantzian, 1990). In addition, negative withdrawal states that emerge with chronic drug use evoke the same feelings that individuals initially try to escape by using drugs. Many of the same neurotransmitter systems implicated in negative emotional states are altered with AUD and other substance use disorders, including DA, gamma-aminobutyric acid (GABA), glucocorticoids, glutamate, corticotropin-releasing factor (CRF), and neuropeptide Y (Gilpin et al., 2011; Gilpin, Misra, & Koob, 2008; Kallupi et al., 2014; Markou, Kosten, & Koob, 1998).

According to the self-medication hypothesis, individuals with AUD prefer a given drug that best helps them to attempt to self-regulate their unique emotional break with homeostasis (Khantzian, 1985). Because different classes of drugs can trigger different subjective feelings (Robbins & Everitt, 1999), drug choice presumably reflects neurochemical actions that reverse or ameliorate, at least temporarily, such negative emotional states. However, chronic use of the preferred drug may lead to the use of another drug to cope with the negative consequences associated with primary drug use (e.g., taking alcohol or opioids to reduce anxious feelings evoked by cocaine intoxication; Khantzian, 1997). Other factors, including availability and cost, significantly impact drug choice and may explain reports that adolescents experiment with cigarettes and alcohol before transitioning to illicit drugs like heroin and cocaine (see above discussion on the gateway hypothesis; Kandel & Faust, 1975). In this way, addicted individuals can develop co-morbid addictions in pursuit of self-medication.

3. Neurobiological interactions

Very little preclinical work has been done in which neurobiological measures have been studied in the context of two drugs interacting in self-administration or by concomitant, experimenter administration, where drugs have been administered or self-administered to alleviate withdrawal, or where drugs have been self-administered following pre-exposure to different drugs. As a result, the following sections focus on given neurotransmitter neurocircuits where drugs do clearly interact or where drugs appear to have common underlying mechanisms that would predict some interaction.

3.1 Glutamate

Glutamate is an excitatory neurotransmitter that regulates synaptic plasticity for long-term learning and memory (Luscher & Malenka, 2012). Glutamatergic connectivity from prefrontal cortex (PFC), amygdala, and hippocampus to striatum are thought to connect motivational states associated with environmental drug-associated stimuli to learned motor responses to trigger drug seeking (Kalivas, 2009). *N*-acetylcysteine, which regulates intracellular glutamate, has emerged as a potential therapeutic to treat relapse in abstinent individuals (Duailibi et al., 2017; Tomko et al., 2018). Under normal circumstances, glial cells convert extra-synaptic cysteine to glutamate through the cysteine-glutamate exchanger. The glutamate released binds to mGluR2/3 to inhibit presynaptic glutamate release. By stimulating the exchanger, *N*-acetylcysteine reduces glutamate release during reinstatement (Kruyer, Chioma, & Kalivas, 2020). The established role for glutamate in relapse propensity across drugs of abuse make *N*-acetylcysteine an attractive candidate drug for individuals suffering with single or polydrug use disorders.

Altered glutamatergic signaling to nucleus accumbens (NAc) medium spiny neurons from cortical and allocortical regions is evident following chronic exposure to drugs of abuse (Kalivas, 2009; Lüscher, 2016; McKetin et al., 2017; Scofield et al., 2016; Turner, Kashima, Manz, Grueter, & Grueter, 2017). Imbalance between synaptic and non-synaptic glutamate in NAc neurons impairs synaptic plasticity and alters input from these regions to NAc (Kalivas, 2009), contributing to relapse propensity and compulsive drug seeking. Chronic drug use produces lasting neuroadaptations in NAc synapses, many of which are consistent across drugs of abuse and reviewed in detail elsewhere (Scofield et al., 2016).

Briefly, blocking post-synaptic glutamatergic receptors that promote synaptic potentiation, including ionotropic NMDA and AMPA receptors and type I metabotropic glutamate receptor 5 (mGluR5), systemically or into NAc core or shell regions reduces reinstatement for cocaine, alcohol, opioids, and amphetamine. However, stimulating components of the synapse that negatively regulate synaptic glutamate levels, including type II mGluR2/3 and the glial glutamate transporter-1 (GLT-1), attenuates reinstatement across drugs of abuse. As a result, multiple components of glutamatergic synapses that provide input to the NAc are dysregulated to drive relapse, and many of these changes are consistent across drugs of abuse.

Polydrug use produces neuroadaptations in the glutamatergic system that are distinct from chronic single drug use, including augmenting the heightened glutamatergic signaling observed following single drug use. Co-administration of alcohol and nicotine increased glutamate levels in NAc shell (Waeiss, Knight, Engleman, Hauser, & Rodd, 2020) or PFC (Deehan Jr. et al., 2015), compared to each drug separately. Similarly, rats with prior voluntary access to 10% alcohol had elevated striatal glutamate after a non-contingent injection of methamphetamine, compared to those given methamphetamine or alcohol alone (Blaker, Moore, & Yamamoto, 2019). Polydrug use also induces changes in glutamatergic signaling that alter excitability of these cells. Chronic nicotine intoxication via vapor inhalation and withdrawal attenuated phosphorylation of the AMPA GluA1 subunit in dorsomedial PFC, dorsal striatum, and central amygdala induced by a non-contingent alcohol injection (McGinn et al., 2016), suggesting that nicotine influences the reinforcing effects of alcohol (Leao et al., 2015). Rats trained on sequential 2-h intravenous cocaine and 6-h alcohol self-administration for 12 days had increased surface expression of the pre-synaptic GLT-1 and reduced glutamate efflux into NAc core during cocaine reinstatement, compared to rats trained on cocaine self-administration alone (Stennett, Padovan-Hernandez, & Knackstedt, 2020). Prenatal and postnatal exposure to alcohol for 4 days over a 6-week period also increased cocaine self-administration and reinstatement in adulthood and altered the AMPA GluA1 to GluA2 ratio in PFC and striatum after reinstatement, indicating long-lasting changes in the excitability of glutamatergic neurons following a history of alcohol exposure (Cantacorps et al., 2020). Thus, these findings suggest that polydrug use may exacerbate dysfunctional glutamatergic signaling observed with single drug use and increase propensity for relapse.

Although there are few studies that have examined the effects of *N*-acetylcysteine in polydrug use and relapse in animal models or humans,

N-acetylcysteine has been shown to attenuate cocaine, nicotine, alcohol, and heroin reinstatement in animal models (Baker et al., 2003; Israel et al., 2019; Namba et al., 2019; Zhou & Kalivas, 2008). In cocaine-dependent humans, higher impulsivity correlated with increased cortical glutamate levels, which was normalized by *N*-acetylcysteine (Schmaal, Veltman, Nederveen, van den Brink, & Goudriaan, 2012). *N*-acetylcysteine similarly reduced craving for cocaine in cocaine-dependent, abstinent individuals (Amen et al., 2011; LaRowe et al., 2013) and reduced craving for drugs in posttraumatic stress disorder (PTSD) patients diagnosed with substance use disorder (Back et al., 2016), but not methamphetamine in individuals with methamphetamine dependence (Grant, Odlaug, & Kim, 2010) or nicotine in heavy smokers during short-term abstinence (Schmaal et al., 2011). Thus, although *N*-acetylcysteine consistently decreases reinstatement in animal models, its application in humans has had mixed results, appears limited to non-drug taking measures, and may depend on drug class and/or history of drug use prior to treatment.

Thus, disruptions to glutamatergic signaling contribute to maladaptive learning and memory processing and relapse in addiction. Although many of these changes are consistent across drugs of abuse, polydrug use produces alterations in glutamate systems that are distinct from single drug use. *N*-acetylcysteine attenuates elevated glutamate and restores glutamatergic tone but to date has some effects on craving in human studies, but not on relapse.

3.2 GABA

GABA is a major inhibitory neurotransmitter that regulates synaptic plasticity and neuronal excitability. GABA negatively regulates brain reward and stress systems, and both the ion-gated $GABA_A$ and ligand-gated $GABA_B$ receptors are necessary for drug seeking/taking behaviors (Engin, Benham, & Rudolph, 2018; Vlachou & Markou, 2010). In particular, VTA DA neurons receive GABAergic input from VTA GABA neurons and from other GABAergic brain regions like the ventral pallidum, NAc, and amygdala (Gilpin, Herman, & Roberto, 2015; Vlachou & Markou, 2010). GABAergic neurons can also inhibit the hypothalamic-pituitary-adrenal (HPA) axis and CRF release (Calogero, Gallucci, Chrousos, & Gold, 1988; Herman, Mueller, & Figueiredo, 2004), possibly through hypothalamic or extended amygdala GABAergic neurons.

Both acute and chronic use of addictive drugs or alcohol impact GABAergic synaptic transmission. Acute exposure to alcohol, nicotine, and cocaine increased GABAergic transmission (Kallupi et al., 2013; Roberto, Madamba, Moore, Tallent, & Siggins, 2003; Zhu & Lovinger, 2006), though there is some evidence to suggest that the mechanism depends on drug type. In the hippocampus, acute ethanol administration enhanced GABA transmission through $GABA_A$ receptor signaling, but only if presynaptic $GABA_B$ receptors were blocked (Wan, Berton, Madamba, Francesconi, & Siggins, 1996). In contrast, nicotine binds to excitatory nicotinic acetylcholine receptors on presynaptic GABA neurons to depolarize GABA cells (D'Souza & Markou, 2013). An increase in GABA transmission, via increased GABA release, was found in in vitro CeA slice preparations from rats that were chronically exposed to alcohol compared with alcohol-naïve rats. The acute application of alcohol increased GABA transmission in alcohol-exposed and naive rats, suggesting a lack of alcohol tolerance for this effect (Roberto, Madamba, Stouffer, Parsons, & Siggins, 2004). In vivo microdialysis studies indicated that the administration of alcohol caused a dose-dependent increase in GABA release in the CeA in both rats chronically treated with alcohol and naive rats (Roberto et al., 2004). Changes in GABA transmission in the CeA were also found with cocaine. GABA transmission was enhanced in the CeA of rats that were chronically exposed to cocaine compared with cocaine-naive rats. Acute cocaine bath application decreased GABA release in both cocaine-treated and cocaine-naïve rats (Kallupi et al., 2013). In contrast, withdrawal from some drugs of abuse, such as morphine and cocaine, inhibits $GABA_B$ receptor-mediated transmission (Shoji, Delfs, & Williams, 1999; Tan, Nuno-Perez, Mameli, & Meye, 2019).

Rodent models of addiction highlighted that GABA receptor signaling inhibits drug consumption and may mediate the reinforcing effects of alcohol and drugs of abuse. Systemic administration of the $GABA_B$ receptor agonist baclofen reduced operant responding for alcohol (Walker & Koob, 2007), nicotine (Fattore, Cossu, Martellotta, & Fratta, 2002; Paterson, Froestl, & Markou, 2004), heroin (Spano, Fattore, Fratta, & Fadda, 2007), and stimulants (Brebner, Phelan, & Roberts, 2000; Di Ciano & Everitt, 2003). Baclofen also dose-dependently reduced the drug-induced increases in DA release in the NAc shell following acute administration of morphine, cocaine, or nicotine (Fadda, Scherma, Fresu, Collu, & Fratta, 2003).

Few rodent studies have examined GABA receptor signaling in withdrawal, though some studies show that it may contribute to both somatic and affective withdrawal symptoms. Baclofen reduced anxiety-like behaviors in the elevated plus maze following naloxone-precipitated withdrawal in morphine dependent mice (Pedron, Varani, & Balerio, 2016). Baclofen also reduced somatic signs of withdrawal, anxiety-like behaviors, and conditioned place aversion following naloxone- or mecamylamine-precipitated withdrawal from nicotine in mice (Varani, Aso, Moutinho, Maldonado, & Balerio, 2014; Varani, Moutinho, Calvo, & Balerio, 2011). However, both the $GABA_B$ receptor agonist CGP44532 and antagonist CGP56433A exacerbated anhedonia-like effects of nicotine withdrawal and elevated intracranial self-stimulation thresholds (Vlachou et al., 2011).

Several clinical studies have shown that baclofen can reduce drug craving, but its effects may be drug specific. In individuals with AUD, baclofen reduced craving and prolonged the time to first relapse (Morley et al., 2018; Reynaud et al., 2017). However, other studies have shown that baclofen did not reduce alcohol drinking or craving (Garbutt, Kampov-Polevoy, Gallop, Kalka-Juhl, & Flannery, 2010; Hauser et al., 2017). Chronic, but not acute, administration of baclofen reduced the number of cigarettes smoked and craving in smokers (Cousins, Stamat, & de Wit, 2001; Franklin et al., 2009). Baclofen also reduced craving in marijuana smokers but did not decrease relapse (Haney et al., 2010). Similarly, in detoxified individuals with OUD, baclofen reduced opioid withdrawal and depressive systems, but did not significantly reduce craving or self-reported opioid use (Assadi, Radgoodarzi, & Ahmadi-Abhari, 2003). Similarly, in individuals with CUD with or without co-occurring opioid use disorder, baclofen only reduced cocaine use in individuals without opioid use disorder (Haney, Hart, & Foltin, 2006), indicating that it may not be effective for treating co-occurring substance use disorders.

3.3 Norepinephrine

The norepinephrine (NE) system is implicated in the regulation of arousal and activated in stressful circumstances (Berridge & Waterhouse, 2003). Norepinephrine neurons originate in several nuclei in the medulla and pons (Dahlstrom & Fuxe, 1964) and are enriched in the locus coeruleus and subcoeruleus (Swanson & Hartman, 1975). Norepinephrine acts on target cells by binding to the α- or β-adrenergic receptors. Although NE projects throughout the brain, the two largest NE outputs are the paraventricular nucleus of hypothalamus (PVN) and bed nucleus of stria terminalis (BNST; Kilts & Anderson, 1986), which innervate the HPA axis to regulate stress responses. Norepinephrine neurons are also stimulated by CRF in response to stress (Krystal & Neumeister, 2009). Norepinerphine neurons have been linked to stress and addiction for alcohol and drugs of abuse, particularly during withdrawal and stress-induced reinstatement (Espana, Schmeichel, & Berridge, 2016; Smith & Aston-Jones, 2008).

Again, although little preclinical work has been done on alcohol self-administration with concomitant administration of self-administration of other drugs of abuse, NE neurons are responsive to acute administration of drugs of abuse and mediate drug-evoked arousal states. Norepinephrine signaling has been implicated in the sensitized locomotor response to certain drugs of abuse, including morphine (Drouin et al., 2002), amphetamine (Vanderschuren, Beemster, & Schoffelmeer, 2003), alcohol (Matchett & Erickson, 1977), and cocaine (Wellman, Ho, Cepeda-Benito, Bellinger, & Nation, 2002), although results are less clear at higher doses (Vanderschuren et al., 2003; Wellman et al., 2002). In humans, the dual-receptor antagonists labetalol and carvedilol reduced sympathetic responses to cocaine without impacting subjective responses like self-reported feelings of craving or euphoria (Sofuoglu, Brown, Babb, Pentel, & Hatsukami, 2000a; Sofuoglu, Brown, Babb, Pentel, & Hatsukami, 2000b). Acute passive administration of alcohol, morphine, cocaine, amphetamine, or nicotine dose-dependently increased NE concentrations in the BNST as measured by microdialysis (Jadzic, Bassareo, Carta, & Carboni, 2019), indicating that NE neurons may contribute to acute drug-induced arousal through interactions with BNST neurons.

Evidence for involvement of NE in alcohol and drug consumption varies by drug class. Early experiments in rats or rhesus monkeys indicated that NE transporter inhibitors or the α_2-agonist clonidine did not impact self-administration of stimulants (Wee et al., 2006; Wee & Woolverton, 2004; Woolverton, 1987; Yokel & Wise, 1978), although antagonists of both receptors decreased cocaine self-administration in rats or squirrel monkeys (Goldberg & Gonzalez, 1976; Harris, Hedaya, Pan, & Kalivas, 1996; Wee, Mandyam, Lekic, & Koob, 2008). In contrast, the α_2-receptor agonists clonidine and guanfacine (Fredriksson et al., 2015; Rasmussen, Alexander, Malone, Federoff, & Froehlich, 2014) and antagonists of the α-receptor (Froehlich, Hausauer, Fischer, Wise, & Rasmussen, 2015; O'Neil, Beckwith, Kincaid, & Rasmussen, 2013; Walker, Rasmussen, Raskind, & Koob, 2008) and β-receptor (Gilpin & Koob, 2010) reduced alcohol consumption in rats. Although few studies have examined NE during opioid self-administration, the results thus far have indicated that it is not necessary. Yohimbine, which increases brain NE levels, did not impact reacquisition of heroin self-administration after extinction in rats (Minhas & Leri, 2014), and clonidine did not prevent withdrawal-associated increases in heroin choice over food in opioid-dependent rhesus monkeys (Negus & Rice, 2009) or attenuate escalation of heroin intake during extended access self-administration in rats (Park et al., 2015). However, in a study of individuals with co-morbid cocaine and OUD on methadone maintenance therapy, treatment with disulfram, which decreases NE synthesis by inhibiting dopamine β-hydroxylase activity, reduced both the frequency and quantity of cocaine use (Petrakis et al., 2000).

Chronic exposure to alcohol and other addictive drugs suppresses NE activity, and acute withdrawal produces a rebound effect of elevating NE signaling (Fitzgerald, 2013). Agonists of the α_2-receptor clonidine and guanfacine reduced circulating NE and suppressed negative symptoms associated with drug abstinence, including acute withdrawal symptoms in alcohol-dependent (Manhem, Nilsson, Moberg, Wadstein, & Hokfelt, 1985; Muzyk, Fowler, Norwood, & Chilipko, 2011) and opioid-dependent human patients (Gold, Pottash, & Extein, 1982; Gold, Redmond Jr., & Kleber, 1979; Kowalczyk et al., 2015), and psychological symptoms, like self-reported feelings of stress and craving in patients with CUD (Fox et al., 2012). However, clonidine did not prevent seizures or delirium that can accompany acute alcohol withdrawal (Albertson, Chenoweth, Ford, Owen, & Sutter, 2014), which may limit its application in patients with AUD (Kosten & O'Connor, 2003). In rodents, antagonists of the β-receptor reduced somatic signs of morphine withdrawal and aversion to a withdrawal-paired environment (Harris & Aston-Jones, 1993a, 1993b), as well as anxiety evoked by chronic exposure to cocaine or morphine (Harris & Aston-Jones, 1993a, 1993b; Wenzel et al., 2014). Clonidine also attenuated acoustic startle response induced by extended access and withdrawal from heroin in rats (Park et al., 2013). However, prazosin, and propranolol were ineffective at reducing mechanical hypersensitivity induced by extended access to heroin in rats (Park et al., 2015). As a result, there may be a role for NE in the affective and somatic signs of withdrawal across drugs of abuse.

Norepinephrine signaling has also been implicated in relapse, particularly in response to stress. Agonists of the α_2-receptor reduced stress-induced reinstatement for cocaine (Erb et al., 2000), heroin (Shaham, Highfield, Delfs, Leung, & Stewart, 2000), a heroin-cocaine mixture (Highfield, Yap, Grimm, Shalev, & Shaham, 2001), and alcohol (Le et al., 2011; Le, Harding, Juzytsch, Funk, & Shaham, 2005) in rats. Atomoxetine, a NE uptake inhibitor, reduced cue-induced reinstatement of heroin and cocaine seeking after abstinence in rats (Economidou, Dalley, & Everitt, 2011). Antagonists of both β_1- and β_2-adrenoceptors blocked stress-induced reinstatement of cocaine (Leri, Flores, Rodaros, & Stewart, 2002) and an α_1-adrenoceptor blocker and an α_{2A}-adrenoceptor activator stress-induced alcohol seeking (Le et al., 2011) in rats, as well as stress- and cue-induced craving for cocaine in human patients (Fox, Morgan, & Sinha, 2014). In humans, lofexidine reduced both physiological and subjective responses to stress and drug cues in opioid-dependent individuals (Sinha, Kimmerling, Doebrick, & Kosten, 2007). Norepinephrine signaling in response to stress may be sex-specific, as men with co-morbid alcohol dependence and CUD have blunted NE plasma levels when viewing drug cue imagery with increased feelings of anxiety and craving, whereas women have elevated NE levels with similar self-reported emotions (Fox et al., 2009). These discrepancies may result from sex differences in stress system function and plasticity in dependence and chronic polydrug use.

Collectively, there is a role for NE in mediating stress in addiction. In particular, therapies targeting the NE system can attenuate the stress response associated with withdrawal and relapse in animal models and humans. Further investigation is also necessary to determine whether the NE system impacts relapse after extended abstinence.

3.4 Dopamine

The effects of individual drugs on the DA system have been extensively investigated (Hauser et al., 2012) (Fig. 3). Virtually all drugs of abuse increase DA (Di Chiara & Imperato, 1988; Solinas, Belujon, Fernagut, Jaber, & Thiriet, 2019). Although the precise mechanisms of alcohol-induced DA release are unknown, studies have reported that alcohol may stimulate glycine receptors to inhibit GABA receptors in the NAc and the ventral tegmental area (VTA; Soderpalm, Lido, & Ericson, 2017), which in turn stimulates DA release. Alcohol also activates serotonin 5HT-3 receptors to excite dopaminergic neurons (Wang et al., 2019) and similarly induce DA release. Nicotine causes DA release in the NAc by activating nicotinic receptors in the VTA. Cocaine inhibits DA transporters (DATs) and prevents DA reuptake from the synaptic cleft, which leads to increased DA signaling in the NAc (McElvain & Schenk, 1992; Missale et al., 1985). Methamphetamine and amphetamines increase extracellular DA levels in the brain (Anglin, Burke, Perrochet, Stamper, & Dawud-Noursi, 2000;

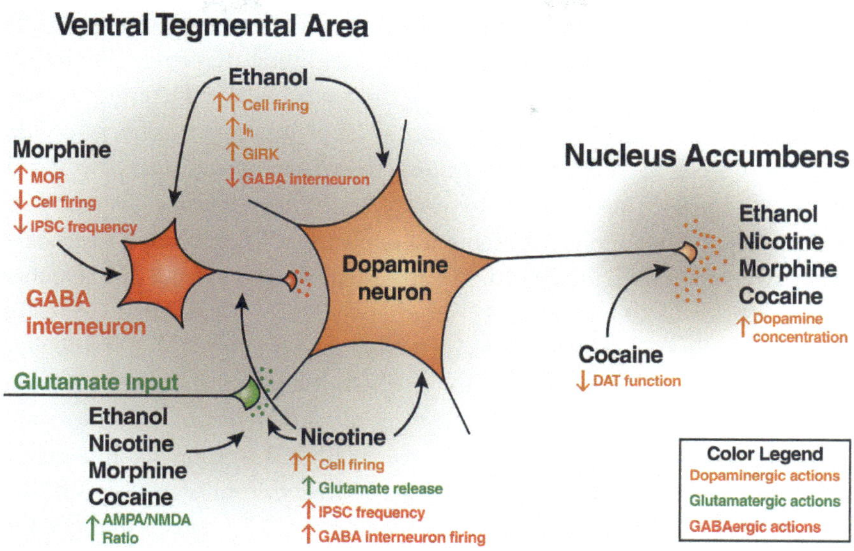

Fig. 3 The converging effects of different drugs of abuse on the DA system in the VTA and NAc. *From Juarez, B., & Han, M. H. (2016). Diversity of dopaminergic neural circuits in response to drug exposure.* Neuropsychopharmacology, *41(10), 2424–2446, with permission.*

Gibson, Leamon, & Flynn, 2002; Rothman et al., 2001) via direct binding to the DAT, causing DA release through the reversal of DAT and accumulation of DA into the synapse (Sulzer, Sonders, Poulsen, & Galli, 2005; Vergo, Johansen, Leist, & Lotharius, 2007; Volz, Hanson, & Fleckenstein, 2007). Mechanisms of opioid-induced increases in DA include activation of μ-opioid receptors expressed on VTA neurons, as well as disinhibition of VTA DA neurons via μ-opioid receptor action in GABAergic interneurons (Juarez & Han, 2016; Tanda, Pontieri, & Di Chiara, 1997). There are several mechanisms implicated in cannabis-mediated increases in DA (Bloomfield, Ashok, Volkow, & Howes, 2016). Δ^9-Tetrahydrocannabinol (THC), the primary psychoactive of cannabis, activates cannabinoid receptors that suppress GABAergic inhibition of VTA dopaminergic neurons (Bloomfield et al., 2016). The μ-opioid receptor also contributes to THC-mediated DA release (Tanda et al., 1997).

Nicotine and alcohol are the most popular combination of drugs among polydrug users. Behavioral experiments using animal models have highlighted that rodents are more highly motivated to obtain both substances than each separately. In studies using an oral alcohol-nicotine self-administration procedure, alcohol-preferring rats preferred the combination of alcohol and nicotine over either of the two substances alone (Hauser et al., 2012; Juarez & Han, 2016). The literature provides evidence for interaction effects of alcohol and nicotine on the DAergic system. However, the impact of nicotine and alcohol interaction effects on the DAergic system varies depending upon the order of exposure, timing of exposure, and the method of exposure to both substances (Van Skike et al., 2016). For example, simultaneous exposure to both nicotine and alcohol caused an additive increase in DA release measured in the NAc (Deehan Jr. et al., 2015; Doyon et al., 2013; Leao et al., 2015; Tizabi, Bai, Copeland Jr., & Taylor, 2007; Tizabi, Copeland Jr., Louis, & Taylor, 2002). However, an acute nicotine injection 3 h prior to alcohol self-administration blunted elevated DA release in the NAc during alcohol self-administration and decreased VTA DA neuron firing (Doyon et al., 2013). One possible mechanism for the neurobiological interactions of these two substances is outlined in Fig. 3. Nicotine activates nicotinic receptors on DA neurons in the VTA, causing DA release in the NAc. Alcohol acts indirectly on VTA and NAc DA neurons by potentiating $GABA_A$ receptor function (Weiner, Zhang, & Carlen, 1994) that inhibits GABAergic interneurons in the substantia nigra reticulata (Mereu & Gessa, 1985). This leads to increased DA release in the NAc (Grace & Bunney, 1985). As nicotine and alcohol increase NAc DA release through distinct processes, the combined administration of the two drugs may further augment NAc DA release and heighten the reinforcing effects of the two drugs when compared to each drug separately. Importantly, the initial intense activation of the DA system by drugs of abuse leads to DA hypofunction, as observed in alcohol- and drug-dependent states, and contributes to the negative emotional state that characterizes substance use disorders (Koob & Volkow, 2016).

The concurrent use of alcohol with psychostimulants is very common; alcohol and cocaine co-abuse is frequent (up to 75%) among individuals with CUD (Barrett, Darredeau, & Pihl, 2006). A few studies have explored how exposure to both cocaine and alcohol affects the reinforcing efficacy of each drug, and the involvement of the dopaminergic system in the interaction of alcohol and cocaine. Rats previously exposed to cocaine via passive intravenous infusions consumed more when allowed to self-administer alcohol the next day than controls that were not exposed to cocaine (Katner et al., 2011). Alcohol pre-exposure using intermittent access to 20% alcohol or alcohol vapor for 7 weeks in rats did not have a significant impact on cocaine self-administration, extinction responding, or reinstatement (Fredriksson et al., 2015). Alcohol and cocaine co-exposure increased DA levels in the NAc (Bardo, 1998; Koob & Bloom, 1988; Koob, Sanna, & Bloom, 1998; Wise & Rompre, 1989) significantly more than either drug separately (Lindholm, Rosin, Dahlin, Georgieva, & Franck, 2001).

Human studies suggest that the co-use of methamphetamine and alcohol has a synergistic effect on intoxication symptoms like euphoria and elevated heart rate (Brecht, Huang, Evans, & Hser, 2008; Bujarski et al., 2014; Kirkpatrick, Gunderson, Levin, Foltin, & Hart, 2012; Martin, Lampinen, & McGhee, 2006; O'Grady, Arria, Fitzelle, & Wish, 2008). However, a study by Winkler et al. showed that methamphetamine self-administration decreased alcohol consumption (Winkler, Greager, Stafford, & Bachtell, 2018), an effect the authors hypothesize may be due to lingering increases in extracellular DA by methamphetamine that blunted the reinforcing properties of alcohol (Winkler et al., 2018).

More than 20% of people who abuse opioids also drink alcohol daily (Peacock et al., 2016), and opioids have been shown to increase alcohol drinking (Corrigan et al., 2015; Corrigan & Hutchinson, 2012; Reid, Delconte, Nichols, Bilsky, & Hubbell, 1991). Rats pretreated with a small dose (2.0 mg/kg) of morphine 30 min prior to self-administration consumed significantly more alcohol than controls (Reid et al., 1991). Studies that investigated the association of alcohol and opioids showed that alcohol administration increases opioidergic neurotransmission, and its ability to increase DA release is may be dependent on the endogenous opioid system (Gianoulakis, 2001; Herz, 1997; Singh, 2019). Notably, opioid receptor antagonism with naloxone (Marfaing-Jallat, Miceli, & Le Magnen, 1983; Weiss, Mitchiner, Bloom, & Koob, 1990), naltrexone (Altshuler, Phillips, & Feinhandler, 1980; Volpicelli, Davis, & Olgin, 1986), and nalmefene (Quelch et al., 2017) decreased alcohol consumption in rodents, and naltrexone is Food and Drug Administration-approved to treat AUD. However, the combined effect of alcohol and opioids on the mesolimbic DA system remains unclear.

3.5 Opioid peptides

The endogenous opioid system plays a major role in opioid, cocaine, and alcohol dependence (Kreek, 1996; Kreek, LaForge, & Butelman, 2002; Xuei et al., 2008). This system comprises the endogenous opioid peptides endorphin, enkephalin, and dynorphin that are generated from precursor proteins encoded by three different genes (Drews & Zimmer, 2010; Kieffer, Befort, Gaveriaux-Ruff, & Hirth, 1992; Kieffer & Evans, 2009).

They all share the common N-terminal sequence Tyr-Gly-Gly-Phe. [Met] enkephalins and [Leu]enkephalins are derived from preproenkephalin (PENK). [Leu]enkephalin and dynorphins derive from preprodynorphin (PDYN). β-Endorphin and the non-opioid hormones adrenocorticotrophic hormone (ACTH), γ-melanocyte stimulating hormone (MSH), and β-lipotropic hormone derive from proopiomelanocortin (POMC). Opioid peptides activate G protein-coupled μ-, δ-, and κ-opioid receptors with different affinities and response profiles. The μ-opioid receptor is activated by β-endorphin and enkephalin, whereas δ-opioid receptors are activated by enkephalin and to a lesser extent by β-endorphin. Dynorphins bind specifically to the κ-opioid receptor (Akil et al., 1984; Kieffer, 1995). Protracted abstinence from morphine and cocaine modulated mRNA expression of PDYN, PENK and OPRD1 (δ-opioid receptor) in opposite directions, in that morphine reduced and cocaine increased gene expression in the extended amygdala (Becker, Kieffer, & Le Merrer, 2017). Alcohol, cocaine and alcohol + cocaine increased the expression of κ-opioid receptor mRNA in the VTA and NAc (Rosin, Lindholm, Franck, & Georgieva, 1999). Alcohol and nicotine increased the expression of enk and cFos in PVN, CeA, and NAc of male rats after 5 days of treatment (Chang, Karatayev, Barson, Liang, & Leibowitz, 2014). Endogenous opioids and opioid receptors are widely expressed throughout the brain, including brain regions involved in drug reward and addiction, such as the extended amygdala, VTA, and NAc (Minami & Satoh, 1995) (Fig. 4), and have been implicated in the hedonic responses of rewarding and aversive stimuli (Kieffer & Gaveriaux-Ruff, 2002; Pan, 1998).

Most studies on brain opioid systems have compared alcohol and drug (opioids, cannabinoids, nicotine, stimulants) independently. There are very few studies that investigate the role of the opioid system in polydrug abuse. Activation of neural circuits that express the μ-opioid receptor is necessary for many, if not all, rewarding drugs to exert their hedonic effect (Berrendero, Kieffer, & Maldonado, 2002; Contarino et al., 2002; Ghozland et al., 2002; Hall, Goeb, Li, Sora, & Uhl, 2004; Zimmer et al., 2001). Cocaine, alcohol, and amphetamine, but not nicotine, caused an increase in endorphin release in the NAc (Olive, Koenig, Nannini, & Hodge, 2001). Naltrexone reduced self-administration of alcohol and D-amphetamine in rhesus monkeys (Jimenez-Gomez, Winger, Dean, Deaver, & Woods, 2011). A low dose of morphine increased alcohol intake, whereas a high dose decreased alcohol intake in male rhesus monkeys. Naltrexone caused a rightward-shift in the morphine doses that suppress

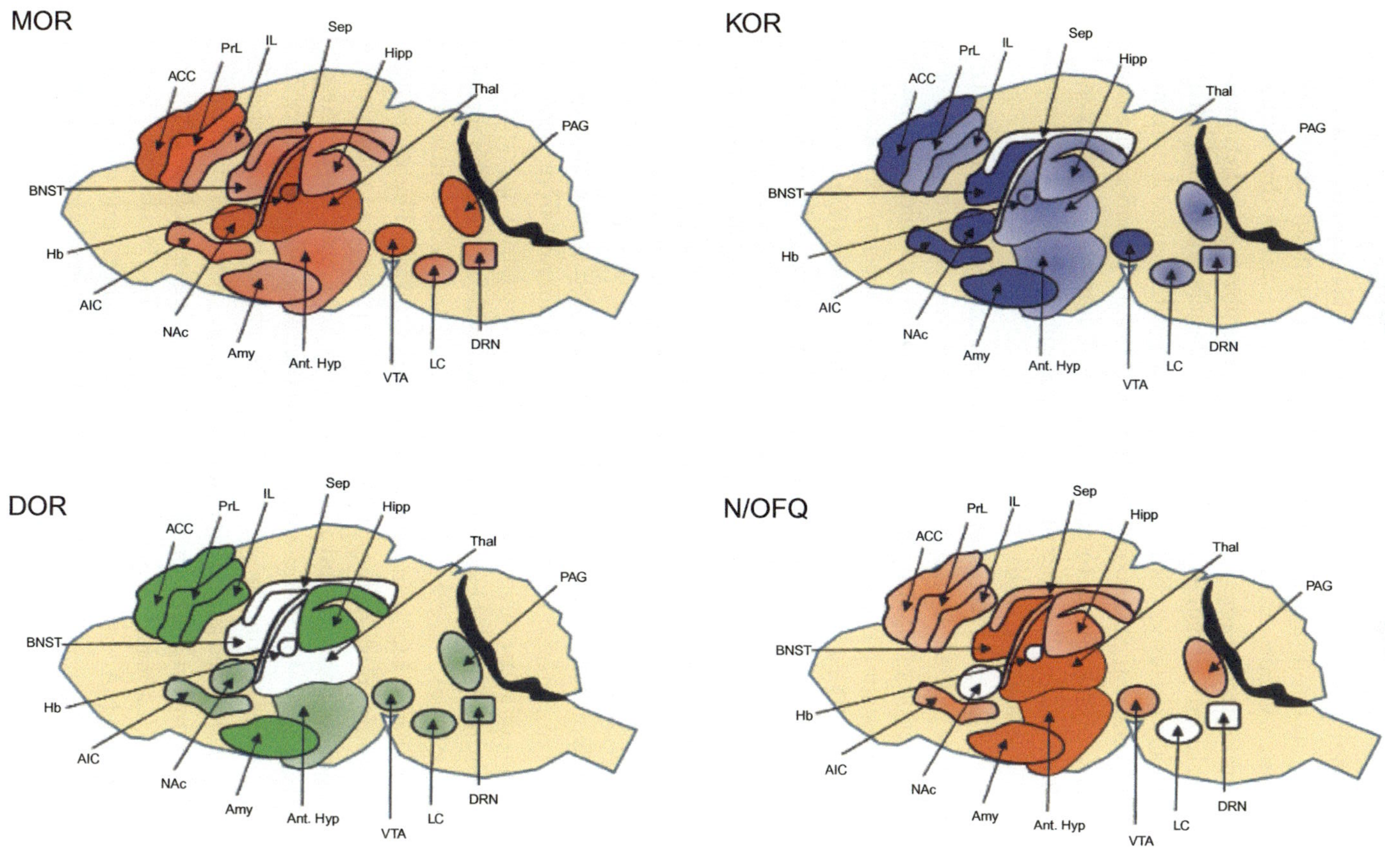

Fig. 4 Expression of μ-opioid receptor (MOR), κ-opioid receptor (KOR), δ-opioid receptor (DOR), and nociception/orphanin FQ receptor (N/OFQ) in brain regions involved in affective states. *From Browne, C. A., & Lucki, I. (2019). Targeting opioid dysregulation in depression for the development of novel therapeutics.* Pharmacology & Therapeutics, 201, *51–76, with permission.*

alcohol intake (Williams, Kane, & Woods, 2001). Combined sub-effective doses of morphine and alcohol produced conditioned place preference that was blocked by naloxone and naltrindole (δ-opioid receptor antagonist; Ise, Mori, Katayama, Nagase, & Suzuki, 2013). However, studies that trained rats to self-administer cocaine or heroin in different phases and then measured the escalation of intake of each drug or the choice of each drug (i.e., in a mutually exclusive choice with a non-drug alternative) found that individual differences in either cocaine or heroin self-administration or escalation or choice do not appear to predict individual differences in the same behavior maintained by the other drug (Lenoir, Guillem, Koob, & Ahmed, 2012; Tunstall, Riley, & Kearns, 2014).

The clinical efficacy of naltrexone for the treatment of AUD and OUD indicates the involvement of the opioid system, in particular the μ-opioid receptor, in the etiology of both of these two disorders. μ-Opioid receptor availability in mesolimbic regions is associated with severity of nicotine dependence in AUD patients going through detoxification (Weerts et al., 2014). Chronic alcohol exposure impairs μ-opioid receptor endocytosis that could contribute to tolerance to opioids (He & Whistler, 2011; Le, Khanna, Kalant, & LeBlanc, 1980). Also, there are three single nucleotide polymorphisms of PENK associated with OUD and two single nucleotide polymorphisms of PMOC associated with polydrug dependence (including alcohol) and OUD (Xuei et al., 2008).

There is a significant literature suggesting that opioid systems also may play a role in the negative reinforcing effects of withdrawal from alcohol and other drugs of abuse, laying the foundation for polydrug interactions. A history of chronic alcohol exposure is associated with an increased sensitivity of NAc κ-opioid receptors in mice, rats, and cynomolgus macaques (Siciliano, Karkhanis, Holleran, Melchior, & Jones, 2018). The κ-opioid receptor antagonist norbinaltorphimine reduced alcohol vapor-induced increases in alcohol consumption and preference in mice, and the κ-opioid receptor agonist U50–488 increased alcohol consumption and preference in alcohol-naïve mice (Rose et al., 2016). Norbinaltorphimine increased responding for combined alcohol and nicotine in male, but not female, adult rats, and did not impact alcohol and nicotine responding in adolescent male or female rats (Larraga, Belluzzi, & Leslie, 2017). Dynorphin was upregulated in the NAc of rats dependent on methamphetamine or heroin, and treatment with norbinaltorphimine blocked the escalation of both methamphetamine and heroin self-administration in male rats (Schlosburg et al., 2013; Whitfield Jr. et al., 2015). Dynorphin and κ-opioid receptors were found to be involved in the dysphoria induced by cannabinoids (Ghozland et al., 2002; Zimmer et al., 2001). One single-nucleotide polymorphism of PDYN has associated with AUD and negative craving, or craving that results from the emergence of negative affect (Karpyak et al., 2013). Bremazocine (κ-opioid receptor agonist, μ-opioid receptor antagonist, δ-opioid receptor agonist) dose-dependently reduced self-administration of alcohol, cocaine, phencyclidine, saccharin, and food in rhesus monkeys (Cosgrove & Carroll, 2002).

3.6 Glucocorticoids

Glucocorticoids are steroid hormones that mediate the response to stress. Glucocorticoids are synthesized in the adrenal glands and are released following activation of the HPA axis by a threat to homeostasis. Glucocorticoids such as cortisol and corticosterone (the predominant glucocorticoid in humans and rats, respectively) bind glucocorticoid receptors (GRs) throughout the periphery to regulate physiological responses to stress. Glucocorticoid receptors are also ubiquitously expressed throughout the brain and mediate the cognitive and affective components of the stress response. Centrally expressed GRs also regulate the HPA axis activity and physiological responses via negative feedback.

There is a vast literature of studies in humans that indicate that the HPA axis is modulated by drugs of abuse and is significantly perturbed by chronic drug use including nicotine, cannabis, psychostimulants, alcohol, and opioids (Wemm & Sinha, 2019). However, limited research has been conducted into the role of central glucocorticoid signaling in drug dependence. The existing research described below indicates that central glucocorticoid activity contributes to the reinforcing properties of alcohol and psychostimulants and is involved in the withdrawal-negative affect stage of the addiction cycle.

In animal studies, a functioning corticosteroid system has been implicated in the reinforcing properties of alcohol. In non-dependent alcohol drinking rats, adrenalectomy decreased alcohol consumption, whereas systemic and intracerebroventricular corticosterone administration restored alcohol consumption (Fahlke, Engel, Eriksson, Hard, & Soderpalm, 1994; Fahlke, Hard, & Hansen, 1996). More recent studies have reported that the glucocorticoid system has a significant role in the withdrawal/negative affect state of alcohol dependence. Repeated activation of the HPA axis by chronic heavy alcohol consumption is hypothesized to drive neuroadaptations in brain glucocorticoid signaling that result in the over-engagement or prolonged activation of this system. Notably, chronic activation of the HPA axis downregulates CRF in the PVN and upregulates CRF in the CeA, and these divergent effects are hypothesized to drive drug intake (Edwards, Little, Richardson, & Vendruscolo, 2015; Vendruscolo & Koob, 2020). Chronic systemic GR antagonism with mifepristone prevented the escalation of alcohol self-administration and the motivation in rats undergoing alcohol vapor exposure to induce alcohol dependence (Vendruscolo et al., 2012). Dysregulated GR activity within the CeA was found to be a significant mechanism underlying alcohol consumption in the alcohol-dependent state. Enhanced GR activity was observed in the CeA of alcohol-dependent rats, as indicated by an increase in GR phosphorylation, and intra-CeA antagonism of GRs with mifepristone reduced alcohol self-administration in alcohol-dependent rats (Vendruscolo et al., 2015). Further, a transcriptional network analysis identified GR as one of the top transcriptional regulators contributing to the altered gene expression profile of the CeA in the alcohol-dependent state (Repunte-Canonigo et al., 2015). Systemic and intra-CeA GR antagonism with mifepristone inhibited yohimbine stress-induced reinstatement of alcohol seeking in rats (Simms, Haass-Koffler, Bito-Onon, Li, & Bartlett, 2012). Mifepristone administration into the ventral striatum decreases alcohol self-administration in alcohol-dependent rats in protracted abstinence (Repunte-Canonigo et al., 2015). Overall, these data indicate that glucocorticoid activity contributes to the maintenance of the alcohol-dependent state.

Similarly, reduction of glucocorticoid function decreases the reinforcing effects of psychostimulants. A reduction of glucocorticoid release by adrenalectomy reduced cocaine self-administration in rats and produced a downward shift of the cocaine dose-response curve, and these effects were reversed by corticosterone supplementation (Deroche, Marinelli, Le Moal, & Piazza, 1997). Systemic administration of the GR antagonist mifepristone reduced the motivation of rats to self-administer cocaine (Deroche-Gamonet et al., 2003). In mice, whole-brain GR deletion and GR deletion in DA D1A receptor-expressing neurons produced a downward shift in the dose-response curve for intravenous cocaine self-administration (Ambroggi et al., 2009; Deroche-Gamonet et al., 2003). This may be attributed to reduced spontaneous activity of VTA DA neurons in these transgenic mice (Ambroggi et al., 2009). In rats trained to self-administer intravenous amphetamine, corticosterone pretreatment enhanced amphetamine self-administration in rats that exhibited low levels of responding for amphetamine, whereas it suppressed self-administration in rats that had high baseline responding levels for amphetamine (Piazza et al., 1991). The effect of intravenous corticosterone treatment on reinstatement of cocaine self-administration in rats exhibited a U-shaped dose-effect curve, such that the exogenous corticosterone doses that recapitulated stress-induced levels of circulating corticosterone enhanced reinstatement, while the dose producing supra-stress levels of circulating corticosterone suppressed reinstatement (Deroche et al., 1997). In line with this, the glucocorticoid synthesis inhibitor metyrapone reduced reinstatement of cocaine seeking (Piazza et al., 1994), and the glucocorticoid synthesis inhibitor ketoconazole suppressed footshock stress-induced reinstatement of cocaine seeking (Mantsch & Goeders, 1999). These data indicate that glucocorticoid signaling facilitates the reinforcing effects of psychostimulants and reinstatement of psychostimulant seeking. While to our knowledge, mifepristone has not been tested in polydrug studies, it may be effective in treating addiction given its role in ameliorating stress and preventing dependence across several drug classes.

3.7 Endocannabinoids

The endocannabinoid system comprises two receptors (CB1 and CB2) and endogenous ligands that are lipid-based, retrograde messengers, found ubiquitously throughout tissues of the body. The arachidonic acid derivatives *N*-arachidonoylethanolamine (AEA) and 2-arachidonoylglycerol (2-AG) are the most well-studied endocannabinoids to date (both are agonists of CB1 and CB2 receptors; Gray et al., 2015). In the context of neurotransmission, the post-synaptic release of endocannabinoids serves to counteract pre-synaptic release of neurotransmitters, thereby returning transmission to homeostasis (i.e., depolarization-induced suppression of inhibition and depolarization-induced suppression of excitation; Diana & Marty, 2004. This is important for example following stress challenges, and it has been hypothesized that hypofunction of the endocannabinoid system contributes to anxiety and PTSD (Hillard, 2018). It is clear that endocannabinoids are involved in acute alcohol reward and also that the endocannabinoid system is dysregulated in alcohol dependence (although the latter is complex; see, for detailed review, Pava & Woodward, 2012). Serrano et al. (2018) demonstrated in rats that during alcohol dependence, especially in withdrawal or in response to either restraint stress or 30 min of alcohol consumption, that there is a deficiency in CeA 2-AG levels compared to non-dependent rats

(Serrano et al., 2018). Furthermore, the inhibition of MAGL to prevent the degradation of 2-AG was able to normalize the anxiety-like behavior and increased drinking in alcohol-dependent rats, and this also worked in normalizing these behaviors in mice exposed to alcohol in a 2-bottle choice paradigm. It has been proposed that the endocannabinoid system, by virtue of its role in maintaining homeostasis in the face of stress, may be a feasible target for treating alcohol dependence (e.g., inhibitors of endocannabinoid degradation to promote CB1 availability or CB1 ligands that serve to normalize function in this system; Tunstall, Carmack, Koob, & Vendruscolo, 2017).

Some direct interactions of alcohol and nicotine have been tested with respect to endocannabinoids. Alcohol and nicotine have generally opposing behavioral effects (e.g., locomotor activation), and a history of nicotine or alcohol treatment causes tolerance (cross-tolerance in the case of nicotine) to alcohol. This is observed in decreased locomotor inhibition by alcohol and an increase in self-administration of alcohol and suggests decreased aversive and increased rewarding effects of alcohol following alcohol or nicotine history (Abburi et al., 2016; Lajtha & Sershen, 2010). Nicotine and alcohol have opposing effects on neurochemistry, and this includes endocannabinoid function (Lajtha & Sershen, 2010). Symmetrical cross-tolerance, such that exposure to either drug enhances tolerance to the other drug occurs in the case of alcohol and THC (Sprague & Craigmill, 1976). Additionally, THC exposure facilitates tolerance to nicotine (Biala & Budzynska, 2010; Sprague & Craigmill, 1976; Valjent, Mitchell, Besson, Caboche, & Maldonado, 2002), and the CB1 receptor has been implicated in the ability of alcohol to cross-sensitize and cross-reinstate nicotine sensitization and nicotine place preference (Biala & Budzynska, 2010). The endogenous ligands for cannabinoid receptors AEA and 2-AG are altered in animals chronically exposed to alcohol and nicotine. Chronic drug exposure caused drug-dependent and regionally-defined alterations in AEA and 2-AG. Notably, both nicotine and alcohol caused increased limbic forebrain AEA content (Gonzalez et al., 2002). Despite the role of endocannabinoids in alcohol reward, chronic exposure to alcohol does not markedly alter the expression of CB1 receptor mRNA or CB1 receptor binding sites in various regions of rat brain (Gonzalez, Fernandez-Ruiz, Sparpaglione, Parolaro, & Ramos, 2002). Slice electrophysiology experiments in the dorsolateral striatum determined that following a history of nicotine or alcohol administration, CB1 agonism no longer increased the paired-pulse ratio of evoked EPSCs as it did in slices from controls. This suggests a role for the endocannabinoid system in the mediation of tolerance and nicotine-to-alcohol cross-tolerance that was observed for alcohol's motor impairing effects and in self-administration of alcohol (Abburi et al., 2016). Dialysates taken from the NAc of rats self-administering alcohol demonstrated that the amount of alcohol self-administered correlated with the increases in 2-AG levels induced by alcohol self-administration (Caille, Alvarez-Jaimes, Polis, Stouffer, & Parsons, 2007). Compounds that alter the activity of the endocannabinoid system (e.g., CB1 receptor agonism with THC; Filev, Engelke, Da Silveira, Mello, & Santos-Junior, 2017; CB1 receptor antagonism with rimonabant; Biala & Budzynska, 2010; Marinho et al., 2015) inhibit expression of alcohol-induced locomotor sensitization. Rimonabant also inhibited alcohol prime of alcohol conditioned place preference reinstatement and locomotor cross-sensitization between nicotine and alcohol. In summary, the endocannabinoid system remains a target for elucidating the etiology of AUD and co-morbid abuse of alcohol and other substances including nicotine and marijuana.

3.8 Corticotropin-releasing factor

The CRF system is a key component of the stress response, as a modulator of physiological, behavioral, and emotional responses to stress. The CRF system is composed of the neuropeptide CRF and the CRF1 receptor, for which CRF has a high affinity (Bale & Vale, 2004). The CRF system also includes the structurally related urocortin peptides 1, 2, and 3, as well as the CRF2 receptor (Fekete & Zorrilla, 2007). Activity of the CRF system is recruited with chronic alcohol and drug use, hypothesized to be a component of a counterregulatory neurobiological process to oppose reward processes and reestablish emotional homeostasis (Koob, 2008). Elevations in brain CRF levels are observed during withdrawal from chronic exposure to alcohol (Funk, Marinelli, & Le, 2006; Merlo Pich et al., 1995; Olive, Koenig, Nannini, & Hodge, 2002), opioids (McNally & Akil, 2002), nicotine (Baiamonte et al., 2014; George et al., 2007), benzodiazepines (Skelton, Nemeroff, & Owens, 2004), and cocaine (Zhou, Spangler, Schlussman, Ho, & Kreek, 2003).

Although there is little evidence in the preclinical animal literature on changes in the CRF system in the acute rewarding effects of alcohol and other drugs, the CRF system contributes to the consumption of alcohol and drugs of abuse in the dependent state. The blockade of CRF system in drug-dependent rodents decreased self-administration of alcohol

(Funk et al., 2006; Funk & Koob, 2007; Valdez, Platt, Rowlett, Ruedi-Bettschen, & Spealman, 2007), opioids (Greenwell et al., 2009; Park et al., 2015), and cocaine (Specio et al., 2008). Specifically, in alcohol-dependent rats, alcohol self-administration is reduced by systemic administration of CRF1 receptor antagonists (Funk & Koob, 2007; Gilpin, Richardson, & Koob, 2008; Roberto et al., 2010; Sabino et al., 2006) and intracerebroventricular or intra-CeA administration of a non-specific CRF1R/CRF2R antagonist (Funk et al., 2006; Valdez et al., 2002). Systemic CRF1 receptor antagonism reduces heroin self-administration in heroin-dependent rats (Greenwell et al., 2009) and chronic systemic CRF1 receptor antagonism prevents escalation of heroin self-administration (Park et al., 2015). In cocaine-dependent rats, systemic CRF1 receptor antagonism reduces cocaine self-administration (Specio et al., 2008).

Blockade of the CRF system reduces aversive stimulus effects and negative affective behaviors in animal models of withdrawal from alcohol, opioids, nicotine, cocaine, and benzodiazepines. Increased anxiety-like behavior of alcohol-dependent rats in withdrawal is reduced by systemic or intra-CeA administration of a CRF1/CRF2 receptor peptide antagonist (Baldwin, Rassnick, Rivier, Koob, & Britton, 1991; Rassnick, Heinrichs, Britton, & Koob, 1993). Intracerebroventricular administration of a CRF1/CRF2 receptor antagonist decreased elevated brain reward thresholds (an index of reward system hypofunction) in withdrawn, alcohol-dependent rats (Bruijnzeel, Small, Pasek, & Yamada, 2010). Additionally, systemic CRF1 receptor antagonism reduced hyperalgesia in alcohol-dependent rats in withdrawal (Edwards et al., 2012). In nicotine-dependent rats, systemic CRF1 receptor antagonism attenuated spontaneous or mecamylamine-precipitated anxiety-like behavior during nicotine withdrawal (Cohen et al., 2015; George et al., 2007). Systemic or intra-CeA administration of a CRF1/CRF2 receptor antagonist or a selective CRF1 receptor antagonist blocked mecamylamine-precipitated nicotine withdrawal-induced elevated brain reward thresholds (Bruijnzeel et al., 2012; Bruijnzeel, Zislis, Wilson, & Gold, 2007; Marcinkiewcz et al., 2009). Systemic CRF1 receptor antagonism decreased hyperalgesia that emerges during withdrawal in nicotine-dependent rats (Baiamonte et al., 2014; Cohen et al., 2015). Systemic administration of a CRF1 receptor antagonist reduced naloxone-precipitated opioid withdrawal conditioned place aversion in rats (Stinus, Cador, Zorrilla, & Koob, 2005), and intra-CeA administration of a CRF1/CRF2 receptor peptide antagonist attenuated this behavior as well (Heinrichs, Menzaghi, Schulteis, Koob, & Stinus, 1995). In addition, precipitated opioid withdrawal conditioned place aversion is decreased in CRF1 receptor-deficient, opioid-dependent mice (Contarino et al., 2002; Garcia-Carmona, Almela, Baroja-Mazo, Milanes, & Laorden, 2012; Garcia-Carmona, Baroja-Mazo, Milanes, & Laorden, 2015). Intra-CeA administration of a CRF1/CRF2 receptor antagonist reduced naloxone-precipitated opioid withdrawal hyperalgesia (McNally & Akil, 2002), and systemic CRF1 receptor antagonism reduces hyperalgesia in spontaneously withdrawn, opioid-dependent rats (Edwards et al., 2012; Park et al., 2015). Increased anxiety-like behavior of benzodiazepine-dependent rats in flumazenil-precipitated withdrawal is attenuated by systemic CRF1 receptor antagonism (Skelton, Gutman, Thrivikraman, Nemeroff, & Owens, 2007). Cocaine withdrawal-associated anxiety-like behavior is reduced by intracerebroventricular administration of a CRF1/CRF2 receptor antagonist (Basso, Spina, Rivier, Vale, & Koob, 1999).

The CRF system is a key contributor to stress-induced relapse across several drug classes. Systemic and intracerebroventricular administration of CRF1/CRF2 receptor antagonists inhibited footshock stress-induced reinstatement of alcohol seeking (Le et al., 2000). Footshock stress-induced reinstatement of alcohol seeking was also attenuated by systemic CRF1 receptor antagonism (Gehlert et al., 2007; Le et al., 2000). Systemic CRF1 receptor antagonism inhibited reinstatement of alcohol seeking following yohimbine-induced stress (Marinelli et al., 2007) or κ-opioid receptor activation (Funk, Coen, & Le, 2014). Footshock stress-induced reinstatement of extinguished nicotine seeking was attenuated by intracerebroventricular administration of a CRF1/CRF2 receptor antagonist (Zislis, Desai, Prado, Shah, & Bruijnzeel, 2007) and systemic administration of a CRF1 receptor antagonist (Bruijnzeel, Prado, & Isaac, 2009). Footshock stress-induced reinstatement of cocaine seeking was reduced by intracerebroventricular administration of a CRF1/CRF2 receptor antagonist (Erb, Shaham, & Stewart, 1998) and systemic administration of a CRF1 receptor antagonist (Shaham, Erb, Leung, Buczek, & Stewart, 1998). Systemic CRF1 receptor antagonism attenuated footshock stress-induced reinstatement of opioid seeking (Shaham et al., 1998).

In summary, the CRF system is a key component of the drug-dependent state, shared among multiple classes of drugs. The CRF system is recruited in withdrawal from alcohol and other addictive drugs, and contributes to excessive alcohol and drug use, potentially by its mediation of the negative affective state of alcohol and drug withdrawal.

3.9 Hypocretin/orexin

Hypocretin 1 and 2 (also called orexin A and B) are hypothalamic neuropeptides recruited during physiological challenge to promote behaviors that are important for homeostasis (de Lecea et al., 1998; Sakurai et al., 1998). Neurons that contain hypocretin/orexin project throughout the brain and act on the G protein-coupled hypocretin 1- or 2-receptors to regulate arousal and motivated behaviors. Although the hypocretin-1 receptor has greater affinity for hypocretin 1, the hypocretin-2 receptor binds to both hypocretin 1 and 2 (Sakurai et al., 1998). Hypocretin/orexin modulates several systems involved in arousal and stress, including the HPA axis and noradrenergic, serotonergic, and dopaminergic systems (Nevárez & de Lecea, 2018). Recently, the dual hypocretin receptor antagonist suvorexant was approved by the US Food and Drug Administration to treat insomnia and may have therapeutic potential for treating substance use disorders (Campbell, Marchant, & Lawrence, 2020; James et al., 2020). Again, although few studies thus far have examined the impact of hypocretin in polydrug use, there are several consistencies in its role in addiction across drug classes (James, Mahler, Moorman, & Aston-Jones, 2017; Matzeu & Martin-Fardon, 2020a, 2020b; Walker & Lawrence, 2017).

Hypocretin/orexin neurons are preferentially involved in behaviors that require a high degree of motivation in animal models of addiction. Expression of hypocretin/orexin or its receptors is upregulated in rodents highly motivated to obtain alcohol (Lawrence, Cowen, Yang, Chen, & Oldfield, 2006), opioids (Fragale, James, & Aston-Jones, 2020; Schmeichel et al., 2015; Thannickal et al., 2018), and psychostimulants (James et al., 2019; Pantazis, James, Bentzley, & Aston-Jones, 2020). Similarly, blocking the hypocretin-1 receptor attenuated motivated operant responding in rodents for stimulants (Bentzley & Aston-Jones, 2015; Borgland et al., 2009; Hollander, Pham, Fowler, & Kenny, 2012), opioids (Fragale, Pantazis, James, & Aston-Jones, 2019; Porter-Stransky, Bentzley, & Aston-Jones, 2017; Smith & Aston-Jones, 2012), and alcohol (Lawrence et al., 2006). However, there are discrepancies in the role of hypocretin/orexin in low-effort responding across drug classes. Blockade of hypocretin/orexin-1 receptor signaling did not impact low-effort responding for cocaine (Espana et al., 2010; Espana, Melchior, Roberts, & Jones, 2011) or fentanyl (Fragale et al., 2019) but did for remifentanil (Porter-Stransky et al., 2017), heroin (Smith & Aston-Jones, 2012), oxycodone (Matzeu & Martin-Fardon, 2020a, 2020b), and alcohol

(Lopez, Moorman, Aston-Jones, & Becker, 2016; Moorman, James, Kilroy, & Aston-Jones, 2017). There are also reported differences in the role of the hypocretin-2 receptor in drug taking. Antagonism of the hypocretin-2 receptor attenuated the consumption of heroin (Schmeichel et al., 2015) and alcohol (Anderson, Becker, Adams, Jesudason, & Rorick-Kehn, 2014; Brown, Khoo, & Lawrence, 2013; Shoblock et al., 2011; though see: Lei, Wegner, Yu, & Hopf, 2016) but not cocaine (Boutrel et al., 2005; Prince, Rau, Yorgason, & Espana, 2015) or oxycodone (Matzeu & Martin-Fardon, 2020a, 2020b).

Hypocretin/orexin neurons are recruited in response to other external stimuli, including stress/withdrawal and drug-associated cues or contexts. Hypocretin expression or activity was upregulated during precipitated and acute but not chronic morphine withdrawal (Georgescu et al., 2003; Sharf, Sarhan, & Dileone, 2008) or naloxone-precipitated cocaine withdrawal (Zhou et al., 2008). Blocking or knocking out hypocretin signaling also attenuated somatic withdrawal symptoms for morphine precipitated by naltrexone (Georgescu et al., 2003) or naloxone (Sharf et al., 2008) and nicotine precipitated by mecamylamine (Plaza-Zabala, Flores, Maldonado, & Berrendero, 2012). Hypocretin-1 receptor antagonism attenuates stress-induced reinstatement of cocaine (Boutrel et al., 2005; Schmeichel, Herman, Roberto, & Koob, 2017) and alcohol (Richards et al., 2008) seeking, as well as cue- or context-induced reinstatement of stimulant (Martin-Fardon & Weiss, 2014a, 2014b; Smith, See, & Aston-Jones, 2009; Smith, Tahsili-Fahadan, & Aston-Jones, 2010), opioid (Fragale et al., 2019; Matzeu & Martin-Fardon, 2020a, 2020b; Smith & Aston-Jones, 2012), and alcohol (Brown et al., 2016; Lawrence et al., 2006; Martin-Fardon & Weiss, 2014a, 2014b) seeking. The hypocretin-2 receptor has been also been shown to be involved in drug-seeking behaviors (Brown et al., 2013; Matzeu & Martin-Fardon, 2020a, 2020b; Smith et al., 2009; though see: Matzeu, Kerr, Weiss, & Martin-Fardon, 2016; Uslaner et al., 2014).

Collectively, these studies point to a role for hypocretin at multiple stages of the addiction cycle. In particular, hypocretin is preferentially involved when motivation for drug is high, such as when drug is difficult to obtain and/or with exposure to stimuli that evoke drug seeking. Hypocretin signaling to areas associated with motivation and arousal may stimulate drug seeking and taking behaviors to cope with stress associated with withdrawal. Importantly, the dual hypocretin receptor antagonist suvorexant has the potential to be repurposed for the treatment of substance use disorder in humans.

3.10 Neuroimmune signaling

Alcohol and other addictive drugs have a substantial impact on the neuroimmune system (Cui, Shurtleff, & Harris, 2014). Chronic exposure to drugs intensifies neuroimmune signaling and is associated with a concomitant deterioration of cognitive function. In addition, studies have reported the contribution of the innate immune response in the enhancement of the mesolimbic dopaminergic system during binge/intoxication, withdrawal, and craving (Coller & Hutchinson, 2012; Mayfield, Ferguson, & Harris, 2013). Again, there is increasing evidence for a role of neuroimmune signaling in mediating aspects of the addiction cycle for individual drugs but little study to date of drug-drug interactions.

Alcohol misuse induces a sustained neuroinflammatory state (He & Crews, 2008). Specifically, alcohol misuse increases the expression of neuroimmune genes and microglial-related transcription (Crews et al., 2015; Mayfield et al., 2013; Robinson et al., 2014) and results in higher levels of cytokines (Achur, Freeman, & Vrana, 2010). Microglia are activated by alcohol, as evidenced by increased microglial markers (IBA1 or CD11B) in mouse and in human AUD postmortem brains (Alfonso-Loeches, Pascual-Lucas, Blanco, Sanchez-Vera, & Guerri, 2010; Barton, Baker, & Leasure, 2017; He & Crews, 2008; Qin & Crews, 2012; Rubio-Araiz et al., 2017; Walter & Crews, 2017; Zhao et al., 2013). Alcohol-induced microglial activity is regulated by several immune receptor families such as Toll-like receptors (TLR; Alfonso-Loeches et al., 2010; Fernandez-Lizarbe, Pascual, & Guerri, 2009), and it has been shown that alcohol induces the release of inflammatory mediators and increases TLR4 and TLR2 in microglial cells (Fernandez-Lizarbe et al., 2009). Neuroinflammation has been shown to enhance alcohol and drug-seeking behavior (Mayfield et al., 2013) and motivate alcohol drinking in mice (Blednov et al., 2011), whereas deletion of neuroimmune genes, such as B2m, Ctss, Il1rn, Cd14, and Il6, reduced alcohol consumption in mice (Blednov et al., 2012). However, genetic and pharmacological manipulation of TLR-4 had very little effect on alcohol drinking in rodent models (Harris et al., 2017; Warden, DaCosta, et al., 2020; Warden, Wolfe, et al., 2020).

Astrocytes are also altered by alcohol-induced neuroinflammation (Fig. 5). A growing number of studies have demonstrated the impact of alcohol on astrocytes and vice-versa (Adermark & Bowers, 2016; Ayers-Ringler, Jia, Qiu, & Choi, 2016). In the superior frontal gyrus and primary motor cortex of postmortem brains of humans with AUD, glial fibrillary acidic

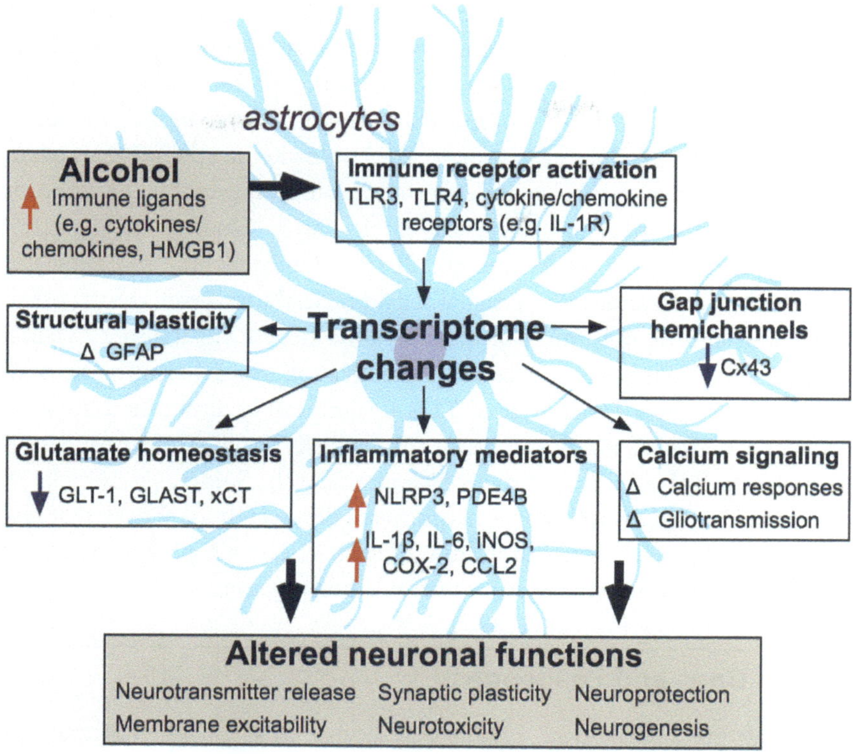

Fig. 5 Effect of alcohol-induced neuroinflammation on astrocytes. From Erickson, E. K., Grantham, E. K., Warden, A. S., & Harris, R. A. (2019). Neuroimmune signaling in alcohol use disorder. *Pharmacology, Biochemistry, and Behavior,* 34–60, with permission.

protein (GFAP) expression was significantly reduced, suggesting atrophy of astrocytes following alcohol-associated proinflammatory activity (Kim, Healey, Sepulveda-Orengo, & Reissner, 2018; Lewohl, Wixey, Harper, & Dodd, 2005). Similar results have been shown in the rat hippocampus, striatum and cortex following 6 weeks of alcohol drinking (Evrard et al., 2006).

Other drugs are known to impact the neuroimmune system (Hofford, Russo, & Kiraly, 2019; Hutchinson et al., 2008; Loftis & Janowsky, 2014; Namba et al., 2019), suggesting a potential involvement of inflammation in polydrug use. Opioids trigger neuroinflammation through TLR4 (Hutchinson et al., 2010; Wang et al., 2012) and TLR2 (Li et al., 2009, 2010; Zhang et al., 2011) activation. When combined with alcohol, they act synergistically to induce sedation through TLR2 activation (Corrigan et al., 2015). Nicotine modulates the neuroimmune system via a unique pathway: α_7-nicotinic acetylcholine receptors (nAChRs; Shytle et al., 2004). Studies have also shown that nicotine consumption disrupts the blood-brain barrier (Hawkins et al., 2004) and activates the protein nuclear factor-κ B (NF-κB; Barr et al., 2007). Cocaine may engage TLR4 to induce an inflammatory response (Kashima & Grueter, 2017; Northcutt et al., 2015). Many clinical studies reported an acute effect of cocaine on peripheral expression of inflammatory mediators (Calipari et al., 2018; Hofford et al., 2019; Moreira et al., 2016; Pedraz et al., 2015). Research also indicates that methamphetamine induces microglial activation in the striatum, rhinal cortex, subiculum, and hippocampus of rats in a dose-dependent manner (Thanos et al., 2016). The inhibition of microglial activation using the antibiotic minocycline or ibudilast (antagonist of TLR4) significantly diminished the reinforcing properties of psychostimulants (Attarzadeh-Yazdi, Arezoomandan, & Haghparast, 2014; Chen, Uz, & Manev, 2009; Snider, Hendrick, & Beardsley, 2013). Given that an emerging area of research has shown that several drugs are proinflammatory, the adverse effects of polydrug use on sustained neuroimmune activation warrants investigation.

4. Gaps and challenges

The epidemiological evidence indicates that the co-morbidity of AUD and other substance use disorders is more a rule than an exception. However, the vast majority of preclinical studies on alcohol and drug abuse investigates the neurobiological changes caused by a single drug in isolation. Given that we are still working to understand the effects of individual drugs, it is challenging to study all possible interactions in polydrug use. However, future research on polydrug use will be important for developing effective treatments for addiction. The literature reviewed above indicates that there are adaptations in several neurotransmitter systems such as dopamine, opioid, cannabinoid, CRF, hypocretin, glucocorticoid, and norepinephrine, as well as the engagement of neuroimmune systems that appear to be shared by multiple drugs. The combination of alcohol with other drugs of abuse appears, in limited examples, to have additive effects on these systems. Whether there are synergistic neuroadaptations caused by co-use of drugs or quantitative and qualitative differences caused by a drug alone versus in combination with alcohol is currently unknown. Moreover, the majority of preclinical studies have been conducted in male subjects, although the prevalence of AUD co-morbid with other substance use disorders in females is also very high. Because the treatment of drug addiction co-morbidities appears to be more difficult to treat than addiction to a single drug, more studies are necessary to understand the neurobiology of polydrug use, misuse, and polysubstance use disorders for the identification of challenges facing their prevention and treatment.

Acknowledgments

The authors thank Michael Arends for assistance with manuscript preparation and editing. The National Institute on Drug Abuse Intramural Research Program supported this work.

References

Abburi, C., Wolfman, S. L., Metz, R. A., Kamber, R., McGehee, D. S., & McDaid, J. (2016). Tolerance to ethanol or nicotine results in increased ethanol self-administration and long-term depression in the dorsolateral striatum. *eNeuro*, *3*(4). https://doi.org/10.1523/ENEURO.0112-15.2016.

Achur, R. N., Freeman, W. M., & Vrana, K. E. (2010). Circulating cytokines as biomarkers of alcohol abuse and alcoholism. *Journal of Neuroimmune Pharmacology*, *5*(1), 83–91.

Adams, S. (2017). Psychopharmacology of tobacco and alcohol comorbidity: A review of current evidence. *Current Addiction Reports*, *4*(1), 25–34.

Adermark, L., & Bowers, M. S. (2016). Disentangling the role of astrocytes in alcohol use disorder. *Alcoholism, Clinical and Experimental Research*, *40*(9), 1802–1816.

Adolescent Brain Cognitive Development (ABCD) Study. (2020). Collaborative Research on Addiction at NIH. Retrieved September 8, 2020, from https://www.addictionresearch.nih.gov/abcd-study.

Akil, H., Watson, S. J., Young, E., Lewis, M. E., Khachaturian, H., & Walker, J. M. (1984). Endogenous opioids: Biology and function. *Annual Review of Neuroscience*, *7*, 223–255.

Albertson, T. E., Chenoweth, J., Ford, J., Owen, K., & Sutter, M. E. (2014). Is it prime time for alpha2-adrenocepter agonists in the treatment of withdrawal syndromes? *Journal of Medical Toxicology*, *10*(4), 369–381.

Alfonso-Loeches, S., Pascual-Lucas, M., Blanco, A. M., Sanchez-Vera, I., & Guerri, C. (2010). Pivotal role of TLR4 receptors in alcohol-induced neuroinflammation and brain damage. *The Journal of Neuroscience*, *30*(24), 8285–8295.

Altshuler, H. L., Phillips, P. E., & Feinhandler, D. A. (1980). Alteration of ethanol self-administration by naltrexone. *Life Sciences*, *26*(9), 679–688.

Ambroggi, F., Turiault, M., Milet, A., Deroche-Gamonet, V., Parnaudeau, S., Balado, E., et al. (2009). Stress and addiction: Glucocorticoid receptor in dopaminoceptive neurons facilitates cocaine seeking. *Nature Neuroscience*, *12*(3), 247–249.

Amen, S. L., Piacentine, L. B., Ahmad, M. E., Li, S. J., Mantsch, J. R., Risinger, R. C., et al. (2011). Repeated N-acetyl cysteine reduces cocaine seeking in rodents and craving in cocaine-dependent humans. *Neuropsychopharmacology*, *36*(4), 871–878.

Anderson, R. I., Becker, H. C., Adams, B. L., Jesudason, C. D., & Rorick-Kehn, L. M. (2014). Orexin-1 and orexin-2 receptor antagonists reduce ethanol self-administration in high-drinking rodent models. *Frontiers in Neuroscience*, *8*, 33.

Anglin, M. D., Burke, C., Perrochet, B., Stamper, E., & Dawud-Noursi, S. (2000). History of the methamphetamine problem. *Journal of Psychoactive Drugs*, *32*(2), 137–141.

Assadi, S. M., Radgoodarzi, R., & Ahmadi-Abhari, S. A. (2003). Baclofen for maintenance treatment of opioid dependence: A randomized double-blind placebo-controlled clinical trial [ISRCTN32121581]. *BMC Psychiatry*, *3*, 16.

Attarzadeh-Yazdi, G., Arezoomandan, R., & Haghparast, A. (2014). Minocycline, an antibiotic with inhibitory effect on microglial activation, attenuates the maintenance and reinstatement of methamphetamine-seeking behavior in rat. *Progress in Neuro-Psychopharmacology & Biological Psychiatry*, *53*, 142–148.

Ayers-Ringler, J. R., Jia, Y. F., Qiu, Y. Y., & Choi, D. S. (2016). Role of astrocytic glutamate transporter in alcohol use disorder. *World Journal of Psychiatry*, *6*(1), 31–42.

Back, S. E., McCauley, J. L., Korte, K. J., Gros, D. F., Leavitt, V., Gray, K. M., et al. (2016). A double-blind, randomized, controlled pilot trial of N-acetylcysteine in Veterans with posttraumatic stress disorder and substance use disorders. *The Journal of Clinical Psychiatry*, 77(11), e1439–e1446.

Baggio, S., Sapin, M., Khazaal, Y., Studer, J., Wolff, H., & Gmel, G. (2018). Comorbidity of symptoms of alcohol and cannabis use disorders among a population-based sample of simultaneous users insight from a network perspective. *International Journal of Environmental Research and Public Health*, *15*(12), 2893.

Baiamonte, B. A., Valenza, M., Roltsch, E. A., Whitaker, A. M., Baynes, B. B., Sabino, V., et al. (2014). Nicotine dependence produces hyperalgesia: Role of corticotropin-releasing factor-1 receptors (CRF1Rs) in the central amygdala (CeA). *Neuropharmacology*, *77*, 217–223.

Baker, D. A., McFarland, K., Lake, R. W., Shen, H., Toda, S., & Kalivas, P. W. (2003). N-acetyl cysteine-induced blockade of cocaine-induced reinstatement. *Annals of the New York Academy of Sciences*, *1003*, 349–351.

Baldwin, H. A., Rassnick, S., Rivier, J., Koob, G. F., & Britton, K. T. (1991). CRF antagonist reverses the "anxiogenic" response to ethanol withdrawal in the rat. *Psychopharmacology*, *103*(2), 227–232.

Bale, T. L., & Vale, W. W. (2004). CRF and CRF receptors: Role in stress responsivity and other behaviors. *Annual Review of Pharmacology and Toxicology*, *44*, 525–557.

Bardo, M. T. (1998). Neuropharmacological mechanisms of drug reward: Beyond dopamine in the nucleus accumbens. *Critical Reviews in Neurobiology*, *12*(1–2), 37–67.

Barr, J., Sharma, C. S., Sarkar, S., Wise, K., Dong, L., Periyakaruppan, A., et al. (2007). Nicotine induces oxidative stress and activates nuclear transcription factor kappa B in rat mesencephalic cells. *Molecular and Cellular Biochemistry*, *297*(1–2), 93–99.

Barrett, S. P., Darredeau, C., & Pihl, R. O. (2006). Patterns of simultaneous polysubstance use in drug using university students. *Human Psychopharmacology*, *21*(4), 255–263.

Barton, E. A., Baker, C., & Leasure, J. L. (2017). Investigation of sex differences in the microglial response to binge ethanol and exercise. *Brain Sciences*, 7(10), 139.

Basso, A. M., Spina, M., Rivier, J., Vale, W., & Koob, G. F. (1999). Corticotropin-releasing factor antagonist attenuates the "anxiogenic-like" effect in the defensive burying paradigm but not in the elevated plus-maze following chronic cocaine in rats. *Psychopharmacology*, *145*(1), 21–30.

Becker, J. A. J., Kieffer, B. L., & Le Merrer, J. (2017). Differential behavioral and molecular alterations upon protracted abstinence from cocaine versus morphine, nicotine, THC and alcohol. *Addiction Biology*, *22*(5), 1205–1217.

Bentzley, B. S., & Aston-Jones, G. (2015). Orexin-1 receptor signaling increases motivation for cocaine-associated cues. *The European Journal of Neuroscience*, *41*(9), 1149–1156.

Berrendero, F., Kieffer, B. L., & Maldonado, R. (2002). Attenuation of nicotine-induced antinociception, rewarding effects, and dependence in mu-opioid receptor knock-out mice. *The Journal of Neuroscience*, *22*(24), 10935–10940.

Berridge, C. W., & Waterhouse, B. D. (2003). The locus coeruleus-noradrenergic system: Modulation of behavioral state and state-dependent cognitive processes. *Brain Research. Brain Research Reviews*, *42*(1), 33–84.

Biala, G., & Budzynska, B. (2010). Rimonabant attenuates sensitization, cross-sensitization and cross-reinstatement of place preference induced by nicotine and ethanol. *Pharmacological Reports*, *62*(5), 797–807.

Blaker, A. L., Moore, E. R., & Yamamoto, B. K. (2019). Serial exposure to ethanol drinking and methamphetamine enhances glutamate excitotoxicity. *Journal of Neurochemistry*, *151*(6), 749–763.

Blednov, Y. A., Benavidez, J. M., Geil, C., Perra, S., Morikawa, H., & Harris, R. A. (2011). Activation of inflammatory signaling by lipopolysaccharide produces a prolonged increase of voluntary alcohol intake in mice. *Brain, Behavior, and Immunity*, *25*(Suppl. 1), S92–S105.

Blednov, Y. A., Ponomarev, I., Geil, C., Bergeson, S., Koob, G. F., & Harris, R. A. (2012). Neuroimmune regulation of alcohol consumption: Behavioral validation of genes obtained from genomic studies. *Addiction Biology*, *17*(1), 108–120.

Bloomfield, M. A., Ashok, A. H., Volkow, N. D., & Howes, O. D. (2016). The effects of delta(9)-tetrahydrocannabinol on the dopamine system. *Nature*, *539*(7629), 369–377.

Borgland, S. L., Chang, S. J., Bowers, M. S., Thompson, J. L., Vittoz, N., Floresco, S. B., et al. (2009). Orexin A/hypocretin-1 selectively promotes motivation for positive reinforcers. *The Journal of Neuroscience*, *29*(36), 11215–11225.

Boutrel, B., Kenny, P. J., Specio, S. E., Martin-Fardon, R., Markou, A., Koob, G. F., et al. (2005). Role for hypocretin in mediating stress-induced reinstatement of cocaine-seeking behavior. *Proceedings of the National Academy of Sciences of the United States of America*, *102*(52), 19168–19173.

Brebner, K., Phelan, R., & Roberts, D. C. (2000). Effect of baclofen on cocaine self-administration in rats reinforced under fixed-ratio 1 and progressive-ratio schedules. *Psychopharmacology*, *148*(3), 314–321.

Brecht, M. L., Huang, D., Evans, E., & Hser, Y. I. (2008). Polydrug use and implications for longitudinal research: Ten-year trajectories for heroin, cocaine, and methamphetamine users. *Drug and Alcohol Dependence*, *96*(3), 193–201.

Brown, R. M., Khoo, S. Y., & Lawrence, A. J. (2013). Central orexin (hypocretin) 2 receptor antagonism reduces ethanol self-administration, but not cue-conditioned ethanol-seeking, in ethanol-preferring rats. *The International Journal of Neuropsychopharmacology*, *16*(9), 2067–2079.

Brown, R. M., Kim, A. K., Khoo, S. Y., Kim, J. H., Jupp, B., & Lawrence, A. J. (2016). Orexin-1 receptor signalling in the prelimbic cortex and ventral tegmental area regulates cue-induced reinstatement of ethanol-seeking in iP rats. *Addiction Biology*, *21*(3), 603–612.

Bruijnzeel, A. W., Ford, J., Rogers, J. A., Scheick, S., Ji, Y., Bishnoi, M., et al. (2012). Blockade of CRF1 receptors in the central nucleus of the amygdala attenuates the dysphoria associated with nicotine withdrawal in rats. *Pharmacology, Biochemistry, and Behavior*, *101*(1), 62–68.

Bruijnzeel, A. W., Prado, M., & Isaac, S. (2009). Corticotropin-releasing factor-1 receptor activation mediates nicotine withdrawal-induced deficit in brain reward function and stress-induced relapse. *Biological Psychiatry*, *66*(2), 110–117.

Bruijnzeel, A. W., Small, E., Pasek, T. M., & Yamada, H. (2010). Corticotropin-releasing factor mediates the dysphoria-like state associated with alcohol withdrawal in rats. *Behavioural Brain Research*, *210*(2), 288–291.

Bruijnzeel, A. W., Zislis, G., Wilson, C., & Gold, M. S. (2007). Antagonism of CRF receptors prevents the deficit in brain reward function associated with precipitated nicotine withdrawal in rats. *Neuropsychopharmacology*, *32*(4), 955–963.

Bujarski, S., Roche, D. J., Lunny, K., Moallem, N. R., Courtney, K. E., Allen, V., et al. (2014). The relationship between methamphetamine and alcohol use in a community sample of methamphetamine users. *Drug and Alcohol Dependence*, *142*, 127–132.

Caille, S., Alvarez-Jaimes, L., Polis, I., Stouffer, D. G., & Parsons, L. H. (2007). Specific alterations of extracellular endocannabinoid levels in the nucleus accumbens by ethanol, heroin, and cocaine self-administration. *The Journal of Neuroscience*, *27*(14), 3695–3702.

Calipari, E. S., Godino, A., Peck, E. G., Salery, M., Mervosh, N. L., Landry, J. A., et al. (2018). Granulocyte-colony stimulating factor controls neural and behavioral plasticity in response to cocaine. *Nature Communications*, *9*(1), 9.

Calogero, A. E., Gallucci, W. T., Chrousos, G. P., & Gold, P. W. (1988). Interaction between GABAergic neurotransmission and rat hypothalamic corticotropin-releasing hormone secretion in vitro. *Brain Research*, *463*(1), 28–36.

Campbell, E. J., Marchant, N. J., & Lawrence, A. J. (2020). A sleeping giant: Suvorexant for the treatment of alcohol use disorder? *Brain Research*, *1731*, 145902.

Cantacorps, L., Montagud-Romero, S., Lujan, M. A., & Valverde, O. (2020). Prenatal and postnatal alcohol exposure increases vulnerability to cocaine addiction in adult mice. *British Journal of Pharmacology*, *177*(5), 1090–1105.

Chang, G. Q., Karatayev, O., Barson, J. R., Liang, S. C., & Leibowitz, S. F. (2014). Common effects of fat, ethanol, and nicotine on enkephalin in discrete areas of the brain. *Neuroscience*, *277*, 665–678.

Chassin, L., Presson, C. C., Sherman, S. J., & Edwards, D. A. (1990). The natural history of cigarette smoking: Predicting young-adult smoking outcomes from adolescent smoking patterns. *Health Psychology*, *9*(6), 701–716.

Chen, H., Uz, T., & Manev, H. (2009). Minocycline affects cocaine sensitization in mice. *Neuroscience Letters*, *452*(3), 258–261.

Choi, S. W., Chon, Y., Bhang, S. Y., Jang, Y. L., Won, W. Y., Choi, J. T., et al. (2014). Nicotine dependence syndrome scale and craving: Comparing nicotine-dependent individuals with and without comorbid alcohol dependence. *Asia-Pacific Psychiatry*, *6*(2), 200–206.

Cohen, A., Treweek, J., Edwards, S., Leao, R. M., Schulteis, G., Koob, G. F., et al. (2015). Extended access to nicotine leads to a CRF1 receptor dependent increase in anxiety-like behavior and hyperalgesia in rats. *Addiction Biology*, *20*(1), 56–68.

Cohn, A. M., Johnson, A. L., Rath, J. M., & Villanti, A. C. (2016). Patterns of the co-use of alcohol, marijuana, and emerging tobacco products in a national sample of young adults. *The American Journal on Addictions*, *25*(8), 634–640.

Cohn, A. M., Johnson, A. L., Rose, S. W., Pearson, J. L., Villanti, A. C., & Stanton, C. (2018). Population-level patterns and mental health and substance use correlates of alcohol, marijuana, and tobacco use and co-use in US young adults and adults: Results from the population assessment for tobacco and health. *The American Journal on Addictions*, *27*(6), 491–500.

Coller, J. K., & Hutchinson, M. R. (2012). Implications of central immune signaling caused by drugs of abuse: Mechanisms, mediators and new therapeutic approaches for prediction and treatment of drug dependence. *Pharmacology & Therapeutics*, *134*(2), 219–245.

Contarino, A., Picetti, R., Matthes, H. W., Koob, G. F., Kieffer, B. L., & Gold, L. H. (2002). Lack of reward and locomotor stimulation induced by heroin in mu-opioid receptor-deficient mice. *European Journal of Pharmacology*, *446*(1–3), 103–109.

Cooney, N. L., Litt, M. D., Sevarino, K. A., Levy, L., Kranitz, L. S., Sackler, H., et al. (2015). Concurrent alcohol and tobacco treatment: Effect on daily process measures of alcohol relapse risk. *Journal of Consulting and Clinical Psychology*, *83*(2), 346–358.

Corrigan, F., & Hutchinson, M. (2012). Are the effects of alcohol on the CNS influenced by Toll-like receptor signaling? *Expert Review of Clinical Immunology*, *8*(3), 201–203.

Corrigan, F., Wu, Y., Tuke, J., Coller, J. K., Rice, K. C., Diener, K. R., et al. (2015). Alcohol-induced sedation and synergistic interactions between alcohol and morphine: A key mechanistic role for Toll-like receptors and MyD88-dependent signaling. *Brain, Behavior, and Immunity*, *45*, 245–252.
Cosgrove, K. P., & Carroll, M. E. (2002). Effects of bremazocine on self-administration of smoked cocaine base and orally delivered ethanol, phencyclidine, saccharin, and food in rhesus monkeys: A behavioral economic analysis. *The Journal of Pharmacology and Experimental Therapeutics*, *301*(3), 993–1002.
Cousins, M. S., Stamat, H. M., & de Wit, H. (2001). Effects of a single dose of baclofen on self-reported subjective effects and tobacco smoking. *Nicotine & Tobacco Research*, *3*(2), 123–129.
Crews, F. T., Sarkar, D. K., Qin, L., Zou, J., Boyadjieva, N., & Vetreno, R. P. (2015). Neuroimmune function and the consequences of alcohol exposure. *Alcohol Research: Current Reviews*, *37*(2), 331–341. 344–351.
Cui, C., Shurtleff, D., & Harris, R. A. (2014). Neuroimmune mechanisms of alcohol and drug addiction. *International Review of Neurobiology*, *118*, 1–12.
Dahlstrom, A., & Fuxe, K. (1964). Localization of monoamines in the lower brain stem. *Experientia*, *20*(7), 398–399.
Dash, G. F., Slutske, W. S., Martin, N. G., Statham, D. J., Agrawal, A., & Lynskey, M. T. (2019). Big five personality traits and alcohol, nicotine, cannabis, and gambling disorder comorbidity. *Psychology of Addictive Behaviors*, *33*(4), 420–429.
Daw, J., Nowotny, K. M., & Boardman, J. D. (2013). Changing patterns of tobacco and alcohol co-use by gender in the United States, 1976–2010. *Demographic Research*, *28*, 637–648.
de Lecea, L., Kilduff, T. S., Peyron, C., Gao, X., Foye, P. E., Danielson, P. E., et al. (1998). The hypocretins: Hypothalamus-specific peptides with neuroexcitatory activity. *Proceedings of the National Academy of Sciences of the United States of America*, *95*(1), 322–327.
Deehan, G. A., Jr., Hauser, S. R., Waeiss, R. A., Knight, C. P., Toalston, J. E., Truitt, W. A., et al. (2015). Co-administration of ethanol and nicotine: The enduring alterations in the rewarding properties of nicotine and glutamate activity within the mesocorticolimbic system of female alcohol-preferring (P) rats. *Psychopharmacology*, *232*(23), 4293–4302.
Deroche, V., Marinelli, M., Le Moal, M., & Piazza, P. V. (1997). Glucocorticoids and behavioral effects of psychostimulants. II: Cocaine intravenous self-administration and reinstatement depend on glucocorticoid levels. *The Journal of Pharmacology and Experimental Therapeutics*, *281*(3), 1401–1407.
Deroche-Gamonet, V., Sillaber, I., Aouizerate, B., Izawa, R., Jaber, M., Ghozland, S., et al. (2003). The glucocorticoid receptor as a potential target to reduce cocaine abuse. *The Journal of Neuroscience*, *23*(11), 4785–4790.
Diana, M. A., & Marty, A. (2004). Endocannabinoid-mediated short-term synaptic plasticity: depolarization-induced suppression of inhibition (DSI) and depolarization-induced suppression of excitation (DSE). *British Journal of Pharmacology*, *142*(1), 9–19.
Di Chiara, G., & Imperato, A. (1988). Drugs abused by humans preferentially increase synaptic dopamine concentrations in the mesolimbic system of freely moving rats. *Proceedings of the National Academy of Sciences of the United States of America*, *85*(14), 5274–5278.
Di Ciano, P., & Everitt, B. J. (2003). The GABA(B) receptor agonist baclofen attenuates cocaine- and heroin-seeking behavior by rats. *Neuropsychopharmacology*, *28*(3), 510–518.
Doyon, W. M., Dong, Y., Ostroumov, A., Thomas, A. M., Zhang, T. A., & Dani, J. A. (2013). Nicotine decreases ethanol-induced dopamine signaling and increases self-administration via stress hormones. *Neuron*, *79*(3), 530–540.
Drews, E., & Zimmer, A. (2010). Modulation of alcohol and nicotine responses through the endogenous opioid system. *Progress in Neurobiology*, *90*(1), 1–15.

Drouin, C., Darracq, L., Trovero, F., Blanc, G., Glowinski, J., Cotecchia, S., et al. (2002). Alpha1b-adrenergic receptors control locomotor and rewarding effects of psychostimulants and opiates. *The Journal of Neuroscience*, *22*(7), 2873–2884.

D'Souza, M. S., & Markou, A. (2013). The "stop" and "go" of nicotine dependence: Role of GABA and glutamate. *Cold Spring Harbor Perspectives in Medicine*, *3*(6), a012146.

Duailibi, M. S., Cordeiro, Q., Brietzke, E., Ribeiro, M., LaRowe, S., Berk, M., et al. (2017). N-acetylcysteine in the treatment of craving in substance use disorders: Systematic review and meta-analysis. *The American Journal on Addictions*, *26*(7), 660–666.

Duncan, S. C., Gau, J. M., Farmer, R. F., Seeley, J. R., Kosty, D. B., & Lewinsohn, P. M. (2015). Comorbidity and temporal relations of alcohol and cannabis use disorders from youth through adulthood. *Drug and Alcohol Dependence*, *149*, 80–86.

Economidou, D., Dalley, J. W., & Everitt, B. J. (2011). Selective norepinephrine reuptake inhibition by atomoxetine prevents cue-induced heroin and cocaine seeking. *Biological Psychiatry*, *69*(3), 266–274.

Edwards, S., Little, H. J., Richardson, H. N., & Vendruscolo, L. F. (2015). Divergent regulation of distinct glucocorticoid systems in alcohol dependence. *Alcohol*, *49*(8), 811–816.

Edwards, S., Vendruscolo, L. F., Gilpin, N. W., Wojnar, M., & Witkiewitz, K. (2020). Alcohol and pain: A translational review of preclinical and clinical findings to inform future treatment strategies. *Alcoholism Clinical and Experimental Research*, *44*(2), 368–383.

Edwards, S., Vendruscolo, L. F., Schlosburg, J. E., Misra, K. K., Wee, S., Park, P. E., et al. (2012). Development of mechanical hypersensitivity in rats during heroin and ethanol dependence: Alleviation by CRF(1) receptor antagonism. *Neuropharmacology*, *62*(2), 1142–1151.

Engin, E., Benham, R. S., & Rudolph, U. (2018). An emerging circuit pharmacology of GABAA receptors. *Trends in Pharmacological Sciences*, *39*(8), 710–732.

Erb, S., Hitchcott, P. K., Rajabi, H., Mueller, D., Shaham, Y., & Stewart, J. (2000). Alpha-2 adrenergic receptor agonists block stress-induced reinstatement of cocaine seeking. *Neuropsychopharmacology*, *23*(2), 138–150.

Erb, S., Shaham, Y., & Stewart, J. (1998). The role of corticotropin-releasing factor and corticosterone in stress- and cocaine-induced relapse to cocaine seeking in rats. *The Journal of Neuroscience*, *18*(14), 5529–5536.

Espana, R. A., Melchior, J. R., Roberts, D. C., & Jones, S. R. (2011). Hypocretin 1/orexin A in the ventral tegmental area enhances dopamine responses to cocaine and promotes cocaine self-administration. *Psychopharmacology*, *214*(2), 415–426.

Espana, R. A., Oleson, E. B., Locke, J. L., Brookshire, B. R., Roberts, D. C., & Jones, S. R. (2010). The hypocretin-orexin system regulates cocaine self-administration via actions on the mesolimbic dopamine system. *The European Journal of Neuroscience*, *31*(2), 336–348.

Espana, R. A., Schmeichel, B. E., & Berridge, C. W. (2016). Norepinephrine at the nexus of arousal, motivation and relapse. *Brain Research*, *1641*(Pt. B), 207–216.

Evrard, S. G., Duhalde-Vega, M., Tagliaferro, P., Mirochnic, S., Caltana, L. R., & Brusco, A. (2006). A low chronic ethanol exposure induces morphological changes in the adolescent rat brain that are not fully recovered even after a long abstinence: An immunohistochemical study. *Experimental Neurology*, *200*(2), 438–459.

Fadda, P., Scherma, M., Fresu, A., Collu, M., & Fratta, W. (2003). Baclofen antagonizes nicotine-, cocaine-, and morphine-induced dopamine release in the nucleus accumbens of rat. *Synapse*, *50*(1), 1–6.

Fagerstrom, K., & Hughes, J. (2008). Varenicline in the treatment of tobacco dependence. *Neuropsychiatric Disease and Treatment*, *4*(2), 353–363.

Fahlke, C., Engel, J. A., Eriksson, C. J., Hard, E., & Soderpalm, B. (1994). Involvement of corticosterone in the modulation of ethanol consumption in the rat. *Alcohol, 11*(3), 195–202.

Fahlke, C., Hard, E., & Hansen, S. (1996). Facilitation of ethanol consumption by intracerebroventricular infusions of corticosterone. *Psychopharmacology (Berlin), 127*(2), 133–139.

Fattore, L., Cossu, G., Martellotta, M. C., & Fratta, W. (2002). Baclofen antagonizes intravenous self-administration of nicotine in mice and rats. *Alcohol and Alcoholism, 37*(5), 495–498.

Fekete, E. M., & Zorrilla, E. P. (2007). Physiology, pharmacology, and therapeutic relevance of urocortins in mammals: Ancient CRF paralogs. *Frontiers in Neuroendocrinology, 28*(1), 1–27.

Fernandez-Lizarbe, S., Pascual, M., & Guerri, C. (2009). Critical role of TLR4 response in the activation of microglia induced by ethanol. *Journal of Immunology, 183*(7), 4733–4744.

Filev, R., Engelke, D. S., Da Silveira, D. X., Mello, L. E., & Santos-Junior, J. G. (2017). THC inhibits the expression of ethanol-induced locomotor sensitization in mice. *Alcohol, 65*, 31–35.

Fitzgerald, P. J. (2013). Elevated norepinephrine may be a unifying etiological factor in the abuse of a broad range of substances: Alcohol, nicotine, marijuana, heroin, cocaine, and caffeine. *Substance Abuse*, 7, 171–183.

Fox, H. C., Hong, K. I., Siedlarz, K. M., Bergquist, K., Anderson, G., Kreek, M. J., et al. (2009). Sex-specific dissociations in autonomic and HPA responses to stress and cues in alcohol-dependent patients with cocaine abuse. *Alcohol and Alcoholism, 44*(6), 575–585.

Fox, H. C., Morgan, P. T., & Sinha, R. (2014). Sex differences in guanfacine effects on drug craving and stress arousal in cocaine-dependent individuals. *Neuropsychopharmacology, 39*(6), 1527–1537.

Fox, H. C., Seo, D., Tuit, K., Hansen, J., Kimmerling, A., Morgan, P. T., et al. (2012). Guanfacine effects on stress, drug craving and prefrontal activation in cocaine dependent individuals: Preliminary findings. *Journal of Psychopharmacology, 26*(7), 958–972.

Fragale, J. E., James, M. H., & Aston-Jones, G. (2020). Intermittent self-administration of fentanyl induces a multifaceted addiction state associated with persistent changes in the orexin system. *Addiction Biology*. https://doi.org/10.1111/adb.12946, e12946.

Fragale, J. E., Pantazis, C. B., James, M. H., & Aston-Jones, G. (2019). The role of orexin-1 receptor signaling in demand for the opioid fentanyl. *Neuropsychopharmacology, 44*(10), 1690–1697.

Franklin, T. R., Harper, D., Kampman, K., Kildea-McCrea, S., Jens, W., Lynch, K. G., et al. (2009). The GABA B agonist baclofen reduces cigarette consumption in a preliminary double-blind placebo-controlled smoking reduction study. *Drug and Alcohol Dependence, 103*(1–2), 30–36.

Fredriksson, I., Adhikary, S., Steensland, P., Vendruscolo, L. F., Bonci, A., Shaham, Y., et al. (2017). Prior exposure to alcohol has no effect on cocaine self-administration and relapse in rats: Evidence from a rat model that does not support the gateway hypothesis. *Neuropsychopharmacology, 42*(5), 1001–1011.

Fredriksson, I., Jayaram-Lindstrom, N., Wirf, M., Nylander, E., Nystrom, E., Jardemark, K., et al. (2015). Evaluation of guanfacine as a potential medication for alcohol use disorder in long-term drinking rats: Behavioral and electrophysiological findings. *Neuropsychopharmacology, 40*(5), 1130–1140.

Froehlich, J. C., Hausauer, B., Fischer, S., Wise, B., & Rasmussen, D. D. (2015). Prazosin reduces alcohol intake in an animal model of alcohol relapse. *Alcoholism, Clinical and Experimental Research, 39*(8), 1538–1546.

Funk, D., Coen, K., & Le, A. D. (2014). The role of kappa opioid receptors in stress-induced reinstatement of alcohol seeking in rats. *Brain and Behavior: A Cognitive Neuroscience Perspective, 4*(3), 356–367.

Funk, C. K., & Koob, G. F. (2007). A CRF(2) agonist administered into the central nucleus of the amygdala decreases ethanol self-administration in ethanol-dependent rats. *Brain Research*, *1155*, 172–178.

Funk, D., Marinelli, P. W., & Le, A. D. (2006). Biological processes underlying co-use of alcohol and nicotine: Neuronal mechanisms, cross-tolerance, and genetic factors. *Alcohol Research & Health*, *29*(3), 186–192.

Garbutt, J. C., Kampov-Polevoy, A. B., Gallop, R., Kalka-Juhl, L., & Flannery, B. A. (2010). Efficacy and safety of baclofen for alcohol dependence: A randomized, double-blind, placebo-controlled trial. *Alcoholism, Clinical and Experimental Research*, *34*(11), 1849–1857.

Garcia-Carmona, J. A., Almela, P., Baroja-Mazo, A., Milanes, M. V., & Laorden, M. L. (2012). Restricted role of CRF1 receptor for the activity of brainstem catecholaminergic neurons in the negative state of morphine withdrawal. *Psychopharmacology*, *220*(2), 379–393.

Garcia-Carmona, J. A., Baroja-Mazo, A., Milanes, M. V., & Laorden, M. L. (2015). Sex differences between CRF1 receptor deficient mice following naloxone-precipitated morphine withdrawal in a conditioned place aversion paradigm: Implication of HPA axis. *PLoS One*, *10*(4), e0121125.

Gehlert, D. R., Cippitelli, A., Thorsell, A., Le, A. D., Hipskind, P. A., Hamdouchi, C., et al. (2007). 3-(4-Chloro-2-morpholin-4-yl-thiazol-5-yl)-8-(1-ethylpropyl)-2,6-dimethyl-imidazo [1,2-b]pyridazine: A novel brain-penetrant, orally available corticotropin-releasing factor receptor 1 antagonist with efficacy in animal models of alcoholism. *The Journal of Neuroscience*, *27*(10), 2718–2726.

George, O., Ghozland, S., Azar, M. R., Cottone, P., Zorrilla, E. P., Parsons, L. H., et al. (2007). CRF-CRF1 system activation mediates withdrawal-induced increases in nicotine self-administration in nicotine-dependent rats. *Proceedings of the National Academy of Sciences of the United States of America*, *104*(43), 17198–17203.

Georgescu, D., Zachariou, V., Barrot, M., Mieda, M., Willie, J. T., Eisch, A. J., et al. (2003). Involvement of the lateral hypothalamic peptide orexin in morphine dependence and withdrawal. *The Journal of Neuroscience*, *23*(8), 3106–3111.

Ghozland, S., Matthes, H. W., Simonin, F., Filliol, D., Kieffer, B. L., & Maldonado, R. (2002). Motivational effects of cannabinoids are mediated by mu-opioid and kappa-opioid receptors. *The Journal of Neuroscience*, *22*(3), 1146–1154.

Gianoulakis, C. (2001). Influence of the endogenous opioid system on high alcohol consumption and genetic predisposition to alcoholism. *Journal of Psychiatry & Neuroscience*, *26*(4), 304–318.

Gibson, D. R., Leamon, M. H., & Flynn, N. (2002). Epidemiology and public health consequences of methamphetamine use in California's Central Valley. *Journal of Psychoactive Drugs*, *34*(3), 313–319.

Gilpin, N. W., Herman, M. A., & Roberto, M. (2015). The central amygdala as an integrative hub for anxiety and alcohol use disorders. *Biological Psychiatry*, 77(10), 859–869.

Gilpin, N. W., & Koob, G. F. (2010). Effects of beta-adrenoceptor antagonists on alcohol drinking by alcohol-dependent rats. *Psychopharmacology*, *212*(3), 431–439.

Gilpin, N. W., Misra, K., Herman, M. A., Cruz, M. T., Koob, G. F., & Roberto, M. (2011). Neuropeptide Y opposes alcohol effects on gamma-aminobutyric acid release in amygdala and blocks the transition to alcohol dependence. *Biological Psychiatry*, *69*(11), 1091–1099.

Gilpin, N. W., Misra, K., & Koob, G. F. (2008). Neuropeptide Y in the central nucleus of the amygdala suppresses dependence-induced increases in alcohol drinking. *Pharmacology, Biochemistry, and Behavior*, *90*(3), 475–480.

Gilpin, N. W., Richardson, H. N., & Koob, G. F. (2008). Effects of CRF1-receptor and opioid-receptor antagonists on dependence-induced increases in alcohol drinking by

alcohol-preferring (P) rats. *Alcoholism, Clinical and Experimental Research, 32*(9), 1535–1542.
Gold, M. S., Pottash, A. L., & Extein, I. (1982). Clonidine: Inpatient studies from 1978 to 1981. *The Journal of Clinical Psychiatry, 43*(6 Pt. 2), 35–38.
Gold, M. S., Redmond, D. E., Jr., & Kleber, H. D. (1979). Noradrenergic hyperactivity in opiate withdrawal supported by clonidine reversal of opiate withdrawal. *The American Journal of Psychiatry, 136*(1), 100–102.
Goldberg, S. R., & Gonzalez, F. A. (1976). Effects of propranolol on behavior maintained under fixed-ratio schedules of cocaine injection or food presentation in squirrel monkeys. *The Journal of Pharmacology and Experimental Therapeutics, 198*(3), 626–634.
Gonzalez, S., Cascio, M. G., Fernandez-Ruiz, J., Fezza, F., Di Marzo, V., & Ramos, J. A. (2002). Changes in endocannabinoid contents in the brain of rats chronically exposed to nicotine, ethanol or cocaine. *Brain Research, 954*(1), 73–81.
Gonzalez, S., Fernandez-Ruiz, J., Sparpaglione, V., Parolaro, D., & Ramos, J. A. (2002). Chronic exposure to morphine, cocaine or ethanol in rats produced different effects in brain cannabinoid CB(1) receptor binding and mRNA levels. *Drug and Alcohol Dependence, 66*(1), 77–84.
Grace, A. A., & Bunney, B. S. (1985). Opposing effects of striatonigral feedback pathways on midbrain dopamine cell activity. *Brain Research, 333*(2), 271–284.
Grant, B. F., & Dawson, D. A. (1998). Age of onset of drug use and its association with DSM-IV drug abuse and dependence: Results from the National Longitudinal Alcohol Epidemiologic Survey. *Journal of Substance Abuse, 10*(2), 163–173.
Grant, B. F., Goldstein, R. B., Saha, T. D., Chou, S. P., Jung, J., Zhang, H., et al. (2015). Epidemiology of DSM-5 alcohol use disorder: Results from the National Epidemiologic Survey on alcohol and related conditions III. *JAMA Psychiatry, 72*(8), 757–766.
Grant, J. E., Odlaug, B. L., & Kim, S. W. (2010). A double-blind, placebo-controlled study of N-acetyl cysteine plus naltrexone for methamphetamine dependence. *European Neuropsychopharmacology, 20*(11), 823–828.
Gray, J. M., Vecchiarelli, H. A., Morena, M., Lee, T. T., Hermanson, D. J., Kim, A. B., et al. (2015). Corticotropin-releasing hormone drives anandamide hydrolysis in the amygdala to promote anxiety. *The Journal of Neuroscience, 35*(9), 3879–3892.
Greenwell, T. N., Funk, C. K., Cottone, P., Richardson, H. N., Chen, S. A., Rice, K. C., et al. (2009). Corticotropin-releasing factor-1 receptor antagonists decrease heroin self-administration in long-but not short-access rats. *Addiction Biology, 14*(2), 130–143.
Gunn, R., Jackson, K., Borsari, B., & Metrik, J. (2019). A longitudinal examination of daily patterns of cannabis and alcohol co-use among medicinal and recreational veteran cannabis users. *Drug and Alcohol Dependence, 205*, 107661.
Hall, F. S., Goeb, M., Li, X. F., Sora, I., & Uhl, G. R. (2004). mu-Opioid receptor knockout mice display reduced cocaine conditioned place preference but enhanced sensitization of cocaine-induced locomotion. *Brain Research. Molecular Brain Research, 121*(1–2), 123–130.
Haney, M., Hart, C. L., & Foltin, R. W. (2006). Effects of baclofen on cocaine self-administration: Opioid- and nonopioid-dependent volunteers. *Neuropsychopharmacology, 31*(8), 1814–1821.
Haney, M., Hart, C. L., Vosburg, S. K., Comer, S. D., Reed, S. C., Cooper, Z. D., et al. (2010). Effects of baclofen and mirtazapine on a laboratory model of marijuana withdrawal and relapse. *Psychopharmacology, 211*(2), 233–244.
Harris, G. C., & Aston-Jones, G. (1993a). Beta-adrenergic antagonists attenuate somatic and aversive signs of opiate withdrawal. *Neuropsychopharmacology, 9*(4), 303–311.
Harris, G. C., & Aston-Jones, G. (1993b). Beta-adrenergic antagonists attenuate withdrawal anxiety in cocaine- and morphine-dependent rats. *Psychopharmacology, 113*(1), 131–136.

Harris, R. A., Bajo, M., Bell, R. L., Blednov, Y. A., Varodayan, F. P., Truitt, J. M., et al. (2017). Genetic and pharmacologic manipulation of TLR4 has minimal impact on ethanol consumption in rodents. *The Journal of Neuroscience*, *37*(5), 1139–1155.

Harris, G. C., Hedaya, M. A., Pan, W. J., & Kalivas, P. (1996). beta-adrenergic antagonism alters the behavioral and neurochemical responses to cocaine. *Neuropsychopharmacology*, *14*(3), 195–204.

Hauser, P., Fuller, B., Ho, S. B., Thuras, P., Kern, S., & Dieperink, E. (2017). The safety and efficacy of baclofen to reduce alcohol use in veterans with chronic hepatitis C: A randomized controlled trial. *Addiction*, *112*(7), 1173–1183.

Hauser, S. R., Katner, S. N., Deehan, G. A., Jr., Ding, Z. M., Toalston, J. E., Scott, B. J., et al. (2012). Development of an oral operant nicotine/ethanol co-use model in alcohol-preferring (p) rats. *Alcoholism, Clinical and Experimental Research*, *36*(11), 1963–1972.

Hautala, D., Sittner, K., & Walls, M. (2019). Onset, comorbidity, and predictors of nicotine, alcohol, and marijuana use disorders among north american indigenous adolescents. *Journal of Abnormal Child Psychology*, *47*(6), 1025–1038.

Hawkins, B. T., Abbruscato, T. J., Egleton, R. D., Brown, R. C., Huber, J. D., Campos, C. R., et al. (2004). Nicotine increases in vivo blood-brain barrier permeability and alters cerebral microvascular tight junction protein distribution. *Brain Research*, *1027*(1–2), 48–58.

Hayaki, J., Anderson, B. J., & Stein, M. D. (2016). Dual cannabis and alcohol use disorders in young adults: Problems magnified. *Substance Abuse*, *37*(4), 579–583.

He, J., & Crews, F. T. (2008). Increased MCP-1 and microglia in various regions of the human alcoholic brain. *Experimental Neurology*, *210*(2), 349–358.

He, L., & Whistler, J. L. (2011). Chronic ethanol consumption in rats produces opioid antinociceptive tolerance through inhibition of mu opioid receptor endocytosis. *PLoS One*, *6*(5), e19372.

Heinrichs, S. C., Menzaghi, F., Schulteis, G., Koob, G. F., & Stinus, L. (1995). Suppression of corticotropin-releasing factor in the amygdala attenuates aversive consequences of morphine withdrawal. *Behavioural Pharmacology*, *6*(1), 74–80.

Herman, J. P., Mueller, N. K., & Figueiredo, H. (2004). Role of GABA and glutamate circuitry in hypothalamo-pituitary-adrenocortical stress integration. *Annals of the New York Academy of Sciences*, *1018*, 35–45.

Hersh, D., Van Kirk, J. R., & Kranzler, H. R. (1998). Naltrexone treatment of comorbid alcohol and cocaine use disorders. *Psychopharmacology*, *139*(1–2), 44–52.

Herz, A. (1997). Endogenous opioid systems and alcohol addiction. *Psychopharmacology*, *129*(2), 99–111.

Highfield, D., Yap, J., Grimm, J. W., Shalev, U., & Shaham, Y. (2001). Repeated lofexidine treatment attenuates stress-induced, but not drug cues-induced reinstatement of a heroin-cocaine mixture (speedball) seeking in rats. *Neuropsychopharmacology*, *25*(3), 320–331.

Hillard, C. J. (2018). Circulating endocannabinoids: From whence do they come and where are they going? *Neuropsychopharmacology*, *43*(1), 155–172.

Hingson, R., Heeren, T., Zakocs, R., Winter, M., & Wechsler, H. (2003). Age of first intoxication, heavy drinking, driving after drinking and risk of unintentional injury among U.S. college students. *Journal of Studies on Alcohol*, *64*(1), 23–31.

Hofford, R. S., Russo, S. J., & Kiraly, D. D. (2019). Neuroimmune mechanisms of psychostimulant and opioid use disorders. *The European Journal of Neuroscience*, *50*(3), 2562–2573.

Hollander, J. A., Pham, D., Fowler, C. D., & Kenny, P. J. (2012). Hypocretin-1 receptors regulate the reinforcing and reward-enhancing effects of cocaine: Pharmacological and behavioral genetics evidence. *Frontiers in Behavioral Neuroscience*, *6*, 47.

Hood, L. E., Leyrer-Jackson, J. M., & Olive, M. F. (2020). Pharmacotherapeutic management of co-morbid alcohol and opioid use. *Expert Opinion on Pharmacotherapy, 21*(7), 823–839.

Hser, Y. I., Mooney, L. J., Saxon, A. J., Miotto, K., Bell, D. S., & Huang, D. (2017). Chronic pain among patients with opioid use disorder: Results from electronic health records data. *Journal of Substance Abuse Treatment, 77*, 26–30.

Hursh, S. R., Raslear, T. G., Shurtleff, D., Bauman, R., & Simmons, L. (1988). A cost-benefit analysis of demand for food. *Journal of the Experimental Analysis of Behavior, 50*(3), 419–440.

Hursh, S. R., & Silberberg, A. (2008). Economic demand and essential value. *Psychological Review, 115*(1), 186.

Hursh, S. R., & Winger, G. (1995). Normalized demand for drugs and other reinforcers. *Journal of the Experimental Analysis of Behavior, 64*(3), 373–384.

Hutchinson, M. R., Coats, B. D., Lewis, S. S., Zhang, Y., Sprunger, D. B., Rezvani, N., et al. (2008). Proinflammatory cytokines oppose opioid-induced acute and chronic analgesia. *Brain, Behavior, and Immunity, 22*(8), 1178–1189.

Hutchinson, M. R., Zhang, Y., Shridhar, M., Evans, J. H., Buchanan, M. M., Zhao, T. X., et al. (2010). Evidence that opioids may have Toll-like receptor 4 and MD-2 effects. *Brain, Behavior, and Immunity, 24*(1), 83–95.

Hutchison, M. A., & Riley, A. L. (2012). Ethanol exposure during either adolescence or adulthood alters the rewarding effects of cocaine in adult rats. *Pharmacology, Biochemistry, and Behavior, 101*(3), 458–464.

Ise, Y., Mori, T., Katayama, S., Nagase, H., & Suzuki, T. (2013). Rewarding effects of ethanol combined with low doses of morphine through dopamine D1 receptors. *Journal of Nippon Medical School, 80*(1), 34–41.

Israel, Y., Quintanilla, M. E., Ezquer, F., Morales, P., Santapau, D., Berrios-Carcamo, P., et al. (2019). Aspirin and N-acetylcysteine co-administration markedly inhibit chronic ethanol intake and block relapse binge drinking: Role of neuroinflammation-oxidative stress self-perpetuation. *Addiction Biology*. https://doi.org/10.1111/adb.12853, e12853.

Jadzic, D., Bassareo, V., Carta, A. R., & Carboni, E. (2019). Nicotine, cocaine, amphetamine, morphine, and ethanol increase norepinephrine output in the bed nucleus of stria terminalis of freely moving rats. *Addiction Biology*, e12864.

James, M. H., Fragale, J. E., Aurora, R. N., Cooperman, N. A., Langleben, D. D., & Aston-Jones, G. (2020). Repurposing the dual orexin receptor antagonist suvorexant for the treatment of opioid use disorder: Why sleep on this any longer? *Neuropsychopharmacology, 45*(5), 717–719.

James, M. H., Mahler, S. V., Moorman, D. E., & Aston-Jones, G. (2017). A decade of orexin/hypocretin and addiction: Where are we now? *Current Topics in Behavioral Neurosciences, 33*, 247–281.

James, M. H., Stopper, C. M., Zimmer, B. A., Koll, N. E., Bowrey, H. E., & Aston-Jones, G. (2019). Increased number and activity of a lateral subpopulation of hypothalamic orexin/hypocretin neurons underlies the expression of an addicted state in rats. *Biological Psychiatry, 85*(11), 925–935.

Jimenez-Gomez, C., Winger, G., Dean, R. L., Deaver, D. R., & Woods, J. H. (2011). Naltrexone decreases D-amphetamine and ethanol self-administration in rhesus monkeys. *Behavioural Pharmacology, 22*(1), 87–90.

Jones, C. M., & McCance-Katz, E. F. (2019). Co-occurring substance use and mental disorders among adults with opioid use disorder. *Drug and Alcohol Dependence, 197*, 78–82.

Juarez, B., & Han, M. H. (2016). Diversity of dopaminergic neural circuits in response to drug exposure. *Neuropsychopharmacology, 41*(10), 2424–2446.

Jung, J., Goldstein, R. B., & Grant, B. F. (2016). Association of respondent psychiatric comorbidity with family history of comorbidity: Results from the National

Epidemiologic Survey on Alcohol and Related Conditions-III. *Comprehensive Psychiatry*, *71*, 49–56.

Kalivas, P. W. (2009). The glutamate homeostasis hypothesis of addiction. *Nature Reviews. Neuroscience*, *10*(8), 561–572.

Kallupi, M., Vendruscolo, L. F., Carmichael, C. Y., George, O., Koob, G. F., & Gilpin, N. W. (2014). Neuropeptide YY(2)R blockade in the central amygdala reduces anxiety-like behavior but not alcohol drinking in alcohol-dependent rats. *Addiction Biology*, *19*(5), 755–757.

Kallupi, M., Wee, S., Edwards, S., Whitfield, T. W., Jr., Oleata, C. S., Luu, G., et al. (2013). Kappa opioid receptor-mediated dysregulation of gamma-aminobutyric acidergic transmission in the central amygdala in cocaine addiction. *Biological Psychiatry*, *74*(7), 520–528.

Kalman, D., Kim, S., DiGirolamo, G., Smelson, D., & Ziedonis, D. (2010). Addressing tobacco use disorder in smokers in early remission from alcohol dependence: The case for integrating smoking cessation services in substance use disorder treatment programs. *Clinical Psychology Review*, *30*(1), 12–24.

Kandel, D. (1975). Stages in adolescent involvement in drug use. *Science*, *190*(4217), 912–914.

Kandel, D., & Faust, R. (1975). Sequence and stages in patterns of adolescent drug use. *Archives of General Psychiatry*, *32*(7), 923–932.

Kandel, D. B., & Kandel, E. R. (2014). A molecular basis for nicotine as a gateway drug. *The New England Journal of Medicine*, *371*(21), 2038–2039.

Kandel, D. B., Yamaguchi, K., & Chen, K. (1992). Stages of progression in drug involvement from adolescence to adulthood: Further evidence for the gateway theory. *Journal of Studies on Alcohol*, *53*(5), 447–457.

Karpyak, V. M., Winham, S. J., Preuss, U. W., Zill, P., Cunningham, J. M., Walker, D. L., et al. (2013). Association of the PDYN gene with alcohol dependence and the propensity to drink in negative emotional states. *The International Journal of Neuropsychopharmacology*, *16*(5), 975–985.

Kashima, D. T., & Grueter, B. A. (2017). Toll-like receptor 4 deficiency alters nucleus accumbens synaptic physiology and drug reward behavior. *Proceedings of the National Academy of Sciences of the United States of America*, *114*(33), 8865–8870.

Katner, S. N., Oster, S. M., Ding, Z. M., Deehan, G. A., Jr., Toalston, J. E., Hauser, S. R., et al. (2011). Alcohol-preferring (P) rats are more sensitive than Wistar rats to the reinforcing effects of cocaine self-administered directly into the nucleus accumbens shell. *Pharmacology, Biochemistry, and Behavior*, *99*(4), 688–695.

Kerridge, B. T., Pickering, R., Chou, P., Saha, T. D., & Hasin, D. S. (2018). DSM-5 cannabis use disorder in the National Epidemiologic Survey on Alcohol and Related Conditions-III: Gender-specific profiles. *Addictive Behaviors*, *76*, 52–60.

Khantzian, E. J. (1985). The self-medication hypothesis of addictive disorders: Focus on heroin and cocaine dependence. *The American Journal of Psychiatry*, *142*(11), 1259–1264.

Khantzian, E. J. (1990). Self-regulation and self-medication factors in alcoholism and the addictions. Similarities and differences. *Recent Developments in Alcoholism*, *8*, 255–271.

Khantzian, E. J. (1997). The self-medication hypothesis of substance use disorders: A reconsideration and recent applications. *Harvard Review of Psychiatry*, *4*(5), 231–244.

Kieffer, B. L. (1995). Recent advances in molecular recognition and signal transduction of active peptides: Receptors for opioid peptides. *Cellular and Molecular Neurobiology*, *15*(6), 615–635.

Kieffer, B. L., Befort, K., Gaveriaux-Ruff, C., & Hirth, C. G. (1992). The delta-opioid receptor: Isolation of a cDNA by expression cloning and pharmacological characterization. *Proceedings of the National Academy of Sciences of the United States of America*, *89*(24), 12048–12052.

Kieffer, B. L., & Evans, C. J. (2009). Opioid receptors: From binding sites to visible molecules in vivo. *Neuropharmacology, 56*(Suppl. 1), 205–212.
Kieffer, B. L., & Gaveriaux-Ruff, C. (2002). Exploring the opioid system by gene knockout. *Progress in Neurobiology, 66*(5), 285–306.
Kilts, C. D., & Anderson, C. M. (1986). The simultaneous quantification of dopamine, norepinephrine and epinephrine in micropunched rat brain nuclei by on-line trace enrichment HPLC with electrochemical detection: Distribution of catecholamines in the limbic system. *Neurochemistry International, 9*(3), 437–445.
Kim, R., Healey, K. L., Sepulveda-Orengo, M. T., & Reissner, K. J. (2018). Astroglial correlates of neuropsychiatric disease: From astrocytopathy to astrogliosis. *Progress in Neuropsychopharmacology and Biological Psychiatry, 87*(Pt. A), 126–146.
Kirkpatrick, M. G., Gunderson, E. W., Levin, F. R., Foltin, R. W., & Hart, C. L. (2012). Acute and residual interactive effects of repeated administrations of oral methamphetamine and alcohol in humans. *Psychopharmacology, 219*(1), 191–204.
Koob, G. F. (2008). A role for brain stress systems in addiction. *Neuron, 59*(1), 11–34.
Koob, G. F., & Bloom, F. E. (1988). Cellular and molecular mechanisms of drug dependence. *Science, 242*(4879), 715–723.
Koob, G. F., Buck, C. L., Cohen, A., Edwards, S., Park, P. E., Schlosburg, J. E., et al. (2014). Addiction as a stress surfeit disorder. *Neuropharmacology, 76*(Pt. B), 370–382.
Koob, G. F., & Le Moal, M. (1997). Drug abuse: Hedonic homeostatic dysregulation. *Science, 278*(5335), 52–58.
Koob, G. F., & Le Moal, M. (2001). Drug addiction, dysregulation of reward, and allostasis. *Neuropsychopharmacology, 24*(2), 97–129.
Koob, G. F., Sanna, P. P., & Bloom, F. E. (1998). Neuroscience of addiction. *Neuron, 21*(3), 467–476.
Koob, G. F., & Schulkin, J. (2019). Addiction and stress: An allostatic view. *Neuroscience and Biobehavioral Reviews, 106*, 245–262.
Koob, G. F., & Volkow, N. D. (2016). Neurobiology of addiction: A neurocircuitry analysis. *Lancet Psychiatry, 3*(8), 760–773.
Kosten, T. R., & O'Connor, P. G. (2003). Management of drug and alcohol withdrawal. *The New England Journal of Medicine, 348*(18), 1786–1795.
Kowalczyk, W. J., Phillips, K. A., Jobes, M. L., Kennedy, A. P., Ghitza, U. E., Agage, D. A., et al. (2015). Clonidine maintenance prolongs opioid abstinence and decouples stress from craving in daily life: A randomized controlled trial with ecological momentary assessment. *The American Journal of Psychiatry, 172*(8), 760–767.
Kreek, M. J. (1996). Opiates, opioids and addiction. *Molecular Psychiatry, 1*(3), 232–254.
Kreek, M. J., LaForge, K. S., & Butelman, E. (2002). Pharmacotherapy of addictions. *Nature Reviews. Drug Discovery, 1*(9), 710–726.
Kruyer, A., Chioma, C., & Kalivas, P. W. (2020). The opioid-addicted tetrapartite synapse. *Biological Psychiatry, 87*(1), 34–43.
Krystal, J. H., & Neumeister, A. (2009). Noradrenergic and serotonergic mechanisms in the neurobiology of posttraumatic stress disorder and resilience. *Brain Research, 1293*, 13–23.
Lajtha, A., & Sershen, H. (2010). Nicotine: Alcohol reward interactions. *Neurochemical Research, 35*(8), 1248–1258.
LaRowe, S. D., Kalivas, P. W., Nicholas, J. S., Randall, P. K., Mardikian, P. N., & Malcolm, R. J. (2013). A double-blind placebo-controlled trial of N-acetylcysteine in the treatment of cocaine dependence. *The American Journal on Addictions, 22*(5), 443–452.
LaRowe, L. R., Powers, J. M., Garey, L., Rogers, A. H., Zvolensky, M. J., & Ditre, J. W. (2020). Pain-related anxiety, sex, and co-use of alcohol and prescription opioids among adults with chronic low back pain. *Drug and Alcohol Dependence, 214*, 108171.
Larraga, A., Belluzzi, J. D., & Leslie, F. M. (2017). Nicotine increases alcohol intake in adolescent male rats. *Frontiers in Behavioral Neuroscience, 11*, 25.

Lawrence, A. J., Cowen, M. S., Yang, H. J., Chen, F., & Oldfield, B. (2006). The orexin system regulates alcohol-seeking in rats. *British Journal of Pharmacology*, *148*(6), 752–759.

Le, A. D., Funk, D., Juzytsch, W., Coen, K., Navarre, B. M., Cifani, C., et al. (2011). Effect of prazosin and guanfacine on stress-induced reinstatement of alcohol and food seeking in rats. *Psychopharmacology*, *218*(1), 89–99.

Le, A. D., Harding, S., Juzytsch, W., Funk, D., & Shaham, Y. (2005). Role of alpha-2 adrenoceptors in stress-induced reinstatement of alcohol seeking and alcohol self-administration in rats. *Psychopharmacology*, *179*(2), 366–373.

Le, A. D., Harding, S., Juzytsch, W., Watchus, J., Shalev, U., & Shaham, Y. (2000). The role of corticotrophin-releasing factor in stress-induced relapse to alcohol-seeking behavior in rats. *Psychopharmacology*, *150*(3), 317–324.

Le, A. D., Khanna, J. M., Kalant, H., & LeBlanc, A. E. (1980). Cross-tolerance between ethanol and morphine. *Advances in Experimental Medicine and Biology*, *132*, 771–777.

Lea, S. (1978). The psychology and economics of demand. *Psychological Bulletin*, *85*(3), 441.

Leao, R. M., Cruz, F. C., Vendruscolo, L. F., de Guglielmo, G., Logrip, M. L., Planeta, C. S., et al. (2015). Chronic nicotine activates stress/reward-related brain regions and facilitates the transition to compulsive alcohol drinking. *The Journal of Neuroscience*, *35*(15), 6241–6253.

Leatherdale, S. T., Hammond, D., & Ahmed, R. (2008). Alcohol, marijuana, and tobacco use patterns among youth in Canada. *Cancer Causes & Control*, *19*(4), 361–369.

Lei, K., Wegner, S. A., Yu, J. H., & Hopf, F. W. (2016). Orexin-1 receptor blockade suppresses compulsive-like alcohol drinking in mice. *Neuropharmacology*, *110*(Pt. A), 431–437.

Leri, F., Bruneau, J., & Stewart, J. (2003). Understanding polydrug use: Review of heroin and cocaine co-use. *Addiction*, *98*(1), 7–22.

Lenoir, M., Guillem, K., Koob, G. F., & Ahmed, S. H. (2012). Drug specificity in extended access cocaine and heroin self-administration. *Addiction Biology*, *17*(6), 964–976.

Leri, F., Flores, J., Rodaros, D., & Stewart, J. (2002). Blockade of stress-induced but not cocaine-induced reinstatement by infusion of noradrenergic antagonists into the bed nucleus of the stria terminalis or the central nucleus of the amygdala. *The Journal of Neuroscience*, *22*(13), 5713–5718.

Levine, A., Huang, Y., Drisaldi, B., Griffin, E. A., Jr., Pollak, D. D., Xu, S., et al. (2011). Molecular mechanism for a gateway drug: Epigenetic changes initiated by nicotine prime gene expression by cocaine. *Science Translational Medicine*, *3*(107), 107ra109.

Lewohl, J. M., Wixey, J., Harper, C. G., & Dodd, P. R. (2005). Expression of MBP, PLP, MAG, CNP, and GFAP in the human alcoholic brain. *Alcoholism, Clinical and Experimental Research*, *29*(9), 1698–1705.

Li, Y., Li, H., Zhang, Y., Sun, X., Hanley, G. A., LeSage, G., et al. (2010). Toll-like receptor 2 is required for opioids-induced neuronal apoptosis. *Biochemical and Biophysical Research Communications*, *391*(1), 426–430.

Li, Y., Sun, X., Zhang, Y., Huang, J., Hanley, G., Ferslew, K. E., et al. (2009). Morphine promotes apoptosis via TLR2, and this is negatively regulated by beta-arrestin 2. *Biochemical and Biophysical Research Communications*, *378*(4), 857–861.

Lindholm, S., Rosin, A., Dahlin, I., Georgieva, J., & Franck, J. (2001). Ethanol administration potentiates cocaine-induced dopamine levels in the rat nucleus accumbens. *Brain Research*, *915*(2), 176–184.

Litten, R. Z., Ryan, M. L., Fertig, J. B., Falk, D. E., Johnson, B., Dunn, K. E., et al. (2013). A double-blind, placebo-controlled trial assessing the efficacy of varenicline tartrate for alcohol dependence. *Journal of Addiction Medicine*, 7(4), 277–286.

Loftis, J. M., & Janowsky, A. (2014). Neuroimmune basis of methamphetamine toxicity. *International Review of Neurobiology*, *118*, 165–197.

Lopez, M. F., Moorman, D. E., Aston-Jones, G., & Becker, H. C. (2016). The highly selective orexin/hypocretin 1 receptor antagonist GSK1059865 potently reduces ethanol drinking in ethanol dependent mice. *Brain Research, 1636*, 74–80.

Lüscher, C. (2016). The emergence of a circuit model for addiction. *Annual Review of Neuroscience, 39*, 257–276.

Luscher, C., & Malenka, R. C. (2012). NMDA receptor-dependent long-term potentiation and long-term depression (LTP/LTD). *Cold Spring Harbor Perspectives in Biology, 4*(6), a005710.

Mackesy-Amiti, M. E., Fendrich, M., & Goldstein, P. J. (1997). Sequence of drug use among serious drug users: Typical vs atypical progression. *Drug and Alcohol Dependence, 45*(3), 185–196.

Manhem, P., Nilsson, L. H., Moberg, A. L., Wadstein, J., & Hokfelt, B. (1985). Alcohol withdrawal: Effects of clonidine treatment on sympathetic activity, the renin-aldosterone system, and clinical symptoms. *Alcoholism, Clinical and Experimental Research, 9*(3), 238–243.

Mantsch, J. R., & Goeders, N. E. (1999). Ketoconazole blocks the stress-induced reinstatement of cocaine-seeking behavior in rats: Relationship to the discriminative stimulus effects of cocaine. *Psychopharmacology, 142*(4), 399–407.

Marcinkiewcz, C. A., Prado, M. M., Isaac, S. K., Marshall, A., Rylkova, D., & Bruijnzeel, A. W. (2009). Corticotropin-releasing factor within the central nucleus of the amygdala and the nucleus accumbens shell mediates the negative affective state of nicotine withdrawal in rats. *Neuropsychopharmacology, 34*(7), 1743–1752.

Marfaing-Jallat, P., Miceli, D., & Le Magnen, J. (1983). Decrease in ethanol consumption by naloxone in naive and dependent rats. *Pharmacology, Biochemistry, and Behavior, 18*(Suppl. 1), 537–539.

Marinelli, P. W., Funk, D., Juzytsch, W., Harding, S., Rice, K. C., Shaham, Y., et al. (2007). The CRF1 receptor antagonist antalarmin attenuates yohimbine-induced increases in operant alcohol self-administration and reinstatement of alcohol seeking in rats. *Psychopharmacology, 195*(3), 345–355.

Marinho, E. A., Oliveira-Lima, A. J., Santos, R., Hollais, A. W., Baldaia, M. A., Wuo-Silva, R., et al. (2015). Effects of rimonabant on the development of single dose-induced behavioral sensitization to ethanol, morphine and cocaine in mice. *Progress in Neuro-Psychopharmacology & Biological Psychiatry, 58*, 22–31.

Markou, A., Kosten, T. R., & Koob, G. F. (1998). Neurobiological similarities in depression and drug dependence: A self-medication hypothesis. *Neuropsychopharmacology, 18*(3), 135–174.

Martin, I., Lampinen, T. M., & McGhee, D. (2006). Methamphetamine use among marginalized youth in British Columbia. *Canadian Journal of Public Health, 97*(4), 320–324.

Martin-Fardon, R., & Weiss, F. (2014a). Blockade of hypocretin receptor-1 preferentially prevents cocaine seeking: Comparison with natural reward seeking. *Neuroreport, 25*(7), 485–488.

Martin-Fardon, R., & Weiss, F. (2014b). N-(2-methyl-6-benzoxazolyl)-N'-1, 5-naphthyridin-4-yl urea (SB334867), a hypocretin receptor-1 antagonist, preferentially prevents ethanol seeking: Comparison with natural reward seeking. *Addiction Biology, 19*(2), 233–236.

Matchett, J. A., & Erickson, C. K. (1977). Alteration of ethanol-induced changes in locomotor activity by adrenergic blockers in mice. *Psychopharmacology, 52*(2), 201–206.

Mateos-Garcia, A., Manzanedo, C., Rodriguez-Arias, M., Aguilar, M. A., Reig-Sanchis, E., Navarro-Frances, C. I., et al. (2015). Sex differences in the long-lasting consequences of adolescent ethanol exposure for the rewarding effects of cocaine in mice. *Psychopharmacology, 232*(16), 2995–3007.

Matzeu, A., Kerr, T. M., Weiss, F., & Martin-Fardon, R. (2016). Orexin-A/hypocretin-1 mediates cocaine-seeking behavior in the posterior paraventricular nucleus of the thalamus via orexin/hypocretin receptor-2. *The Journal of Pharmacology and Experimental Therapeutics*, *359*(2), 273–279.

Matzeu, A., & Martin-Fardon, R. (2020a). Targeting the orexin system for prescription opioid use disorder. *Brain Sciences*, *10*(4), 226.

Matzeu, A., & Martin-Fardon, R. (2020b). Targeting the orexin system for prescription opioid use disorder: Orexin-1 receptor blockade prevents oxycodone taking and seeking in rats. *Neuropharmacology*, *164*, 107906.

Mayfield, J., Ferguson, L., & Harris, R. A. (2013). Neuroimmune signaling: A key component of alcohol abuse. *Current Opinion in Neurobiology*, *23*(4), 513–520.

McClure, E. A., Baker, N. L., Hood, C. O., Tomko, R. L., Squeglia, L. M., Flanagan, J. C., et al. (2020). Cannabis and alcohol co-use in a smoking cessation pharmacotherapy trial for adolescents and emerging adults. *Nicotine & Tobacco Research*, *22*(8), 1374–1382.

McConnell, S. A., Brandner, A. J., Blank, B. A., Kearns, D. A., Koob, G. F., Vendruscolo, L. F., et al. (2021). Demand for fentanyl becomes inelastic following extended access to fentanyl vapor self-administration. *Neuropharmacology*, *182*, 108355.

McElvain, J. S., & Schenk, J. O. (1992). A multisubstrate mechanism of striatal dopamine uptake and its inhibition by cocaine. *Biochemical Pharmacology*, *43*(10), 2189–2199.

McGinn, M. A., Edwards, K. N., & Edwards, S. (2020). Chronic inflammatory pain alters alcohol-regulated frontocortical signaling and associations between alcohol drinking and thermal sensitivity. *Neurobiology of Pain*, *8*, 100052.

McGinn, M. A., Paulsen, R. I., Itoga, C. A., Farooq, M. A., Reppel, J. E., Edwards, K. N., et al. (2016). Withdrawal from chronic nicotine exposure produces region-specific tolerance to alcohol-stimulated GluA1 phosphorylation. *Alcoholism, Clinical and Experimental Research*, *40*(12), 2537–2547.

McKetin, R., Dean, O. M., Baker, A. L., Carter, G., Turner, A., Kelly, P. J., et al. (2017). A potential role for N-acetylcysteine in the management of methamphetamine dependence. *Drug and Alcohol Review*, *36*(2), 153–159.

McNally, G. P., & Akil, H. (2002). Role of corticotropin-releasing hormone in the amygdala and bed nucleus of the stria terminalis in the behavioral, pain modulatory, and endocrine consequences of opiate withdrawal. *Neuroscience*, *112*(3), 605–617.

McPherson, S. M., Burduli, E., Smith, C. L., Brooks, O., Orr, M. F., Barbosa-Leiker, C., et al. (2018). Predictors of tobacco and alcohol co-use from ages 15 to 32: The Amsterdam Growth and Health Longitudinal Study. *Experimental and Clinical Psychopharmacology*, *26*(6), 549–559.

Mereu, G., & Gessa, G. L. (1985). Low doses of ethanol inhibit the firing of neurons in the substantia nigra, pars reticulata: A GABAergic effect? *Brain Research*, *360*(1–2), 325–330.

Merlo Pich, E., Lorang, M., Yeganeh, M., Rodriguez de Fonseca, F., Raber, J., Koob, G. F., et al. (1995). Increase of extracellular corticotropin-releasing factor-like immunoreactivity levels in the amygdala of awake rats during restraint stress and ethanol withdrawal as measured by microdialysis. *The Journal of Neuroscience*, *15*(8), 5439–5447.

Metrik, J., Gunn, R. L., Jackson, K. M., Sokolovsky, A. W., & Borsari, B. (2018). Daily patterns of marijuana and alcohol co-use among individuals with alcohol and cannabis use disorders. *Alcoholism, Clinical and Experimental Research*, *42*(6), 1096–1104.

Minami, M., & Satoh, M. (1995). Molecular biology of the opioid receptors: Structures, functions and distributions. *Neuroscience Research*, *23*(2), 121–145.

Minhas, M., & Leri, F. (2014). The effect of heroin dependence on resumption of heroin self-administration in rats. *Drug and Alcohol Dependence*, *138*, 24–31.

Missale, C., Castelletti, L., Govoni, S., Spano, P. F., Trabucchi, M., & Hanbauer, I. (1985). Dopamine uptake is differentially regulated in rat striatum and nucleus accumbens. *Journal of Neurochemistry*, *45*(1), 51–56.

Mitchell, J. M., Teague, C. H., Kayser, A. S., Bartlett, S. E., & Fields, H. L. (2012). Varenicline decreases alcohol consumption in heavy-drinking smokers. *Psychopharmacology, 223*(3), 299–306.

Molet, J., Herve, D., Thiebot, M. H., Hamon, M., & Lanfumey, L. (2013). Juvenile ethanol exposure increases rewarding properties of cocaine and morphine in adult DBA/2J mice. *European Neuropsychopharmacology, 23*(12), 1816–1825.

Montanari, C., Secci, M. E., Driskell, A., McDonald, K. O., Schratz, C. L., & Gilpin, N. W. (2020). Chronic nicotine increases alcohol self-administration in adult male Wistar rats. *Psychopharmacology*. https://doi.org/10.1007/s00213-020-05669-8.

Moorman, D. E., James, M. H., Kilroy, E. A., & Aston-Jones, G. (2017). Orexin/hypocretin-1 receptor antagonism reduces ethanol self-administration and reinstatement selectively in highly-motivated rats. *Brain Research, 1654*(Pt. A), 34–42.

Moreira, F. P., Medeiros, J. R., Lhullier, A. C., Souza, L. D., Jansen, K., Portela, L. V., et al. (2016). Cocaine abuse and effects in the serum levels of cytokines IL-6 and IL-10. *Drug and Alcohol Dependence, 158*, 181–185.

Morley, K. C., Baillie, A., Fraser, I., Furneaux-Bate, A., Dore, G., Roberts, M., et al. (2018). Baclofen in the treatment of alcohol dependence with or without liver disease: Multisite, randomised, double-blind, placebo-controlled trial. *The British Journal of Psychiatry, 212*(6), 362–369.

Muzyk, A. J., Fowler, J. A., Norwood, D. K., & Chilipko, A. (2011). Role of alpha2-agonists in the treatment of acute alcohol withdrawal. *The Annals of Pharmacotherapy, 45*(5), 649–657.

Namba, M. D., Kupchik, Y. M., Spencer, S. M., Garcia-Keller, C., Goenaga, J. G., Powell, G. L., et al. (2019). Accumbens neuroimmune signaling and dysregulation of astrocytic glutamate transport underlie conditioned nicotine-seeking behavior. *Addiction Biology, 25*(5), e12797.

Negus, S. S., & Rice, K. C. (2009). Mechanisms of withdrawal-associated increases in heroin self-administration: Pharmacologic modulation of heroin vs food choice in heroin-dependent rhesus monkeys. *Neuropsychopharmacology, 34*(4), 899–911.

Nevárez, N., & de Lecea, L. (2018). Recent advances in understanding the roles of hypocretin/orexin in arousal, affect, and motivation [version 1; peer review: 3 approved]. *F1000Research*, 7(F1000 Faculty Rev). https://doi.org/10.12688/f1000research.15097.1, 1421.

Northcutt, A. L., Hutchinson, M. R., Wang, X., Baratta, M. V., Hiranita, T., Cochran, T. A., et al. (2015). DAT isn't all that: Cocaine reward and reinforcement require Toll-like receptor 4 signaling. *Molecular Psychiatry, 20*(12), 1525–1537.

O'Grady, K. E., Arria, A. M., Fitzelle, D. M., & Wish, E. D. (2008). Heavy drinking and polydrug use among college students. *Journal of Drug Issues, 38*(2), 445–466.

Olive, M. F., Koenig, H. N., Nannini, M. A., & Hodge, C. W. (2001). Stimulation of endorphin neurotransmission in the nucleus accumbens by ethanol, cocaine, and amphetamine. *The Journal of Neuroscience, 21*(23), RC184.

Olive, M. F., Koenig, H. N., Nannini, M. A., & Hodge, C. W. (2002). Elevated extracellular CRF levels in the bed nucleus of the stria terminalis during ethanol withdrawal and reduction by subsequent ethanol intake. *Pharmacology, Biochemistry, and Behavior, 72*(1–2), 213–220.

O'Neil, M. L., Beckwith, L. E., Kincaid, C. L., & Rasmussen, D. D. (2013). The alpha1-adrenergic receptor antagonist, doxazosin, reduces alcohol drinking in alcohol-preferring (P) rats. *Alcoholism, Clinical and Experimental Research, 37*(2), 202–212.

Pahng, A. R., Paulsen, R. I., McGinn, M. A., Edwards, K. N., & Edwards, S. (2017). Neurobiological correlates of pain avoidance-like behavior in morphine-dependent and non-dependent rats. *Neuroscience, 366*, 1–14.

Pan, Z. Z. (1998). mu-Opposing actions of the kappa-opioid receptor. *Trends in Pharmacological Sciences*, *19*(3), 94–98.

Pantazis, C. B., James, M. H., Bentzley, B. S., & Aston-Jones, G. (2020). The number of lateral hypothalamus orexin/hypocretin neurons contributes to individual differences in cocaine demand. *Addiction Biology*, *25*(4), e12795.

Park, P. E., Schlosburg, J. E., Vendruscolo, L. F., Schulteis, G., Edwards, S., & Koob, G. F. (2015). Chronic CRF1 receptor blockade reduces heroin intake escalation and dependence-induced hyperalgesia. *Addiction Biology*, *20*(2), 275–284.

Park, P. E., Vendruscolo, L. F., Schlosburg, J. E., Edwards, S., Schulteis, G., & Koob, G. F. (2013). Corticotropin-releasing factor (CRF) and alpha 2 adrenergic receptors mediate heroin withdrawal-potentiated startle in rats. *The International Journal of Neuropsychopharmacology*, *16*(8), 1867–1875.

Paterson, N. E., Froestl, W., & Markou, A. (2004). The GABAB receptor agonists baclofen and CGP44532 decreased nicotine self-administration in the rat. *Psychopharmacology*, *172*(2), 179–186.

Pava, M. J., & Woodward, J. J. (2012). A review of the interactions between alcohol and the endocannabinoid system: Implications for alcohol dependence and future directions for research. *Alcohol*, *46*(3), 185–204.

Peacock, A., Bruno, R., Larance, B., Lintzeris, N., Nielsen, S., Ali, R., et al. (2016). Same-day use of opioids and other central nervous system depressants amongst people who tamper with pharmaceutical opioids: A retrospective 7-day diary study. *Drug and Alcohol Dependence*, *166*, 125–133.

Pedraz, M., Araos, P., Garcia-Marchena, N., Serrano, A., Romero-Sanchiz, P., Suarez, J., et al. (2015). Sex differences in psychiatric comorbidity and plasma biomarkers for cocaine addiction in abstinent cocaine-addicted subjects in outpatient settings. *Frontiers in Psychiatry*, *6*, 17.

Pedron, V. T., Varani, A. P., & Balerio, G. N. (2016). Baclofen prevents the elevated plus maze behavior and BDNF expression during naloxone precipitated morphine withdrawal in male and female mice. *Synapse*, *70*(5), 187–197.

Petrakis, I. L., Carroll, K. M., Nich, C., Gordon, L. T., McCance-Katz, E. F., Frankforter, T., et al. (2000). Disulfiram treatment for cocaine dependence in methadone-maintained opioid addicts. *Addiction*, *95*(2), 219–228.

Petry, N. M. (2001). A behavioral economic analysis of polydrug abuse in alcoholics: Asymmetrical substitution of alcohol and cocaine. *Drug and Alcohol Dependence*, *62*(1), 31–39.

Piasecki, T. M., Jahng, S., Wood, P. K., Robertson, B. M., Epler, A. J., Cronk, N. J., et al. (2011). The subjective effects of alcohol-tobacco co-use: An ecological momentary assessment investigation. *Journal of Abnormal Psychology*, *120*(3), 557–571.

Piazza, P. V., Maccari, S., Deminiere, J. M., Le Moal, M., Mormede, P., & Simon, H. (1991). Corticosterone levels determine individual vulnerability to amphetamine self-administration. *Proceedings of the National Academy of Sciences of the United States of America*, *88*(6), 2088–2092.

Piazza, P. V., Marinelli, M., Jodogne, C., Deroche, V., Rouge-Pont, F., Maccari, S., et al. (1994). Inhibition of corticosterone synthesis by metyrapone decreases cocaine-induced locomotion and relapse of cocaine self-administration. *Brain Research*, *658*(1–2), 259–264.

Plaza-Zabala, A., Flores, A., Maldonado, R., & Berrendero, F. (2012). Hypocretin/orexin signaling in the hypothalamic paraventricular nucleus is essential for the expression of nicotine withdrawal. *Biological Psychiatry*, *71*(3), 214–223.

Porter-Stransky, K. A., Bentzley, B. S., & Aston-Jones, G. (2017). Individual differences in orexin-I receptor modulation of motivation for the opioid remifentanil. *Addiction Biology*, *22*(2), 303–317.

Preston, K. L., Jobes, M. L., Phillips, K. A., & Epstein, D. H. (2016). Real-time assessment of alcohol drinking and drug use in opioid-dependent polydrug users. *Behavioural Pharmacology*, *27*(7), 579–584.

Prince, C. D., Rau, A. R., Yorgason, J. T., & Espana, R. A. (2015). Hypocretin/orexin regulation of dopamine signaling and cocaine self-administration is mediated predominantly by hypocretin receptor 1. *ACS Chemical Neuroscience*, *6*(1), 138–146.

Qin, L., & Crews, F. T. (2012). NADPH oxidase and reactive oxygen species contribute to alcohol-induced microglial activation and neurodegeneration. *Journal of Neuroinflammation*, *9*, 5.

Quelch, D. R., Mick, I., McGonigle, J., Ramos, A. C., Flechais, R. S. A., Bolstridge, M., et al. (2017). Nalmefene reduces reward anticipation in alcohol dependence: An experimental functional magnetic resonance imaging study. *Biological Psychiatry*, *81*(11), 941–948.

Rasmussen, D. D., Alexander, L., Malone, J., Federoff, D., & Froehlich, J. C. (2014). The alpha2-adrenergic receptor agonist, clonidine, reduces alcohol drinking in alcohol-preferring (P) rats. *Alcohol*, *48*(6), 543–549.

Rassnick, S., Heinrichs, S. C., Britton, K. T., & Koob, G. F. (1993). Microinjection of a corticotropin-releasing factor antagonist into the central nucleus of the amygdala reverses anxiogenic-like effects of ethanol withdrawal. *Brain Research*, *605*(1), 25–32.

Reid, L. D., Delconte, J. D., Nichols, M. L., Bilsky, E. J., & Hubbell, C. L. (1991). Tests of opioid deficiency hypotheses of alcoholism. *Alcohol*, *8*(4), 247–257.

Repunte-Canonigo, V., Shin, W., Vendruscolo, L. F., Lefebvre, C., van der Stap, L., Kawamura, T., et al. (2015). Identifying candidate drivers of alcohol dependence-induced excessive drinking by assembly and interrogation of brain-specific regulatory networks. *Genome Biology*, *16*, 68.

Reynaud, M., Aubin, H. J., Trinquet, F., Zakine, B., Dano, C., Dematteis, M., et al. (2017). A randomized, placebo-controlled study of high-dose baclofen in alcohol-dependent patients—The ALPADIR study. *Alcohol and Alcoholism*, *52*(4), 439–446.

Richards, J. K., Simms, J. A., Steensland, P., Taha, S. A., Borgland, S. L., Bonci, A., et al. (2008). Inhibition of orexin-1/hypocretin-1 receptors inhibits yohimbine-induced reinstatement of ethanol and sucrose seeking in Long-Evans rats. *Psychopharmacology*, *199*(1), 109–117.

Robbins, T. W., & Everitt, B. J. (1999). Drug addiction: Bad habits add up. *Nature*, *398*(6728), 567–570.

Roberto, M., Cruz, M. T., Gilpin, N. W., Sabino, V., Schweitzer, P., Bajo, M., et al. (2010). Corticotropin releasing factor-induced amygdala gamma-aminobutyric acid release plays a key role in alcohol dependence. *Biological Psychiatry*, *67*(9), 831–839.

Roberto, M., Madamba, S. G., Moore, S. D., Tallent, M. K., & Siggins, G. R. (2003). Ethanol increases GABAergic transmission at both pre- and postsynaptic sites in rat central amygdala neurons. *Proceedings of the National Academy of Sciences of the United States of America*, *100*(4), 2053–2058.

Roberto, M., Madamba, S. G., Stouffer, D. G., Parsons, L. H., & Siggins, G. R. (2004). Increased GABA release in the central amygdala of ethanol-dependent rats. *The Journal of Neuroscience*, *24*(45), 10159–10166.

Robinson, G., Most, D., Ferguson, L. B., Mayfield, J., Harris, R. A., & Blednov, Y. A. (2014). Neuroimmune pathways in alcohol consumption: Evidence from behavioral and genetic studies in rodents and humans. *International Review of Neurobiology*, *118*, 13–39.

Romberger, D. J., & Grant, K. (2004). Alcohol consumption and smoking status: The role of smoking cessation. *Biomedicine & Pharmacotherapy*, *58*(2), 77–83.

Rose, J. H., Karkhanis, A. N., Chen, R., Gioia, D., Lopez, M. F., Becker, H. C., et al. (2016). Supersensitive kappa opioid receptors promotes ethanol withdrawal-related behaviors and reduce dopamine signaling in the nucleus accumbens. *The International Journal of Neuropsychopharmacology*, *19*(5), pyv127.

Rosin, A., Lindholm, S., Franck, J., & Georgieva, J. (1999). Downregulation of kappa opioid receptor mRNA levels by chronic ethanol and repetitive cocaine in rat ventral tegmentum and nucleus accumbens. *Neuroscience Letters*, *275*(1), 1–4.

Rothman, R. B., Baumann, M. H., Dersch, C. M., Romero, D. V., Rice, K. C., Carroll, F. I., et al. (2001). Amphetamine-type central nervous system stimulants release norepinephrine more potently than they release dopamine and serotonin. *Synapse*, *39*(1), 32–41.

Rubio-Araiz, A., Porcu, F., Perez-Hernandez, M., Garcia-Gutierrez, M. S., Aracil-Fernandez, M. A., Gutierrez-Lopez, M. D., et al. (2017). Disruption of blood-brain barrier integrity in postmortem alcoholic brain: Preclinical evidence of TLR4 involvement from a binge-like drinking model. *Addiction Biology*, *22*(4), 1103–1116.

Sabino, V., Cottone, P., Koob, G. F., Steardo, L., Lee, M. J., Rice, K. C., et al. (2006). Dissociation between opioid and CRF1 antagonist sensitive drinking in Sardinian alcohol-preferring rats. *Psychopharmacology*, *189*(2), 175–186.

Saha, T. D., Grant, B. F., Chou, S. P., Kerridge, B. T., Pickering, R. P., & Ruan, W. J. (2018). Concurrent use of alcohol with other drugs and DSM-5 alcohol use disorder comorbid with other drug use disorders: Sociodemographic characteristics, severity, and psychopathology. *Drug and Alcohol Dependence*, *187*, 261–269.

Sakurai, T., Amemiya, A., Ishii, M., Matsuzaki, I., Chemelli, R. M., Tanaka, H., et al. (1998). Orexins and orexin receptors: A family of hypothalamic neuropeptides and G protein-coupled receptors that regulate feeding behavior. *Cell*, *92*(4), 573–585.

Schlosburg, J. E., Whitfield, T. W., Jr., Park, P. E., Crawford, E. F., George, O., Vendruscolo, L. F., et al. (2013). Long-term antagonism of kappa opioid receptors prevents escalation of and increased motivation for heroin intake. *The Journal of Neuroscience*, *33*(49), 19384–19392.

Schmaal, L., Berk, L., Hulstijn, K. P., Cousijn, J., Wiers, R. W., & van den Brink, W. (2011). Efficacy of N-acetylcysteine in the treatment of nicotine dependence: A double-blind placebo-controlled pilot study. *European Addiction Research*, *17*(4), 211–216.

Schmaal, L., Veltman, D. J., Nederveen, A., van den Brink, W., & Goudriaan, A. E. (2012). N-acetylcysteine normalizes glutamate levels in cocaine-dependent patients: A randomized crossover magnetic resonance spectroscopy study. *Neuropsychopharmacology*, *37*(9), 2143–2152.

Schmeichel, B. E., Barbier, E., Misra, K. K., Contet, C., Schlosburg, J. E., Grigoriadis, D., et al. (2015). Hypocretin receptor 2 antagonism dose-dependently reduces escalated heroin self-administration in rats. *Neuropsychopharmacology*, *40*(5), 1123–1129.

Schmeichel, B. E., Herman, M. A., Roberto, M., & Koob, G. F. (2017). Hypocretin neurotransmission within the central amygdala mediates escalated cocaine self-administration and stress-induced reinstatement in rats. *Biological Psychiatry*, *81*(7), 606–615.

Scofield, M. D., Heinsbroek, J. A., Gipson, C. D., Kupchik, Y. M., Spencer, S., Smith, A. C., et al. (2016). The nucleus accumbens: Mechanisms of addiction across drug classes reflect the importance of glutamate homeostasis. *Pharmacological Reviews*, *68*(3), 816–871.

Serrano, A., Pavon, F. J., Buczynski, M. W., Schlosburg, J., Natividad, L. A., Polis, I. Y., et al. (2018). Deficient endocannabinoid signaling in the central amygdala contributes to alcohol dependence-related anxiety-like behavior and excessive alcohol intake. *Neuropsychopharmacology*, *43*(9), 1840–1850.

Shaham, Y., Erb, S., Leung, S., Buczek, Y., & Stewart, J. (1998). CP-154,526, a selective, non-peptide antagonist of the corticotropin-releasing factor1 receptor attenuates stress-induced relapse to drug seeking in cocaine- and heroin-trained rats. *Psychopharmacology*, *137*(2), 184–190.

Shaham, Y., Highfield, D., Delfs, J., Leung, S., & Stewart, J. (2000). Clonidine blocks stress-induced reinstatement of heroin seeking in rats: An effect independent of locus coeruleus noradrenergic neurons. *The European Journal of Neuroscience*, *12*(1), 292–302.

Sharf, R., Sarhan, M., & Dileone, R. J. (2008). Orexin mediates the expression of precipitated morphine withdrawal and concurrent activation of the nucleus accumbens shell. *Biological Psychiatry*, *64*(3), 175–183.

Shoblock, J. R., Welty, N., Aluisio, L., Fraser, I., Motley, S. T., Morton, K., et al. (2011). Selective blockade of the orexin-2 receptor attenuates ethanol self-administration, place preference, and reinstatement. *Psychopharmacology*, *215*(1), 191–203.

Shoji, Y., Delfs, J., & Williams, J. T. (1999). Presynaptic inhibition of GABA(B)-mediated synaptic potentials in the ventral tegmental area during morphine withdrawal. *The Journal of Neuroscience*, *19*(6), 2347–2355.

Shytle, R. D., Mori, T., Townsend, K., Vendrame, M., Sun, N., Zeng, J., et al. (2004). Cholinergic modulation of microglial activation by alpha 7 nicotinic receptors. *Journal of Neurochemistry*, *89*(2), 337–343.

Siciliano, C. A., Karkhanis, A. N., Holleran, K. M., Melchior, J. R., & Jones, S. R. (2018). Cross-species alterations in synaptic dopamine regulation after chronic alcohol exposure. *Handbook of Experimental Pharmacology*, *248*, 213–238.

Simms, J. A., Haass-Koffler, C. L., Bito-Onon, J., Li, R., & Bartlett, S. E. (2012). Mifepristone in the central nucleus of the amygdala reduces yohimbine stress-induced reinstatement of ethanol-seeking. *Neuropsychopharmacology*, *37*(4), 906–918.

Singh, A. K. (2019). Alcohol interaction with cocaine, methamphetamine, opioids, nicotine, cannabis, and gamma-hydroxybutyric acid. *Biomedicine*, 7(1), 16.

Sinha, R., Kimmerling, A., Doebrick, C., & Kosten, T. R. (2007). Effects of lofexidine on stress-induced and cue-induced opioid craving and opioid abstinence rates: Preliminary findings. *Psychopharmacology*, *190*(4), 569–574.

Skelton, K. H., Gutman, D. A., Thrivikraman, K. V., Nemeroff, C. B., & Owens, M. J. (2007). The CRF1 receptor antagonist R121919 attenuates the neuroendocrine and behavioral effects of precipitated lorazepam withdrawal. *Psychopharmacology*, *192*(3), 385–396.

Skelton, K. H., Nemeroff, C. B., & Owens, M. J. (2004). Spontaneous withdrawal from the triazolobenzodiazepine alprazolam increases cortical corticotropin-releasing factor mRNA expression. *The Journal of Neuroscience*, *24*(42), 9303–9312.

Smith, R. J., & Aston-Jones, G. (2008). Noradrenergic transmission in the extended amygdala: Role in increased drug-seeking and relapse during protracted drug abstinence. *Brain Structure & Function*, *213*(1–2), 43–61.

Smith, R. J., & Aston-Jones, G. (2012). Orexin/hypocretin 1 receptor antagonist reduces heroin self-administration and cue-induced heroin seeking. *The European Journal of Neuroscience*, *35*(5), 798–804.

Smith, B. R., Horan, J. T., Gaskin, S., & Amit, Z. (1999). Exposure to nicotine enhances acquisition of ethanol drinking by laboratory rats in a limited access paradigm. *Psychopharmacology*, *142*(4), 408–412.

Smith, R. J., See, R. E., & Aston-Jones, G. (2009). Orexin/hypocretin signaling at the orexin 1 receptor regulates cue-elicited cocaine-seeking. *The European Journal of Neuroscience*, *30*(3), 493–503.

Smith, R. J., Tahsili-Fahadan, P., & Aston-Jones, G. (2010). Orexin/hypocretin is necessary for context-driven cocaine-seeking. *Neuropharmacology*, *58*(1), 179–184.

Snider, S. E., Hendrick, E. S., & Beardsley, P. M. (2013). Glial cell modulators attenuate methamphetamine self-administration in the rat. *European Journal of Pharmacology*, *701*(1–3), 124–130.

Soderpalm, B., Lido, H. H., & Ericson, M. (2017). The glycine receptor-A functionally important primary brain target of ethanol. *Alcoholism, Clinical and Experimental Research*, *41*(11), 1816–1830.

Sofuoglu, M., Brown, S., Babb, D. A., Pentel, P. R., & Hatsukami, D. K. (2000a). Carvedilol affects the physiological and behavioral response to smoked cocaine in humans. *Drug and Alcohol Dependence*, *60*(1), 69–76.

Sofuoglu, M., Brown, S., Babb, D. A., Pentel, P. R., & Hatsukami, D. K. (2000b). Effects of labetalol treatment on the physiological and subjective response to smoked cocaine. *Pharmacology, Biochemistry, and Behavior*, *65*(2), 255–259.

Sokolovsky, A. W., Gunn, R. L., Micalizzi, L., White, H. R., & Jackson, K. M. (2020). Alcohol and marijuana co-use: Consequences, subjective intoxication, and the operationalization of simultaneous use. *Drug and Alcohol Dependence*, *212*, 107986.

Solinas, M., Belujon, P., Fernagut, P. O., Jaber, M., & Thiriet, N. (2019). Dopamine and addiction: What have we learned from 40 years of research. *Journal of Neural Transmission (Vienna)*, *126*(4), 481–516.

Spano, M. S., Fattore, L., Fratta, W., & Fadda, P. (2007). The GABAB receptor agonist baclofen prevents heroin-induced reinstatement of heroin-seeking behavior in rats. *Neuropharmacology*, *52*(7), 1555–1562.

Specio, S. E., Wee, S., O'Dell, L. E., Boutrel, B., Zorrilla, E. P., & Koob, G. F. (2008). CRF (1) receptor antagonists attenuate escalated cocaine self-administration in rats. *Psychopharmacology*, *196*(3), 473–482.

Sprague, G. L., & Craigmill, A. L. (1976). Ethanol and delta-9-tetrahydrocannabinol: Mechanism for cross-tolerance in mice. *Pharmacology, Biochemistry, and Behavior*, *5*(4), 409–415.

Stennett, B. A., Padovan-Hernandez, Y., & Knackstedt, L. A. (2020). Sequential cocaine-alcohol self-administration produces adaptations in rat nucleus accumbens core glutamate homeostasis that are distinct from those produced by cocaine self-administration alone. *Neuropsychopharmacology*, *45*(3), 441–450.

Stinus, L., Cador, M., Zorrilla, E. P., & Koob, G. F. (2005). Buprenorphine and a CRF1 antagonist block the acquisition of opiate withdrawal-induced conditioned place aversion in rats. *Neuropsychopharmacology*, *30*(1), 90–98.

Subbaraman, M. S., & Kerr, W. C. (2020). Subgroup trends in alcohol and cannabis co-use and related harms during the rollout of recreational cannabis legalization in Washington state. *The International Journal on Drug Policy*, *75*, 102508.

Substance Abuse and Mental Health Services Administration (SAMHSA). (2018). *2018 National Survey on Drug Use and Health (NSDUH). Table 5.4A: Alcohol use disorder in past year among persons aged 12 or older, by age group and demographic characteristics: Numbers in thousands, 2017 and 2018*. Retrieved September 13, 2020, from https://www.samhsa.gov/data/sites/default/files/cbhsq-reports/NSDUHDetailedTabs2018R2/NSDUHDetTabsSect5pe2018.htm#tab5-4a.

Sulzer, D., Sonders, M. S., Poulsen, N. W., & Galli, A. (2005). Mechanisms of neurotransmitter release by amphetamines: A review. *Progress in Neurobiology*, *75*(6), 406–433.

Swanson, L. W., & Hartman, B. K. (1975). The central adrenergic system. An immunofluorescence study of the location of cell bodies and their efferent connections in the rat utilizing dopamine-beta-hydroxylase as a marker. *The Journal of Comparative Neurology*, *163*(4), 467–505.

Tan, D., Nuno-Perez, A., Mameli, M., & Meye, F. J. (2019). Cocaine withdrawal reduces GABAB R transmission at entopeduncular nucleus—Lateral habenula synapses. *The European Journal of Neuroscience*, *50*(3), 2124–2133.

Tanda, G., Pontieri, F. E., & Di Chiara, G. (1997). Cannabinoid and heroin activation of mesolimbic dopamine transmission by a common mu1 opioid receptor mechanism. *Science*, *276*(5321), 2048–2050.

Tarter, R. E., Vanyukov, M., Kirisci, L., Reynolds, M., & Clark, D. B. (2006). Predictors of marijuana use in adolescents before and after licit drug use: Examination of the gateway hypothesis. *The American Journal of Psychiatry*, *163*(12), 2134–2140.

Thannickal, T. C., John, J., Shan, L., Swaab, D. F., Wu, M. F., Ramanathan, L., et al. (2018). Opiates increase the number of hypocretin-producing cells in human and mouse brain and reverse cataplexy in a mouse model of narcolepsy. *Science Translational Medicine*, *10*(447), eaao4953.

Thanos, P. K., Kim, R., Delis, F., Ananth, M., Chachati, G., Rocco, M. J., et al. (2016). Chronic methamphetamine effects on brain structure and function in rats. *PLoS One*, *11*(6), e0155457.

Thomas, A. M., Ostroumov, A., Kimmey, B. A., Taormina, M. B., Holden, W. M., Kim, K., et al. (2018). Adolescent nicotine exposure alters GABAA receptor signaling in the ventral tegmental area and increases adult ethanol self-administration. *Cell Reports*, *23*(1), 68–77.

Tizabi, Y., Bai, L., Copeland, R. L., Jr., & Taylor, R. E. (2007). Combined effects of systemic alcohol and nicotine on dopamine release in the nucleus accumbens shell. *Alcohol and Alcoholism*, *42*(5), 413–416.

Tizabi, Y., Copeland, R. L., Jr., Louis, V. A., & Taylor, R. E. (2002). Effects of combined systemic alcohol and central nicotine administration into ventral tegmental area on dopamine release in the nucleus accumbens. *Alcoholism, Clinical and Experimental Research*, *26*(3), 394–399.

Tomko, R. L., Jones, J. L., Gilmore, A. K., Brady, K. T., Back, S. E., & Gray, K. M. (2018). N-acetylcysteine: A potential treatment for substance use disorders. *Current Psychiatry*, *17*(6), 30–36. 41–42, 55.

Tunstall, B. J., Carmack, S. A., Koob, G. F., & Vendruscolo, L. F. (2017). Dysregulation of brain stress systems mediates compulsive alcohol drinking. *Current Opinion in Behavioral Sciences*, *13*, 85–90.

Tunstall, B. J., Riley, A. L., & Kearns, D. N. (2014). Drug specificity in drug versus food choice in male rats. *Experimental and Clinical Psychopharmacology*, *22*(4), 364–372.

Tunstall, B., Vendruscolo, L., & Worthington, K. A. (2020). Rat models of alcohol use disorder. *The laboratory rat* (3rd, pp. 967–986). Academic Press.

Turner, B. D., Kashima, D. T., Manz, K. M., Grueter, C. A., & Grueter, B. A. (2017). Synaptic plasticity in the nucleus accumbens: Lessons learned from experience. *ACS Chemical Neuroscience*, *9*(9), 2114–2126.

Uslaner, J. M., Winrow, C. J., Gotter, A. L., Roecker, A. J., Coleman, P. J., Hutson, P. H., et al. (2014). Selective orexin 2 receptor antagonism blocks cue-induced reinstatement, but not nicotine self-administration or nicotine-induced reinstatement. *Behavioural Brain Research*, *269*, 61–65.

Valdez, G. R., Platt, D. M., Rowlett, J. K., Ruedi-Bettschen, D., & Spealman, R. D. (2007). Kappa agonist-induced reinstatement of cocaine seeking in squirrel monkeys: A role for opioid and stress-related mechanisms. *The Journal of Pharmacology and Experimental Therapeutics*, *323*(2), 525–533.

Valdez, G. R., Roberts, A. J., Chan, K., Davis, H., Brennan, M., Zorrilla, E. P., et al. (2002). Increased ethanol self-administration and anxiety-like behavior during acute ethanol withdrawal and protracted abstinence: Regulation by corticotropin-releasing factor. *Alcoholism, Clinical and Experimental Research*, *26*(10), 1494–1501.

Valjent, E., Mitchell, J. M., Besson, M. J., Caboche, J., & Maldonado, R. (2002). Behavioural and biochemical evidence for interactions between delta 9-tetrahydrocannabinol and nicotine. *British Journal of Pharmacology*, *135*(2), 564–578.

Van Skike, C. E., Maggio, S. E., Reynolds, A. R., Casey, E. M., Bardo, M. T., Dwoskin, L. P., et al. (2016). Critical needs in drug discovery for cessation of alcohol and nicotine polysubstance abuse. *Progress in Neuro-Psychopharmacology & Biological Psychiatry*, *65*, 269–287.

Vanderschuren, L. J., Beemster, P., & Schoffelmeer, A. N. (2003). On the role of noradrenaline in psychostimulant-induced psychomotor activity and sensitization. *Psychopharmacology*, *169*(2), 176–185.

Vanyukov, M. M., Tarter, R. E., Kirillova, G. P., Kirisci, L., Reynolds, M. D., Kreek, M. J., et al. (2012). Common liability to addiction and "gateway hypothesis": Theoretical, empirical and evolutionary perspective. *Drug and Alcohol Dependence*, *123*(Suppl. 1), S3–17.

Varani, A. P., Aso, E., Moutinho, L. M., Maldonado, R., & Balerio, G. N. (2014). Attenuation by baclofen of nicotine rewarding properties and nicotine withdrawal manifestations. *Psychopharmacology*, *231*(15), 3031–3040.

Varani, A. P., Moutinho, L. M., Calvo, M., & Balerio, G. N. (2011). Ability of baclofen to prevent somatic manifestations and neurochemical changes during nicotine withdrawal. *Drug and Alcohol Dependence*, *119*(1–2), e5–12.

Vendruscolo, L. F., Barbier, E., Schlosburg, J. E., Misra, K. K., Whitfield, T. W., Jr., Logrip, M. L., et al. (2012). Corticosteroid-dependent plasticity mediates compulsive alcohol drinking in rats. *The Journal of Neuroscience*, *32*(22), 7563–7571.

Vendruscolo, L. F., Estey, D., Goodell, V., Macshane, L. G., Logrip, M. L., Schlosburg, J. E., et al. (2015). Glucocorticoid receptor antagonism decreases alcohol seeking in alcohol-dependent individuals. *The Journal of Clinical Investigation*, *125*(8), 3193–3197.

Vendruscolo, L. F., & Koob, G. F. (2020). Alcohol dependence conceptualized as a stress disorder. In *The Oxford handbook of stress and mental health* Oxford University Press.

Vergo, S., Johansen, J. L., Leist, M., & Lotharius, J. (2007). Vesicular monoamine transporter 2 regulates the sensitivity of rat dopaminergic neurons to disturbed cytosolic dopamine levels. *Brain Research*, *1185*, 18–32.

Vlachou, S., & Markou, A. (2010). GABAB receptors in reward processes. *Advances in Pharmacology*, *58*, 315–371.

Vlachou, S., Paterson, N. E., Guery, S., Kaupmann, K., Froestl, W., Banerjee, D., et al. (2011). Both GABA(B) receptor activation and blockade exacerbated anhedonic aspects of nicotine withdrawal in rats. *European Journal of Pharmacology*, *655*(1–3), 52–58.

Volpicelli, J. R., Davis, M. A., & Olgin, J. E. (1986). Naltrexone blocks the post-shock increase of ethanol consumption. *Life Sciences*, *38*(9), 841–847.

Volz, T. J., Hanson, G. R., & Fleckenstein, A. E. (2007). The role of the plasmalemmal dopamine and vesicular monoamine transporters in methamphetamine-induced dopaminergic deficits. *Journal of Neurochemistry*, *101*(4), 883–888.

Waeiss, R. A., Knight, C. P., Engleman, E. A., Hauser, S. R., & Rodd, Z. A. (2020). Co-administration of ethanol and nicotine heightens sensitivity to ethanol reward within the nucleus accumbens (NAc) shell and increasing NAc shell BDNF is sufficient to enhance ethanol reward in naive Wistar rats. *Journal of Neurochemistry*, *152*(5), 556–569.

Walker, B. M., & Koob, G. F. (2007). The gamma-aminobutyric acid-B receptor agonist baclofen attenuates responding for ethanol in ethanol-dependent rats. *Alcoholism, Clinical and Experimental Research*, *31*(1), 11–18.

Walker, L. C., & Lawrence, A. J. (2017). The role of orexins/hypocretins in alcohol use and abuse. *Current Topics in Behavioral Neurosciences*, *33*, 221–246.

Walker, B. M., Rasmussen, D. D., Raskind, M. A., & Koob, G. F. (2008). Alpha1-noradrenergic receptor antagonism blocks dependence-induced increases in responding for ethanol. *Alcohol*, *42*(2), 91–97.

Walter, T. J., & Crews, F. T. (2017). Microglial depletion alters the brain neuroimmune response to acute binge ethanol withdrawal. *Journal of Neuroinflammation*, *14*(1), 86.

Wan, F. J., Berton, F., Madamba, S. G., Francesconi, W., & Siggins, G. R. (1996). Low ethanol concentrations enhance GABAergic inhibitory postsynaptic potentials in hippocampal pyramidal neurons only after block of GABAB receptors. *Proceedings of the National Academy of Sciences of the United States of America*, *93*(10), 5049–5054.

Wang, X., Loram, L. C., Ramos, K., de Jesus, A. J., Thomas, J., Cheng, K., et al. (2012). Morphine activates neuroinflammation in a manner parallel to endotoxin. *Proceedings of the National Academy of Sciences of the United States of America, 109*(16), 6325–6330.

Wang, H. L., Zhang, S., Qi, J., Wang, H., Cachope, R., Mejias-Aponte, C. A., et al. (2019). Dorsal raphe dual serotonin-glutamate neurons drive reward by establishing excitatory synapses on VTA mesoaccumbens dopamine neurons. *Cell Reports, 26*(5), 1128–1142. e1127.

Warden, A. S., DaCosta, A., Mason, S., Blednov, Y. A., Mayfield, R. D., & Harris, R. A. (2020). Inbred substrain differences influence neuroimmune response and drinking behavior. *Alcoholism, Clinical and Experimental Research, 44*(9), 1760–1768.

Warden, A. S., Wolfe, S. A., Khom, S., Varodayan, F. P., Patel, R. R., Steinman, M. Q., et al. (2020). Microglia control escalation of drinking in alcohol-dependent mice: Genomic and synaptic drivers. *Biological Psychiatry, 88*(12), 910–921.

Wee, S., Mandyam, C. D., Lekic, D. M., & Koob, G. F. (2008). Alpha 1-noradrenergic system role in increased motivation for cocaine intake in rats with prolonged access. *European Neuropsychopharmacology, 18*(4), 303–311.

Wee, S., Wang, Z., He, R., Zhou, J., Kozikowski, A. P., & Woolverton, W. L. (2006). Role of the increased noradrenergic neurotransmission in drug self-administration. *Drug and Alcohol Dependence, 82*(2), 151–157.

Wee, S., & Woolverton, W. L. (2004). Evaluation of the reinforcing effects of atomoxetine in monkeys: Comparison to methylphenidate and desipramine. *Drug and Alcohol Dependence, 75*(3), 271–276.

Weerts, E. M., Wand, G. S., Kuwabara, H., Xu, X., Frost, J. J., Wong, D. F., et al. (2014). Association of smoking with mu-opioid receptor availability before and during naltrexone blockade in alcohol-dependent subjects. *Addiction Biology, 19*(4), 733–742.

Weiner, J. L., Zhang, L., & Carlen, P. L. (1994). Potentiation of GABAA-mediated synaptic current by ethanol in hippocampal CA1 neurons: Possible role of protein kinase C. *The Journal of Pharmacology and Experimental Therapeutics, 268*(3), 1388–1395.

Weiss, F., Mitchiner, M., Bloom, F. E., & Koob, G. F. (1990). Free-choice responding for ethanol versus water in alcohol preferring (P) and unselected Wistar rats is differentially modified by naloxone, bromocriptine, and methysergide. *Psychopharmacology, 101*(2), 178–186.

Wellman, P., Ho, D., Cepeda-Benito, A., Bellinger, L., & Nation, J. (2002). Cocaine-induced hypophagia and hyperlocomotion in rats are attenuated by prazosin. *European Journal of Pharmacology, 455*(2–3), 117–126.

Wemm, S. E., & Sinha, R. (2019). Drug-induced stress responses and addiction risk and relapse. *Neurobiology of Stress, 10*, 100148.

Wenzel, J. M., Cotten, S. W., Dominguez, H. M., Lane, J. E., Shelton, K., Su, Z. I., et al. (2014). Noradrenergic beta-receptor antagonism within the central nucleus of the amygdala or bed nucleus of the stria terminalis attenuates the negative/anxiogenic effects of cocaine. *The Journal of Neuroscience, 34*(10), 3467–3474.

White, J., Walton, D., & Walker, N. (2015). Exploring comorbid use of marijuana, tobacco, and alcohol among 14 to 15-year-olds: Findings from a national survey on adolescent substance use. *BMC Public Health, 15*, 233.

Whitfield, T. W., Jr., Schlosburg, J. E., Wee, S., Gould, A., George, O., Grant, Y., et al. (2015). Kappa opioid receptors in the nucleus accumbens shell mediate escalation of methamphetamine intake. *The Journal of Neuroscience, 35*(10), 4296–4305.

Williams, K. L., Kane, E. C., & Woods, J. H. (2001). Interaction of morphine and naltrexone on oral ethanol self-administration in rhesus monkeys. *Behavioural Pharmacology, 12*(5), 325–333.

Winkler, M. C., Greager, E. M., Stafford, J., & Bachtell, R. K. (2018). Methamphetamine self-administration reduces alcohol consumption and preference in alcohol-preferring P rats. *Addiction Biology*, *23*(1), 90–101.

Wise, R. A., & Rompre, P. P. (1989). Brain dopamine and reward. *Annual Review of Psychology*, *40*, 191–225.

Witkiewitz, K., & Vowles, K. E. (2018). Alcohol and opioid use, co-use, and chronic pain in the context of the opioid epidemic: A critical review. *Alcoholism, Clinical and Experimental Research*, *42*(3), 478–488.

Woolverton, W. L. (1987). Evaluation of the role of norepinephrine in the reinforcing effects of psychomotor stimulants in rhesus monkeys. *Pharmacology, Biochemistry, and Behavior*, *26*(4), 835–839.

Wu, L. T., Pilowsky, D. J., & Schlenger, W. E. (2005). High prevalence of substance use disorders among adolescents who use marijuana and inhalants. *Drug and Alcohol Dependence*, *78*(1), 23–32.

Xuei, X., Flury-Wetherill, L., Almasy, L., Bierut, L., Tischfield, J., Schuckit, M., et al. (2008). Association analysis of genes encoding the nociceptin receptor (OPRL1) and its endogenous ligand (PNOC) with alcohol or illicit drug dependence. *Addiction Biology*, *13*(1), 80–87.

Yokel, R. A., & Wise, R. A. (1978). Amphetamine-type reinforcement by dopaminergic agonists in the rat. *Psychopharmacology*, *58*(3), 289–296.

Zhang, Y., Li, H., Li, Y., Sun, X., Zhu, M., Hanley, G., et al. (2011). Essential role of Toll-like receptor 2 in morphine-induced microglia activation in mice. *Neuroscience Letters*, *489*(1), 43–47.

Zhao, Y. N., Wang, F., Fan, Y. X., Ping, G. F., Yang, J. Y., & Wu, C. F. (2013). Activated microglia are implicated in cognitive deficits, neuronal death, and successful recovery following intermittent ethanol exposure. *Behavioural Brain Research*, *236*(1), 270–282.

Zhou, Y., Cui, C. L., Schlussman, S. D., Choi, J. C., Ho, A., Han, J. S., et al. (2008). Effects of cocaine place conditioning, chronic escalating-dose "binge" pattern cocaine administration and acute withdrawal on orexin/hypocretin and preprodynorphin gene expressions in lateral hypothalamus of Fischer and Sprague-Dawley rats. *Neuroscience*, *153*(4), 1225–1234.

Zhou, W., & Kalivas, P. W. (2008). N-acetylcysteine reduces extinction responding and induces enduring reductions in cue- and heroin-induced drug-seeking. *Biological Psychiatry*, *63*(3), 338–340.

Zhou, Y., Spangler, R., Schlussman, S. D., Ho, A., & Kreek, M. J. (2003). Alterations in hypothalamic-pituitary-adrenal axis activity and in levels of proopiomelanocortin and corticotropin-releasing hormone-receptor 1 mRNAs in the pituitary and hypothalamus of the rat during chronic 'binge' cocaine and withdrawal. *Brain Research*, *964*(2), 187–199.

Zhu, P. J., & Lovinger, D. M. (2006). Ethanol potentiates GABAergic synaptic transmission in a postsynaptic neuron/synaptic bouton preparation from basolateral amygdala. *Journal of Neurophysiology*, *96*(1), 433–441.

Zimmer, A., Valjent, E., Konig, M., Zimmer, A. M., Robledo, P., Hahn, H., et al. (2001). Absence of delta-9-tetrahydrocannabinol dysphoric effects in dynorphin-deficient mice. *The Journal of Neuroscience*, *21*(23), 9499–9505.

Zislis, G., Desai, T. V., Prado, M., Shah, H. P., & Bruijnzeel, A. W. (2007). Effects of the CRF receptor antagonist D-Phe CRF(12-41) and the alpha2-adrenergic receptor agonist clonidine on stress-induced reinstatement of nicotine-seeking behavior in rats. *Neuropharmacology*, *53*(8), 958–966.

Kolibri 360 GmbH
Ferdinand-Jühlke-Straße 7
99095 Erfurt, Deutschland
produktsicherheit@kolibri360.de